WITHDRAWN
D0686793

DEFECTS, FRACTURE AND FATIGUE

Defects, Fracture and Fatigue

Proceedings of the Second International Symposium,
held at Mont Gabriel, Canada, May 30–June 5, 1982

edited by

G.C. SIH

Institute of Fracture and Solid Mechanics
Lehigh University, Bethlehem, Pennsylvania, USA

J.W. PROVAN

Mechanical Engineering Department
McGill University, Montreal, P.Q., Canada

1983

MARTINUS NIJHOFF PUBLISHERS
THE HAGUE / BOSTON / LONDON

Distributors:

for the United States and Canada

Kluwer Boston, Inc.
190 Old Derby Street
Hingham, MA 02043
USA

for all other countries

Kluwer Academic Publishers Group
Distribution Center
P.O. Box 322
3300 AH Dordrecht
The Netherlands

Library of Congress Cataloging in Publication Data CIP

International Symposium on Defects, Fracture, and Fatigue (2nd : 1982 : Mont Gabriel, Québec)
Proceedings of Second International Symposium on Defects, Fracture, and Fatigue.

1. Materials--Defects--Congresses. 2. Fracture mechanics--Congresses. 3. Materials--Fatigue--Congresses. I. Sih, G. C. (George C.) II. Provan, J. W. III. Title. IV. Title: Defects, fracture, and fatigue.
TA409.I555 1982 620.1'123 82-24576
ISBN 90-247-2804-5

ISBN 90-247-2804-5

PRINTED IN THE NETHERLANDS

CONTENTS

PREFACE

The Second International Symposium on Defects, Fracture and Fatigue took place at Mont Gabriel, Quebec, Canada, May 30 to June 5, 1982, and was organized by the Mechanical Engineering Department of McGill University and Institute of Fracture and Solid Mechanics, Lehigh University. The Co-Chairmen of the Symposium were Professor G. C. Sih of Lehigh University and Professor J. W. Provan of McGill University. Among those who served on the Organizing Committee were G. C. Sih (Co-Chairman), J. W. Provan (Co-Chairman), H. Mughrabi, H. Zorski, R. Bullough, M. Matczyński, G. Barenblatt and G. Caglioti. As a result of the interest expressed at the First Symposium that was held in October 1980, in Poland, the need for a follow-up meeting to further explore the phenomena of material damage became apparent. Among the areas considered were dislocations, persistent-slip-bands, void creation, microcracking, microstructure effects, micro/macro fracture mechanics, ductile fracture criteria, fatigue crack initiation and propagation, stress and failure analysis, deterministic and statistical crack models, and fracture control. This wide spectrum of topics attracted researchers and engineers in solid state physics, continuum mechanics, applied mathematics, metallurgy and fracture mechanics from many different countries. This spectrum is also indicative of the interdisciplinary character of material damage that must be addressed at the atomic, microscopic and macroscopic scale level.

Mont Gabriel, situated just 39 miles north of Montreal, provided an ideal environment for a small group of specialists to gather and interact on a personal basis. Intense discussions between the physical metallurgists and fracture mechanicians are necessary for reconciliating the details of the microstructure with the assumed continua of the computational methods and toughness assessment techniques. Both groups benefited and developed a wider appreciation of the difference in viewpoints. A unification of concepts, however, can only be accomplished if effective communications are maintained continuously among researchers and engineers. To this end, the Organizing Committee has agreed to hold a third Symposium two years from now in Europe.

The Co-Chairmen wish to take this opportunity to thank the Canadian Committee for Research on the Strength and Fracture of Materials for supporting the Symposium and those who have contributed towards the success of the meeting. In particular, Mrs. Evelyn Schliecker of McGill University should be recognized for her attendance at the registration desk and for taking care of many arrangements. The efforts of Messrs. Louis Archard and Yves Theriault are also appreciated for providing local transportation.

Finally, the credit for retyping all the manuscripts goes to Mrs. Barbara DeLazaro and Mrs. Constance Weaver of the Institute of Fracture and Solid Mechanics at Lehigh University. Their quality of work is evidenced in this book.

Mont Gabriel, Canada
June 1982

G. C. Sih
J. W. Provan
Co-Chairmen

GENERAL REMARKS

The theme of this Symposium follows a long standing notion that material failure initiates from defects. Perhaps, the ancient Chinese transcript "Mo-Ching" in the -4th Century has said it all:

It is upon evenness or continuity that breaking or not-breaking depend.

Let a weight hang on hair. The hair will break because it is not truly even or continuous. If it were, it would not break.

In recent times, much effort has been made to identify movement of imperfections at the earliest possible stage and to follow their growth in details. The results unavoidably have tended to be qualitative and expressed in terms of microscopic entities and their interactions such as dislocations, inclusions, slip-bands, grain boundary cracking, environmental effects, etc. This is mainly because of the difficulties associated with constructing a quantitative theory that can consistently assess material damage at the different scale levels ranging from the atomic, say 10^{-8} cm in linear dimension, to the macroscopic, say 10^{-2} cm. This involves at least six orders of magnitude or more. While the electron microscope has enabled a detailed description of the sequence of events which culminate in material damage, the mathematical models still lag far behind not being able to express the microscopic variables in terms of useful macroscopic parameters. It is precisely due to this lack of interaction between the physical metallurgists and fracture mechanicians that this Symposium has been organized to reconcile the intricate details of the microstructure with the assumed mechanics of continua and computational methods.

Major difficulties still perplex the practitioners who attempt to estimate fatigue lives of structural components with or without the added complexities of hostile environments. Judgement and experience must often be exercised to provide provision against premature failure. The discipline of fracture mechanics, however, has provided a drastic improvement over the conventional methods that do not account for the presence of initial defects. Learning to design with defects or cracks is the key for making prediction with increased accuracy. Undoubtedly, consistent description of the qualitative observations can further improve the agreement of theory and experiment. Material damage at the atomic, microscopic and macroscopic scale level needs to be related mathematically in a unified manner.

Another important objective of this Symposium is its international character that encourages a closer cooperation among the researchers and engineers from different nations. This joint effort to develop better concepts and methods in design concerning with failure prevention should always be kept in mind when organizing future meetings.

Mont Gabriel
June 1982

G. C. Sih
Lehigh University

NIGHTMARE

With severe migrane
went to sleep
in a crack tip
at boundary of a grain
in my brain

Deep in sleep
I could not see
but could hear
Head Mechanics Sih

He said you are hazy
your JAY is lousy
your KAY is crazy
either pump strain energy
or see Herr Dr. in Metallurgy

You have slip bands
said Herr Dr. Extrusion, inclusions
Defects in your facts
Galloping dislocations
Spying disclinations
Unscrew your head
before you are dead
you must get rid
sources of Frank Read

He started to whistle
with an inverted thistle
He called it crack
Materials are not local
Nor are they non-local
To see them clearly
you must wear bifocal

Some regions have no locations
Cracks take no vacations
You *fail* to understand
and you understand to fail
for you have too many
dislocations

I woke up hazy
A little more wise
A little more crazy.

Mont Gabriel
June 4, 1982

A. C. Eringen

A happy group of delegates

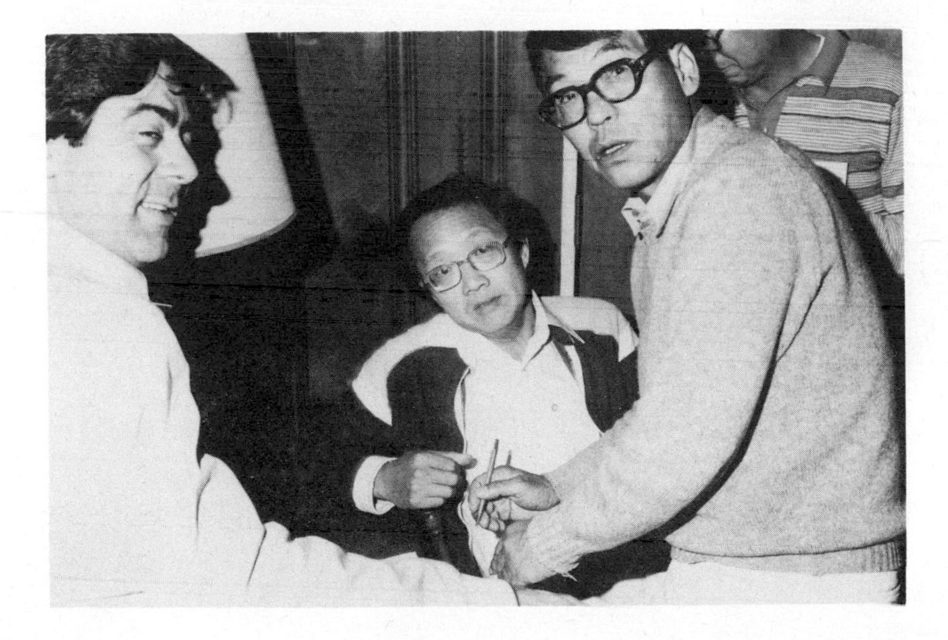

E. Gdoutos, G. Sih and T. Mura

C. Eringen, S. M. Ohr and H. Shirai

K. Bandyopadhyay, B. Tyson and J. Provan

J. Provan receiving a plaque from G. Sih

H. Mughrabi and G. Sih receiving Canadian Northwest Indian
copper boxes from J. Provan

H. Mughrabi and G. Sih admiring their gifts
as Jim and Barbara Provan look on

After banquet talk by G. Sih

After banquet talk by V. Gerold

After banquet poem by C. Eringen

Conference participants

SECTION I

DISLOCATION BEHAVIOR AND INFLUENCE

DIRECT OBSERVATIONS OF CRACK TIP DISLOCATION BEHAVIOR DURING TENSILE AND CYCLIC DEFORMATION*

S. M. Ohr, J. A. Horton

Oak Ridge National Laboratory
Oak Ridge, Tennessee 37830

and

S.-J. Chang

Union Carbide Corporation
Oak Ridge, Tennessee 37830

ABSTRACT

Direct observations have been made of the emission and distribution of dislocations at the crack tip of propagating cracks in various metals during tensile and cyclic deformation in an electron microscope. The results show a number of new findings regarding the dislocation behavior near the crack tip and its relationship to the propagation of cracks. Depending on the loading geometry, it was possible to observe the propagation of shear cracks of Mode III type as well as tensile cracks of Mode I type. Some of the observations were made using a video recording system. For shear cracks of Mode III type, the crack tip generated screw dislocations on planes which were coplanar with the cracks. The dislocations were generally in the form of an inverse pileup but a dislocation-free zone was present near the crack tip. For tensile cracks of Mode I type, edge dislocations were generated at the crack tip on planes which were inclined to crack. As the dislocations were emitted, the crack tip was blunted and crack propagation ceased. Under cyclic loading conditions, some of the dislocations near the crack tip reversed their direction and returned to the crack tip and disappeared as the stress was reduced. In order to explain the physical origin of the dislocation-free zone near the crack tip, the dislocation theory of fracture was extended and the dependence of the stress intensity factor on the number of dislocations in the plastic zone was calculated. It was found that the dislocation-free zone was expected if there was a critical stress intensity factor required for the generation of dislocations at the crack tip. The magnitudes of the critical stress intensity factor for dislocation generations were estimated for various metals from the dislocation theory and these values were compared

*Research sponsored by the Division of Materials Science, U.S. Department of Energy under Contract W-7405-eng-26 with Union Carbide Corporation.

with the values determined from the electron microscope fracture experiments. The problem of a linear array of dislocations on a plane which is inclined to the crack plane was also considered for Mode III geometry.

INTRODUCTION

It is well established that fracture toughness of structural materials is generally much greater than that expected from the fracture surface energy because of plastic deformation that occurs at a crack tip. Under extreme temperatures and harsh chemical or radiation environments, the fracture toughness is reduced considerably and the materials fracture in a brittle manner under relatively low levels of external stress. In order to understand the fracture behavior of these materials, it is therefore necessary to clarify the mechanisms by which the materials deform plastically at the crack tip. In the present study, direct observations have been made of the emission and distribution of dislocations near the crack tip of propagating cracks in various metals, including stainless steel, copper, nickel, aluminum, niobium, molybdenum and tungsten, during in situ tensile and cyclic deformation in an electron microscope [1-5]. Depending on the orientation of the specimen, it was possible to observe the propagation of shear cracks of Mode III type as well as tensile cracks of Mode I type.

The experiments have shown that the distribution of dislocations in the plastic zone was in good agreement with the prediction of the theory of fracture proposed by Bilby, Cottrell and Swinden (BCS) [6]. The only exception was the presence of a dislocation-free zone (DFZ) near the crack tip. In order to understand the physical origin of the DFZ and its relationship to fracture criteria, the BCS theory of fracture was extended to include the DFZ between the crack tip and the plastic zone [7,8]. The theoretical analysis has led to the definition of a critical stress intensity factor K_g for dislocation generation at the crack tip. This parameter, along with the applied stress, determines the extent of plastic deformation occurring at the crack tip. The present paper reviews the highlights of experimental observations and the development of the dislocation-free zone model of fracture. The paper also includes some recent results of the experimenta and theoretical studies of crack tip dislocation behavior.

EXPERIMENTAL

Fracture experiments were performed during deformation inside an electron microscope. Figure 1 shows the deformation stage which was built for a Philip EM 400T [9]. Sheet tensile specimens (3 x 6.5 x 0.025 mm) were spark cut and an area about 2 mm in diameter was electropolished in a Struers Tenupol until perforation. The specimen was mounted between two guide pins and the load was applied by retracting the movable pulling rod. The pulling force on the specimen was applied either by turning a differentially threaded screw or by using a hydraulic control system. As the stress was applied, the cracks were initiated at the edge of the polishing hole and propagated into the specimen. In the electropolished thin area, the cracks propagated as Mode III shear cracks by emitting dislocations on a plane coplanar to the crack planes. As the cracks propagated into thicker areas, the area ahead of the crack tip was thinned by plastic deformation and became electron transparent. In these originally thicker areas, the fracture mode observed was predominantly Mode I and the cracks emitted dislocations on planes inclined to the crack plane. All of the micrographs were taken in the stressed condition.

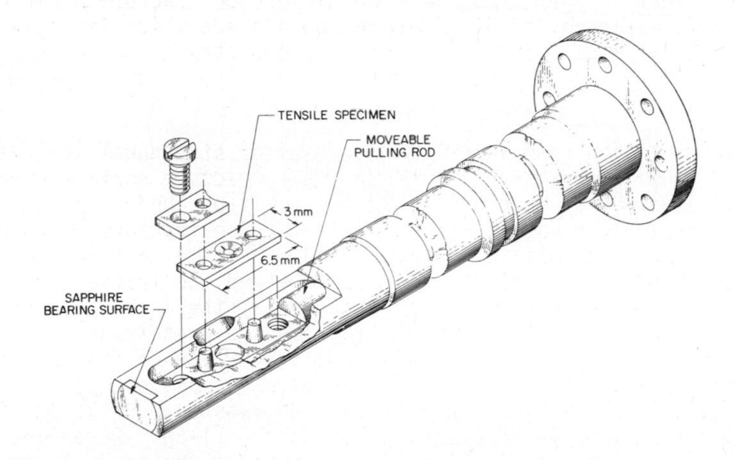

Fig. (1) - Deformation stage for a Philip EM 400T electron microscope

EXPERIMENTAL RESULTS

Figure 2 shows an electron micrograph of the crack tip area taken in a single crystal of pure stainless steel (15% Ni, 15% Cr) deformed in tension. It can be

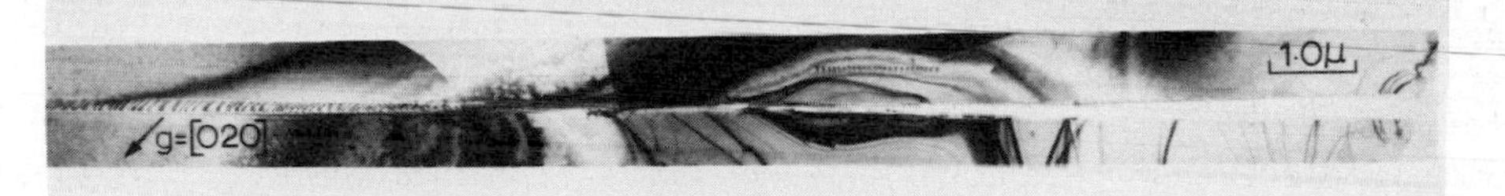

Fig. (2) - Electron micrograph showing a Mode III shear crack and its plastic zone in a (001) foil of stainless steel single crystal

seen that the crack, which originated at the edge of the polishing hole, has propagated into the specimen and the plastic zone, consisting of a linear array of dislocations, has been created on a plane which is coplanar with the crack. Figure 3 shows the dislocation density in the plastic zone of Figure 2 plotted against the distance from the crack tip. The dislocation density reached its maximum near the crack tip, indicating an inverse pileup, but the area immediately ahead of the crack tip, approximately 1 μm in length, was free of dislocations. According to contrast analysis, the dislocations in the plastic zone were pure screw dislocations with Burgers vector a/2 [011] and they were split into two partials of a/6 [112] type. The crack therefore represents an antiplane shear crack of Mode III type. There were approximately 300 dislocations in the plastic zone so that their contribution to the crack opening displacement (COD) was nearly 850 A. This was almost equal to the thickness of the foil at the crack tip which indicates that there were sufficient number of dislocations in the plastic zone to provide the COD needed to propagate the crack. These observations are in good agreement with the plastic zone model of Bilby, Cottrell and Swinden [6].

Figure 4 shows a similar crack tip geometry observed in copper which was also deformed in tension in the electron microscope. The plastic zone consisted of a

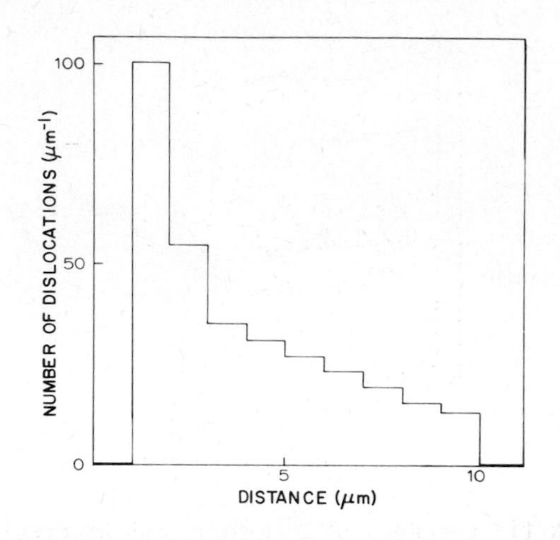

Fig. (3) - Dislocation density from Figure 2 in the plastic zone as a function of the distance from the crack tip showing an inverse pileup

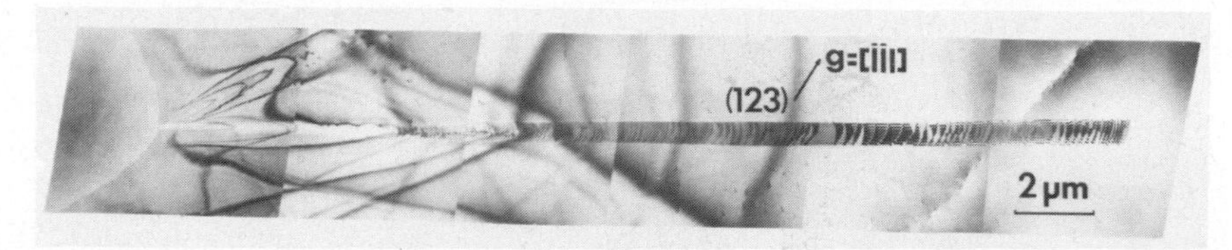

Fig. (4) - Electron micrograph of a Mode III shear crack and its plastic zone in copper. There is a dislocation-free zone between the crack tip and the plastic zone

linear array of screw dislocations that were split into partials. In this micrograph, the dislocation-free zone was approximately 5 μm in length and its presence can be clearly seen. The slip plane was identified as (111) and the slip system represents the system of maximum resolved shear stress. Figure 5 shows schematically the Mode III crack geometry observed. The crack tip is located at c and the DFZ is present between c and e. The plastic zone, consisting of an inverse pileup of screw dislocations, extends from e to a. The dislocations in the plastic zone have contributed δ_p = bN to the COD where b is the Burgers vector and N is the total number of dislocations in the plastic zone. The crack propagates through the DFZ without emitting any more dislocations.

Figure 6 is an electron micrograph taken from a crack tip area of aluminum [10]. The area was originally too thick to be electron transparent. It shows an array of edge dislocations that were emitted from the crack tip on a plane which was nearly perpendicular to the plane of the crack. The crack geometry therefore corresponds to that of Mode I crack propagation. It could be seen that as the edge dislocations were emitted the crack tip was blunted. The blunted cracks did not propagate until the area ahead of the crack tip was thinned by plastic deformation. The cracks then propagated abruptly without emitting dislocations. The se-

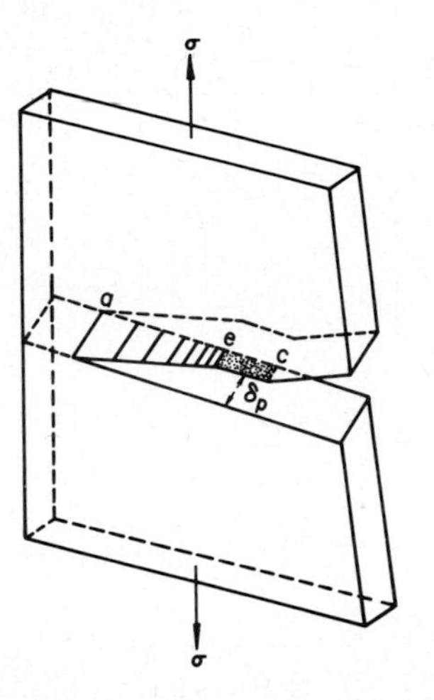

Fig. (5) - Crack tip geometry of a Mode III shear crack observed in the electron microscope. The dislocation-free zone (DFZ) is present between the crack tip c and the plastic zone (e to a)

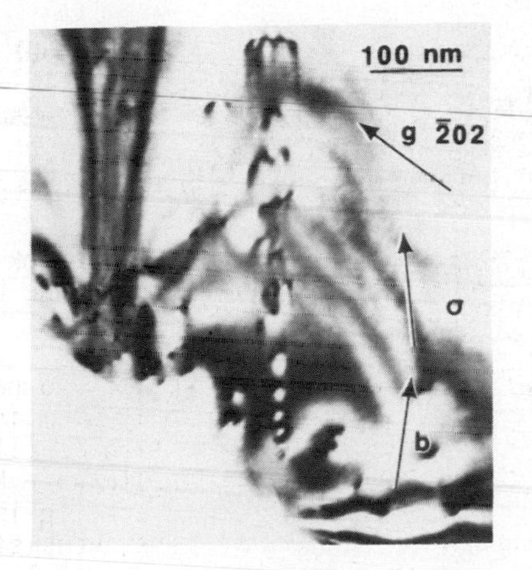

Fig. (6) - Electron micrograph of edge dislocations emitted from a crack tip during Mode I crack propagation in aluminum

quence of crack blunting and the nucleation of a sharp crack at the crack front is sketched in Figure 7(a).

Figure 8 shows a crack tip geometry different than Figure 7 which was observed in molybdenum but was also seen in other metals [11]. In this geometry, a crack propagates in a zig-zag manner by emitting edge dislocations at the crack tip along the plane of the crack. After emitting a number of dislocations, the crack ceases to emit dislocations and then it changes its direction and propagates along this new direction by emitting dislocations. The process is schematically shown

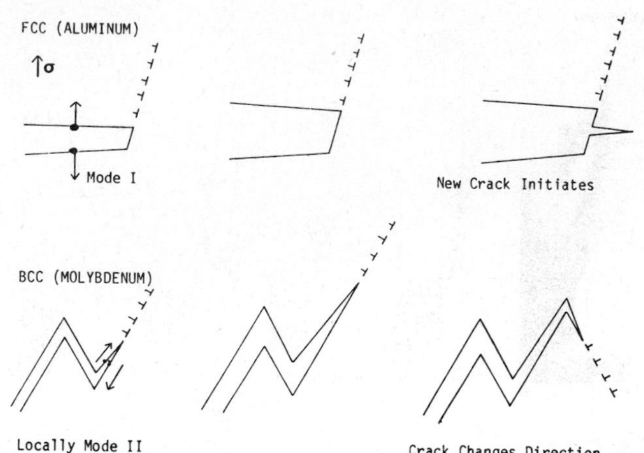

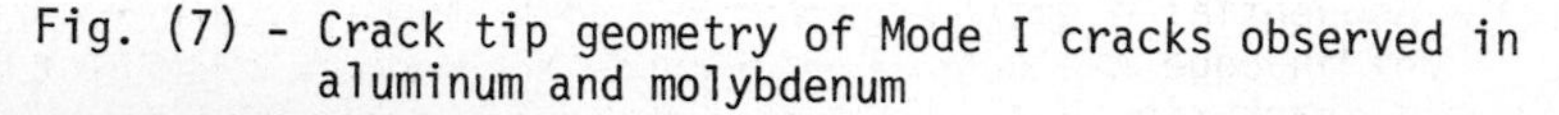
Fig. (7) - Crack tip geometry of Mode I cracks observed in aluminum and molybdenum

Fig. (8) - Crack propagation along a zig-zag path observed in molybdenum

in Figure 7(b) where it can be seen that the crack propagates by Mode II locally. The Mode II crack geometry, combined with that of Mode I, is expected to give a zig-zag crack path.

Figure 9 shows a sequence of electron micrographs taken near a crack tip of aluminum specimen as the load was reduced from its peak value [5]. It can be seen that a dislocation, marked by the arrow, returned to the crack tip and finally vanished at the crack tip. As the load was increased, the dislocation reemitted from the crack tip and moved away along the identical slip plane. A load reduction of 20% was sufficient to bring some of the dislocations in the plastic zone back to the crack.

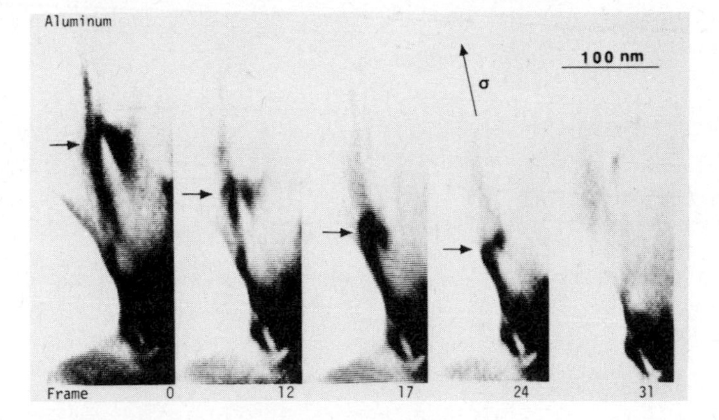

Fig. (9) - Sequential electron micrographs made from a video recording of an edge dislocation, marked by the arrow, returning to a crack tip in aluminum as the load is reduced

DISLOCATION-FREE ZONE MODEL OF FRACTURE

The behavior of crack tip dislocations observed during deformation in the electron microscope is generally in good agreement with the plastic zone model proposed by Bilby, Cottrell and Swinden [6]. Their plastic zone consists of an inverse pileup of dislocations on a plane which is coplanar with the crack. The dislocations in the plastic zone have contributed to the crack tip opening displacement. The experimental observation which is not expected from the BCS theory is the presence of a DFZ at the crack tip. The concept of a highly stressed elastic core at the crack tip was proposed recently by Thomson [12] and Weertman [13]. The DFZ observed in the present study is in many ways similar to the elastic core region in their model.

In order to examine in detail the physical origin of the DFZ, we have extended the BCS theory of fracture to include the DFZ as part of the crack tip equilibrium geometry [7,8]. A singular integral equation, describing Mode III crack geometry with the DFZ present, can be written as

$$A \left[\int_{-a}^{-e} + \int_{-c}^{c} + \int_{e}^{a} \right] \frac{f(x')dx'}{x-x'} + \sigma_a = 0, \quad |x|<c \qquad (1)$$
$$= \sigma_f, \quad e<|x|<a$$

where $A = \mu b/2\pi$, μ is the shear modulus, b is the Burgers vector, f(x) is the distribution function for the dislocations, σ_a is the applied stress and σ_f is the fraction stress. The symbols, c, e and a were defined in Figure 5. Equation (1) was inverted by the method of Muskhelishvili [14] to obtain the distribution function f(x). It is found that f(x) can be expressed in terms of complete and incomplete elliptic integrals of the first kind $F(\beta,k)$ and third kind $II(\beta,\alpha^2,k)$,

$$f(x) = G[x, F(\beta,k), II(\beta,\alpha^2,k)] \qquad (2)$$

where $\alpha^2 = (a^2-e^2)/(a^2-c^2)$ and $k^2 = \alpha^2 c^2/e^2$. As the necessary condition for the solution (12) to exist, we get

$$\frac{\pi\sigma_a}{2\sigma_f} = \frac{(e^2-c^2)}{e(a^2-c^2)^{1/2}} \, \Pi(\pi/2,\alpha^2,k) \tag{3}$$

We have shown that equations (2) and (3) are reduced to the results of the BCS theory as e approaches a, i.e., for a vanishingly small DFZ [15].

From the distribution function f(x), we have obtained an expression for the stress intensity factor K at the crack tip using the definition given by Bilby and Eshelby [16]. It was found that

$$\frac{K}{\sigma_f(\pi c)^{1/2}} = \frac{2(e^2-c^2)^{1/2}}{\pi e} F(\pi/2,k) \tag{4}$$

Figure 10 shows a plot of K versus the number of dislocations N in the plastic zone, which is obtained from equations (2), (3) and (4). The number of dislocations N is obtained by integrating f(x) from e to a. For a given value of σ_a/σ_f, K decreases from the elastic value ($(\pi c)^{1/2}\ \sigma_a$) as N increases and eventually

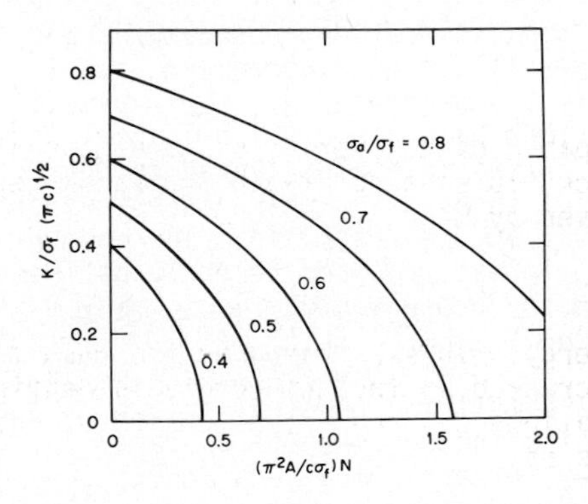

Fig. (10) - Dependence of stress intensity factor K of an elastic-plastic crack as a function of the number of dislocations N in the plastic zone

reaches zero. The maximum number of dislocations in the plastic zone N_{max} corresponding to K=0 is found to be

$$N_{max} = \frac{2c\sigma_f}{\pi^2 A} \ln(a/c) \tag{5}$$

which is identical to the number of dislocations expected in the plastic zone of a BCS crack. The BCS crack therefore corresponds to a crack which is completely relaxed (K=0) by plastic deformation. Between the elastic crack (N=0) and the plastic crack ($N=N_{max}$), we expect a wide range of elastic-plastic cracks which are partly elastic with a stress singularity ($K \neq 0$) and partly plastic with a plastic zone.

The fact that the DFZ is present in many of the cracks observed in the electron microscope indicates that the crack tip was unable to generate a sufficient number of dislocations to eliminate the stress concentration at the crack tip. In order to understand the difficulty associated with the generation of dislocations at the crack tip, we have considered the process of dislocation nucleation based on the model of Rice and Thomson [17]. When a dislocation is generated close to the crack tip of a Mode III crack, the shear stress on the dislocation is given by

$$\sigma_{yz}(r) = \frac{K}{(2\pi r)^{1/2}} - \frac{A}{2r} - \sigma_f \tag{6}$$

where r is the distance between the crack tip and the dislocation. The first term represents the elastic crack stress, the second term the image stress and the third the friction stress. At an equilibrium distance r_e from the crack tip, the stress on the dislocation vanishes, i.e., $\sigma_{yz}(r_e) = 0$. The condition for spontaneous generation of dislocations is given by $r_e < r_c$ where r_c is the core radius of a dislocation. According to equation (6), this condition is equivalent to

$$\sigma_{yz}(r_c) > 0 \tag{7}$$

For materials that are capable of generating dislocations at the crack tip, we propose that K at the crack tip does not reach the value required for brittle fracture, K_c, which is given by

$$K_c = 4\mu\gamma \tag{8}$$

where γ is the surface energy. This is based on the result shown in Figure 10 which indicated that K decreased as the number of dislocations N increased. It is then possible from equations (6) and (7) to derive a condition for spontaneous generation of dislocations as

$$K > K_g \tag{9}$$

where K_g is the critical stress intensity factor required for the generation of dislocations at the crack tip and is given by

$$K_g = A(\pi/2r_c)^{1/2} + \sigma_f(2\pi r_c)^{1/2} \tag{10}$$

K_g defines quantitatively the difficulty imposed on a crack tip in generating a dislocation and is a material constant. The physical factors involved are the image stress and the friction stress which act together against the plastic crack stress in the generation of dislocations.

As the applied stress is increased, K at the crack tip increases. The dislocations are not generated until K reaches K_g. For $K > K_g$, the dislocations are generated and as they form a plastic zone they reduce the value of K at the crack tip. This process is described by Majumdar and Burns [18] as the plastic zone shielding the crack tip from the applied stress. The magnitude of K is reduced until K reaches K_g and the generation of dislocations will cease. As the applied stress is increased further, more dislocations will be generated but K will remain at K_g as can be seen in Figure 10. The number of dislocations N in the plastic zone and hence the extent of plastic deformation at the crack tip will be decided by the values of K_g and the applied stress σ_a. The relative magnitudes of K_g and K_c may be used as a criterion of ductile versus brittle fracture of a material. We see that

$K_c < K_g$ brittle fracture

$K_c = K_g$ semi-brittle fracture

$K_c > K_g$ ductile fracture

$K_g = 0$ completely ductile fracture.

In order to test the validity of the DFZ model of fracture developed here, we have determined the values of K experimentally from the crack tip geometry observed in the electron microscope in various metals using equation (4). These values were then compared with the theoretical values K_g and K_c defined by equations (8) and (10). For the values of σ_f, we used the yield stress in shear determined for the materials studied in the fracture experiments. We have found that in the expression for K_g given in equation (10), the contribution of the friction stress can be neglected compared to that of the image stress. Table 1 shows the measured values of K and σ_f and the predicted values of K_g and K_c for

TABLE 1 - COMPARISON OF STRESS INTENSITY FACTORS BETWEEN EXPERIMENT AND THEORY

	Stress Intensity Factor			
	Experiment		Theory	
Material	σ_f	K	K_g	K_c
S. Steel	43	1.6	1.8	7.6
Copper	15	0.9	0.9	5.2
Nickel	34	1.8	1.7	7.2
Aluminum	20	0.8	0.6	2.9
Niobium	25	1.7	1.6	7.9
Molybdenum	240	11.0	5.0	13.0
Tungsten	160	12.0	6.5	16.5
M_gO			8.0	7.5
	($M P_a$)	(10^5 $N/m^{3/2}$)		

the seven metals studied. It can be seen that for all of the metals studied, $K_g < K_c$. For the fcc metals and niobium, the measured K values are in very good agreement with the theoretical values of K_g. In semi-brittle bcc metals, the measured values of K are greater than K_g but are less than K_c. For these semi-brittle metals, there must be additional difficulties besides the image and friction stresses in generating the dislocations at the crack tip. The general agreement between experiment and theory must be considered as strong evidence in favor of the DFZ model of fracture, and in particular, the concept of the critical stress intensity factor K_g required for the generation of dislocations at the crack tip.

DISLOCATIONS ON AN INCLINED PLANE

As was shown in Figure 6, a linear array of dislocations has been frequently observed in front of a crack tip on a plane which is inclined to the crack plane when the specimen is loaded under Mode I conditions. In order to study the equilibrium distribution of dislocations on an inclined plane, we have treated this problem by formulating an integral equation. Owing to the mathematical difficulties involved, we will only present the results for the pileup of screw dislocations in the present paper. The results for the edge dislocation pileup are incomplete and they will be presented elsewhere.

Figure 11 shows the geometry of the problem in which an array of screw dislocations are located on a plane which makes an angle of ϕ with the crack plane. The Cartesian coordinate system (x,y) is set up with its origin at the tip of a semi-infinite crack. Antiplane shear stress is applied which results in the

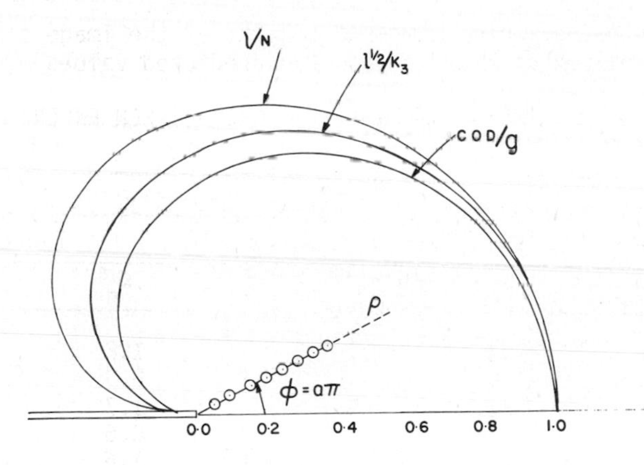

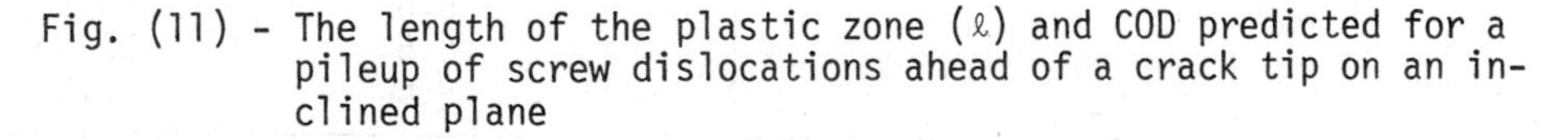
Fig. (11) - The length of the plastic zone (ℓ) and COD predicted for a pileup of screw dislocations ahead of a crack tip on an inclined plane

stress distribution

$$\sigma_{yz} + i\sigma_{xz} = \frac{K_3}{\sqrt{2\pi z}} \tag{12}$$

where σ_{yz} and σ_{xz} are the components of shear stress, K_3 is the Mode III stress intensity factor, $z = x + iy = re^{i\zeta}$ with $\zeta<\pi$. By considering the balance of force on each dislocation, we obtain the pileup integral equation

$$\frac{K_3}{\sqrt{2\pi r}} \cos(\zeta/2) - \sigma_f = \frac{A}{4\sqrt{r}} \int_0^{\ell} f(r')\left(\frac{2}{\sqrt{r'}-\sqrt{r}} + \frac{1}{\sqrt{r}+\sqrt{r'}e^{i\zeta}} + \frac{1}{\sqrt{r}+\sqrt{r'}e^{-i\zeta}}\right)dr' \quad (13)$$

where f is the distribution function for dislocations and σ_f is the friction stress. Equation (13) is solved by the Wiener-Hopf technique after applying the Mellin transform to the equation [19].

The condition of finite stress at the end of the plastic zone, sometimes referred to as the BCS (or Dugdale) condition, is obtained in a simple analytical form as

$$\frac{2\sigma_f\sqrt{2\pi\ell}}{\pi K_3} = \frac{\sin(a\pi)}{a\pi} (1+a)^{\frac{1+a}{2}} (1-a)^{\frac{1-a}{2}} \quad (14)$$

where ℓ is the length of the pileup and $a = \zeta/\pi$. The total number of dislocations N in the plastic zone divided by the elastic energy release rate $g = K_3^2/2\mu$, subject to the BCS condition, is also simply given by

$$\frac{\sigma_f bN}{g} = \frac{\sin(a\pi)}{a\pi} \quad (15)$$

Numerical results for these equations and also the ratio ℓ/N are plotted in Figure 11 in polar coordinate as a function of the angle of inclination. They indicate that for a fixed applied force, K_3 = const., the pileup length ℓ, as well as the number of dislocations N, decreases as the angle of inclination ζ increases. It is noted that both ℓ and N decrease rapidly for $\zeta > \pi/2$. On the other hand, the amount of energy, $\sigma_f bN$, associated with the inclined pileup is nearly independent of ζ for $\zeta < \pi/3$. Furthermore, the integral equation (13) may not represent a realistic physical situation for large $\zeta(> 3\pi/4)$ because the screw dislocations in this region are attracted strongly to the crack by the image stress so that they may be annihilated by cross-slipping into the crack surface.

REFERENCES

[1] Ohr, S. M. and Narayan, J., Phil. Mag. A, Vol. 41, p. 81, 1980.

[2] Kobayashi, S. and Ohr, S. M., Phil. Mag. A, Vol. 42, p. 763, 1980.

[3] Ohr, S. M. and Kobayashi, S., J. Metals, Vol. 32(5), p. 35, 1980.

[4] Kobayashi, S. and Ohr, S. M., Scripta Metall., Vol. 15, p. 343, 1981.

[5] Horton, J. A. and Ohr, S. M., Scripta Metall., Vol. 16, p. 621, 1982.

[6] Bilby, B. A., Cottrell, A. H. and Swinden, K. H., Proc. R. Soc. London, Ser. A, Vol. 272, p. 304, 1963.

[7] Chang, S.-J. and Ohr, S. M., Dislocation Modelling of Physical Systems, M. F. Ashby, R. Bullough, C. S. Hartley and J. P. Hirth, eds., Pergamon Press, New York, pp. 23-27, 1981.

[8] Chang, S.-J. and Ohr, S. M., J. Appl. Phys., Vol. 52, p. 7174, 1981.

[9] Horton, J. A., Proc. 40th Ann. Meeting of EMSA, G. W. Bailey, ed., 1982.

[10] Horton, J. A. and Ohr, S. M., J. Mater. Sci., (in press).

[11] Horton, J. A., to be published.

[12] Thomson, R., J. Mater. Sci., Vol. 13, p. 128, 1978.

[13] Weertman, J., Acta Metall., Vol. 26, p. 1731, 1978.

[14] Muskhelishvili, N. I., Singular Integral Equations, Noordhoff, Groningen, 1953.

[15] Chang, S.-J., Ohr, S. M. and Horton, J. A., Int. J. Fracture, (in press).

[16] Bilby, B. A. and Eshelby, J. D., Fracture, H. Liebowitz, ed., Academic Press, New York, Vol. 1, pp. 99-182, 1968.

[17] Rice, J. R. and Thomson, R., Phil. Mag., Vol. 29, p. 73, 1974.

[18] Majumdar, B. S. and Burns, S. J., Acta Metall., Vol. 29, p. 579, 1981.

[19] Chang, S.-J. and Ohr, S. M., to be published.

THE BEHAVIOR OF DISLOCATIONS AND THE FORMATION OF WALL STRUCTURES OBSERVED BY IN SITU HIGH VOLTAGE ELECTRON MICROSCOPY

T. Imura and A. Yamamoto

Department of Metallurgy, Faculty of Engineering, Nagoya University
Furo-cho, Chikusa-ku, Nagoya 464, Japan

ABSTRACT

The aim of the present experiments is to observe directly the behavior of dislocations and the dynamic formation of wall structures in connection with surface markings, i.e., the persistent slip bands (PSBs) during the cyclic deformation of Cu-2at%Al single crystals by carrying out in situ fatigue experiments in our 1000 kV HVEM in order to obtain the information useful for elucidating the formation mechanism of the wall structure and the PSBs.

EXPERIMENTAL PROCEDURE

The size and the geometry of the single cyrstals used are shown in Figure 1. Cyclic deformation was performed at first with bulk specimens in

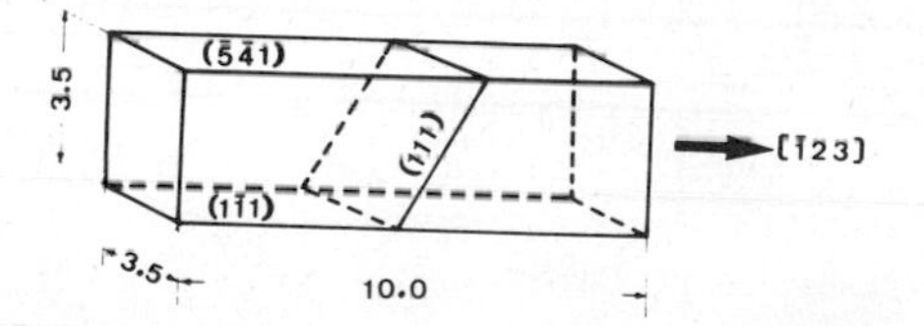

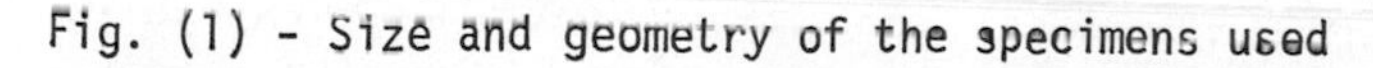
Fig. (1) - Size and geometry of the specimens used

push-pull (1Hz) at a constant plastic shear strain amplitude of ±0.4% for 500, 1000, 2000 and 4000 cycles. Each of these fatigued specimens was cut into slices parallel to (1) the cross slip plane $(1\bar{1}\bar{1})$, (2) the primary slip plane (111), and (3) one of the free surfaces $(54\bar{1})$ by turns at 0.4 mm spacing by a multi-wire saw and thinned down by electro-polishing to prepare foil specimens for the in situ HVEM observations. In the present experiments, the in situ push-pull straining device of bending type for use with a high voltage electron microscope developed by the authors was used [1]. Schematic drawings of the device and the way of mounting and straining the specimen are illustrated in Figure 2 and Figure 3, respectively.

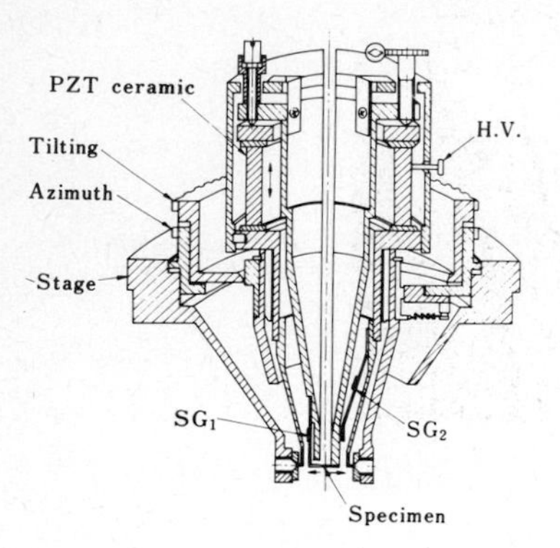

Fig. (2) - In situ fatigue device

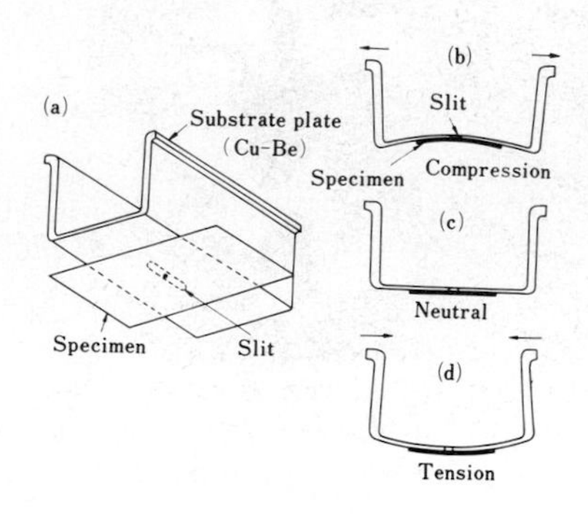

Fig. (3) - Specimen mounting and straining

RESULTS AND DISCUSSIONS

The behavior of dislocations in the foil specimen prepared from the pre-fatigued crystals by slicing it parallel to $(1\bar{1}1)$ is illustrated in Figure 4. In the rapid-hardening stage (500 cycles), the primary slips are homogeneously observed throughout the specimen (Figure 4A), but in the specimen fatigued for 1000 cycles (transition stage), an inhomogeneous deformation started to occur already. It seems that at the beginning of the transition stage, the dislocation substructures included in the volumes undergoing of deformation begin to change into the wall structures (Figure 4B). In the saturation-hardening stage (4000 cycles), slips are observed only in the wall structure (Figure 4C).

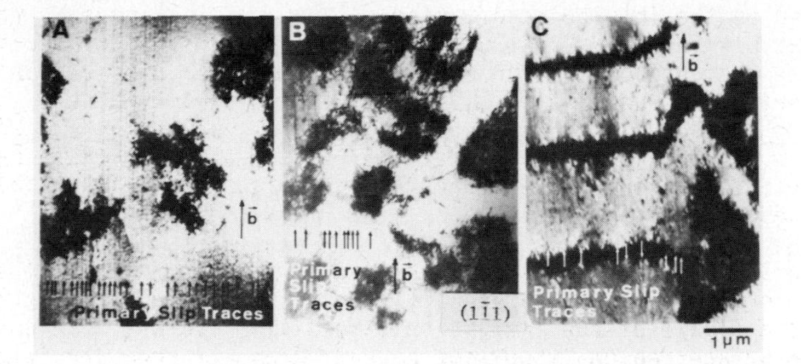

Fig. (4) - The behavior of dislocations observed in the $(1\bar{1}1)$ slice; A: 500 cycles, B: 1000 cycles, C: 4000 cycles

The typical dislocation substructures observed at the saturation-hardening stage (4000 cycles) in the thin foils whose surfaces are parallel to the cross slip plane $(1\bar{1}1)$ and the primary slip plane (111) are illustrated in Figure 5 and Figure 6, respectively.

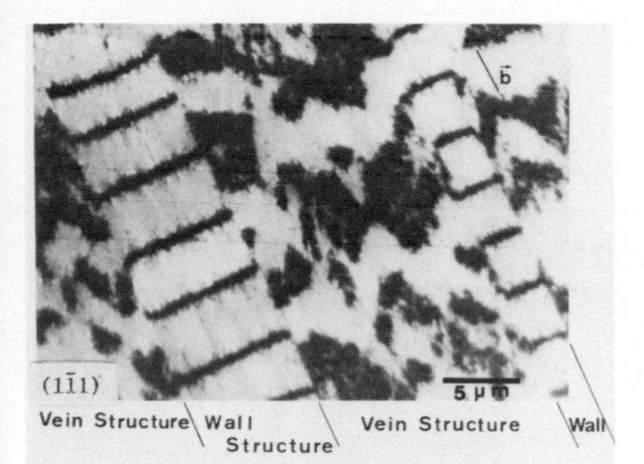

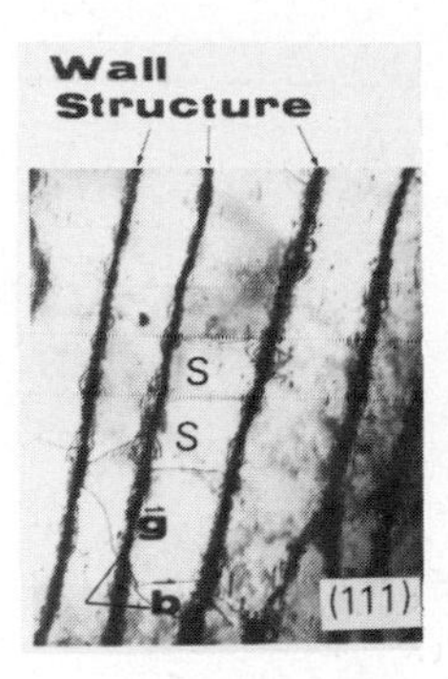

Fig. (5) - Typical dislocation substructure observed in (1$\bar{1}$1) slice at saturation-hardening stage (4000 cycles)

Fig. (6) - Dislocation substructure in (111) slice after 4000 cycles

Figure 7 indicates a three-dimensional structure of dislocation configurations developed in the specimen fatigued up to saturation hardening. Walls lined up fairly regularly at a constant spacing perpendicularly to the primary Burgers vector b=[10$\bar{1}$] on the primary slip plane in the wall structure. On the other hand, in the matrix, veins were observed, consisting of bundles of dislocations and having a slight tendency to line up.

Being based on the in situ observation illustrated in Figure 8, the behavior of dislocations in the wall structure is schematically indicated in Figure 9. Edge segment bows out to form a half loop from the wall and reaches to the neighboring wall under a stress, accompanying screws. These screws move to and fro under cyclic stresses, making occasionally jogs or cross slips which result in forming jogs and edge dipoles. The motion of these screws tends to contribute to form wall structures and voids (inside) and the PSBs (on the surfaces) [2,3].

Figure 10 illustrates the results obtained by x-ray topography (Berg-Barrett type) on the slices prepared from the Cu-2at%Al single crystal fatigued up to the middle of the transition stage (1700 cycles). 1A and 1B are from the near-surface region (front and back), 2A and 2B from the intermediate region and 3A and 3B from the center of the specimen. Slip bands are well observed in 2A, 2B, 3A and 3B, but not so clearly in 1A and 1B. The observation is supported by the tensile tests of the slices prepared from the near-surface region and the central region [4].

It is found that the dislocation substructures developed faster in central region and that those in the near-surface region retarded most probably by the escape of dislocations to the surfaces. The escape of edge components moving near the one of the free surfaces (54$\bar{1}$) plane, results in forming

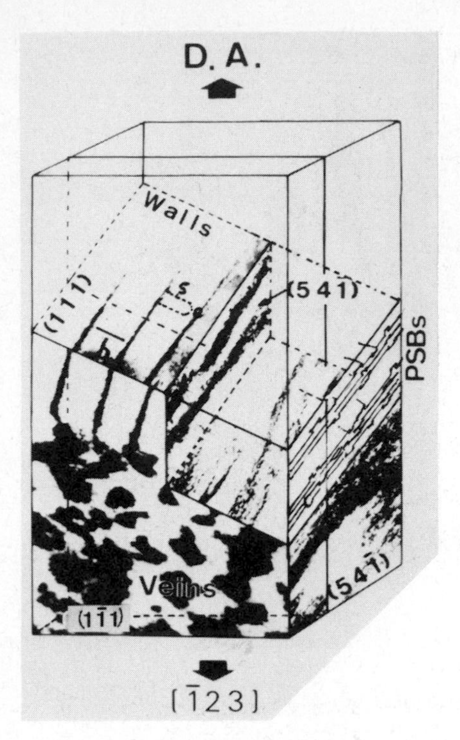

Fig. (7) - Three dimensional representation of dislocation configuration developed after cyclic deformation of 4000 cycles

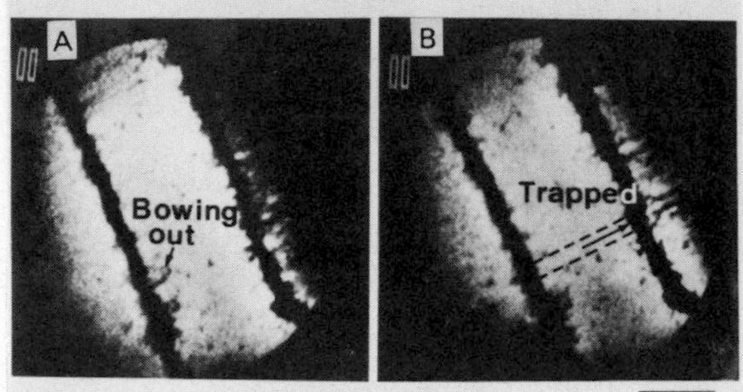

Fig. (8) - Video image of an edge segment bowing out from the wall (1Hz)

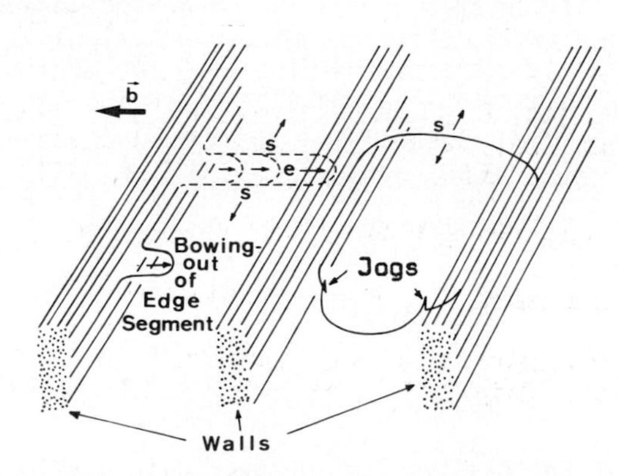

Fig. (9) - Schematic representation of the behavior of dislocations in the wall structure

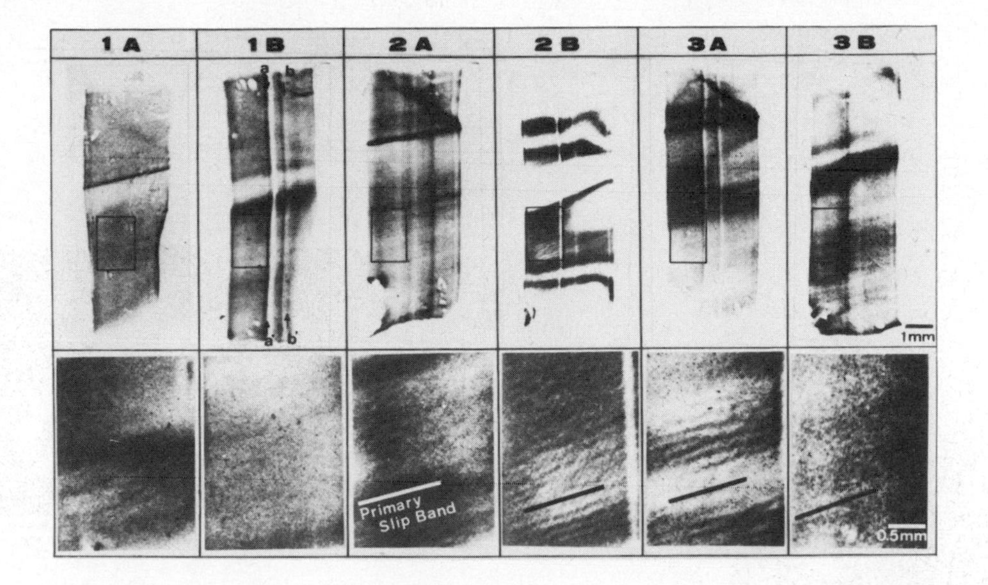

Fig. (10) - Slip bands observed by Berg-Barrett method at the near-surface region (1A, 1B), intermediate region (2A, 2B), and central region (3A, 3B) after 1700 cycles

screws which emerge at the free surface. The cyclic motion of these screws contributes to form surface markings, i.e., PSBs, in addition to the afore-mentioned edge components. Here, attention is directed to the paper by U. Essmann, U. Gösele and H. Mughrabi showing the production of vacancies by the annihilation of edge dislocations which leads to an elongation of the PSBs parallel to the slip vector.

REFERENCES

[1] Yamamoto, A., Morita, C., Tono, T., Saimoto, S., Saka, S. and Imura, T., Proc. Fifth Int. Conf. on High Voltage Electron Microscopy, Kyoto, (edited by T. Imura and H. Hashimoto), p. 133, 1977.

[2] Yamamoto, A., Saka, H. and Imura, T., Kristall Tech., 14, p. 1275, 1979.

[3] Yamamoto, A. and Imura, T., Electron Microscopy, p. 368, 1980.

[4] Yamamoto, A. and Imura, T., Proc. Int. Conf. on Dislocation Modelling of Physical Systems, Gainesville, Florida, U.S.A., 1980.

DISLOCATION DYNAMICS IN ALUMINUM AND IN ALUMINUM-BASED ALLOYS INVESTIGATED BY TEM AND NMR TECHNIQUES

J. Th. M. De Hosson and W. H. M. Alsem

Department of Applied Physics, Materials Science Centre,
University of Groningen, Nijenborgh 18, 9747 AG Groningen,
The Netherlands

H. Tamler and O. Kanert

Institute of Physics, Experimental Physics III,
University of Dortmund, 46 Dortmund 50, W. Germany

ABSTRACT

A new attempt has been made to evaluate the yield stress of an alloy containing non-deforming plate-like precipitates. It is based on an application of nuclear magnetic resonance techniques. Since the process of dislocation motion consists of atomic movements nuclear magnetic resonance techniques offer a possibility to determine the manner in which dislocations progress through the solid as a function of time. The spin lattice relaxation rate in the rotating frame, $T_{1\rho}^{-1}$ of ^{27}Al has been measured as a function of the plastic strain rate $\dot{\varepsilon}$ at 77 K. From $T_{1\rho}$-measurements which were performed on pure Al, Al-0.1 at% Cu in solid solution and on Al-1 at% Cu containing θ' plates, the mean jump distance of a mobile dislocation is found. Further, the mean diameter and mean thickness of the θ' plates were obtained from transmission electron microscopic observations. The yield stress of the Al-1 at% Cu alloy could be predicted by using the NMR and TEM data in a parallel plate model.

INTRODUCTION

The principal reason for the great interest in precipitation from solid solution is due to the startling increase in yield stress which usually accompanies it. To understand this increase in yield strength one must try to understand the way in which moving dislocations interact with precipitate particles of a second phase. In this paper, a nuclear magnetic resonance study on the dislocation dynamics in Al and in Al-Cu alloys is presented in terms of the mean jump distance of dislocations. The NMR method of determining characteristics of dislocations is essentially based on the interaction between nuclear electric quadrupole moments and electric field gradients at the nucleus. Around a dislocation the cubic symmetry is destroyed and consequently there exists an interaction between nuclear electric quadrupole moments and electric field gradients. In order to apply the NMR technique for studies on dislocations, it is essential to have a nucleus with a non-zero quadrupole moment.

In the following we will focus on plastic deformation experiments with a constant strain rate $\dot{\varepsilon}$. This type of experiment is governed by the Orowan equation:

$$\dot{\varepsilon} = \phi \cdot b \cdot \rho_m \cdot \frac{L}{\tau_w} \tag{1}$$

assuming a thermally activated, jerky motion of mobile dislocations of density ρ_m. The motion may be considered to be jerky-like if the actual jump time τ_j is small compared to the mean time of stay τ_w at an obstacle. In Equation (1) ϕ denotes a geometrical factor, b symbolizes the magnitude of the Burgers vector and L is the mean jump distance between obstacles which are considered to be uniform.

THEORETICAL BACKGROUND

While deforming a sample with a constant strain rate $\dot{\varepsilon}$ the spin-lattice relaxation rate in a weak rotating field H_1 ("locking field"), $1/T_{1\rho}$, of the resonant nuclei in the sample is enhanced due to the motion of dislocations. The resulting total relaxation rate may be decomposed into a background relaxation rate $(1/T_{1\rho})_0$ and the contribution $(1/T_{1\rho})_D$ which is governed by the mechanism of dislocation motion, i.e. by Equation (1):

$$(T_{1\rho}^{-1}) = (T_{1\rho}^{-1})_0 + (T_{1\rho}^{-1})_D \tag{2}$$

In metals and alloys, $(1/T_{1\rho})_0$ is due to fluctuations in the conduction electron-nucleus interaction leading to the Korringa relation $(T_{1\rho})_0 \cdot T = c$, where the magnitude of the constant c depends slightly on the strength of the locking field H_1 [1]. At a finite plastic strain rate, dislocations move in the crystal, i.e. causing time fluctuations both of the quadrupolar and dipolar spin Hamiltonian for spins with $I > 1/2$. However, the dipolar effects on the resulting nuclear spin relaxation due to dislocation motion are negligible and quadrupolar interactions dominate the relaxation behavior. The resulting expression for the relaxation rate induced by dislocation motion is given by [2,3].

$$\left(\frac{1}{T_{1\rho}}\right)_D = \frac{\delta_Q}{H_1^2 + H_{L\rho}^2} \cdot \langle V^2 \rangle \, g_Q(L) \cdot \frac{\rho_m}{\tau_w} \tag{3}$$

δ_Q is a quadrupole coupling constant and $\langle V^2 \rangle$ denotes the second moment of the electric field gradient due to the stress field of a dislocation of unit length [1]. $H_{L\rho}$ is the mean local field in the rotating frame determined by the local dipolar field $H_{D\rho}$ and the local quadrupolar field $H_{Q\rho}$. The quadrupolar geometry factor g_Q in Equation (3) which depends on the mean jump distance L approaches to one if L is of the order of 0.1-1 μm [3].

For ^{27}Al in aluminum, the quadrupole coupling constant δ_Q (Equation (3)) has the value 2.85 10^{-25} G^2 $dyne^{-1}cm^4$. The value of $\langle V^2 \rangle$ may either be determined theoretically by means of the corresponding theoretical expression of $V(r,\theta)$ for dislocations as given by Kanert and Mehring [4] or derived experimentally from an analysis of the line shape of the NMR signal of the sample which is quadrupole distorted by a known number of dislocations. In both cases $A_Q = \delta_Q \langle V^2 \rangle$ appeared to be about 3.10^{-10} G^2cm^2 [3].

Finally, combining Orowan's relation (1) with Equation (3) one obtains:

$$\left(\frac{1}{T_{1\rho}}\right)_D = \frac{A_Q}{H_1^2 + H_L^2} \cdot \frac{1}{\phi \cdot b} \cdot \frac{g_Q(L)}{L\rho} \quad \dot{\varepsilon} \tag{4}$$

Hence, for a given plastic strain rate $\dot{\varepsilon}$ the dislocation induced spin relaxation rate is proportional to the inverse of the mean jump distance L. This relationship is used in the experiments discussed below to determine L.

EXPERIMENTAL DETAILS

A. Sample-Preparation and Transmission Electron Microscopic Measurements

For the investigation, polycrystalline samples with a grain of the order of 100-200 μm were used. To avoid skin effect distortions of the NMR signal each sample consisted of a single rectangular foil with a length of 27 mm, a width of 12 mm, and a thickness of about 50 μm. The starting material for the samples was (a) 5N aluminum, (b) 5N Al:0.1 at% Cu, (c) 5N Al:1 at% Cu. After a homogenizing procedure at 550°C for 2.5 days the material was rolled out to the thin foils with a thickness of about 50 μm and has then been cut by spark erosion to the sample size given above. The ultrapure aluminum samples were annealed a second time at 290°C for 1 hour.

In order to get samples of solid solution the foils were annealed at 550°C for 2.5 h and then quenched to 20°C. Some of these samples were exposed to a third heat treatment (200°C for 1 day) in order to produce platelike precipitates of copper in these samples (θ'-phase in the Al:Cu phase diagram). Transmission electron micrographs were taken by using JEM 200 CX operating at 160 keV.

B. NMR Measurements and Deformation Experiments

In the NMR experiment, the sample under investigation is plastically deformed by a servo-hydraulic tensile machine (ZONIC Technical Lab. Inc. Cincinnati) of which the exciter head XCI TE 1105 moves a driving rod with a constant velocity. The movement is controlled by a digital function generator which serves the Master controller of the exciter head. While the specimen was deforming, ^{27}Al nuclear spin measurements were carried out by means of a BRUKER pulse spectrometer SXP 4-100 operating at 15.7 MHz corresponding to a magnetic field of 1.4 T controlled by an NMR stabilizer (BRUKER B-SN 15). The NMR head of the spectrometer and the frame in which the rod moves formed a unit which was inserted between the pole pieces of the electromagnet of the spectrometer. The unit could be temperature-controlled between 77 K and 550 K. The spectrometer was triggered by the electronic control of the tensile machine. The trigger starts the nuclear spin relaxation experiment at a definite time during deformation. The NMR measurements discussed here were carried out at 77 K. At such a low temperature nuclear spin relaxation effects due to diffusive atomic motions are negligible. Calculations of the correlation times for Al and Cu [3] show that in fact Al and Cu are immobile at 77 K. Therefore, an observable contribution of diffusive atomic motion to the measured relaxation rates does not occur. The local fields $H_{L\rho}$ in ultrapure Al and in Al-Cu were determined as follows: According to Equation (4), a plot of the dislocation-induced contribution to $T_{1\rho}$ vs. H_1^2 will yield a straight line which can be extrapolated to find the abscissa-intercept at $H_1^2 = -H_{L\rho}^2$. The local field $H_{L\rho}$ thus obtained in ultra pure Al is 3.16 G. In order to obtain the static local fields $H_{D\rho}$ and $H_{Q\rho}$ of samples (b) (c) the ^{27}Al spin echo signal was measured. $H_{L\rho}$ was found to be 3.3 G in Al-0.1 at% Cu and 4.4 G in Al-1 at% Cu.

Figure 1 exhibits deformation curves of some of the samples measured at 77 K. In particular, the data demonstrate the different plastic behavior of the Al:0.1 at% Cu (solid solution) sample and of Al:1 at% Cu (θ' phase). The increase of the yield strength of the Al:1 at% Cu sample compared to Al:0.1 at% Cu and pure aluminum is about 70 MPa.

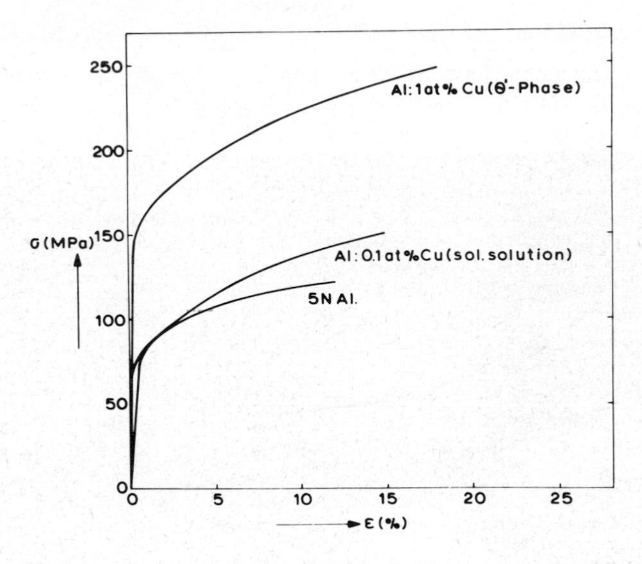

Fig. (1) - Experimental stress-strain curves at 77K

RESULTS AND DISCUSSION

A. Mean Jump Distance of Moving Dislocations in Al

In Figure 2 the mean jump distance measured by NMR in pure Al is illustrated as a function of strain. The mean jump distance L measured by NMR in Al has to be interpreted with care in terms of mean slip distance and statistical slip length (Λ_{st}). As commonly found in annealed f.c.c. metals, a cell

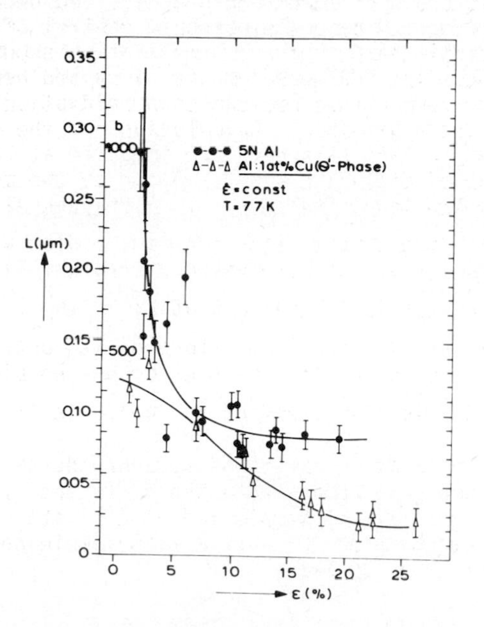

Fig. (2) - The mean jump distance measured by NMR as a function of strain ε in Al and in Al:1 at% Cu

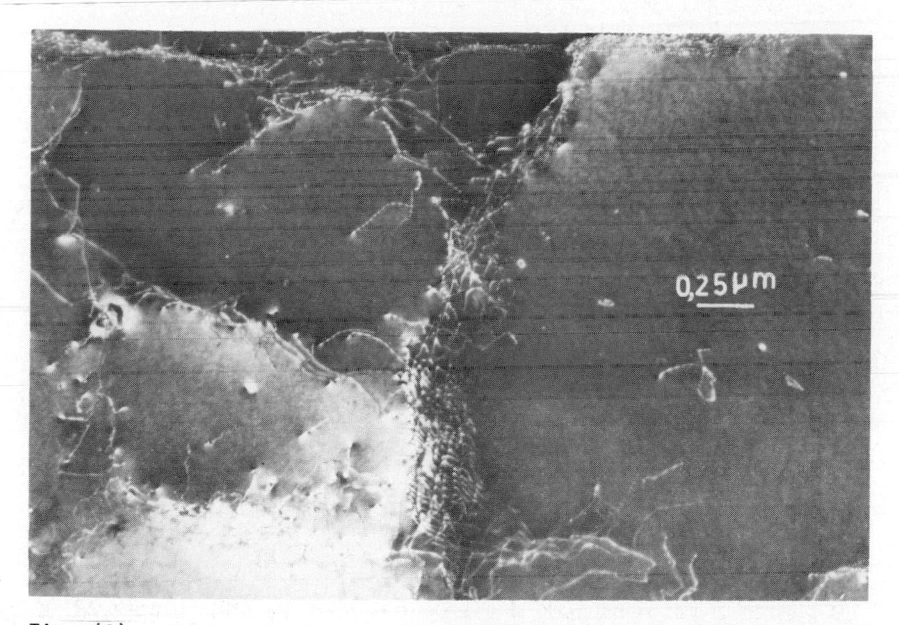

Fig. (3) - Cell structure of Al-deformed 15% at 77 K. Dark field/ weak beam image, [100] orientation, $\vec{g}$ = [002]

structure is formed in Al after deformation at 77 K. An electron micrograph illustrating the cell structure of deformed Al at 15% strain is shown in Figure 3. As a result, the mean slip distance of dislocations is mainly determined by the cell size when the cell structure is well developed. The statistical slip length Λ_{st} will be of the same order of magnitude as the cell size ($\simeq$ 1-2 μm) [5], i.e. much larger than the mean jump distance measured by NMR ($\simeq$ 0.1 μm for $\varepsilon \gtrsim 7\%$). A plausible explanation for this difference is that all moving dislocations, present both in the cell boundary and in the interior region of the cell, affect the spin lattice relaxation rate. The mean jump distance of dislocations measured by NMR is possibly related to the spacing of the dislocation tangles near the cell boundary ranging from 0.01 μm to 0.1 μm. The model that we assume is schematically depicted in Figure 4.

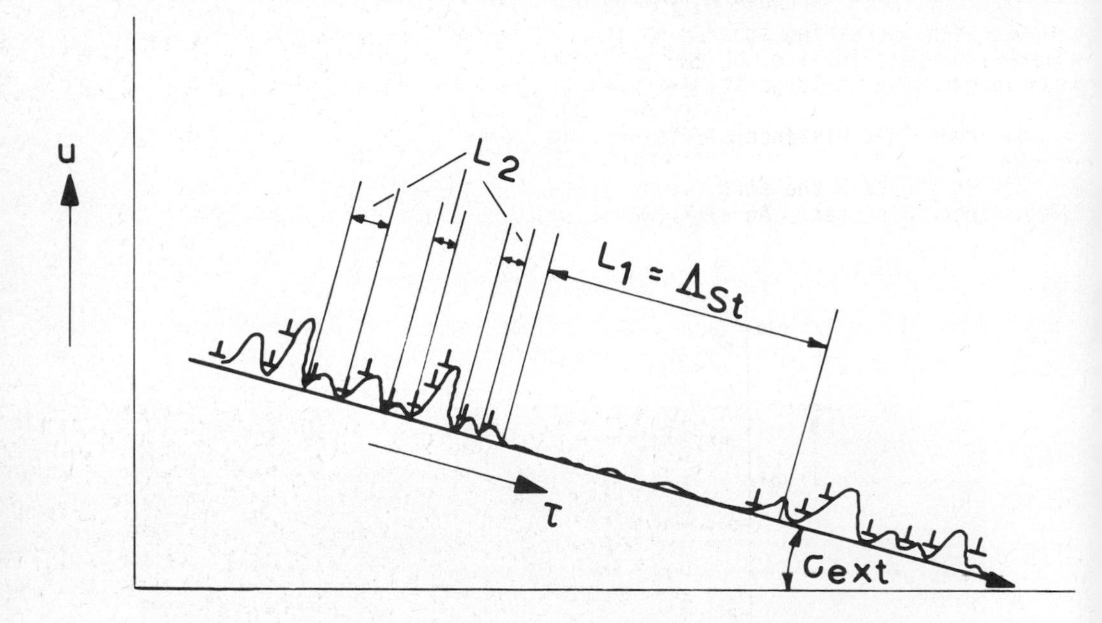

Fig. (4) - Schematical representation of the barrier potential in the case of a cell structure

A mobile dislocation crosses a cell by one step, i.e. the corresponding jump distance L_1 is of the order of the cell diameter ($\simeq \Lambda_{st}$). Subsequently, many short jumps occur with a distance L_2 ($\simeq$ spacing of the tangles 0.01-0.1 μm). Assuming two different sets of corresponding mobile dislocation densities: ρ_1 in the interior of the cell and ρ_2 inside the cell wall the total spin lattice relaxation rate can be written as:

$$\left(\frac{1}{T_{1\rho}}\right)_D = \left(\frac{1}{T_{1\rho}}\right)_D^{(1)} + \left(\frac{1}{T_{1\rho}}\right)_D^{(2)} \tag{5}$$

Further, since $L_1 >> L_2$ and $g_Q(L_1) \simeq 1$ and $g_Q(L_2) \simeq 0.6$ the total spin lattice relaxation rate measured by NMR is largely determined by the jump distance inside the cell wall: $(T_{1\rho}^{-1})_D \simeq (T_{1\rho}^{-1})_D^{(2)}$. Another explanation for the difference between the jump distance L measured by NMR and the mean slip distance Λ_{st} can be based on a dislocation mechanism in which the dislocation free path Λ_{st} depends on a number of dislocation intersections. If the dislocations are delayed at each of the intersections during a period of time $\tau_c > 10^{-4}$s, spin-lattice relaxation takes place. As a result $T_{1\rho}^{-1}$ is determined by the waiting time τ_c at each intersection. The dislocation mean jump distance L thus obtained is much smaller than the actual mean jump distance Λ_{st}. Since L decreases with increasing strains up to ≃ 7% (Figure 2), the formed model is somewhat preferable; i.e. before cells are being established, L is found to be much larger than at large strains when distinct cells are being formed.

B. Mean Jump Distance of Dislocations in Al-0.1 at% Cu (Solid Solution)

In Figure 5 the strain dependence of L in Al-0.1 at% Cu as measured by NMR has been depicted. An electron micrograph illustrating Al-0.1 at% Cu de-

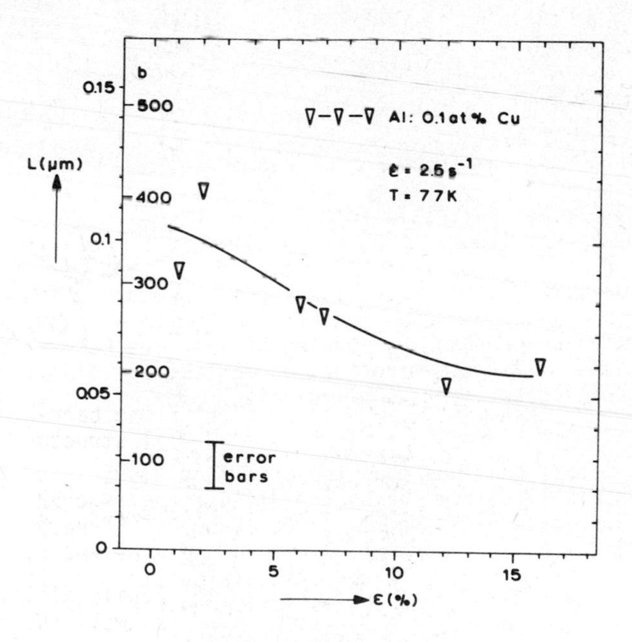

Fig. (5) - The mean jump distance measured by NMR as a function of strain ε in Al:0.1 at% Cu formed at 77 K (ε = 16%) is shown in Figure 6. In comparison with pure aluminum (Figure 3) there are more dislocations inside the cells indicating a certain amount of dislocation pinning by the matrix. The cells seem to have more ragged cell walls due to lowering of the stacking fault energy in Al upon adding Cu. A lowering of the stacking fault energy impedes cross slip and the

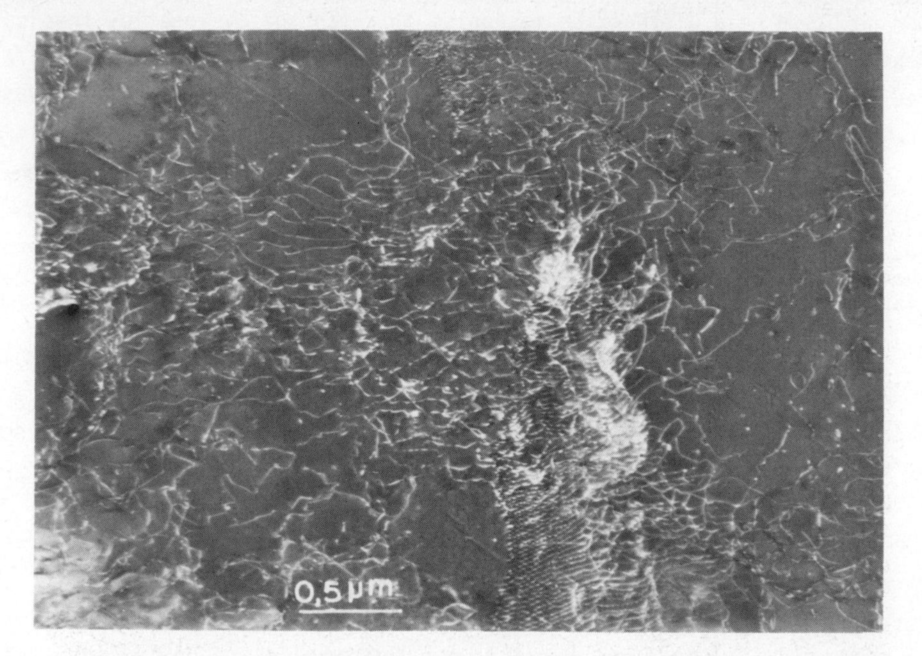

Fig. (6) - Al-0.1 at% Cu (solid solution) deformed at 77 K. Dark field/ weak beam image, [100] orientation, $\vec{g}$ = [002]

cell structure resulting from cross slipping screw dislocations is less readily formed. The mean jump distance measured by NMR ($\simeq$ 0.06 µm) is about one order of magnitude larger than the spacing between the Cu particles as calculated in the Friedel limit assuming that during the yielding process the dislocation takes up a steady state configuration. The mean jump distance would be 0.005 µm.

Therefore, we must conclude that the mean jump distance of dislocations in Al-0.1 at% Cu is mainly determined by the dislocation density and dislocation distribution. The latter is directly affected by Cu, impeding the cross slip process. Assuming a parabolic relation $\sigma \sim \sqrt{\rho}$ one expects from the experimental stress-strain curve (Figure 1) that the dislocation density for the same strain in Al-0.1 at% Cu is increased as compared to pure Al leading to a smaller value of the mean jump distance in Al-0.1 at% Cu. At large strains L in Al-0.1 at% Cu has been found to be 0.06 µm whereas in pure Al a value of 0.09 µm has been measured.

C. Mean Jump Distance of Dislocations in Al-1 at% Cu (θ')

In Figure 2 the NMR results are displayed in the case of Al-1 at% Cu (θ'-phase). Figure 7 shows an electron micrograph of Al-1 at% Cu. The typical microstructure shows plate-like θ' precipitates. According to Russell and Ashby [6] the interplate spacing in the slip direction (Λ_p) sets an upper limit for the slip distance, i.e. the actual distance traversed before it gets stuck.

Fig. (7) - Al-1 at% Cu (θ' phase), 26% deformation. Bright field/ strong beam image, [100] orientation, $\vec{g}$ = [002]

The mean jump distance of the dislocations measured by NMR can be associated with the effective interparticle spacing obtained from the transmission electron micrographs. The precipitates have a definite angle with the glide plane (54° 44'). In the glide plane, their mean length of the obstacles D' is $(\pi/4)D$ and their mean thickness d' is $d\sqrt{3/2}$ [7,8]. The mean centre-to-centre spacing is then given by:

$$\lambda' = \sqrt{\frac{D'\,d'}{f}} \tag{5}$$

where f is the volume fraction of the precipitates. Following Foreman et al. [9] a distribution of linear parallel obstacles of length S = D' - d' will give a hardening that is $1 + S/\lambda'$ times greater than for the associated distribution of point obstacles. The latter applies to aligned, line obstacles only. If we take into account the finite thickness of plate shaped particles the effective separation between the precipitates is

$$\Lambda_p = \frac{\lambda'}{1+S/\lambda'} - d' \qquad (6)$$

From the transmission electronmicroscopic observations follows: D = 0.36 μm, d = 0.005 μm and f = 0.03. Substituting the values for D' and d' into Equation (5) and Equation (6) leads to Λ_p = 0.11 μm. The effective particle spacing obtained from the transmission electronmicroscopic observations is in good agreement with the mean jump distance of dislocations measured by NMR (L_{NMR} = 0.12 μm, Figure 2). The hardening is thus controlled by the microstructure at the beginning of deformation. The dislocation configuration after an elongation of 20% at 77 K in the same material has been depicted in Figure 8. It should be noted at this point that Λ_p, defined by Ashby [10] as a constant, is independent of the shear strain a. Thompson et al. [5] proposed a modification to the Ashby view point where Λ_p sets an upper limit to Λ_{st} and $\Lambda_{st} \simeq \Lambda_p$ at yield. Then at small strain the slip distance is $\simeq \Lambda_p$ and at a large strain the slip distance is $\simeq \Lambda_{st}$. This situation has been found in our experiments. At the beginning of deformation $L_{NMR} \simeq \Lambda_p$. At larger strains the mean distance between the statistical dislocations is about 0.023 μm (determined from area indicated in Figure 8) whereas $L_{NMR} \simeq$ 0.03 μm (Figure 2).

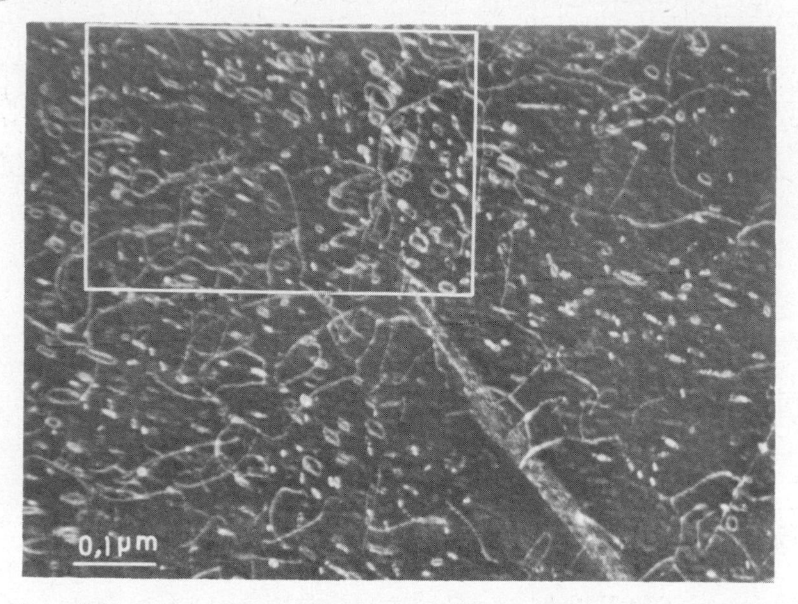

Fig. (8) - Al-1 at% Cu (θ' phase), 20% deformation at 77 K. Dark field/weak beam image. [100] orientation, $\vec{g}$ = [002]

For a theoretical evaluation of the yield stress, which is based on the NMR data, and for comparision with the deformation experiment the following theoretical expression of the yield stress is assumed:

$$\sigma = \sigma_0 + \frac{k_1 k_2}{2\pi\lambda} \mu \, b \ln \left(\frac{\lambda}{b}\right) \qquad (7)$$

where λ is the mean end-to-end spacing of the precipitates in the glide plane, k_1 is a statistical factor and k_2 is a factor depending on the character of the dislocation. σ_0 is considered to be the contribution to flow stress due to the residual solid solution ($\simeq$ 0.09 at% Cu). Equation (7) is similar to the expression proposed by Ashby [11]. If in the case of Orowan by-passing the bowed-out dislocation between the θ' obstacles are not much affected by dipole interactions the constant k_2 can be calculated on the basis of a line tension approximation [7,8]. The maximum value thus obtained is 1.5 and the minimum value of k_2 is 1. Substituting the values in Equation (7) and taking for $\lambda = L_{NMR} = 0.12$ μm, $\mu = 2.8 \; 10^{11}$ dyn/cm^2 and $k_1 = 0.85$ [8] we find that the minimum of the Orowan stress is 54.5 MPa and the maximum value is found to be 81.8 MPa. It means that the yield stress due to the hardening effect of the precipitates lies in the range of 54.5 MPa to 81.8 MPa. From Equation (7) it follows that this hardening effect has to be added to the contribution to flow stress due to the residual solid solution ($\simeq$ 0.09 at% Cu). This is in good agreement with the experimental stress-strain curves (Figure 1): the increase of the yield strength of Al:1 at% Cu samples compared to Al:0.1 at% Cu in solid solution is about 70 MPa.

The agreement between the experimentally observed hardening effect due to the precipitates and the theoretically predicted increase of the yield stress based on transmission electronmicroscopic observations would be even better, assuming that the glide plane is randomly oriented with respect to the precipitates. According to Kelly [8] the Orowan stress is then given by:

$$\tau = \tau_0 + B \frac{C}{D(1-\pi d C/2D)} \ln \left(\frac{2D}{\pi r_0}\right) \qquad (8)$$

where

$$B = \frac{k_1 \mu b}{2\pi\sqrt{1-\nu}}$$

and

$$C = \sqrt{\left(\frac{fD}{d}\right) + \left(\frac{2}{\pi} - \frac{\pi d}{2D}\right) \frac{fD}{d}}$$

Substituting the aforementioned values for f, D and d obtained from transmission electronmicroscopic observations in Equation (8), and taking k_1 equal to 0.85 and the core radius equal to b as is usual for a metallic system, we calculate for the hardening due to an array of parallel precipitates 73.4 MPa which is in good agreement with experiments (Figure 1).

CONCLUSIONS

The conclusion may be drawn that pulsed nuclear magnetic resonance is a complementary new technique for the study of dislocations in metallic systems. Since the process of dislocation motion is made up of atomic movements nuclear magnetic resonance technique offers a possibility to determine the manner in which dislocations progress through the crystal as a function of time utilizing nuclear spin relaxation as a tool.

It turned out that from the pulsed nuclear magnetic resonance experiments on Al and Al-Cu alloys the mean jump distance of moving dislocations can be deduced and subsequently the hardening effect due to precipitates in Al-1 at% Cu (θ') can be predicted. The mean jump distance of moving dislocations detected are:

(a) Al: ranging from 0.28 μm at small strains to 0.09 μm large strains (Figure 2).

(b) Al-0.1 at% Cu (solid solution), ranging from 0.11 μm at small strains to 0.06 μm at large strains (Figure 5).

(c) Al-1 at% Cu (θ' phase): varying from 0.12 μm at small strains to 0.03 μm at large strains (Figure 2).

ACKNOWLEDGEMENTS

The authors wish to thank Dr. H. J. Hackelöer for his technical assistance in the NMR measurements and Dr. G. J. L. Van Der Wegen for preparing the electronmicrographs. Particular thanks are due to Mr. H. J. Bron, Mr. J. Harkema and Mr. U. B. Nieborg for their technical assistance in preparing and analyzing the specimens.

This work is part of the research program of the Foundation for Fundamental Research on Matter (F.O.M. - Utrecht) and has been made possible by financial support from the Netherlands Organization for the Advancement of Pure Research (Z.W.O. - The Hague) and the Deutsche Forschungsgemeinschaft, W. Germany.

REFERENCES

[1] Wolf, D., "Spin Temperature and Nuclear Spin Relaxation in Matter", Clarendon Press, Oxford, 1979.

[2] Tamler, H., Hackelöer, H. J., Kanert, O., Alsem, W. H. M., De Hosson, J. Th. M., "Nuclear and Electron Resonance Spectroscopies Applied to Materials Science" (Eds. E. N. Kaufman, G. K. Shenoy), North Holland, 421, 1981.

[3] Tamler, H., Kanert, O., Alsem, W. H. M., De Hosson, J. Th. M., Acta Met. 30, 1982, in press.

[4] Kanert, O., Mehring, M., "Static Quadrupole Effects in Disordered Cubic Solids" (Springer-Verlag, Berlin) NMR, Vol. 3, pp. 40 ff. 1971.

[5] Thompson, A. W., Baskes, M. I., Flanagan, W. F., Acta Met. 25, p. 1017, 1973.

[6] Russell, K. C. and Ashby, M. F., Acta Met. 18, p. 891, 1970.

[7] Merle, P., Fouquet, F., Merlin, J., Mat. Sci. & Engin. 50, 215, 1981.

[8] Kelly, P. M., Scripta Met. 6, p. 647, 1972.

[9] Foreman, A. J. E., Hirsch, P. B., Humphreys, F. J., NBS Spec. Publ. 317, Vol. 2, p. 1083, 1970.

[10] Ashby, M. F., Phil. Mag. 21, p. 399, 1970.

[11] Ashby, M. F., Acta Met. 14, 679, 1966.

THE CYCLIC DEFORMATION OF TITANIUM: DISLOCATION SUBSTRUCTURES AND EFFECTIVE AND INTERNAL STRESSES

L. Handfield and J. I. Dickson

Ecole Polytechnique
Montreal, Quebec, Canada H3C 3A7

ABSTRACT

The dislocation substructures produced during the cyclic deformation at constant total strain amplitudes of polycrystalline commercial-purity titanium resembles substructures obtained in f.c.c. and b.c.c. metals. The cyclic softening stage in annealed titanium is associated with an increased mobility of screw dislocations and cross-slip. The cyclic stress amplitude, $\Delta\sigma$, is decomposed into an effective, $\Delta\sigma^*$, and an internal, $\Delta\sigma_i$, stress component and the different cyclic hardening or softening stages are related to changes in these stress components. The cyclic softening stage corresponds essentially to a decrease in $\Delta\sigma^*$, which result is discussed in terms of the mechanism by which oxygen in solution hardens titanium and zirconium.

INTRODUCTION

Transmission electron microscopy (TEM) observations of cyclically deformed polycrystals and their comparison with dislocation microstructures produced in single crystals have been of considerable recent interest [1-3]. The primary objective of this paper is to present TEM observations obtained on a polycrystalline commercial-purity titanium (Ti-40) and to compare these with observations obtained on mono- and polycrystals of other metals.

The room temperature cyclic behavior of Ti-40 has been described previously [4,5] and can be summarized as follows: For material initially in the annealed state, the stress amplitude $\Delta\sigma$, taken as the average of the tensile and compressive peak stresses, shows for reversed total strain amplitudes, $\Delta\varepsilon_t$, between (±) 0.15% and 0.75%, initial cyclic hardening followed by cyclic softening, Figure 1, and occasionally, if the fatigue life at $\Delta\varepsilon_t$ = 0.25 or 0.35% is sufficiently long, cyclic rehardening. For $\Delta\varepsilon_t$ = 1%, the softening observed can be slight or, as in Figure 1, absent. For material prestrained in tension, the relaxation of the residual stress influences the σ_{max}-log N curve and other cyclic effects can be seen by analyzing the σ_{min}-log N curve. A sufficient prestrain (> 8%) prevents the cyclic softening stage observed in the annealed material. The cy-

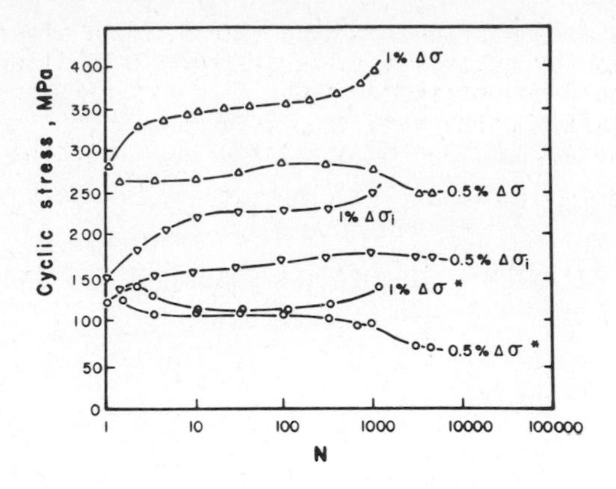

Fig. (1) - Curves of $\Delta\sigma$, $\Delta\sigma_i$ and $\Delta\sigma^*$ versus log N for annealed samples cycled at $\Delta\varepsilon_t$ = 0.5 and 1.0%

clic behavior of other grades of commercial-purity titanium [6-9] is similar, although the initial hardening stage may be absent [6,7]. In contrast, only cyclic hardening was observed in higher purity titanium [10]. A second objective of the present study is to present results in which $\Delta\sigma$ has been decomposed into its components, $\Delta\sigma^*$ and $\Delta\sigma_i$, in order to help understand the relationships between the $\Delta\sigma$-log N curves and the changes in dislocation substructures.

EXPERIMENTAL PROCEDURE

The mechanical testing procedure employed has been described previously [4]. The material contains approximately 1200 wt ppm of oxygen. The average grain size was, for most samples, approximately 75 μm. Fully reversed pull-push strain cycling at constant $\Delta\varepsilon_t$ between ±0.15 - ±1.0% was employed generally with a sinusoidal waveform but occasionally with a symmetric sawtooth waveform. The product of frequency in Hertz and $\Delta\varepsilon_t$ in % was 0.35.

The thin foils were prepared from sections cut perpendicular to the tensile axes and examined in an older AEI electron microscope, or in a JEOL 100CX microscope. As in the study of Stevenson and Breedis [8], considerable use was made of the fact that since slip occurs essentially on $\{10\bar{1}0\}$ $\langle 1\bar{2}10\rangle$ slip systems [5,8,11] for foils viewed in directions approximately perpendicular to the c-axis, dislocations parallel and perpendicular to the trace of the basal plane can be considered to be screw and edge dislocations, respectively. Observations with foils viewed in directions close to the c-axis permit to study the dislocation arrangements on the different $\{10\bar{1}0\}$ slip planes or on their $\{1\bar{2}10\}$ polygonization planes. As well, the dipole loops produced during cycling are elongated in the direction of edge dislocations.

For the separation of $\Delta\sigma$ into its components, the procedure, originally suggested by Cottrell [12] and employed by Kuhlmann-Wilsdorf and Laird [13] to obtain back and flow stresses, was attempted. However, especially for tests in which a sinusoidal waveform was employed, the truly elastic portion of the hysteresis loop became very short during cycling. Noting that the logic behind this procedure is similar to that behind the method of determining internal and effective stresses during monotonic deformation by unloading to zero stress relaxation [14] and that zero stress relaxation should be obtained, as verified experimentally, close to the middle of the short elastic portion of the hysteresis loop, it was found preferable to simply determine $\Delta\sigma_i$ and $\Delta\sigma^*$ by obtaining the mid-point of this elastic portion as schematized in Figure 2. In titanium, it has been shown [15] that

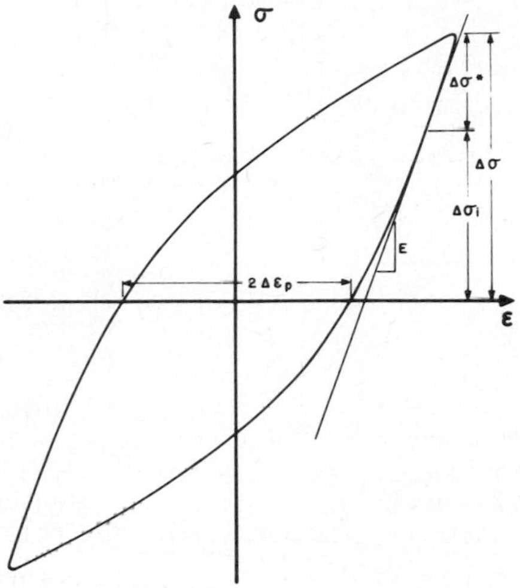

Fig. (2) - Schema showing the separation of $\Delta\sigma$ into $\Delta\sigma^*$ and $\Delta\sigma_i$

the unloading to zero stress relaxation method gives results similar to the double strain rate change method [16], and $\Delta\sigma^*$ can be considered as the thermal component of $\Delta\sigma$.

RESULTS AND OBSERVATIONS

A. Effective and Internal Stresses

The scatter in the separation of $\Delta\sigma$ into its component $\Delta\sigma_i$ and $\Delta\sigma^*$ depended on the quality of the recordings of the hysteresis loops. Good quality recordings were especially important in order to follow variations during individual tests. The general trend observed from the hysteresis loop recordings which were of better quality is shown in Figures 1 and 3. Figure 1 is for two annealed samples and Figure 3 shows results for a prestrained and an annealed sample cyclically strained at room temperature and 150°C respectively. Initial cyclic hardening in annealed samples corresponds to an increase in $\Delta\sigma_i$, cyclic softening to a decrease in $\Delta\sigma^*$, and cyclic rehardening at least partially to a reincrease in $\Delta\sigma^*$. Initial cyclic softening in prestrained samples corresponds to a decrease in $\Delta\sigma_i$. Similar trends have been observed on zirconium [17]. Figure 4 presents the curves for the room-temperature (295 K) cyclic stress amplitude $\Delta\sigma$ and its two components measured at mid-life versus $\Delta\varepsilon_t$ for annealed samples and for samples prestrained 10%, at 77, 295 and 468 K. For the three results for annealed samples at $\Delta\varepsilon_t = \pm 1.0\%$, the larg-

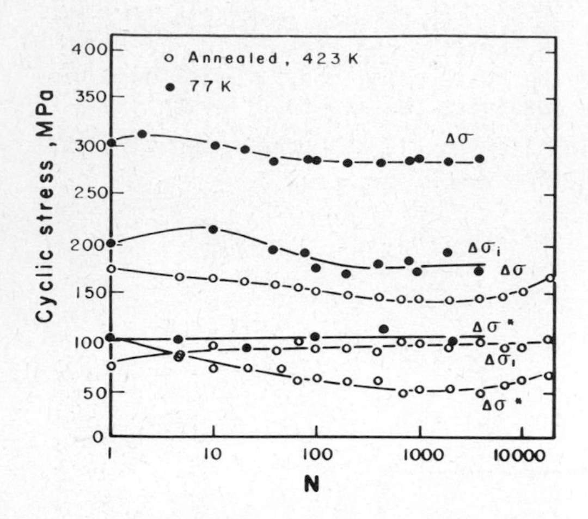

Fig. (3) - Curves of $\Delta\sigma$, $\Delta\sigma_i$ and $\Delta\sigma^*$ versus log N for a sample prestrained at 77K and for an annealed sample cycled at 423K, $\Delta\varepsilon_t = 0.5\%$

Fig. (4) - Curves of the mid-life values of $\Delta\sigma$, $\Delta\sigma^*$, $\Delta\sigma_i$ versus $\Delta\varepsilon_t$ for annealed and prestrained samples. Prestraining temperatures indicated

est internal stress corresponds to a sample which showed continuous cyclic hardening, while the other two samples showed initial hardening followed by slight softening. For the sample that showed continuous cyclic hardening, Figure 1 suggests that cyclic softening occurred in the σ^*-log N curve.

B. Dislocation Substructures

The dislocation substructures observed in initially annealed titanium cycled to the end or close to the end of the fatigue life were studied as a function of strain amplitude. At the lowest strain amplitudes employed, $\Delta\varepsilon_t = 0.15\%$, loose arrangements of dipoles, debris and dislocations can be found, Figure 5, similar to observations on copper polycrystals [1]. As well, walls of edge dislocation dipole loops perpendicular to the trace of the basal plane with a high density of screw dislocations in the channels between the walls can be found, Figure 6. Edge dislocations can at times be seen to bow out from walls, as observed in copper [18]. As the amplitude increases, the dipoles loops are arranged in patches, roughly perpendicular to the trace of the basal plane, and then into denser narrower walls, generally more perpendicular to the trace of the basal plane. As well, for observations made at the end of the fatigue life and therefore at a decreasing N with increasing $\Delta\varepsilon_t$, the width of the channels between the walls or patches of dipole loops increases. Screw dislocations can generally be found in these channels. The walls and patches appear from the different observations to tend to be perpendicular to the primary <1$\bar{2}$10> Burgers vector. At an amplitude of $\Delta\varepsilon_t = \pm 0.75\%$, series of still narrower parallel walls can form,

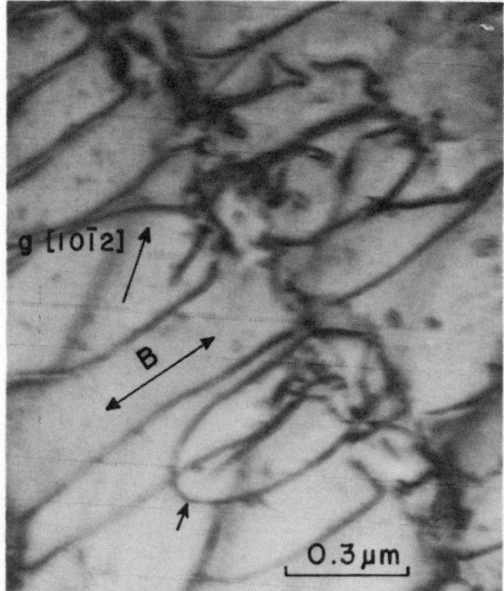

Fig. (5) - Loose arrangements of dipole loops and debris, $\Delta\varepsilon_t$ = 0.15%, N = 30,000 cycles, B = basal trace

Fig. (6) - Edge dislocation bowing out of a wall of dipole loops, $\Delta\varepsilon_t$ = 0.15%, N = 167,000

similar to those in the ladder structure of persistent slip bands, although actual narrow persistent slip bands were not observed by TEM. Screw dislocations perpendicular to these walls were at times detected in the channels between these narrow walls, and some of the observations, Figure 7, suggested alignment of screw dislocation across walls. Although the misorientation across most walls is small, significant misorientations can be found across some of these walls. From foils viewed closer to the c-axis, Figure 8, it can be seen that most walls have the same orientation but a few short walls of two other orientation can be present, with the angles between walls suggesting the 120° between the $\{1\bar{2}10\}$ polygonization planes. The indication is that each wall orientation contains principally dislocations of one slip system and that these wall structures form when the slip is principally on the primary $\{10\bar{1}0\}$ slip plane but that limited amounts of slip can occur on the other $\{10\bar{1}0\}\langle 1\bar{2}10\rangle$ slip systems.

At higher amplitudes ($\Delta\varepsilon_t \simeq 1\%$), cells appear with significant misorientations across most of the cell boundaries. These cells are elongated parallel to the trace of the basal plane indicating that these boundaries are twist boundaries. The presence of twist boundaries indicate that an important amount of dislocation activity is now occurring on at least a second $\{10\bar{1}0\}\langle 1\bar{2}10\rangle$ slip system. The cell walls in the perpendicular direction often appear to be polygonized boundaries. Because three $\{1\bar{2}10\}$ polygonization planes are possible, one for each Burgers vector, three-dimensional cells can be enclosed by such a combination of twist and polygonized boundaries.

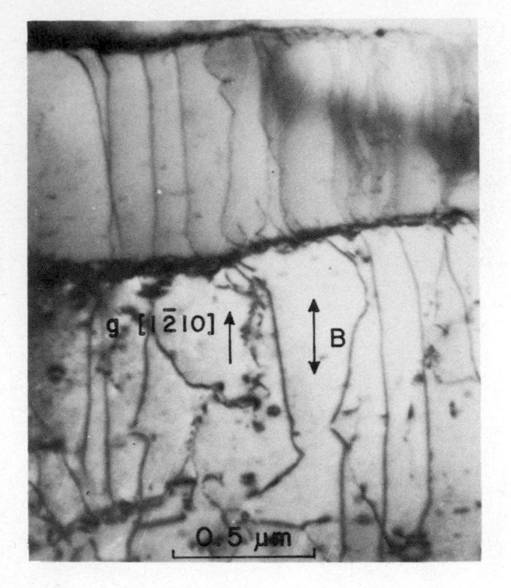

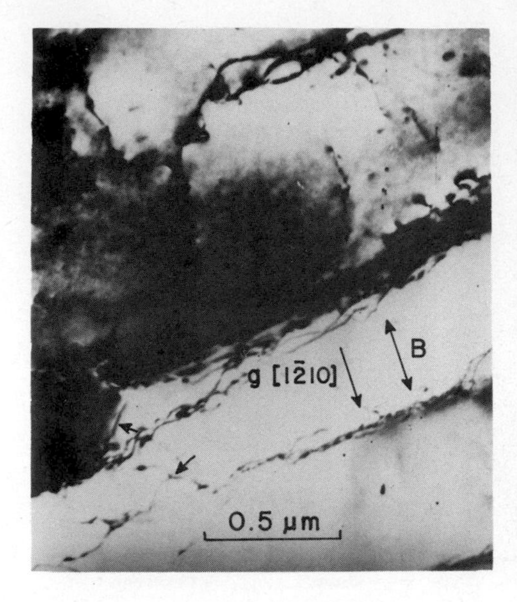

Fig. (7) - Screw dislocations gliding between narrow polygonized walls, $\Delta\varepsilon_t$ = 0.75%, N = 2100

Fig. (8) - Series of parallel narrow walls. Walls of different orientations are indicated by arrows. $\Delta\varepsilon_t$ = 0.75%, N = 2100

Tests were interrupted at different stages of the cyclic behavior curve particularly for $\Delta\varepsilon_t$ = 0.25 and 0.35%, in order to observe the dislocation substructures associated with the different stages of the $\Delta\sigma$ - log N curve. Just prior to the start of cyclic softening, the substructure consists primarily of long screw dislocations, Figure 9, although the dislocation density was lower at $\Delta\varepsilon_t$ = 0.15%, where less cyclic hardening occurred. As cyclic softening proceeds, the screw dislocations gradually become less numerous. As well, they become jogged, Figure 10, indicating that their mobility has increased and they show some tendency to pair off in dislocations of opposite signs (jogs in opposite directions). As well, dipole loops, which appear to form from the jogs, become visible as well as loose clusters of dipoles loops which begin to form perpendicular to the screw dislocations. As cycling proceeds, the long screw dislocations gradually disappear. When the minimum stress amplitude is reached, the dislocation substructure consists of loose patches of dipole loops, Figure 11, with screw dislocations in the wide channels between patches. As cycling proceeds, the dipole loops become compressed into narrower denser walls or veins. The few samples which showed cyclic rehardening showed more numerous veins or walls and narrower channels between the walls, Figure 12. Similar observations have been made for a commercial purity zirconium [20] at $\Delta\varepsilon_t$ = 0.35%, for which the cyclic rehardening stage was very prominent. The only difference was that the initial cyclic hardening stage was absent and screw dislocations formed at the start of cycling but did not build up in density. In a similar commercial purity titanium (Ti-50)

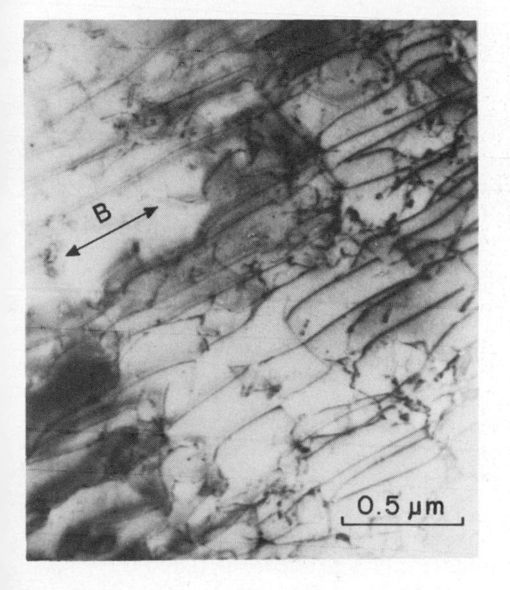

Fig. (9) - Long screw dislocations observed near the end of the cyclic hardening stage, $\Delta\varepsilon_t$ = 0.25%, N = 200

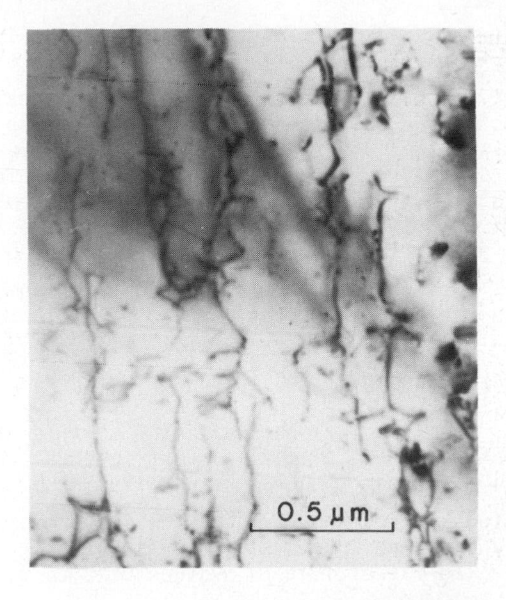

Fig. (10) - Jogged screw dislocation observed during the softening stage, $\Delta\varepsilon_t$ = 0.25%, N = 500

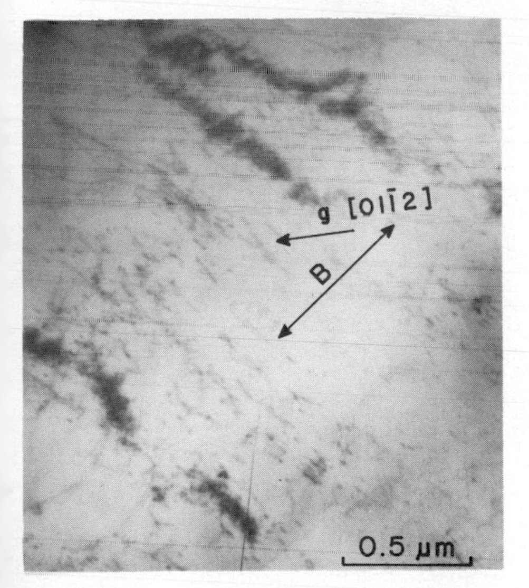

Fig. (11) - Loose patches of dipole loops observed toward the end of the softening stage, $\Delta\varepsilon_t$ = 0.25%, N = 5000

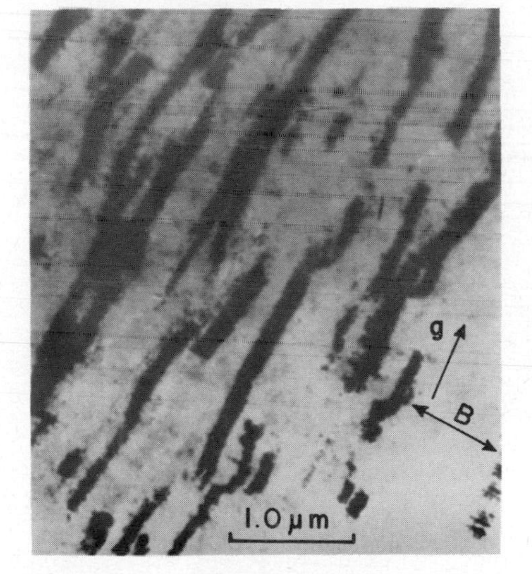

Fig. (12) - Dense walls of dipoles with narrow channels between walls observed during rehardening stage, $\Delta\varepsilon_t$ = 0.25%, N = 40,000

Stevenson and Breedis [8] observed the initial build-up of the substructure of long screw dislocations even for a plastic strain amplitude, $\Delta\varepsilon_p$, of 1%, where it eventually transformed into an elongated cellular substructure without cyclic softening being observed on the $\Delta\sigma$ - log N curve.

Because of the highly anisotropic plastic strain behavior of the h.c.p. lattice, the dislocation substructures observed varied from grain to grain. The dislocations substructure described are those that appeared most typical. Within a grain, the substructure often appeared uniform. At higher values of $\Delta\varepsilon_t$, the regions in the immediate vicinity of some grain boundaries and twins appeared to be preferential sites for the formation of cells. At low values of $\Delta\varepsilon_t$, occasional zones at grain boundaries had much lower dislocation densities, somewhat similar to observations in copper polycrystals [2].

The influence of a 10% tensile prestrain at temperatures of 77, 295 and 468K on the dislocation substructures produced during subsequent cyclic deformation at room temperature was also studied. For $\Delta\varepsilon_t > 0.5\%$, many of the dislocations introduced during a tensile prestrain can be fairly rapidly annihilated and transformed into dislocation substructures typical of those produced in the annealed material at $\Delta\varepsilon_t$ values 0.25 to 0.5% higher. For example, Figure 13 shows

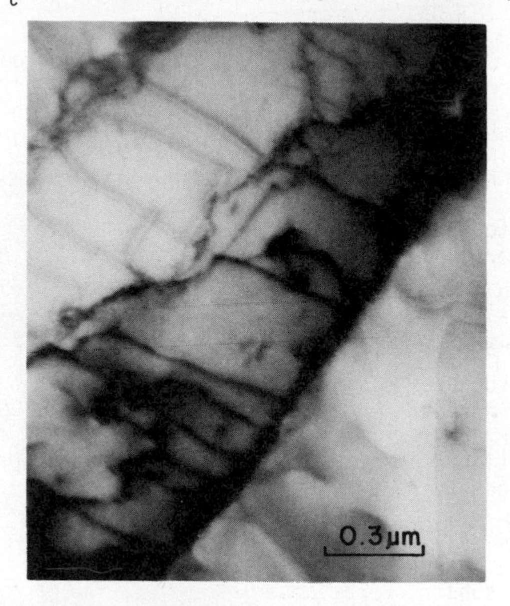

Fig. (13) - Aligned screw dislocations across narrow wall, $\Delta\varepsilon_t = 0.5\%$, prestrained 10% at 295K, N = 7620

aligned screw dislocations across narrow walls which can be compared to Figure 7. The presence of cells was a frequent feature of samples cycled after prestraining Prestraining at 468K caused a particularly strong tendency to form cells even at $\Delta\varepsilon_t = 0.5\%$, but this appears at least partially associated with the smaller grain sizes ($\approx$ 25 μm) of these samples [4]. For $\Delta\varepsilon_t = 1\%$, the sample prestrained at

77K showed cells, some of which were elongated parallel to the trace of the basal plane and some of which were elongated perpendicular to this trace, Figure 14, in contrast to the cells in the sample prestrained at room temperature which were

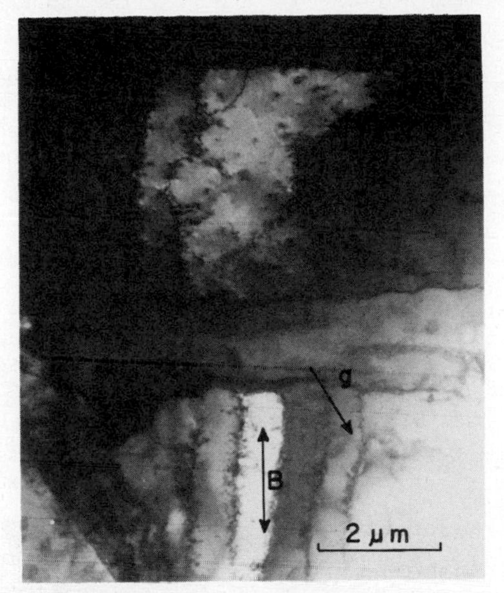

Fig. (14) - Cells elongated parallel and perpendicular to the trace of the basal plane, $\Delta\varepsilon_t$ = 1.0%, prestrained 10% at 77K, N = 478

elongated parallel to the trace of the basal plane, Figure 15. Cells elongated in two directions correspond to the "mosaic" substructure observed for $\Delta\varepsilon_p$ = 1% in annealed Ti-50 [8]. Within many individual cells, mainly edge dislocations were found, although within other cells, screw dislocations dominated or edge and screw dislocations were almost equally prominent. For low values of $\Delta\varepsilon_t$ (0.15 - 0.35%), the substructures observed in some regions approached those typical of cyclic deformation in that the dense veins or walls of dipole loops formed; however, especially for the lower amplitudes within this range, many regions contained high dislocation densities between the veins more typical of the dislocation substructures observed directly after tensile prestraining.

DISCUSSION

The dislocation substructures observed after a large number of cycles in Ti-40 resemble those observed previously in Ti-50 [8] as well as in other metals [1-3, 20-22]. At low amplitudes, patches, veins and walls of dipole loops form perpendicular to the primary Burgers vector. At higher amplitudes, more regular, narrower walls form, which are polygonized and largely dipolar, and which appear similar to the walls in the ladder structure of persistent slip bands. At even higher amplitudes, cells form generally elongated in the direction of twist boundaries. In the less anisotropic f.c.c. and b.c.c. metals, the cells have little tendency to be crystallographic.

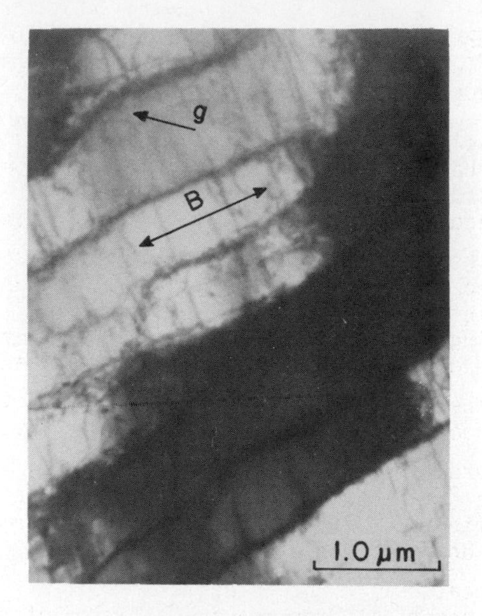

Fig. (15) - Cells elongated parallel to the trace of the basal plane, $\Delta\varepsilon_t$ = 1.0% prestrained 10% at 295K, N = 280

It is clear from the TEM observations on commercial-purity titanium, that the initial cyclic hardening in annealed titanium corresponds to an increase in the density of long screw dislocations. During this hardening stage, the screw dislocations have little mobility, as indicated by the absence of jogs, and the glide of edge dislocations assures the plastic strain. The subsequent cyclic softening is associated with an increase in the mobility of the screw dislocations and with cross-slip. This is indicated by the observation of pairs of jogged screw dislocations of opposite signs, Figure 10, which presumably would annihilate after a few more cycles. As the screw dislocations become more mobile, edge dislocation dipole loops start to form from jogs on the screw dislocations. The softened substructures are dipolar polygonized arrangements of loops and tilt boundaries with screw dislocations observed in the channels between these walls. At least at lower amplitudes, these substructures evolved during cycling as the dipole loops become more tightly compressed in the boundaries and as new walls form decreasing the width of the channels. The latter effect results in cyclic rehardening. Because of this continuous evolution of the dislocation substructures, it is inappropriate to refer to saturated stresses for this material, at least for low amplitudes.

The indications of cooperative glide on opposite sides of the narrow polygonized boundaries support the proposals of Kuhlmann-Wilsdorf and Laird [21]. The curvature of the dislocations in the vicinity of and on opposite sides of the boundary shown in Figure 7 appears consistent with the cooperative glide mechanism of their Figure 6b [21], which they considered as less probable. Such indications for cooperative glide were not found by Grosskreutz and Mughrabi [20] in their neutron-irradiated thin foils of copper. When cross-slip is easy, screw dislocation annihilation may prevent cooperative glide from operating since the required arrangements of screw dislocations do not develop. In the present study, significant misorientations were found across some polygonized walls indicating that the required cooperative glide to maintain the dipolar nature of these walls operated imperfectly. Increasing amplitudes or cycles may then tend to cause increasing misalignment across such walls.

That the initial cyclic hardening in the annealed samples is related largely to an increase in internal stress is in keeping with the increase in the density of long, screw dislocations. That the cyclic softening after a tensile prestrain is associated with a reduction in $\Delta\sigma_i$ is also expected, since during this softening the density of dislocations between walls is reduced and the dislocation arrangements, as indicated from the relaxation of the residual stress [4], are gradually transformed into the more strongly dipolar arrangements typical of cyclic deformation. During monotonic deformation, Williams et al [23] found that a particular high value of σ_i was associated with the arrangement of long, screw dislocations. It was therefore assumed that the cyclic softening stage in the annealed samples corresponded to a decrease in $\Delta\sigma_i$ [4,19]. The absence of cyclic softening when misoriented cells were produced could also be explained on this basis. Miura and Umeda [24], employing double strain rate changes in zirconium, had found that this softening stage corresponded to a decrease in $\Delta\sigma^*$. That the cyclic softening stage does not correspond to a decrease in $\Delta\sigma_i$ suggests that the presence of screw dislocations in the channels as well as the presence of the walls and loop patches compensate for the annihilation of the long screw dislocations.

In zirconium and titanium, the rate controlling process for thermally activated flow in the temperature-range of interest is generally considered to be the overcoming of the oxygen atoms [25,26] and less frequently a Peierl's stress mechanism [27]. A psuedo Peierl's stress mechanism with the screw dislocations dissociated on more than one plane has also been suggested [28] although convincing experimental support is lacking. The influence of decreasing temperature and increasing oxygen concentration on favoring formation of arrangements of long screw dislocations [23] after monotonic deformation is in qualitative agreement with a Peierl's stress or pseudo-Peierl's stress mechanism. Two possibilities therefore must be considered to explain the decrease in $\Delta\sigma^*$. The first is that during cyclic oxygen becomes less effective in causing solid solution hardening. Scavenging of oxygen from solid solution appears improbable. The arguments against this possibility are 1) that at the start of cyclic softening, the main traps for oxygen appear to be long screw dislocations and scavenging should therefore not cause the mobility of these dislocations to increase and 2) that the temperature for strain-aging effects related to the oxygen atom to begin is of the order of 150°C during interrupted tensile tests [29]. The high solubility limit of oxygen in titanium and zirconium also appears to make a scavenging explanation improbable. A second possibility appears if the strong solution hardening were related to short-range ordering of the oxygen atoms, which has been suggested [30] but which explanation has received little support in the literature [9]. Cyclic straining could be expected to destroy any short range order [31]. For a Peierl's stress or pseudo-Peierl's stress mechanism to explain the results, the nucleation of double kinks would have to become facilitated during cycling. It can be speculated that this could be related to the presence of the dipole loops formed after a few screw dislocations have been forced to undergo glide. This could cause an avalanche effect and could explain rapid cyclic softening such as observed at 150°C, where $\Delta\sigma^*$ decreases approximately 20% in the first ten cycles. The TEM observations strongly indicate that cyclic softening is associated with an increased mobility of the screw dislocations and with cross-slip.

These observations are certainly compatible with a Peierl's stress or pseudo-Peierl's stress mechanism if cycling causes the nucleation of pairs of double kinks to become facilitated. The oxygen atoms would then have an indirect role, related to their influence on the Peierl's stress. The reincrease in $\Delta\sigma^*$ that occurs when cyclic rehardening takes place would be related to strain hardening rather than to a reincrease in Peierl's stress.

As already pointed out by Williams et al [23] and as can be seen by comparing Figures 1 and 3, the internal stress is strongly influenced by the deformation temperature, presumably since the dislocation substructures formed depend on the temperature. For this reason, many of the thermally activated deformation studies performed on titanium and zirconium can be criticized, since an internal stress almost independent of temperature has been assumed.

Figure 16 compares semi-logarithmic stress-strain curves for different grades of titanium. The Ti-50 employed by Stevenson and Breedis [8] who obtained similar cyclic hardening-softening behavior appears slightly purer. The IMI 115 employed by Munz [6,7] appears significantly purer. This material did not show the initial cyclic hardening stage, which result appears related to the greater mobility of screw dislocations in the higher purity material. Munz [7], however, also observed

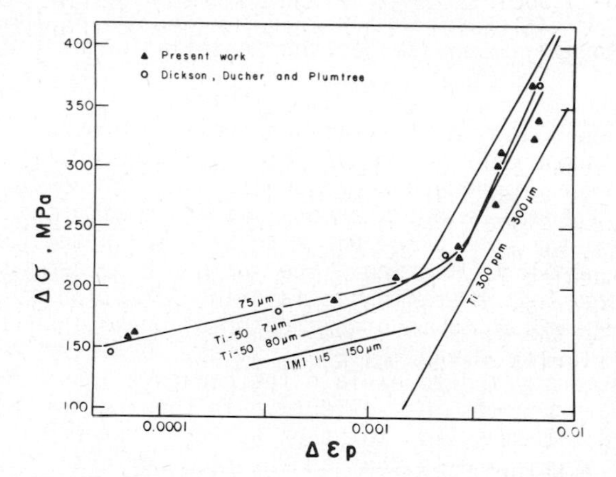

Fig. (16) - Comparison of mid-life cyclic stress-strain curve with previous studies [5-8, 10]

a similar behavior in a lower purity material, which is more difficult to explain. The comparison in Figure 16 with the curve for the highest purity titanium studied is particularly interesting, in that even neglecting the grain size effect, there is at higher values of $\Delta\varepsilon_p$ only ≃17% difference in $\Delta\sigma$ compared to ≃400% difference in yield stress. This clearly indicates the effect of cyclic straining in the low cycle fatigue regime on reducing the influence of oxygen atoms on the flow stresses.

That the start of cyclic softening is related to an increased mobility of screw dislocations and to the onset of cross-slip, which are thermally activated mechanisms, can explain the decrease in the number of cycles to the start of the cyclic softening stage with increasing cycling temperature, increasing strain amplitude [4,5] and decreasing frequency [7]. For sufficiently high purity material, the cyclic softening stage is not observed [10], since the screw dislocations are sufficiently mobile from the start of cycling. It has previously been shown [11] that the cyclic softening stage corresponds, at least for $\Delta\varepsilon_t = 0.25\%$, to the first appearance of persistent slip bands revealed by etching the external surfaces. Presumably, the occurrence of cross-slip triggers their nucleation simultaneously with the onset of cyclic softening.

That the $\Delta\sigma_i$ - $\Delta\varepsilon_t$ curves, Figure 4, tend to be parallel to the $\Delta\sigma$ - $\Delta\varepsilon_t$ curves agree with results obtained during tensile deformation of Ti-40 [29]. That the $\Delta\sigma^*$ values for the prestrained samples tend to be larger than for the annealed samples for $\Delta\varepsilon_t = 0.5$ and 0.75% agrees with the previous conclusion [4] that this tensile prestrain prevents the occurrence of the cyclic softening that occurs in annealed Ti-40. For $\Delta\varepsilon_t > 0.75\%$, $\Delta\sigma^*$ increases with $\Delta\varepsilon_t$ which may also reflect the reduced amount of cyclic softening at high amplitudes.

The $\Delta\sigma_i$ and $\Delta\sigma^*$ versus $\Delta\varepsilon_t$ results are similar to those obtained on iron [21] employing strain rate changes. Other similarities are the initial low mobilities of screw dislocations at low temperatures and high strain rates, which for iron is well-known to be associated with a pseudo-Peierl's stress mechanism. It should be noted that b.c.c. β-titanium alloys as well as α-β alloys can also show cyclic softening stages [9,32].

CONCLUSIONS

It can be concluded that the dislocation substructures produced after cyclically deforming Ti-40 to a large N resemble those produced in other metals. The low mobilities of screw dislocations in Ti-40 permit observations of cooperative glide across dipolar walls as well as dislocations bowing out of walls. The cyclic softening stage in annealed Ti-40 corresponds to an increase in the mobility of screw dislocations and in cross-slip as well as to a decrease in $\Delta\sigma^*$. The results appear consistent with the glide of screw dislocations controlled by a Peierl's stress or pseudo-Peierl's stress mechanism.

ACKNOWLEDGEMENTS

Support from the Natural Sciences and Engineering Research Council of Canada and the Ministry of Education of Quebec (FCAC program) is acknowledged.

REFERENCES

[1] Figueroa, J. C., Bhat, S. P., De LaVeaux, R., Murzenski, S. and Laird, C., Acta Met., Vol. 29, p. 1667, 1981.

[2] Winter, A. T., Pedersen, O. B. and Rasmussen, K. V., Acta Met., Vol. 29, p. 735, 1981.

[3] Pohl, K., Mayr, P. and Macherauch, E., Inter. J. of Fract., Vol. 17, p. 221, 1981.

[4] Handfield, L. and Dickson, J. I., Can. Metall. Quart., Vol. 20, p. 331, 1981.

[5] Dickson, J. I., Ducher, J. and Plumtree, A., Met. Trans., Vol. 7A, p. 1559, 1976.

[6] Munz, D., Scripta Met., Vol. 6, p. 815, 1972.

[7] Munz, D., Eng. Fract. Mech., Vol. 5, p. 353, 1973.

[8] Stevenson, R. and Breedis, J. F., Acta Met., Vol. 23, p. 1419, 1975.

[9] Mahajan, Y. and Margolin, H., Met. Trans., Vol. 13A, p. 269, 1982.

[10] Dickson, J. I., Owens, J. P. and Plumtree, A., "Titanium Science and Technology", Vol. 2, R. I. Jaffee and H. M. Burte, eds., Plenum Press, New York, pp. 1231-1243, 1973.

[11] Ducher, J., M.A.Sc. Thesis, Ecole Polytechnique de Montréal, 1974.

[12] Cottrell, A. H., "Dislocations and Plastic Flow in Crystal", Oxford Univ. Press, p. 111, 1953.

[13] Kuhlmann-Wilsdorf, D. and Laird, C., Mater. Sci. Eng., Vol. 37, p. 111, 1977.

[14] MacEwen, S. R., Kupcis, O. A. and Ramaswami, B., Scripta Met., Vol. 3, p. 441, 1969.

[15] Conrad, H. and Okazaki, K., Scripta Met., Vol. 4, p. 259, 1970.

[16] Li, J. C. M., Can J. Phys., Vol. 45, p. 493, 1969.

[17] Handfield, L. and Dickson, J. I., to be published.

[18] Mughrabi, H., in Constitutive Equations in Plasticity, A. S. Argon, ed., MIT Press, Cambridge, Mass., pp. 199-250, 1975.

[19] Handfield, L. and Dickson, J. I., in Advances in Fracture Research, ICF5 Proc., Vol. 3, D. Francois, ed., Pergamon Press, pp. 1411-1418, 1981.

[20] Grosskreutz, J. C. and Mughrabi, H., in Constitutive Equations in Plasticity, A. S. Argon, ed., MIT Press, Cambridge, Mass., pp. 251-326, 1975.

[21] Kuhlmann-Wilsdorf, D. and Laird, C., Mater. Sci. Eng., Vol. 27, p. 137, 1977.

[22] Mughrabi, H., Kerz, K. and Stark, X., Intern. J. Fract., Vol. 17, p. 193, 1981.

[23] Williams, J. C., Sommer, A. W. and Tung, P. P., Met. Trans., Vol. 3, p. 2979, 1972.

[24] Miura, S. and Umeda, K., Scripta Met., Vol. 7, p. 337, 1973.

[25] Conrad, H., Doner, M. and DeMeester, B., Titanium Science and Technology, Vol. 2, R. I. Jaffee and H. M. Burte, eds., Plenum Press, pp. 969-1005, 1973.

[26] Mills, D. and Craig, G. B., TMS-AIME, Vol. 242, p. 1881, 1968.

[27] Sastry, D. H. and Vasu, K. I., Acta Met., Vol. 20, p. 399, 1972.

[28] Sob, M., Kratochvil, J. and Kroupa, F., Czech. J. Phys., Vol. B25, p. 872, 1975.

[29] Malik, L. M. and Dickson, J. I., Mater. Sci. Eng., Vol. 17, p. 67, 1975.

[30] Weissmann, S. and Shrier, A., in The Science, Applications and Technology of Titanium, R. I. Jaffee and N. Promisel, eds., Pergamon Press, pp. 441-451, 1970.

[31] Calabrese, C. and Laird, C., Mater. Sci. Eng., Vol. 13, p. 141, 1974.

[32] Saleh, Y. and Margolin, H., Met. Trans., Vol. 11A, p. 1295, 1980.

FATIGUE SOFTENING IN PRECIPITATION HARDENED COPPER-COBALT SINGLE CRYSTALS

D. Steiner and V. Gerold

Max-Planck-Institut fuer Metallforschung, Institut fuer Werkstoff-Wissenschaften, Stuttgart, W.-Germany

ABSTRACT

The phenomenon of low cycle fatigue softening in a precipitation hardened alloy and the resulting damage has been studied on single crystals of a Cu-2at.% Co alloy. The amount of softening after passing a maximum stress amplitude is nearly equivalent to the amount of precipitation hardening, $\Delta\tau_0$. This is the result of the total destruction of precipitates in persistent slip bands (PSBs). The latter have formed at the maximum stress amplitude before softening sets in. These bands are very narrow and lead to the nucleation of fatigue cracks. Under constant stress amplitude conditions the same PSBs form also gradually at stress amplitudes around τ_s where τ_s is the amplitude after the softening process has ceased in a strain controlled experiment. At stress amplitude close to the CRSS micro-PSBs are found which may also lead to failure in a high cycle experiment. From these results conclusions for polycrystals are drawn.

INTRODUCTION

Since the first observation of fatigue softening in a Ni base superalloy [1] this phenomenon has been observed quite often. It is connected with the occurrence of persistent slip bands (PSBs) and the resulting strain localization in these bands. The purpose of the present study is to investigate the amount of softening and the final saturation stress amplitude τ_s in relationship to the amount of precipitation hardening. Also of interest was the surface damage which follows the PSB formation. In addition, the question was raised what would happen if the sample is fatigued at a constant stress amplitude below or close to τ_s. In order to get a better macroscopic response to microstructural changes in the sample the experiments were undertaken with single crystals oriented for single slip.

The results discussed in this paper will be restricted to underaged and peak-aged states of a simple alloy system as Cu-Co. Part of these results and some results for the overaged condition have been published elsewhere [2].

EXPERIMENTAL DETAILS

From a copper alloy containing 2at.% cobalt single crystals were grown by the Bridgman technique. The cylindrical samples had a diameter of 6 mm and were oriented for single glide. Gauge sections of 15 mm length and 3 mm diameter were prepared by spark erosion. The heat treatment [2] gave spherical precipitates with radii varying from 6 to 75 Å. The aged single crystals were fatigued in a servohydraulic machine with closed loop control. Either the plastic shear strain amplitude $\gamma_{p\ell}$ or the shear stress amplitude τ of the primary slip system was kept constant. In order to avoid large plastic strains the stress amplitudes necessary for control were produced by a preceding strain controlled experiment. Most tests were undertaken with $\gamma_{p\ell} = 10^{-3}$ and in a vacuum of 10^{-4} mbar. The latter considerably improved the life time of the samples. The age hardening state was characterized by the increase of the CRSS, $\Delta\tau_o$.

Optical surface observations were possible during the experiments. The damage of the samples was investigated by scanning electron microscopy (SEM).

EXPERIMENTAL RESULTS

A. Experiments with Constant Plastic Strain Amplitude

Single crystals containing particle sizes with radii R between 6 and 74 Å were fatigued at a constant plastic shear strain amplitude of about $\gamma_{p\ell} = 10^{-3}$. The resulting stress response as a function of cumulative resolved shear strain ($\gamma_{cum} = 4\ N\ \gamma_{p\ell}$, N = number of cycles) is plotted on Figure 1. All curves are characterized by a hardening stage up to a maximum stress amplitude τ_m followed by a softening stage which finally leads to a constant stress amplitude τ_s (saturation stress). The only exception is the curve for the solid solution (R = 0) which only hardens to a saturation stress amplitude τ_s.

At the beginning of fatigue the observed stress amplitude increases with increasing particle size in the same way as the CRSS does in a tensile experiment. The hardening rate in fatigue increases also which again is comparable to the observation of corresponding tensile experiments [3]. In Figure 2 the observed maximum stress amplitude τ_m is plotted as a function of the increase of the CRSS, $\Delta\tau_o$, due to precipitation hardening. A linear relationship is observed.

$$\tau_m = \tau_s^M + 1.5\ \Delta\tau_o, \qquad (1)$$

where τ_s^M is the saturation stress amplitude of the matrix (solid solution).

The following softening stage can be observed to its end only for the crystals containing smaller particles. For the crystals close to peak-age hardening the samples fail relatively early as will be discussed later. In

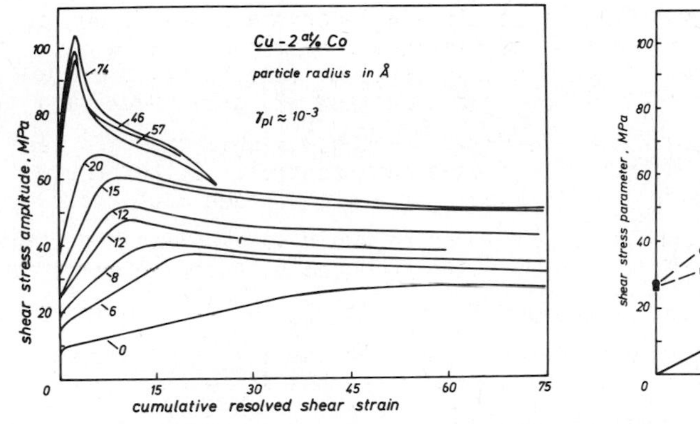

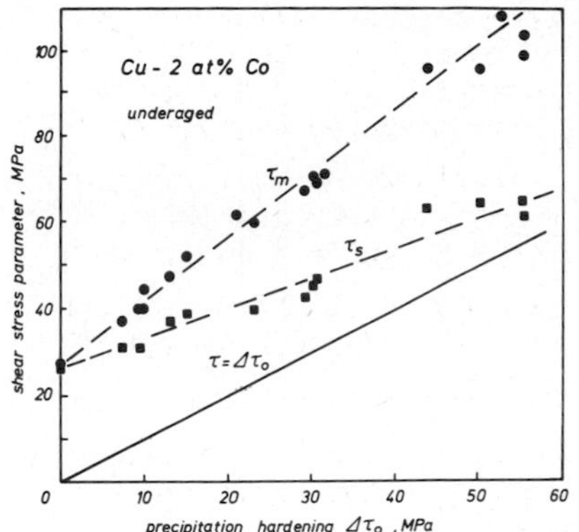

Fig. (1) - Stress response of fatigued single crystals for various sizes of precipitated cobalt particles (underaged peak-aged conditions)

Fig. (2) - Characteristic stress amplitudes τ_m and τ_s as a function of precipitation hardening, $\Delta\tau_o$

that case the final saturation stress amplitude τ_s can be estimated only. The resulting saturation stresses τ_s are plotted on Figure 2 and follow also a straight line

$$\tau_s = \tau_s^M + 0.7\ \Delta\tau_o. \tag{2}$$

The amount of softening is called $\Delta\tau_E$. From Equations (1) and (2) it follows that

$$\Delta\tau_E = 0.8\ \Delta\tau_o. \tag{3}$$

If the $\Delta\tau_E$ values are taken from the individual experiments and plotted as a function of $\Delta\tau_o$ a slightly curved line results as plotted on Figure 3. The corresponding values $\Delta\tau_E$ plotted for the overaged crystals have been discussed elsewhere [2].

The important feature in fatigue is the occurrence of persistent slip bands (PSB) which can easily be observed by light microscopy during the experiment [4]. They start to develop shortly before the maximum stress amplitude τ_m (resp. τ_s^M for the solid solution) is reached and get their final volume fraction immediately after softening has started. Their surface appearance has been studied by SEM techniques. It depends on the precipitation state of the crystal.

The solid solution shows an inhomogeneous distribution of PSBs which cluster together into groups and result in a wavy surface as shown on

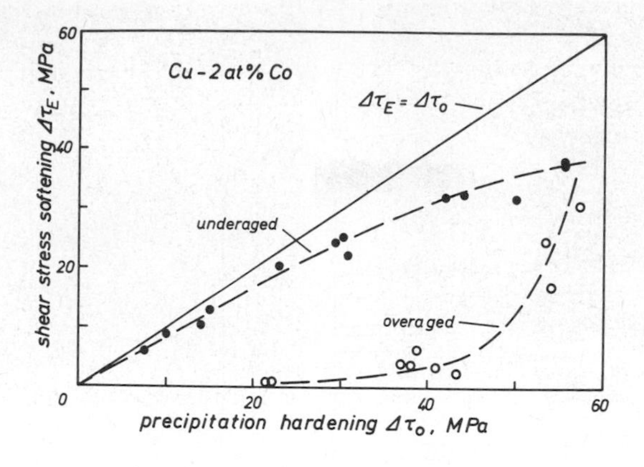

Fig. (3) - The softening $\Delta\tau_E = \tau_m - \tau_s$ as a function of $\Delta\tau_o$

Figure 4a. After prolonged fatigue up to γ_{cum} = 680 the surface roughness becomes more pronounced and voluminous extrusions and intrusions can be found (Figure 4b). This type of roughness is observed only on such parts of the surface which are not parallel to the primary slip vector.

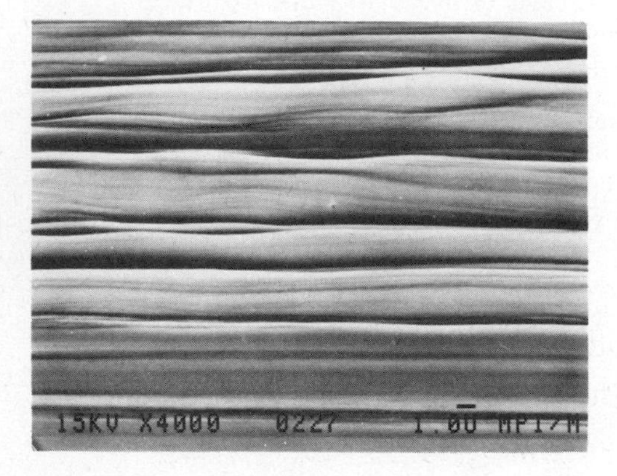

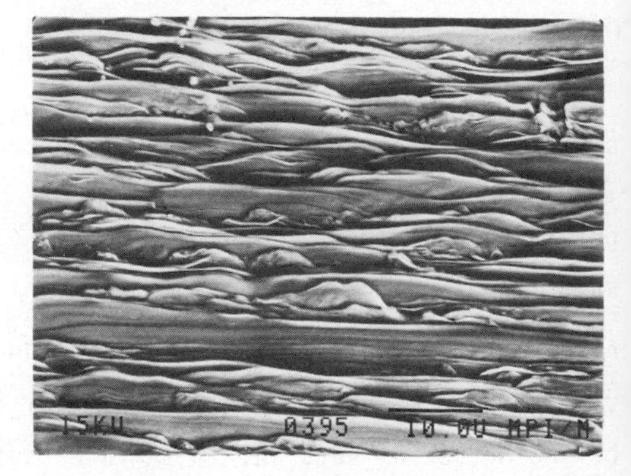

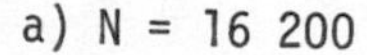
a) N = 16 200

b) N = 172 000

Fig. (4) - Surface profiles of PSBs in a solid solution single crystal

With increasing particle size the appearance of the PSBs changes gradually into very sharp bands as shown in Figure 5a. These bands produce very thin flakes with a thickness less than 100 nm. They have been extruded 2 to 10 μm out of the surface and even more than that (Figure 5b). A closer inspection of these extrusions gives the impression of rest lines comparable to

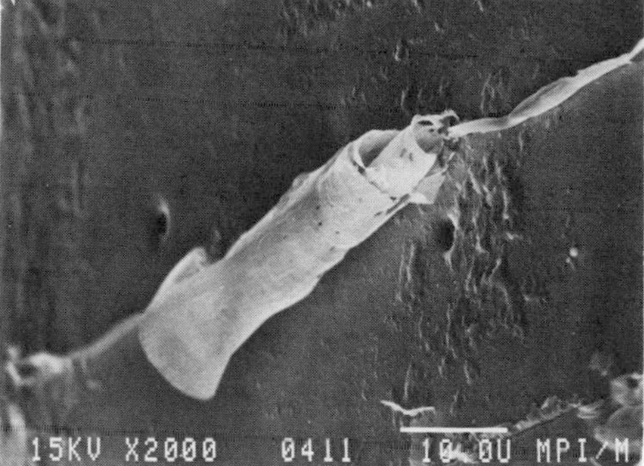

a) typical pattern

b) large extrusion

Fig. (5) - Surface profiles of PSBs in an underaged single crystal. Particle radius 46 Å. Fatigued at various $\gamma_{p\ell}$ from 8.7×10^{-4} to 3.4×10^{-3}. $\gamma_{cum} = 133$

striations on fatigue fracture surfaces. Again, this phenomenon is observed only on surfaces which are not parallel to the primary slip vector.

The marked softening after the appearance of the PSBs is the result of strain localization in these bands and obviously a consequence of particle destruction [4]. The possibility of this effect has been controversely discussed in the literature as reviewed by Calabrese and Laird [5]. In order to get further proof on this effect the following additional experiment was undertaken: an underaged single crystal (particle radius 12 Å) was fatigued well into the softening stage (Figure 6, curve I). At point A the sample was aged again the same amount as before and a second experiment was started (Figure 6, curve II). The stress amplitude at B is about the same as at A. The sample hardens again up to a τ_m value comparable to that of curve I before softening sets in. A surface investigation gave the same distribution of PSBs at points A and C of Figure 6.

B. Experiments with Constant Stress Amplitude

Some experiments were undertaken to investigate the fatigue behavior for controlled stress amplitudes below the maximum amplitude τ_m. The crystals contained precipitation particles with a radius of R = 46 Å which gave the following characteristic stresses:

$\tau_o \approx$ 50 MPa (CRSS)

$\tau_m \approx$ 60 MPa (saturation stress amplitude)

$\tau_s \approx$ 95 MPa (maximum stress amplitude)

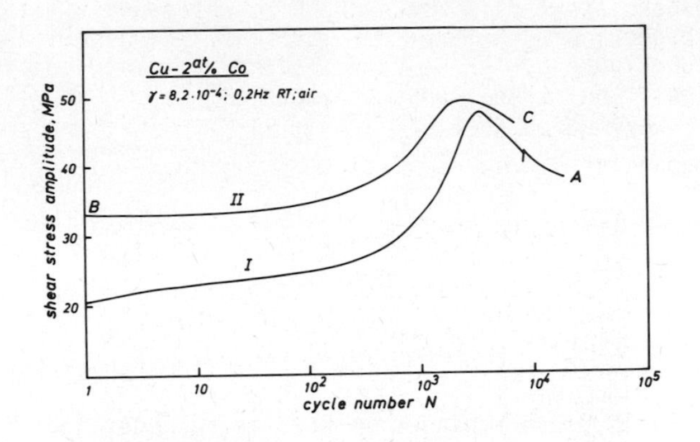

Fig. (6) - Dependence of the stress amplitude on the cycle number for an underaged crystal (curve I) reheated at A and cycled again (curve II)

As an example, one of these crystals was fatigued for the first 100 cycles at a constant plastic strain amplitude $\gamma_{p\ell} = 1.8 \times 10^{-4}$. During this cycling the stress amplitude raised from 48 MPa to 52 MPa which is nearly the saturation stress for this small strain amplitude (Figure 7a). In this case no PSBs

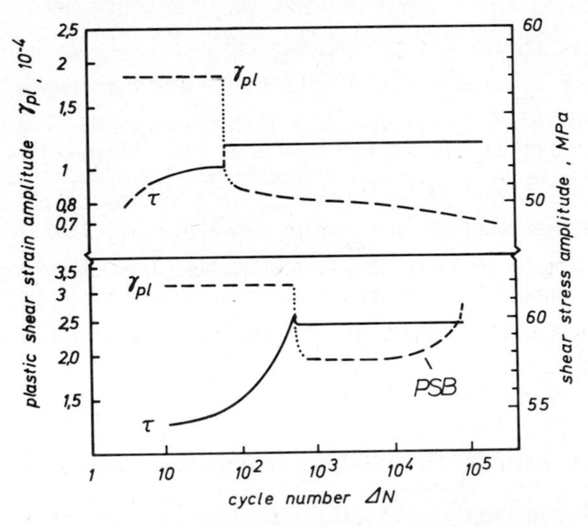

Fig. (7) - Fatigue experiment with varying control conditions (see text)

are formed. Thereafter, the stress amplitude was kept constant at 53 MPa for about 10^5 cycles. This stress value is in between τ_o and τ_s and close to τ_o. In response the plastic strain amplitude decreases continuously to $\gamma_{p\ell} = 0.7 \times 10^{-4}$ which signalizes further hardening of the sample and no strain localization. No persistent slip bands could be observed by light microscopy. However, closer inspection by SEM reveals localized microbands as on the micrograph shown on Figure 8. These bands have only a length of the order 20 μm. Nevertheless, they show the same narrow extrusions as the PSBs on Figures 5a and b. Obviously, they are localized micro-PSBs and may be able to lead to further damage at high cycle fatigue ($N > 10^7$).

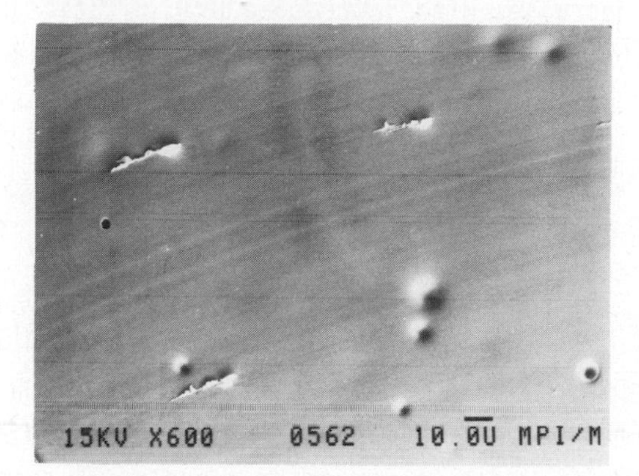

Fig. (8) - Micro-PSBs on the surface after the fatigue experiment shown on Figure 7a

Thereafter, the sample was fatigued again, at first at a higher plastic strain amplitude of $\gamma_{p\ell} = 3.1 \times 10^{-4}$ which was kept constant for another 500 cycles (Figure 7b). As a consequence, the stress amplitude raised to 60 MPa which is the value of the saturation stress τ_s. Then, the stress amplitude was made constant (τ = 59.5 MPa) for another 80 000 cycles. During this time the strain amplitude raised from $\gamma_{p\ell} = 2 \times 10^{-4}$ to 2.7×10^{-4} indicating a softening of the sample at this stress amplitude. The following surface inspection by SEM showed the existence of PSBs which could not be distinguished from bands developed with a constant plastic strain amplitude of 10^{-3} (Figures 5a and b).

DISCUSSION

A. Experiments with Constant Plastic Strain Amplitude

The present strain controlled experiments have shown that progressing precipitation hardening from the underaged to the peak-aged state leads to a progressive resistance of the sample against fatigue deformation. The stress amplitude at the beginning of fatigue deformation as well as the observed fatigue hardening increases with increasing particle size.

The formation of more mobile dislocation structures occurs at a characteristic stress amplitude τ_m which depends only on the amount of precipitation hardening. According to Wilhelm [4] τ_m is independent of the controlled strain amplitude $\gamma_{p\ell}$ if it is in the range from 3×10^{-4} to 3×10^{-3}. For smaller amplitudes only a hardening to a saturation stress smaller than τ_m is observed without PSB formation whereas for larger amplitudes more severe deformation occurs in the bulk which leads to higher stress amplitudes than τ_m.

The value of τ_m increases with increasing precipitation hardening, $\Delta\tau_o$. Even the difference, $\tau_m - \tau_o$ increases from 18 MPa for the solid solution to about 50 MPa for the peak-aged case as can be deduced from Figure 2. This also reflects the increasing resistance of precipitation hardened material to withstand a fast formation of PSBs.

After the PSBs have formed a considerable softening $\Delta\tau_E$ occurs which increases with increasing precipitation hardening, $\Delta\tau_o$. This softening is the result of a continuous destruction of the precipitated particles inside the PSBs. From the thickness and the number of PSBs a volume fraction can be estimated for the bands which is of the order of 1% in the present case ($\gamma_{p\ell} = 10^3$). That leads to an estimated local shear strain amplitude in these bands of the order of 10% which is larger by a factor 10 compared to the situation for the solid solution.

The destruction of precipitates in fatigue bands has been discussed several times in the literature. A clear proof for this has been given by Vogel et al. [6] in their TEM studies of Al-Zn-Mg single crystals. Recently, Laupheimer et al. [7] were able to demonstrate the same for a Cu-Co crystal. Inside the bands the strain contrast of the particles has disappeared whereas other contrast remained still visible as that of radiation damage (due to irradiation with high voltage electrons in the microscope). Another proof was given by neutron small angle scattering [8]. There it was demonstrated that the volume fraction of precipitates had decreased after fatigue by an amount which corresponded to the volume fraction of PSBs. Another indirect proof is given by the experiment reproduced in Figure 8. The reheating of the sample after the end of experiment I led to (1) recovery processes in the bulk and to (2) reprecipitation of the dissolved particles in the PSBs. As a result, the bulk softened and the PSBs hardened leading to the medium stress amplitude at the beginning of experiment II. In order to get again strain localization in the PSBs the bulk of the sample first has to harden again to the same τ_m value as was observed in experiment I. Only at this stress amplitude the new occurrence of PSBs could be observed. These bands were more or less the same as those after experiment I.

After the destruction of precipitates inside the PSBs the fatigue deformation continues at the saturation stress amplitude, τ_s. This amplitude increases also with increasing precipitation hardening but this increase is more moderate and only 70% of $\Delta\tau_o$ (see Equation (2)). The difference between τ_s and τ_o decreases with increasing $\Delta\tau_o$ and becomes nearly zero for peak-age-hardened crystals (Figure 2.) The increase of τ_s is an indication for different dislocation structures inside the PSBs. Mughrabi [9] and Brown [10] derived a relationship between τ_s for pure metals and the distance d between dislocation walls (the ladder structure) inside the PSBs where τ_s is found to be inversely proportional to d. Experiments on copper [11] and on magnesium [12] have shown that the width of a single PSB has about the same value as the wall distance d. In the present case it has been found that the width of a PSB decreases from 1.5 μm for the solid solution to values between 0.1 and 0.2 μm for the peak-aged state. If the experimental results on Cu and Mg are applied to Cu-Co a considerable change of the dislocation structure is expected. Because of the narrowness of the bands no direct observation of the dislocation structure of an age-hardened material is not known as yet.

The continuous decrease of the width of the PSBs and the increasing stress level (τ_m or τ_s) lead to an increasing localization of the deformation as mentioned before and an increasing sensitivity to crack formation at the PSBs. Therefore, the τ_s level never could be reached for peak-aged crystals because they fractured already during softening.

B. Experiments with Controlled Stress Amplitude

The marked softening during fatigue led to the question if localized deformation and damage of the crystals could be observed also at lower stress levels which could be kept constant by stress control. Such experiments led to the conclusion that for amplitudes below the CRSS, τ_o, no damage is expected. For stress amplitudes above τ_o but below τ_s localized short micro-PSBs are developed which may be able to lead also to final damage at high cycles. For stress amplitudes at and above τ_s but below τ_m the same type of PSBs is observed as in a strain controlled experiment which certainly leads to failure. Probably, these PSBs develop gradually from microbands and penetrate slowly the whole cross section of the sample. At the low stress amplitudes the PSBs can only be active if the precipitates have been destroyed already. Therefore, it is suggested that in this case the precipitates have to be destroyed first at the front of the penetrating band. This is in contrast to the strain controlled experiments where at first the bands spread over the whole cross section and the destruction of the precipitates in the bands occurs afterwards.

Recently, these findings could be stated [13]. In addition, it was found that in the stress range τ_o to τ_s the localized damage is increasing with increasing stress amplitude, i.e., the length of micro-PSBs is steadily increasing and microcracks within these PSBs could be found also. These results are independent of the way the constant stress amplitude has been reached.

If these results are generalized for all underaged and peak-aged states and if they are applied to polycrystalline specimens the following conclusions can be drawn: The first surface damage will occur as soon as the shear stress amplitude τ_o is reached for some grains. This will be at a macroscopic applied stress amplitude $\sigma = 2\,\tau_o$. This amount is far below the yield stress $\sigma_o = 3.06\,\tau_o$ where 3.06 is the Taylor factor for a fcc lattice. Therefore, the first damage should occur at stress amplitudes exceeding 0.67 σ_o. Micro-PSBs will form which may lead to final failure at high cycles.

In the case of peak-aged polycrystals one has to take into account that τ_o is close to τ_s where macrobands will be formed. In this case, an earlier fracture at stress amplitudes 0.67 σ_o is predicted.

REFERENCES

[1] Wells, C. H. and Sullivan, C. P., Trans. ASM, Vol. 57, p. 841, 1964.

[2] Gerold, V. and Steiner, D., Scripta Met., in press.

[3] Hartmann, K.-G., Z. f. Metallkunde, Vol. 62, p. 736, 1971.

[4] Wilhelm, M., Mater. Sci. Eng., Vol. 48, p. 91, 1981.

[5] Calabrese, C. and Laird, C., Mater. Sci. Eng., Vol. 13, p. 141, 1974.

[6] Vogel, W., Wilhelm, M. and Gerold, V., Acta Metallurg., Vol. 30, p. 21, 1982.

[7] Laupheimer, A., Ungár, T. and Mughrabi, H., private communication.

[8] Beddoe, R., Gerold, V., Kostorz, G., Schmelczer, R. and Steiner, D., to be published.

[9] Mughrabi, H., 4th Int. Conf. on Continuum Models of Discrete Systems, Stockholm 1981, eds. O. Brulin and R. K. T. Hsieh. North Holland Publ. Comp., Amsterdam 1982.

[10] Brown, L. M., see Ref. [9].

[11] Basinski, Z. S., Korbel, A. S. and Basinski, S. J., Acta Metallurg., Vol. 28, p. 191, 1980.

[12] Kwoadjo, R. and Brown, L. M., Acta Metallurg., Vol. 26, p. 1117, 1978.

[13] Mueller, W. and Steiner, D., to be published.

SECTION II
MODELING OF DISLOCATIONS AND MICROCRACKING

THE ELASTIC STRAIN ENERGY OF DISLOCATION STRUCTURES IN FATIGUED METALS

T. Mura, H. Shirai and J. R. Weertman

Northwestern University
Evanston, Illinois 60201

ABSTRACT

The elastic strain energy is calculated for persistent slip band structures and for vein structures of dislocations in cyclically loaded metals. The evolution of dislocation structures during fatigue of metals is discussed in terms of the elastic strain energy of the structures. The ladders in persistent slip bands and the tangling dislocation dipoles in vein structures are simulated by the Orowan loops surrounding inclusions which consist of slip bands with a multiple shell structure.

INTRODUCTION

The elastic strain energy is calculated for vein structures of dislocations and for persistent slip band structures in cyclically loaded metals. Since the elastic strain energy is a mechanical part of Gibbs' free energy, the evolution of dislocation structures during fatigue testing may be discussed in terms of the elastic strain energy of the structures.

As an analytical tool, the inclusion method of Eshelby [1] is employed since the simplest distribution of dislocations can be simulated by an inclusion in which a uniform plastic distortion is defined.

Preparatory to the following sections, some fundamental equations are introduced in this section.

If a free body of volume V contains a distribution of plastic strain ε^p_{ij}, its elastic strain energy W is evaluated to be

$$W = -(1/2) \int_V \sigma_{ij}\varepsilon^p_{ij} dV \tag{1}$$

This is proved as follows:

$$W = (1/2) \int_V \sigma_{ij}e_{ij} dV = (1/2) \int_V \sigma_{ij}(u_{i,j} - \varepsilon^p_{ij}) dV = -(1/2) \int_V \sigma_{ij}\varepsilon^p_{ij} dV$$
$$+ (1/2)\left(\int_S \sigma_{ij}n_j u_i dS - \int_V \sigma_{ij,j}u_i dV\right) \tag{2}$$

where σ_{ij} is the stress, e_{ij} the elastic strain, u_i the displacement and $_{,j}$ is the derivative with respect to the coordinate x_j. The term "free body" means that

$$\sigma_{ij}n_j = 0 \quad \text{on } S$$

$$\sigma_{ij,j} = 0 \quad \text{in } V \tag{3}$$

$$(1/2)(u_{i,j} + u_{j,i}) = e_{ij} + \varepsilon^p_{ij}$$

where n_i is the normal to the surface S of the body V. Because of (3), equation (2) is reduced to (1).

If a body of infinite extent contains an ellipsoidal inclusion (a sub-domain of material having the same elastic moduli as those of the matrix but a uniform eigenstrain ε^p_{ij}), the stress in the inclusion becomes constant,

$$\sigma_{ij} = C_{ijk\ell}(S_{k\ell mn}\varepsilon^p_{mn} - \varepsilon^p_{k\ell}) \tag{4}$$

where $C_{ijk\ell}$ is the elastic modulus and $S_{k\ell mn}$ is Eshelby's tensor (constant), which depends on the shape of the ellipsoid but is independent of its volume Ω. Then,

$$W = -(1/2)C_{ijk\ell}(S_{k\ell mn}\varepsilon^p_{mn} - \varepsilon^p_{k\ell})\varepsilon^p_{ij}\Omega \tag{5}$$

The stress field σ_{ij} is complicated in the region outside of the inclusion. However, Tanaka and Mori [2] proved that

$$\int_{\Delta\Omega} \sigma_{ij}dV = 0 \tag{6}$$

where $\Delta\Omega$ is the region outside of the ellipsoid Ω bounded by another ellipsoid whose shape is similar to that of Ω (see Figure 1).

If a uniform plastic strain ε^p_{ij} exists between two similar ellipsoids Ω_1 and Ω_2 as shown in Figure 2, the elastic strain energy W is expressed as

$$W = -(1/2) \int_{\Delta\Omega} \sigma_{ij}\varepsilon^p_{ij}dV \tag{7}$$

with

$$\sigma_{ij} = \sigma^{I}_{ij} + \sigma^{II}_{ij} \tag{8}$$

where σ^I_{ij} is the stress due to ε^p_{ij} in Ω_1 and σ^{II}_{ij} is the stress due to $-\varepsilon^p_{ij}$ in Ω_2. It should be noted that points in $\Delta\Omega$ are in the interior of Ω_1 and exterior to Ω_2. Then from (6)

$$\int_{\Delta\Omega} \sigma^{II}_{ij}dV = 0 \tag{9}$$

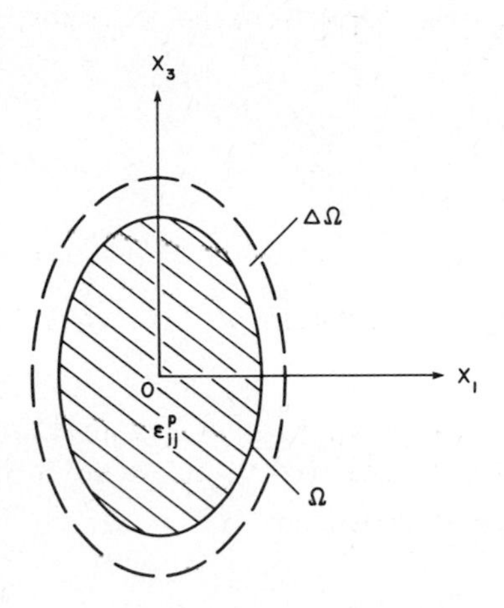

Fig. (1) - Ellipsoidal inclusion

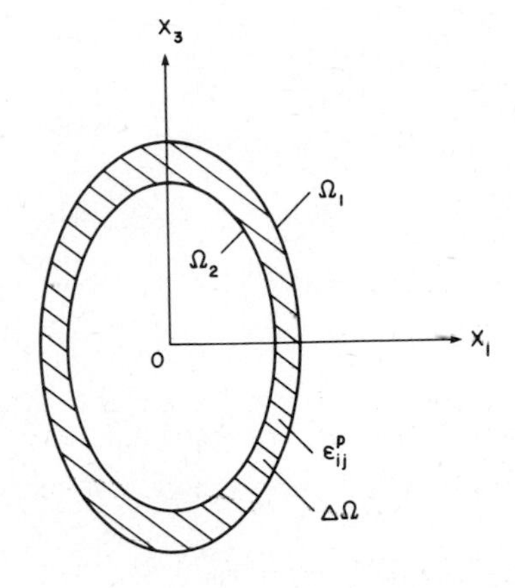

Fig. (2) - Plastic deformation ε^p_{31} in the band region $\Delta\Omega$

therefore,

$$W = -(1/2)\sigma^I_{ij}\varepsilon^p_{ij}\Delta\Omega \tag{10}$$

where σ^I_{ij} is exactly the same as (4) if the shape of Ω_1 is similar to that of Ω.

Since, in Ω_2, $\sigma^{II}_{ij} = -\sigma^I_{ij}$, we have $\sigma_{ij} = 0$ in Ω_2. Similarly, if the plastic domain is defined by the band structure bounded by the similarly-shaped ellipsoids as shown in Figure 3, the elastic strain energy is evaluated to be

$$W = -(1/2)\sigma_{ij}\varepsilon^p_{ij}\Sigma\Delta\Omega \tag{11}$$

where $\Sigma\Delta\Omega$ is the volume of shaded regions in Figure 3 and σ_{ij} is given by (4).

If a body is finite and contains many inclusions which constitute a fracture f of the volume, Mori and Tanaka [3] have found that

$$\begin{aligned} \langle\sigma_{ij}\rangle_I &= (1-f)\sigma^\infty_{ij} \\ \langle\sigma_{ij}\rangle_M &= -f\sigma^\infty_{ij} \end{aligned} \tag{12}$$

where $\langle\sigma_{ij}\rangle_I$ and $\langle\sigma_{ij}\rangle_M$ are the average stresses in the inclusions and the matrix, respectively, and σ^∞_{ij} is the interior stress, given by (4), in an isolate inclusion. The above useful result comes from the fact that the average stress in the whole volume is zero,

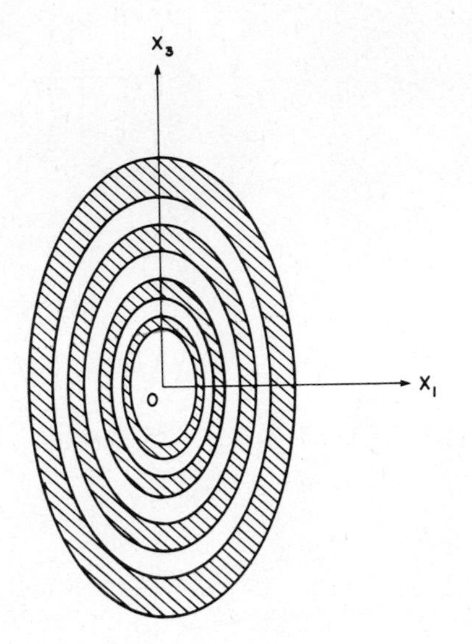

Fig. (3) - Residual plastic strain $\Delta\varepsilon_{31}^{p}$ remains in the shaded region

$$\int_V \sigma_{ij} dV = \int_V \sigma_{ik}\delta_{kj} dV = \int_V \sigma_{ik}\partial x_j/\partial x_k dV = \int_S \sigma_{ik} n_k x_j dS - \int_V \sigma_{ik,k} x_j dV = 0 \tag{13}$$

where δ_{kj} is the Kronecker delta. Therefore,

$$(1/V)\int_V \sigma_{ij} dV = f\langle\sigma_{ij}\rangle_I + (1-f)\langle\sigma_{ij}\rangle_M = 0 \tag{14}$$

On the other hand,

$$\langle\sigma_{ij}\rangle_I = \sigma_{ij}^{\infty} + \langle\sigma_{ij}\rangle_M \tag{15}$$

in which $\langle\sigma_{ij}\rangle_M$ can be considered as the sum of the image stress and the interaction stress. Equations (14) and (15) yield (12).

PERSISTENT SLIP BANDS

Let us consider a persistent slip band Ω as shown in Figure 4. The domains Ω^* are assemblies of dislocation dipoles and constitute the ladder structure as reported by Laird and Duquette [4] and Mughrabi [5] among others. Under an applied stress $\sigma_{31}^{A} = \tau_1$, plastic deformation ε_{31}^{p} takes place in the domain $\Omega-\Omega^*$. The strain ε_{31}^{p} is determined by the balance of stresses acting on mobile dislocations

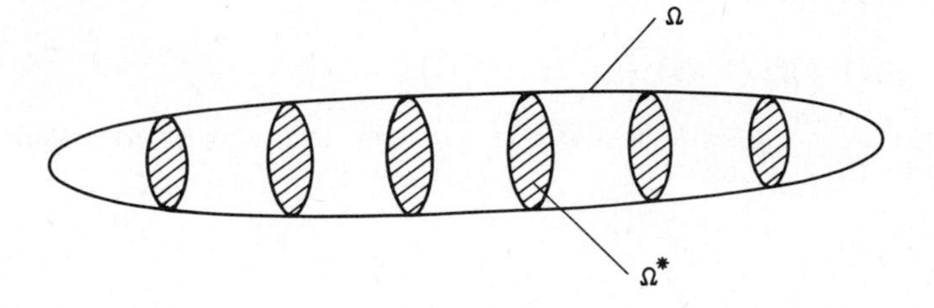

Fig. (4) - A persistent slip band. Micro-structure of Ω^* is Figure 3

in $\Omega-\Omega^*$,

$$\tau_1 + \sigma_1^D - k = 0 \tag{16}$$

where k is the frictional stress for dislocation motion and σ_1^D is the dislocation stress. The dislocation stress is the stress caused by the uniform plastic strain ε_{31}^p in the domain $\Omega-\Omega^*$, which is the sum of stresses σ_{31}^1 and σ_{31}^2. σ_{31}^1 is the stress caused by ε_{31}^p in Ω and σ_{31}^2 is the stress caused by $-\varepsilon_{31}^p$ in Ω^*. We assume that Ω and Ω^* are ellipsoidal. Then, Eshelby [1] gives

$$\sigma_{31}^1 = -2\mu(1-2S)\varepsilon_{31}^p \tag{17}$$

where μ is the shear modulus and S is Eshelby's tensor S_{3131}. The value of Eshelby's tensor is

$$S = (1/2)\{1 + (\nu-2)\pi a_3/4(1-\nu)a_1\} \tag{18}$$

when Ω is a flat thin disk $a_1 = a_2 >> a_3$, and

$$S = (4-5\nu)/15(1-\nu) \tag{19}$$

when Ω is a sphere $a_1 = a_2 = a_3$, where ν is Poisson's ratio, and a_1, a_2, a_3 are the semi-axes of the ellipsoid Ω. Expression (17) is modified according to (12),

$$\sigma_{31}^1 = -(1-f)2\mu(1-2S)\varepsilon_{31}^p \tag{20}$$

where f is the volume fraction of Ω (total volume of Ω-type inclusions in a unit volume). The other stress σ_{31}^2 is not uniform in $\Omega-\Omega^*$, however, it is approximated by the mean value $\langle\sigma_{31}^2\rangle_M$ defined by (12).

$$\sigma_{31}^2 = -f^* 2\mu(1-2S^*)\varepsilon_{31}^p \tag{21}$$

where f^* is the volume fraction of Ω^* (total volume of Ω^*-type inclusions in a unit volume) and $S^* = S_{3131}$ for an ellipsoid with the semi-axes a_1^*, a_2^*, a_3^*. When (20) and (21) are substituted into (16), we have

$$\varepsilon_{31}^p = (\tau_1 - k)/\{(1-f)2\mu(1-2S) + f^* 2\mu(1-2S^*)\} \tag{22}$$

When the applied stress is reduced to τ_2, the balance of stresses on the mobile dislocations in $\Omega-\Omega^*$ becomes

$$\tau_2 + \sigma_1^D + \sigma_2^D + k = 0 \tag{23}$$

where σ_2^D is the dislocation stress caused by a decrease of $\Delta\varepsilon_{31}^p$ in the plastic strain in $\Omega-\Omega^*$. Namely, $\sigma_2^D = \sigma_{31}^3 + \sigma_{31}^4$ and

$$\sigma_{31}^3 = (1-f)2\mu(1-2S)\Delta\varepsilon_{31}^p,\ \sigma_{31}^4 = f^* 2\mu(1-2S^*)\Delta\varepsilon_{31}^p \tag{24}$$

From (16) and (23), we have

$$-\Delta\tau + \sigma_2^D + 2k = 0,\ \Delta\tau = \tau_1 - \tau_2 \tag{25}$$

When (24) is substituted into (25),

$$\Delta\varepsilon_{31}^p = (\Delta\tau - 2k)/\{(1-f)2\mu(1-S) + f^* 2\mu(1-2S^*)\} \tag{26}$$

When the material is loaded again to τ_1, domain $\Omega-\Omega^*$ is deformed by $\Delta\varepsilon_{31}^p$. From the stress balance equation, $\Delta\varepsilon_{31}^p$ is evaluated and it is found to be the same as (26). This regularity continues and positive and negative plastic deformations whose magnitude is given by (26) are alternately created in domain $\Omega-\Omega^*$. However, a slight modification is necessary here in order to show a model of damage accumulation. We assume that the plastic domain $\Omega-\Omega^*$ shrinks after each half cycle of loading and unloading and this shrinking is converted to an expansion of Ω^*. It is also assumed that the expansion of Ω^* is small enough not to affect the magnitude of f^*. The micro-structure of Ω^* after n cycles is shown in Figure 3. Since $\Omega-\Omega^*$ shrinks after each cycle, the positive plastic strain $\Delta\varepsilon_{31}^p$ remains as a residue in the band domains which are shaded in Figure 3. The ellipsoids constituting the band structure are assumed to be similar in shape as considered in connection with equation (11). Then, the elastic strain energy W after n cycles of loading and unloading is evaluated from (11),

$$W = -\sigma_{31}\Delta\varepsilon_{31}^p \Sigma\Delta\Omega \tag{27}$$

where

$$\sigma_{31} = -2\mu(1-2S^*)\Delta\varepsilon_{31}^p \tag{28}$$

and $\Delta\varepsilon_{31}^p$ is given by (26). $\Sigma\Delta\Omega$ is the total shaded volume of all Ω^* inclusions in a unit volume. The number of Ω^* inclusions per unit volume is $f^*/(4/3)\pi a_1^* a_2^* a_3^*$. If the semi-axes of ellipsoidal band boundaries of Ω^* are denoted by (a_1^*, a_2^*, a_3^*), (ta_1^*, ta_2^*, ta_3^*), $(t^2a_1^*, t^2a_2^*, t^2a_3^*)(t^n a_1^*, t^n a_2^*, t^n a_3^*)$, $(t>1)$, simple geometric calculation leads to

$$\Sigma\Delta\Omega \simeq 2\pi a_1^* a_2^* a_3^* nhf^* / (4/3)\pi a_1^* a_2^* a_3^* \simeq (3/2)nhf^* \tag{29}$$

where

$$h = t-1 << 1 \tag{30}$$

Then, (27) becomes

$$W = \frac{(3/2)(1-2S^*)(\Delta\tau-2k)^2 nhf^*}{2\mu\{(1-f)(1-S)+f^*(1-2S^*)\}^2} \tag{31}$$

Due to the band structure of Ω^*, the interaction between Ω^* inclusions can be neglected since the long range stress caused by the band structure vanishes.

The band structure corresponds to distributions of dislocation dipoles. The dislocations are continuously distributed along the boundaries of ellipsoids and the dislocation density tensor is given by

$$\alpha_{hi} = \varepsilon_{h\ell k}\varepsilon^p_{ki} n_\ell \tag{32}$$

where $\varepsilon_{h\ell k}$ is the permutation tensor and n_i is the outer normal to the shaded band in Figure 3 (see [6]). For $\varepsilon^p_{ki} = \Delta\varepsilon^p_{31}$,

$$\begin{aligned} \alpha_{11} &= 2\Delta\varepsilon^p_{31} n_2 = \pm 2\Delta\varepsilon^p_{31}(x_2/a_2^{*2})/(x_1^2/a_1^{*4} + x_2^2/a_2^{*4} + x_3^2/a_3^{*4})^{1/2} \\ \alpha_{21} &= -2\Delta\varepsilon^p_{31} n_1 = \mp 2\Delta\varepsilon^p_{31}(x_1/a_1^{*2})/(x_1^2/a_1^{*4} + x_2^2/a_2^{*4} + x_3^2/a_3^{*4})^{1/2} \end{aligned} \tag{33}$$

The dislocation distribution (33) represents the Orowan loops as shown in Figure 5. The Burgers vector is in the x_1-direction and the dislocation line direc-

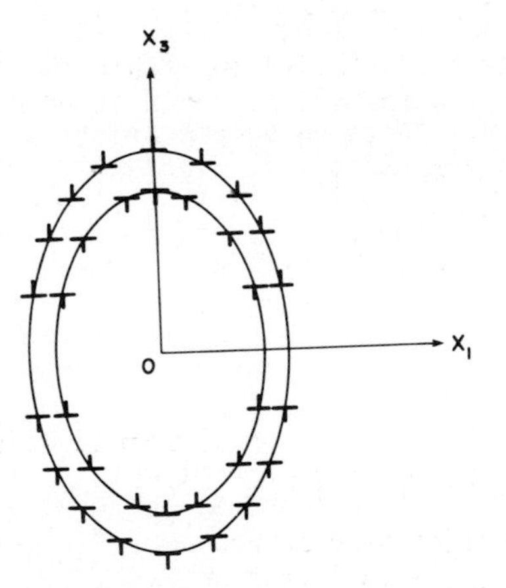

Fig. (5) - Dislocation dipoles caused by the plastic deformation as in Figure 2

tions have no components in the x_3-direction. Since the vector n_i along the inner boundary of the band has the opposite sign to the vector n_i along the outer boundary, expression (33) gives distributions of dislocation dipoles.

The elastic strain energy (31) increases with the cycle number n and is proportioned to $(\Delta\tau-2k)^2$. This characteristic has also been found by Tanaka and Mura [7-9] and Mura [10] in the derivation of the Coffin-Manson law.

VEIN STRUCTURE

The elastic strain energy per unit volume of material consisting of a vein structure of dislocations is obtained as a special case of (31) by taking f = 1. The ellipsoid Ω in this case is all of space. The energy is

$$W = (3/2)(\Delta\tau-2k)^2 nh/f^*2\mu(1-2S^*) \quad (34)$$

The distribution of Ω^* inclusions is a random distribution in the vein structure, in contrast to the ladder distribution in the persistent slip band.

Since S, $S^* \leq 1/2$ and f, $f^* \leq 1$, the energy (31) is lower than (34) and equal when S = 1/2 or f = 1. This means that the persistent slip band structure is energetically favorable compared to the vein structure. The evolution of dislocation structures from the vein structure to the persistent slip band structure is quite natural in view of the elastic strain energy. If this change requires a continuous change of energy, it may be achieved at S = 1/2, which exactly corresponds to the infinitely thin disk-shaped Ω $(a_3/a_1 \to 0)$.

ACKNOWLEDGEMENT

This research was supported by NSF-MRL Grant DMR-07923573 and partially by U.S. Army Research Grant DAAG29-81-K0090.

REFERENCES

[1] Eshelby, J. D., "The determination of the elastic field of an ellipsoidal inclusion, and related problems", Proc. Roy. Soc. A241, p. 376, 1957.

[2] Tanaka, K. and Mori, T., "Note on volume integrals of the elastic field around an ellipsoidal inclusion", J. Elasticity, Vol. 2, p. 199, 1972.

[3] Mori, T. and Tanaka, K., "Average stress in matrix and average energy of materials with misfitting inclusions", Acta Metall., Vol. 21, p. 571, 1973.

[4] Laird, C. and Duquette, D. J., "Mechanisms of fatigue crack and nucleation", Corrosion Fatigue, O. F. Deuereux, A. J. McEvily and R. W. Staehle, eds., NACE, Vol. 2, p. 88, 1972.

[5] Mughrabi, H., "Microscopic mechanisms of metal fatigue", Strength of Metals and Alloys, P. Haasen, V. Gerold and G. Kostorz, eds., Pergamon Press, Vol. 3, p. 1615, 1980.

[6] Mura, T., Micromechanics of Defects in Solids, Martinus Nijhoff, p. 292, 1982.

[7] Tanaka, K. and Mura, T., "A dislocation model for fatigue crack initiation", J. Appl. Mech., Vol. 48, p. 97, 1981.

[8] Tanaka, K. and Mura, T., "A theory of fatigue crack initiation at inclusions", Metall. Trans. A., Vol. 13, p. 117, 1982.

[9] Tanaka, K. and Mura, T., "Micromechanical theory of fatigue crack initiation from notches", Mech. of Materials, Vol. 1, p. 63, 1982.

[10] Mura, T., "Accumulation of elastic strain energy during cyclic loading", Scripta Metall., June, 1982.

DISLOCATION KINETICS AND THE FORMATION OF DEFORMATION BANDS

E. C. Aifantis

University of Minnesota
Minneapolis, Minnesota 55455

ABSTRACT

By viewing material states as superpositions of perfect lattice and dislocated substates, we can adopt the approach of continuum mechanics to arrive at a system of differential equations describing the interaction effects between stress, dislocations and deformation. The various arguments that we utilize are broadly related to ideas earlier advanced in theories of continuous distribution of dislocations and in models of dislocation dynamics. Following a brief account of the general theoretical structure in the introduction, we discuss in the first part of this paper the ability of the present method to recover previous phenomenological models of material behavior, such as macroscopic theories of plasticity. In the second part of the paper, we illustrate the ability of the present method to predict dislocation phenomena associated with the formation of plastic zones, deformation bands and dislocation patterns.

INTRODUCTION

We view a material state as a superposition of a perfect lattice substate L and a dislocated substate D. Each substate can be assigned with mechanical fields (such as densities, velocities, stresses and exchange terms of mass and momentum) which must be compatible with the balance laws of mass and momentum. Moreover, the stresses and the exchange terms can be related to the dislocation densities, elastic stress and the directions of dislocation motion by means of appropriate constitutive equations. Thus, in general, we have the following formal structure

$$\dot{\rho}_L + \mathrm{div}\underset{\sim}{j}_L = \hat{c}_L,\ [\dot{\rho}_D] + \mathrm{div}[\underset{\sim}{j}_D] = [\hat{c}_D] \tag{1}$$

$$\mathrm{div}\underset{\sim}{T}_L^T + \hat{\underset{\sim}{f}}_L = \rho_L\underset{\sim}{a}_L,\ \mathrm{div}[\underset{\sim}{T}_D^T] + [\hat{\underset{\sim}{f}}_D] = [\rho_D\underset{\sim}{a}_D] \tag{2}$$

$$\underset{\sim}{T}_L - \underset{\sim}{T}_L^T = \hat{\underset{\sim}{\Lambda}}_L,\ [\underset{\sim}{T}_D] - [\underset{\sim}{T}_D^T] = [\hat{\underset{\sim}{\Lambda}}_D] \tag{3}$$

$$\{\underset{\sim}{T}_L, \hat{c}_L, \hat{\underset{\sim}{f}}_L, \hat{\underset{\sim}{\Lambda}}_L;\ [\underset{\sim}{T}_D], [\hat{c}_D], [\hat{\underset{\sim}{f}}_D], [\hat{\underset{\sim}{\Lambda}}_D]\} \xrightarrow{\text{fct.'s}} \{\underset{\sim}{E};\ [\rho_D], [\underset{\sim}{j}_D], [\hat{\underset{\sim}{n}}_D], [\hat{\underset{\sim}{\nu}}_D]\} \tag{4}$$

Equations (1) express conservation of mass with ($\rho_L, \underset{\sim}{j}_L$) denoting density and flux of the perfect lattice substate, and ($[\rho_D], [\underset{\sim}{j}_D]$) denoting column matrices with entries the corresponding density and flux associated with particular families and/or types of dislocations. The terms $\hat{c}_L$ and $[\hat{c}_D]$ model the mobilization and/or immobilization, production and/or annihilation, and other transformations occurring between the various families and types of dislocations. Equations (2) express conservation of linear momentum with $(\underset{\sim}{T}_L, \underset{\sim}{a}_L)$ denoting stress and acceleration of the perfect lattice substate, and $([\underset{\sim}{T}_D],[\underset{\sim}{a}_D])$ denoting column matrices wit entries the corresponding stress and acceleration associated with particular families and/or types of dislocations; the symbol "T" denotes transposition so that $[\rho_D]^T$ is a row matrix. The terms $\hat{\underset{\sim}{f}}_L$ and $[\hat{\underset{\sim}{f}}_D]$ denote internal body forces generate by the exchange of linear momentum between the perfect lattice and dislocated substates. For example, these terms will normally include the Peach-Koehler force, the drag exerted by the lattice on a moving dislocation, etc. Similarly, equation (3) express conservation of angular momentum for the lattice and dislocated substates with the terms $\hat{\underset{\sim}{\Lambda}}$'s denoting material tensors modeling the exchange of internal body couples between the states. Finally, equations (4) are the appropriat constitutive equations. The tensor $\underset{\sim}{E}$ is the lattice strain which can be identifie with the elastic strain. The column matrices $[\hat{\underset{\sim}{n}}_D]$ and $[\hat{\underset{\sim}{\nu}}_D]$ have as entries the unit vectors $\hat{\underset{\sim}{n}}^{\alpha}$ and $\hat{\underset{\sim}{\nu}}^{\alpha}$, $\alpha = 1,2,\ldots$, where α denotes the number of distinct families and types of dislocations, $\hat{\underset{\sim}{n}}$ denotes the unit vector perpendicular to the sli plane and $\hat{\underset{\sim}{\nu}}$ denotes the unit vector in the slip direction. The very general balance laws (1)-(3) can be simplified somewhat by neglecting non-local effects

$$(\hat{c}_L + [1]^T[\hat{c}_D] = \hat{\underset{\sim}{f}}_L + [1]^T[\hat{\underset{\sim}{f}}_D] = \hat{\underset{\sim}{\Lambda}}_L + [1]^T[\hat{\underset{\sim}{\Lambda}}_D] = 0,\ [1]^T \equiv [1,1,\ldots])$$

by neglecting inertia effects $(\underset{\sim}{a}_L = [\underset{\sim}{a}_D] = 0)$, by neglecting body couple exchanges $(\hat{\underset{\sim}{\Lambda}}_L = [\hat{\underset{\sim}{\Lambda}}_D] = 0)$, and finally by considering quasi-static processes where the velocity of the lattice state is vanishingly small as compared to the average velocity of moving dislocation states.

To establish contact between the overall macroscopic deformation and the particular microscopic displacements associated with the motion and production of individual families and types of dislocations, we also propose a constitutive equation for the rate of the plastic strain tensor $\dot{\underset{\sim}{E}}_p$. It can be shown that, by assuming macroscopic plastic incompressibility and neglecting the contribution of climbing dislocation motions to the overall plastic strain, the appropriate constitutive equation is of the form

$$\dot{\underset{\sim}{E}}_p = [1]^T[(\zeta_D j_D^g + \xi_D \dot{\rho}_D)\underset{\sim}{M}_D] \tag{5}$$

where (ζ_D, ξ_D) are constants measuring respectively the contributions of gliding dislocation motion and dislocation production associated with the various familie and types of dislocations, j_D^g denote the components of the corresponding fluxes i the slip direction (glide motion), and $[\underset{\sim}{M}_D]$ denotes a column matrix with entries appropriate orientation tensors of the form

$$\underset{\sim}{M}^{\alpha} = \frac{1}{2}(\hat{\underset{\sim}{n}}^{\alpha} \otimes \hat{\underset{\sim}{\nu}}^{\alpha} + \hat{\underset{\sim}{\nu}}^{\alpha} \otimes \hat{\underset{\sim}{n}}^{\alpha}) \tag{5a}$$

where α represents the number of the various families and types of dislocations under consideration. In the next section we discuss the ability of the above the retical structure to recover and generalize various macroscopic models of material behavior, earlier proposed in the literature. Finally, in the last section, we illustrate the ability of the theory to predict various microscopic phenomena associated with dislocation motion and production, and which are delivered at the macroscale with the occurrence of plastic zones, slip bands and other shear localizations.

MACROSCOPIC THEORIES OF PLASTIC DEFORMATION

The theoretical structure discussed in the earlier section can be simplified appreciably if only one family and type of dislocations is considered. By assuming, in addition, negligible non-local effects, inertia and body couple exchanges, the balance laws (1)-(3) are simplified as follows for quasi-static processes

$$\dot{\rho} + \mathrm{div}\mathbf{j} = \hat{c}, \quad \dot{\rho}_L + \mathrm{div}\mathbf{j}_L = -\hat{c} \tag{6}$$

$$\mathrm{div}\mathbf{T} + \hat{\mathbf{f}} = 0, \quad \mathrm{div}\mathbf{T}_L - \hat{\mathbf{f}} = 0 \tag{7}$$

with $\mathbf{T}$ and $\mathbf{T}_L$ symmetric. A relatively simple but physically motivated candidate for the constitutive model (4) is as follows

$$\mathbf{T} = \hat{\pi}\mathbf{M}, \quad \mathbf{T}_L = \hat{\lambda}(\mathrm{tr}\mathbf{E})\mathbf{1}+2\hat{G}\mathbf{E} \tag{8}$$

$$\hat{\mathbf{f}} = -(\hat{\alpha}+\hat{\beta}j^g-\hat{\gamma}\tau)\hat{\boldsymbol{\nu}}, \quad \hat{c} = \hat{c}_0 + \hat{c}_1\rho + \hat{c}_2\rho^2 + \hat{c}_3\rho^3 \tag{9}$$

with the definitions

$$j^g \equiv \mathbf{j}\cdot\hat{\boldsymbol{\nu}}, \quad \tau \equiv \mathrm{tr}(\mathbf{SM}), \quad \mathbf{S} \equiv \mathbf{T}+\mathbf{T}_L \tag{10}$$

The functions $\hat{\pi}$, $\hat{\lambda}$, $\hat{G}$ depend, in general, non-linearly on ρ, and the same is true for the functions $\hat{\alpha}$, $\hat{\beta}$, and $\hat{\gamma}$; $\hat{\gamma}$ measures the effect of the Peach-Koehler force, $\hat{\beta}$ the drag, and $\hat{\alpha}$ measures the effect of all other microstructural factors but dislocations. Interaction stresses and anisotropic effects on the lattice stress were not assumed, and the effect of climbing dislocation motion on the internal body force was neglected. Roughly speaking, the quantities $\hat{c}_0$, $\hat{c}_1$, $\hat{c}_2$ and $\hat{c}_3$, being generally functions of j^g and τ, measure respectively the probabilities leading to dislocation production by activation of fixed number of sources, by self-multiplication or decomposition of other dislocations, by interaction of two dislocations (e.g., annihilation or formation of dipoles), and by interaction of three dislocations (e.g., interaction of a dipole and another dislocation). The constitutive model expressed by (8) and (9) can be easily generalized but we will discuss this issue elsewhere. Consistent with (8) and (9), the appropriate formula for the plastic strain rate is

$$\dot{\mathbf{E}}_p = (\zeta j^g+\xi\dot{\rho})\mathbf{M} \tag{11}$$

Ideally, the system of equations (6)-(11) can be utilized to describe the stress, the dislocation distributions and the macroscopic deformation that occur in a monocrystal characterized by a single family and type of dislocations. It can be shown for example, that under certain suitable assumptions, the temporal and spatial evolution of the dislocated state can be described by the equation

$$\dot{\rho} = D\mathrm{tr}(\mathbf{M}\ \mathrm{grad}^2\rho) + c_1\rho + c_2\rho^2 \tag{12}$$

which, as discussed in [1,2] has essentially been used by Pilecki. He has shown how equation (12) with the first term of the right-hand side replaced by $D\nabla^2\rho$ can serve to interpret fatigue processes in cases where a chaotic motion of dislocations take place.

Physically, yielding occurs when the resistance to the dislocation motion in the slip direction is overcome. Within the present formulation, a yield condition can be obtained by computing the component of the momentum balance for the dislocated state in the slip direction. By neglecting spatial nonuniformities and fluxes associated with the dislocation distribution, this computation yields with $\tau_o \equiv \hat{\alpha}/\hat{\gamma}$

$$tr(\underset{\sim}{S}\underset{\sim}{M}) = \tau_o \tag{13}$$

where τ_o has the meaning of yield stress in pure shear. By evaluating equation (13) at incipient yielding (where it can be assumed that $\underset{\sim}{T}=0$), we have

$$tr(\underset{\sim}{T}_L\underset{\sim}{M}) = \tau_o \tag{14}$$

which is essentially the Tresca yield condition for monocrystals with a single slip system. The evolution of the initial yield surface as defined by equation (14) can easily be traced by means of equation (13) which, with the use of equation $(10)_3$ and a linearized version of equation $(8)_1$, becomes

$$tr([\underset{\sim}{T}_L - \pi\rho\underset{\sim}{M}]\underset{\sim}{M}) = \tau_o \tag{15}$$

Equation (15) may be viewed as substantiating Prager's kinematic hardening condition for monocrystals with ρ determined by a balance law and appropriate constitutive equations.

For polycrystals, the directions $\hat{n}$ and $\hat{\nu}$ cannot be assumed as prescribed by crystallography. Instead, we take the point of view that these directions should be determined in conjunction with directions of the critical resolved shear lattice stress τ_L which is essentially responsible for activating the relevant slip processes. With this motivation, it is natural to seek for the extrema of the critical resolved shear stress

$$\tau_L = tr(T_L\underset{\sim}{M}) \tag{16}$$

subject to the obvious geometric constraints

$$tr\underset{\sim}{M} = 0, \quad tr\underset{\sim}{M}^2 = \frac{1}{2} \tag{17}$$

The result is

$$\underset{\sim}{M} = \frac{\underset{\sim}{S}^d}{2\sqrt{J_2}} \tag{18}$$

where

$$\underset{\sim}{S}^d \equiv \underset{\sim}{T}_L - \frac{1}{3}\, tr\underset{\sim}{T}_L, \quad J_2 = \frac{1}{2}\, tr(\underset{\sim}{S}^d)^2 \tag{19}$$

Interestingly, in equations (18) and (19), familiar quantities from classical macroscopic plasticity appear as a result of average trends in the behavior of dislocations. This connection between macroscopic plasticity and dislocation theory becomes more apparent upon substitution of equation (18) into equation (14). The result is

$$J_2 = \tau_o^2 \tag{20}$$

i.e., the Von-Mises yield condition. Moreover, by utilizing equation (11), we obtain the Prandtl-Reuss relationships of classical plasticity

$$\dot{\underset{\sim}{E}}_p = \dot{\lambda} \frac{\underset{\sim}{S}^d}{2\sqrt{J_2}}, \quad \dot{\lambda} \equiv \Sigma(\zeta j^g + \xi\dot{\rho}) \tag{21}$$

where the sum is over the eight families of dislocations defined by the unit normal to the octahedral planes $\hat{n}$ and the unit in the direction of the traction vector on the octahedral planes $\underset{\sim}{\hat{\nu}}$.

If instead of the constraints (17) we impose directly the constraints

$$\underset{\sim}{\hat{n}}\cdot\underset{\sim}{\hat{n}} = \underset{\sim}{\hat{\nu}}\cdot\underset{\sim}{\hat{\nu}} = 1, \quad \underset{\sim}{\hat{n}}\cdot\underset{\sim}{\hat{\nu}} = 0 \tag{22}$$

we find

$$\underset{\sim}{\hat{n}} = \frac{\pm\hat{e}_1 \pm \hat{e}_3}{\sqrt{2}}, \quad \underset{\sim}{\hat{\nu}} = \frac{\pm\hat{e}_1 \mp \hat{e}_2}{\sqrt{2}}, \quad \underset{\sim}{M} = \frac{1}{2}\,(\hat{e}_1 \otimes \hat{e}_1 - \hat{e}_3 \otimes \hat{e}_3) \tag{23}$$

where $\hat{e}_1$, $\hat{e}_2$ and $\hat{e}_3$ denote the direction of principal stresses. Then, substitution of $(23)_3$ into (14) and use of the spectral decomposition

$$\underset{\sim}{T}_L = \sigma_1 \hat{e}_1 \otimes \hat{e}_1 + \sigma_2 \hat{e}_2 \otimes \hat{e}_2 + \sigma_3 \hat{e}_3 \otimes \hat{e}_3$$

where σ_1, σ_2 and σ_3 denote principal stresses, yields

$$\sigma_1 - \sigma_3 = 2\tau_o \tag{24}$$

which is Tresca's yield condition. Moreover, use of (11) yields

$$\dot{\underset{\sim}{E}}_p = \dot{\lambda} \frac{(\hat{e}_1 \otimes \hat{e}_1 - \hat{e}_3 \otimes \hat{e}_3)}{2}, \quad \dot{\lambda} = \Sigma(\zeta j^g + \xi\dot{\rho}) \tag{25}$$

where now the sum is over the four families of dislocations which, as indicated by $(23)_{1,2}$ are associated with the planes of maximum shear.

The ability of the present theoretical structure to substantiate and generalize other phenomenological models of macroscopic plastic deformation will be discussed elsewhere. The more complex situation where one allows for the possibility of two types of dislocations (that is, mobile and immobile dislocation states) has been considered briefly in [1] and elaborated in more detail in [2]. It was thus shown in [2] that various models of cyclic plasticity can be obtained within the framework of the present approach. In addition, phenomena such as latent hardening and the

Bauschinger effect can be modeled by the present theory. In the next section, we illustrate briefly the ability of the present structure to predict the formation, the character and evolution of regions of shear localization, such as plastic zones and slip bands.

SHEAR LOCALIZATION

In this final part of the paper, we apply the theory of previous section to consider localization of shear in certain elementary but typical configurations. Roughly speaking, localization of shear is taken to correspond to appreciable increase of dislocation density, as indicated by (11). For simplicity and purposes of illustration, we confine attention to situations which can approximately be treated within one-dimensional considerations only.

We first consider an ad hoc treatment of the problem of formation of deformation bands in properly oriented monocrystals and plastic zones in cracked plane-stressed polycrystals. To illustrate the formation of deformation, we consider properly oriented so that its slip direction is parallel to the x-axis, while the direction of the load is parallel to the y-axis as shown in Figure 1. It then

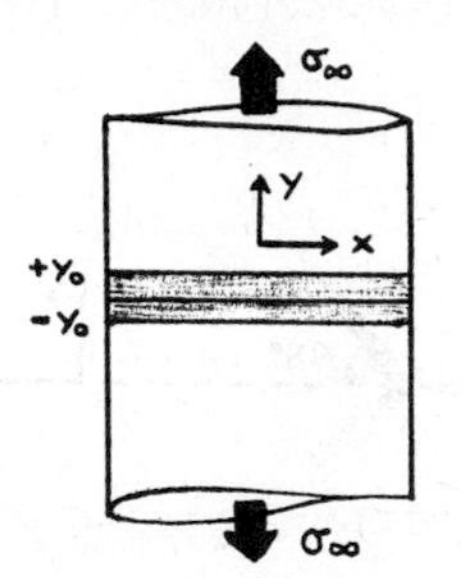

Fig. (1) - Deformation band

turns out that equation (12) becomes

$$\dot{\rho} = \frac{1}{2} D\rho_{xy} + c_1\rho + c_2\rho^2 \tag{26}$$

We confine attention to stationary states where $\dot{\rho}=0$ and we also consider solutions which are uniform in x. Furthermore, we assume that due to an inhomogeneity place at y=0, the stress distribution is given approximately by

$$\sigma = \sigma_\infty + \sigma_0 h(y_0+y)h(y_0-y) = \begin{cases} \sigma_\infty, & |y|>y_0 \\ \sigma_\star \equiv \sigma_\infty+\sigma_0, & |y|<y_0 \end{cases} \tag{27}$$

where $h(\cdot)$ denotes the usual unit Heaviside function and σ_0, the uniform effect of inhomogeneity on the stress distribution within a small region $y<|y_0|$. Finally, the kinetic coefficients c_1 and c_2 in equation (26) are assumed to be respectively independent and linearly dependent on the stress σ; that is,

$$c_1 = \mathring{c}_1,\ c_2 = -\mathring{c}_2 h(\sigma-\sigma_c) \tag{28}$$

where $\overset{\circ}{c}_1$ and $\overset{\circ}{c}_2$ denote positive constants and σ_c is a critical stress (a material parameter) which relates to the present loading configuration according to the condition $\sigma_\infty < \sigma_c < \sigma_*$. Under these conditions, the solution to (26) is easily found to be

$$\rho(y) = \begin{cases} 0\ , & |y| > y_0 \\ \rho_*, & |y| < y_0 \end{cases} \tag{29}$$

with $\rho_* \equiv \overset{\circ}{c}_1/\overset{\circ}{c}_2$; that is, in the small region $|y| < y_0$ the dislocation density can be appreciably large and this can lead to macroscopic shear localization.

By employing similar arguments, we can illustrate the formation of plastic zones in cracked polycrystals loaded under plane-deformation conditions with tensile stresses σ_∞ at infinity. From equations (23), we conclude that slip occurs on planes forming 45° angles with the plane of the crack as illustrated in Figure 2. It then turns out that equation (12) becomes

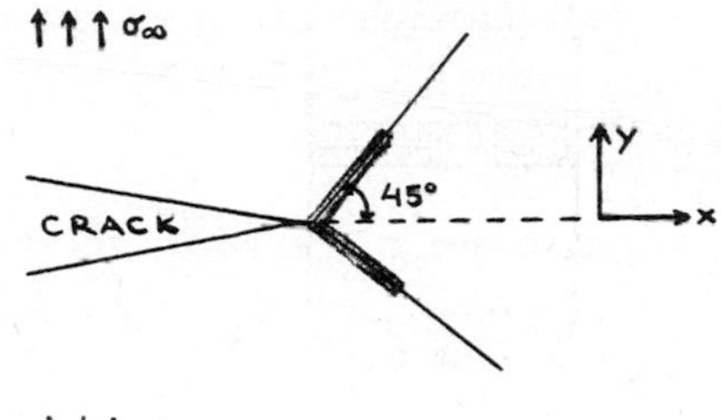

Fig. (2) - Plastic zone

$$\dot{\rho} = \frac{1}{2} D(\rho_{yy} - \rho_{xx}) + c_1\rho + c_2\rho^2 \tag{30}$$

As before, we confine attention to stationary states and consider solutions which are uniform in x+y with the exception of a small region about the line x-y which extends up to a distance r_0 ahead of the crack tip. We have thus reduced approximately the problem into confining the plastic zones along the lines x-y and we seek to determine their size; due to symmetry we elaborate for y>0. The appropriate stress which controls the interactions of dislocations on the slip plane is given by

$$\tau = tr(\underset{\sim}{T}_L \underset{\sim}{M}) \tag{31}$$

which for the present configuration turns out to be

$$\tau = \sigma_{yy} - \sigma_{xx} = \frac{k}{\sqrt{r}} \tag{32}$$

where k is a constant ($\sigma_\infty\sqrt{\alpha/2}$ sin 67.5°, α = half crack length) as determined by linear elastic fracture mechanics. The kinetic coefficients c_1 and c_2 are assumed as before to be of the form

$$c_1 = \overset{\circ}{c}_1,\ c_2 = -\overset{\circ}{c}_2 h(\tau-\tau_o) \tag{33}$$

where τ_o denotes the yield stress. Under these conditions, the solution to (30) is easily found to be

$$\rho(r) = \begin{cases} 0\ ,\ r>r_o \\ \rho_*,\ r<r_o \end{cases} \tag{34}$$

where ρ_* is as in (29) and r_o denotes the length of plastic zone determined via (32) as $r_o \equiv k^2/\tau_o^2$.

Next, we consider a more precise analysis on the possibility of formation of stationary one-dimensional spatially distributed patterns of dislocations. By assuming that average dislocation motion occurs along a single direction only (say x) and by also considering dislocation interactions up to third order, we can show that the appropriate differential equation governing the distribution of dislocations is

$$\rho_{xx} = a\rho - b\rho^2 + c\rho^3 \equiv g(\rho) \tag{35}$$

where the kinetic coefficients a, b and c are taken for simplicity as constants. We assume

$$b^2 > 4ac \tag{35a}$$

so that the graph of $g(\rho)$ is as in Figure 3 below.

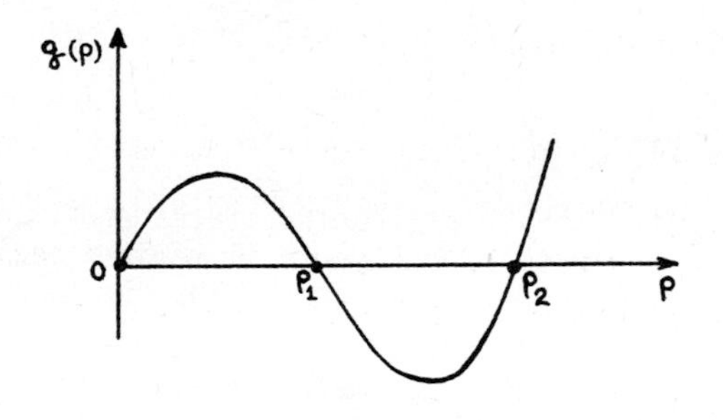

Fig. (3) - Graph of $g(\rho)$

The roots ρ_1 and ρ_2 are given by

$$\rho_1 = \frac{b-\sqrt{b^2-4ac}}{2c},\ \rho_2 = \frac{b+\sqrt{b^2+4ac}}{2c} \tag{36}$$

Then it can be shown that for infinite regions the only possible solutions of (35) are either of the forms depicted in Figure 4. The solutions of type (i)

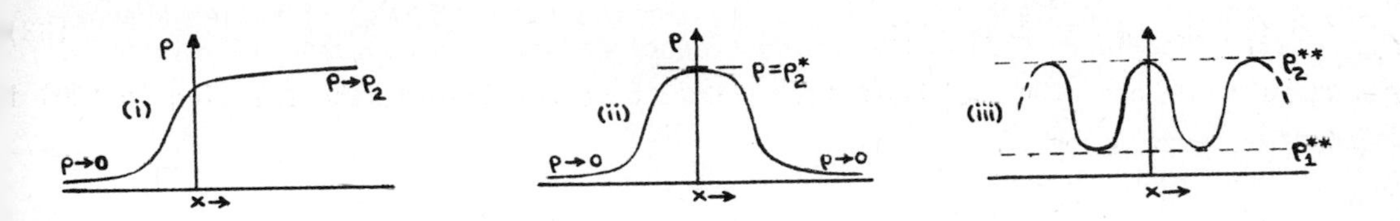

Fig. (4) - Types of solution

are uniquely determined (up to choice of x_0) by the explicit representation

$$x = x_0 + \int_{\rho(x_0)}^{\rho(x)} \frac{d\rho}{\sqrt{2G(\rho)}}, \quad G(\rho) \equiv \int_0^{\rho} g(\rho)d\rho \tag{37}$$

and the following equal area condition must be satisfied

$$\int_0^{\rho_2} g(\rho)d\rho = 0; \; g(0) = g(\rho_2) = 0 \tag{38}$$

If condition (38) is not fulfilled, then only solutions of types (ii) and (iii) can be possible. The solutions of type (ii) have also the representation (37) for $-\infty<x\leq\bar{x}$ where $\rho(\bar{x}) = \rho_2^* = \max\rho(x)$; the graph of $\rho(x)$ is symmetric about $\bar{x}$. The appropriate area condition that must now be satisfied is

$$\int_0^{\rho_2^*} g(\rho)d\rho = 0; \; g(0) = 0, \; g(\rho_2^*) < 0 \tag{39}$$

Finally, the solutions of type (iii) also have the representation (37) for $\bar{\bar{x}}\leq x\leq\bar{x}$ where $\bar{\bar{x}}$ and $\bar{x}$ are any two consecutive minima and maxima of $\rho(x)$ with $\rho(\bar{\bar{x}}) = \rho_1^{**} = \min\rho(x)$ and $\rho(\bar{x}) = \rho_2^{**} = \max\rho(x)$; the graph of $\rho(x)$ is periodic. The appropriate area condition for existence of solutions is now given by

$$\int_{\rho_1^{**}}^{\rho_2^{**}} g(\rho)d\rho = 0; \; g(\rho_2^{**}) < 0 < g(\rho_1^{**}) \tag{40}$$

All solutions depicted in Figure 4 can easily be taken to correspond qualitatively to many physical situations where shear localization occurs. The above results are obtained when the condition (35a) is satisfied. In general, the kinetic coefficients a, b and c depend on the state of stress and it is conceivable that for values of stress smaller than a critical value, $g(\rho)$ will only have one trivial root. Then, localization does not occur and this fact, which will be discussed in detail elsewhere, is in accord with intuition.

Finally, we attempt to provide briefly a qualitative but formal reasoning of the formation of persistent slip bands during fatigue. As it will be discussed in detail elsewhere, if the resolved shear strain amplitude γ^p is approximately proportional to the dislocation density ρ, and the cyclic stress τ^s varies approximately linearly with the dislocation production $g(\rho)$, we can cast (35) in the form

$$\gamma^p_{xx} = \tau^s(\gamma^p) - \bar{\tau}^s \qquad (41)$$

where $\tau^s(\gamma^p)$ has a graph as in Figure 5 and $\bar{\tau}^s$ is a constant having the meaning

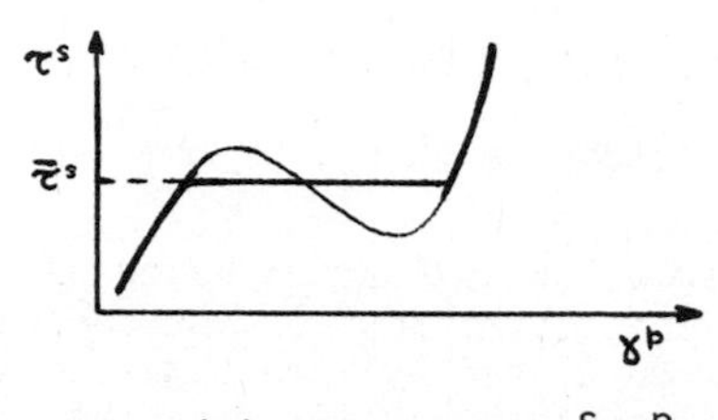

Fig. (5) - Graph of $\tau^s(\gamma^p)$

of a saturation stress. As in the problem of phase transitions, states with $d\tau^s/d\gamma^p < 0$ are inaccessible; instead, a saturation stress $\bar{\tau}^s$ is drawn such as to cut-off equal areas. Moreover, the spatial distribution of γ^p is determined through (41) by applying techniques similar to those used for the solution of (35). Less formally, these analogies between persistent slip bands and phase transitions were also pointed out in [3] where, in addition, some comparisons with experiments were made.

ACKNOWLEDGEMENT

Support of the Solid Mechanics Program of the National Science Foundation and the Corrosion Center of the University of Minnesota is gratefully acknowledged.

REFERENCES

[1] Aifantis, E. C., "Elementary physiochemical degradation processes", Mechanics of Structured Media, Part A, A.P.S. Selvadurai, ed., Elsevier, Amsterdam, pp. 301-317, 1981.

[2] Bamman, D. J. and Aifantis, E. C., "On a proposal for a continuum with microstructure", Acta Mechanica, 1982.

[3] Mughrabi, H., "Cyclic plasticity of matrix and persistent slip bands in fatigued metals", Continuum Models of Discrete Systems 4, O. Brulin and R.K.T. Hsieh, eds., North-Holland, Amsterdam, pp. 241-257, 1981.

INTERGRANULAR FRACTURE CRITERIA AND INTERFACIAL THERMODYNAMIC PROPERTIES

O. A. Bamiro

University of Ibadan
Ibadan, Nigeria

ABSTRACT

A theory of intergranular crack nucleation based upon the stress concentration set up by double slip bands coincident on the boundary has been presented. The propagation criteria are developed by investigating the stability of such nucleated crack against blunting by nucleated dislocations. The cohesion potential, emanating from the crack nucleation and propagation criteria, is developed in terms of the thermodynamic description of Gibbs with particular reference to separations that occur too rapidly for any significant redistribution of segregant to take place and separations that are slow enough to allow full adsorption equilibrium. An atomistic model has been developed in order to evaluate the interfacial potential at ordinary temperatures. The prediction of the model is in very good agreement with the experimental observations on the embrittlement of Cu by Bi.

INTRODUCTION

The intergranular fracture of several materials has in the recent literature been attributed to the segregation of certain impurities to the grain boundaries. Intergranular cracking has been observed in alloy steel due to phosphorous, antinomy and tin [1-3]; and in copper alloys with traces of bismuth [4]. This has been achieved by utilizing the technique of the Auger Electron Spectroscopy to analyze the chemical composition of the fracture surfaces immediately after the fracture has taken place [5]. Thus, experimental data of the degree of segregation to the grain boundary for an alloy such as Cu-Bi as well as the corresponding fracture stress at room temperature have been obtained [4]. The goal of any theory of intergranular fracture, therefore, is to relate the fracture stress to the bulk impurity concentration. So far, no analytical theory has been able to achieve this, although, some developed theories have been able to predict whether intergranular fracture will occur or not [4,6-8]. This paper addresses further some of the fundamental problems of theoretical formulation.

In line with the previous analysis, the intergranular fracture process is conceptually modelled as one involving crack nucleation followed by propagation of the nucleated crack. The theoretical stress required to initiate a crack is obtained. Such nucleated crack may propagate to cause a brittle fracture or be non-propagating as a result of preferential nucleation of dislocation which blunts the nucleated crack. The necessary and sufficient condition for the crack to remain atomically sharp and be stable against spontaneous blunting is examined in the sense of Rice [6], Mason [7] and Bamiro [8].

The theoretical analysis is shown to culminate in expressions involving some basic thermodynamic properties of the grain boundary. However, there is no experimental technique at present to determine these properties unambiguously. Recourse is therefore made to theoretical evaluation which is discussed in the paper.

ANALYSIS

A. Thermodynamics of Fracture Process

Let us consider a general bicrystal, with a grain boundary normal to the tensile axis, and undergoing the separation process caused by the locally applied normal stress, σ_{22}, as shown in Figure 1a. Let us suppose that the two crystals

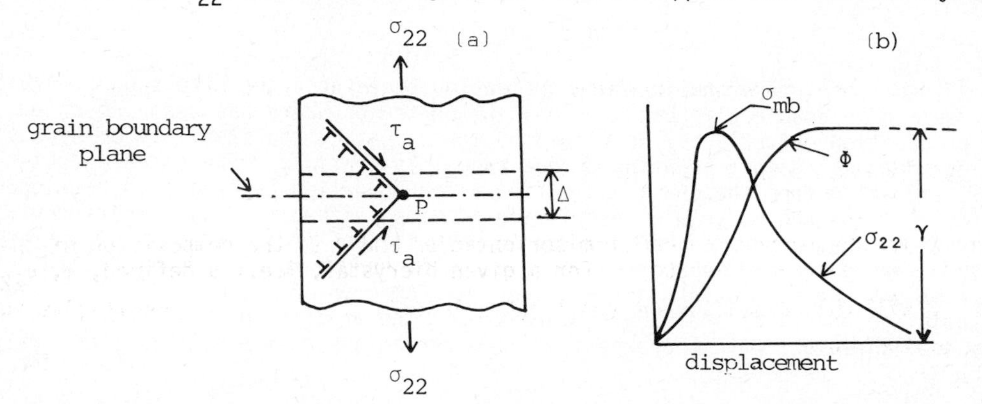

Fig. (1a) - Double-slip band mechanism subjected to externally applied shear stress τ_a; (b) - Interfacial potential Φ and stress-displacement relation for a separating interface

are initially at a mean equilibrium separation λ_o (whose determination is discussed later on) and the separation process takes place isothermally and reversibly. The free energy change with relative displacement δ from the equilibrium configuration is assumed to vary as shown in Figure 2. The stress σ_{22} required for the extension is therefore given by

$$\sigma_{22} = \frac{\partial \Phi}{\partial \delta} \tag{1}$$

with possible variation also shown in Figure 1b. The maximum value of σ_{22} given by σ_{mb} defines the cohesive stress to the grain boundary. It is reasonable to expect Φ, and hence, σ_{22} and σ_{mb} depend also upon the composition and structure of the grain boundary. Thus, in general,

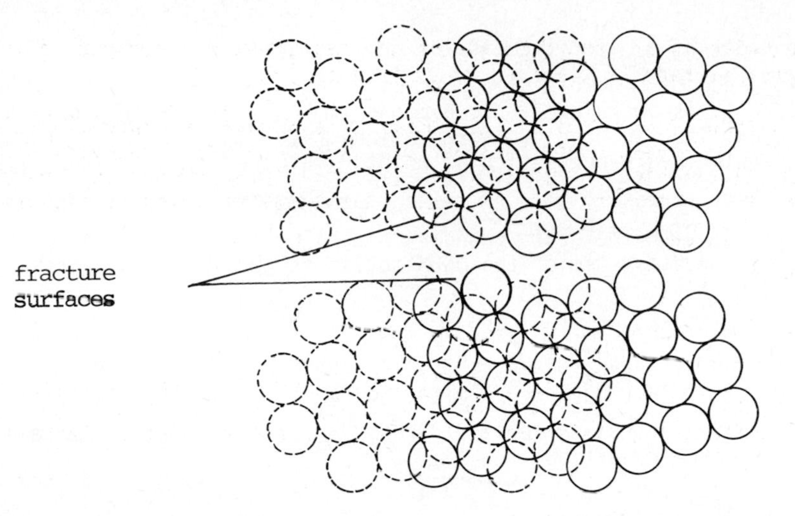

Fig. (2) - Fracture surfaces of the 36.9° grain boundary in copper

$$\Phi = \Phi(\delta,\underset{\sim}{\theta},\Gamma);\ \sigma_{22} = \sigma_{22}(\delta,\underset{\sim}{\theta},\Gamma);\ \sigma_{mb} = \sigma_{mb}(\delta,\underset{\sim}{\theta},\Gamma) \tag{2}$$

where $\underset{\sim}{\theta}$ is the bicrystal angular misorientation and Γ is the composition of the impurity in the grain boundary. For a given bicrystal, i.e., $\underset{\sim}{\theta}$ defined, Φ, σ_{22} and σ_{mb} will then depend upon δ and Γ only, leading thereby to a constitutive equation of the form:

$$\sigma_{22} = \sigma_{22}(\delta,\Gamma) \tag{3}$$

The reversible work required to separate the bicrystal γ, is given by

$$\gamma = \Phi(\infty,\Gamma) \tag{4}$$

Evaluation of Φ at $\delta=\infty$ as per equation (4) serves to accentuate the fact that the bicrystal is expected to have been completely separated into two non-interacting crystals. Physically, δ is of the order of the atomic radius. Determination of γ requires the specification of the relaxation process that will occur during separation as discussed later.

B. Theory of Intergranular Crack Nucleation

The process of crack nucleation at a point in a material involves the breakage of atomic bonds at the point. No matter the degree of local decohesion or strength lowering, the stress required to accomplish this task is still much higher than any externally applied fracture stress. The implication of this is obvious: before a crack can nucleate at a point, there must be a mechanism available to locally raise the small externally applied stress to a much higher value of the order of σ_{mb} for the case of an intergranular fracture. In this context, the existence of a double-slip band dislocation mechanism (DSBM) such as shown in Figure 1a is postulated. Such mechanism was originally proposed by Cottrell [9] and utilized in previous formulations [4,6-8]. For the particular case of intergranular crack nucleation at point P, Figure 1a, the external stress σ_a exerts a shear stress τ_a (= $\sigma_a/2$, since the symmetric bands are assumed to make angle of 45° with the grain boundary plane) on the dislocation pile-up such that the local stress system of the DSBM satisfies the local bond-breakage condition:

$$\sigma_{22} \geq \sigma_{mb} \tag{5}$$

The solution of the stress system of the DSBM under an externally applied stress obtained by Stroh [10] and by Smith and Barnby [11], has been utilized [8] to show that the externally applied stress, σ_a^N, required to nucleate a crack at P, is given by the condition:

$$\sigma^N \geq \beta\left[\frac{E_b\gamma}{d}\right]^{1/2} \tag{6}$$

where $\beta = [3\pi/196(1-\nu^2)]^{1/2}$; d is the grain diameter; ν is the Poisson's ratio; and E_b is the Young's modulus of the grain boundary. E_b, which describes the variation of the stress σ_{22}, Figure 1b, in the elastic region, i.e., for very small displacement δ from the equilibrium configuration, is given by:

$$E_b = \lambda_o \left.\frac{\partial^2\Phi}{\partial^2\delta^2}\right|_{\delta=0} \tag{7}$$

The equilibrium interplanar spacing at the grain boundary plane λ_o, was arbitrarily equated to $\Delta/2$ (Δ = grain boundary thickness) in the previous analysis [8]. Assuming that in the elastic region the interfacial potential function is parabolic and of the form:

$$\Phi = C_1\delta^2 - C_2\delta^3; \quad \text{(for small } \delta) \tag{8}$$

where C_1 and C_2 are positive quantities. In the present analysis, we demand that Φ give the linear thermal coefficient of expansion, α, of the bulk material. Following the analysis of Kittle [12] and utilizing equation (8):

$$\lambda_o = \left[\frac{3KC_2}{4\alpha C_1^2}\right]^{1/3} \tag{9}$$

where K is the Boltzmann constant. From equations (8) and (9), equation (7) becomes

$$E_b = 2\lambda_o C_1 \tag{10}$$

The crack nucleation condition (6) has been derived assuming no frictional stress opposing dislocation movement in the slip plane. In real materials, static friction stress acts on the glide dislocations in the band. In general, the friction stress is anisotropic and dependent upon the bulk concentration of solute [13]. Assuming isotropic condition and following the analysis of Evans [14], it can be shown that the friction stress σ_f, in a glide plane is given by:

$$\sigma_f \geq 2nbE/\pi d(1-\nu^2) \tag{11}$$

where n is the number of dislocation; b is the Burger's vector of the edge dislocation; and E is the elastic modulus of the crystal lattice in the region of DSBM. From Eshelby et al [15], n is given by:

$$n = \frac{\pi d(1-\nu^2)\sigma^N}{2bE} \tag{12}$$

Substitution of n from equation (12) in equation (11) leads to an estimate of the friction stress given by:

$$\sigma_f \geq \sigma^N \tag{13}$$

Combining equations (6) and (13), the total externally applied stress required to nucleate a crack, $\sigma_a^N (\sigma_a^N = \sigma^N + \sigma_f)$ can therefore be expressed in the form:

$$\sigma_a^N \geq (q+\beta)[E_b\gamma/d]^{1/2} \tag{14}$$

q is the lattice friction factor which on the basis of the above analysis is given by $q \geq \beta$.

C. Crack Propagation Criteria

Non-propagating cracks in alloys which have failed due to intergranular fracture have been experimentally observed [4]. This can be interpreted to mean that while the condition for crack nucleation, such as derived in the previous section may be met in a locality, the propagation of such crack is subjected to other conditions being satisfied. In other words, the high stresses near the atomically sharp crack of sufficient magnitude to propagate it according to the Griffith's theory may be relieved by a stress-attenuating process. In this context, it is assumed in this study that the atomically sharp crack can become non-propagating as a result of the alternative process of dislocation nucleation which causes blunting. Thus, the derivation of the crack propagation criterion is based upon an examination of the stability of the crack against blunting by nucleated dislocations.

The above criterion was first applied by Rice and Thomson [6] to cracks in single crystals. Rice [6] later extended the analysis to intergranular crack. An alternative approach, modifying the theoretical analysis of Rice, was developed by Mason [7]. Although the results based upon the two different models of Rice and Mason have been shown by Bamiro [18] to give essentially the same results, the approach of Mason, which gives solution in closed form, is adopted in this paper.

Following the analysis of Mason, the total energy required to form a dislocation loop of radius rb (b = Burger's vector) is given by:

$$U(r) = \frac{\mu b^3 r(2-\nu)}{8(1-\nu)} \ln\left(\frac{8r}{e^2\zeta_o}\right) + \frac{2}{\beta^1}\gamma_s b^2(r-\zeta_o) - \frac{1.395}{\beta}\frac{b^2}{(1-\nu)^{1/2}} \times$$

$$\times \left(\frac{\Omega bG}{2}\right)^{1/2} (r^{3/2} - \zeta_o^{3/2}); \ \frac{1}{\beta^1} = \sin\phi\cos\lambda, \ 1/\beta = 1/\beta^1\cos\phi/2 \qquad (15)$$

Ω is the shear modulus. λ is the angle of inclination to the crack propagation direction. ϕ is the angle between the Burger's vector of the dislocation and the normal to the crack surface. γ_s is the specific energy of the new surface to form a "step" or "ledge" left behind at the crack tip after nucleation of the dislocation. $\zeta_o = r_o/b$, where r_o is the dislocation core cut-off in the sense of Hirth and Lothe [17]. $G = (1-\nu)K^2/2\Omega$, is the energy release rate corresponding to the Griffith load condition. K is the general stress intensity factor which increases the load on the crack. The crack propagation is based upon the following interpretation: if, as K increases, the K_g corresponding to the Griffith load is first reached, then the crack propagates in a brittle manner; but, if K_c corresponding to a load at which dislocation nucleation is possible is first reached, then the crack is blunted. Thus, blunting occurs if $(dU/dr) = 0$ and $(d^2U/dr^2) = 0$. These two conditions lead to the critical energy release rate:

$$G_c = \frac{(1-\nu)K_c^2}{2\Omega} = \frac{2\beta^2}{9(0.986)^2 e^3}\frac{(2-\nu)^2}{(1-\nu)}\frac{\Omega b}{\zeta_o}\exp(5S) \qquad (16)$$

On the other hand, a Griffith microcrack exists if the energy release rate:

$$G_g = \frac{(1-\nu)K_g^2}{2\Omega} = 2\gamma_{int} = \gamma \qquad (17)$$

with γ as defined before. Utilizing equations (16) and (17), the ductile-brittle transition given by $G_c = G_g$ can be shown to lead to:

$$R_o^* = 0.058\ S^{-1}\exp(5S) \qquad (18)$$

where the non-dimensional parameters R_o and S are given by:

$$R_o = \frac{16\beta^1}{5(2-\nu)\beta^2}\zeta_o\frac{\gamma_{int}}{\gamma_s}; \ S = \frac{16(1-\nu)}{5\beta^1(2-\nu)}\frac{\gamma_s}{\Omega b} \qquad (19)$$

If $R_o < R_o^*$, the crack is a Griffith microcrack. Since $G_g = \frac{(1-\nu^2)K_g^2}{E} = \gamma$, and $K_g = \sigma_a^p\sqrt{\pi c}$, the stress σ_a^p, required to cause brittle propagation of the Griffith crack length (2c), is then given by:

$$\sigma_a^p = \left[\frac{E_b\gamma}{\pi(1-\nu^2)c}\right]^{1/2} \qquad (20)$$

Thus, the nucleated crack, which must have at least satisfied the nucleation condition (14), becomes a Griffith microcrack if the ductile-brittle transition condition (18) is satisfied as well as equation (20) for σ_a^p. It can readily be appreciated that an evaluation of all the above three conditions depends critically on our knowledge of γ which, in turn, depends upon the details of the assumed separation process.

D. Interfacial Potential and the Thermodynamics of Interfacial Separation

There are two limiting cases in which the bicrystal in Figure 1a can be separated. The process of separation can be accomplished so fast that there is no time for diffusion of the segregant to the grain boundary, i.e., at constant excess segregant Γ. Alternatively, the separation can proceed very slowly such that there is diffusion of the segregant to the grain boundary. The latter case will ensure equilibrium between the grain boundary composition and the bulk lattice, i.e., at constant potential, μ. As pointed out by Rice [6], kinetic consideration will determine which of the two cases is more probable in some given circumstance. However, at ordinary temperatures, separation at constant Γ is more applicable due to low mobility while at high temperatures, either case may apply.

Seah [19] first presented an analysis of the two cases which had to be modified in the light of later work by Rice [6]. The problem has similarly been extensively addressed by Asaro [20], Hirth [21], and by Hirth and Rice [22]. In this paper, one is concerned with the analysis of a bicrystal containing only one alloying element (for simplicity of presentation as generalization to multicomponent system can be obtained) undergoing reversible isothermal separation under the two limiting conditions indicated above.

For separation at constant Γ, the interfacial potential function Φ, the stress σ_{22} (henceforth represented by σ), the boundary cohesive stress σ_{mb} are given by:

$$\bar{\Phi} = \bar{\Phi}(\delta,\Gamma)_{\Gamma const}; \ \sigma = \frac{d\bar{\Phi}}{d\delta}; \ \bar{\sigma}_{mb} = \bar{\sigma}|_{\delta_m} \tag{21}$$

where δ_m is the δ at maximum σ. The work of fracture $\bar{\gamma}$ is given by:

$$\bar{\gamma} = \bar{\Phi}(\infty,\Gamma) \tag{22}$$

The corresponding values of Φ, σ, σ_{mb} and γ for the separation at constant μ designated respectively by $\hat{\Phi}$, $\hat{\sigma}$, $\hat{\sigma}_{mb}$ and $\hat{\gamma}$, can be similarly obtained by replacing Γ in equations (21) and (22) by μ.

In accordance with Gibbs [23], the change in the grain boundary internal energy u, is given by:

$$du = Tds + \sigma d\delta + \mu d\Gamma \tag{23}$$

where s is the excess entropy and T is temperature. u, s and Γ are excess quantities per unit area of the grain boundary. The reversible work of fracture is given in the differential form by:

$$d\Phi + \sigma d\delta = d(u-Ts) - \mu d\Gamma \tag{24}$$

The grain boundary excess free energy, f, is given by:

$$f = u-Ts \tag{25}$$

For separation at constant Γ, i.e., $d\Gamma = 0$, equation (24) reduces to

$$d\bar{\Phi} = \bar{\sigma} d\delta = d\bar{f} \tag{26}$$

Hence, the work of fracture $\bar{\gamma}$ is given by:

$$\bar{\gamma} = \bar{\Phi}(\infty,\Gamma) = f_\infty(\Gamma) - f_b(\Gamma) \tag{27}$$

as obtained also by Asaro [20]. $f_\infty(\Gamma)$ is the Helmholtz free energy of the two fracture surfaces, $f_b(\Gamma)$ is the equilibrium free energy (i.e., at δ=0) of the grain boundary.

For separation at constant μ, integration of equation (29) yields

$$\hat{\gamma} = \hat{f}_\infty(\mu) - \hat{f}_b(\mu) - \mu[\Gamma_\infty(\mu) - \Gamma_b(\mu)] \tag{28}$$

where the parameters with subscript ∞ represent quantities for the two fracture surfaces, and those with subscripts b are quantities for the underformed grain boundary.

Let us consider an initially unstressed grain boundary which is in equilibrium with the lattices. In this case, $f_b(\mu) = f_b(\Gamma)$. Following the procedure of Asaro or Rice, one can easily show that

$$\bar{\gamma} - \hat{\gamma} = \int_b^{\Gamma_\infty(\mu)} [\mu_\infty(\Gamma^\infty) - \mu_\infty(\Gamma)]d\Gamma \tag{29}$$

Asserting, as Rice [6] did, that Γ on the free surfaces increases monotonically with μ, so that the integral in equation (29) is non-negative, then

$$\bar{\gamma} > \hat{\gamma} \tag{30}$$

with strict inequality holding whenever $\Gamma_\infty(\mu)$ differs from Γ_b. Thus, fracture is easier under the condition of slow separation compared with fast separation.

DETERMINATION OF WORK OF FRACTURE

The question that logically follows the above analysis is, what is the fracture stress and under what condition will intergranular fracture occur? In order to answer this question, equations (14), (18) and (20) must be evaluated. Each of these equations involves the work of fracture γ. Evaluation of γ is the most difficult aspect of the various models of intergranular fracture. In most cases, γ is often evaluated by the cohesion equation:

$$\gamma = 2\gamma_s - \gamma_b \tag{31}$$

where γ_s and γ_b are the surface energies of the created surfaces and the grain boundary. Equation (28) can be utilized, as done by Asaro [20] to show that equation (31) applies to the equilibrium separation at constant μ, i.e., $\hat{\gamma}$ leading thereby to the lower value of γ since from equation (30), $\bar{\gamma} > \hat{\gamma}$. Unfortunately, one is interested in fracture at ordinary temperatures which will involve the calculation of $\bar{\gamma}$ since it is more physically applicable. The problem is complicated further by the fact that γ as per equation (31) can be evaluated only at high temperatures at which the experimental measurements of γ_s and γ_b take place. Experimental values of γ_s or γ_b are unavailable at ordinary temperatures and can be obtained from the high temperature values if the temperature coefficients, i.e., the entropies of both the surfaces and the boundary are known. Only the theoretical estimates of such entropies exist in the literature and these are even mainly for pure conditions, i.e., no alloying element. Since direct experimental values are not available, recourse is made to theoretical evaluation and that also, as discussed below, is beset with certain fundamental problems.

The proposed theoretical model is based upon the realization of the fact that the process of cracking takes place on an atomic scale during which some atomic bonds are stretched while some are broken. Thus, the process of crack extension is one of progressive breakage of atomic bonds in the front of advancing crack in the sense of Barenblatt [24]. Thus, the energetics of the cracking process can be evaluated if there is a model capable of accounting for the bond energies of all the atoms reasonably close to the grain boundary plane as the process of separation takes place. In this context, the free energy f of a group of N atoms, say, can be written as

$$f = \frac{1}{2} \sum_{i=1}^{N} \sum_{j=1}^{Q} \psi(r_{ij}) - T \sum_{i=1}^{N} S_i \tag{32}$$

where $\psi(r_{ij})$ is the inter-atomic potential between atoms i and j separated by distance r_{ij}. Q is the total number of atoms that interact with atom i while S_i is the specific entropy of atom i. The first term in equation (32) corresponds to the total internal energy of the N atoms. Although inter-nuclear forces are, in general, many-body forces rather than pair forces, central pair-wise forces such as above are almost always assumed because of the computational simplicity they introduce. The problem reduces to finding a potential, $\psi(r_{ij})$, to describe in a physically reasonable manner the inter-atomic forces between the different species of atoms. Also, a model for calculating the entropy must be developed. Bamiro [25] developed a model based upon the central pair-wise inter-atomic potential to calculate the fundamental frequencies of an atom which bibrates in the potential well created by the surrounding atoms. The calculated frequencies, υ_j, are utilized to calculate the entropy of the atom as:

$$S_i = K \sum_{j=1}^{3} \left[\frac{\beta_j e^{\beta_j}}{(e^{\beta_j}-1)} - \ln(e^{\beta_j}-1) + \frac{K^2 g^2 T}{2\pi^2 m} \left(\frac{1}{\upsilon_j^2} - \frac{1}{\upsilon_o^2}\right)\right]; \quad \beta_j = \frac{h\upsilon_j}{KT} \tag{33}$$

h is the Planck's constant, K is the Boltzmann constant, m is the atomic mass, g is a parameter of the potential well of the atom, υ_o is the frequency of vibration of an atom in the perfect lattice. Application of this model to the calculation of the equilibrium energies of grain boundaries in copper has been shown to give values in reasonable agreement with the experimental observations [25]. Furthermore, the calculated entropies of the boundaries are in excellent agreement with the results obtained by Hashimoto et al [26] based upon the calculation of the local density of states (LDS) of individual atoms at the grain boundary using the recursion method. Thus, Bamiro's model can be extended to the treatment of the energetics of the separation process. This will conceptually involve first determining the equilibrium structure of the grain boundary in the presence of the segregant. With such equilibrium structure as the starting basis, the two crystals are given incremental dilatational displacements. At each stage, the atoms in the boundary are allowed to relax and the free energy is calculated. Separation at constant Γ is easier since at any stage in the separation process, the segregant only relaxes within the grain boundary region. And, of course, it is assumed that a potential function $\psi(r_{ij})$ exists.

In most simulations of defects, the pair-wise potentials such as the Morse and Lennard Jones have been utilized for the treatment of pure crystals [25,26]. Potentials which can describe the interaction between the host atom and the segregant are generally unavailable. In an unpublished work, Bamiro and Weins used the Lennard Jones 6-12 potential of the form:

$$\psi(r_{ij}) = -\frac{A^{qp}}{|r_{ij}|^m} + \frac{B^{qp}}{|r_{ij}|^n}$$

$$A^{qp} = \frac{2L^{qp}(h^{qp})^6}{(1-\frac{m}{n})C_m}; \quad B^{qp} = \frac{2L^{qp}(h^{qp})^{12}}{(\frac{m}{n}-1)C_n} \qquad (34)$$

where the subscripts q, p denote the type of atoms interacting: 1 for one atom and 2 for the other atom. The heats of sublimation L for the three different types of interaction (11, 22 and 12 or 21) are given by

$$L^{12} = L^{21} = (L^{11}\cdot L^{22})^{1/2}; \quad h^{12} = h^{21} = \frac{1}{2}(h^{11}+h^{22})$$

C_{m}, n is a geometric factor. Inconclusive results were obtained when applied to the analysis of gold-copper alloys. Further work is required to construct more realistic potentials to reflect the interatomic forces between different species. In the absence of such potential, the fracture properties of pure boundaries have been simulated using the Morse potential. This serves to test the developed model by examining whether such pure boundaries will cleave or not.

NUMERICAL RESULTS AND DISCUSSION

The equilibrium structures and the free energies of coincidence grain boundaries in copper have been simulated using the Morse potential function:

$$\psi(r_{ij}) = D\{\exp[-2\xi(r_{ij}-r_o)] - 2\exp[-\xi(r_{ij}-r_o)]\} \qquad (35)$$

with the constants D, ξ and r_o representing a characteristic energy, a measure of the compressibility, and a constant of the crystalline lattice, determined on the vacancy energy basis and considering the nearest neighbor interactions [25, 27]. The separation process was simulated by giving the bicrystal an incremental relative displacement of 0.1Å. At each stage, atoms were allowed to relax to minimize the calculated free energy. The fracture surfaces of the θ = 36.9° coincidence boundary in copper are shown in Figure 2. Solid lines indicate atoms in one plane while dashed line show atoms in another plane. The variation of free energy with relative displacement, i.e., Φ for this grain boundary is shown in Figure 3a. Φ for the θ = 28.1° coincidence boundary in copper is also shown in Figure 3b. Since one is interested in the stress calculation, i.e., the slope of Φ, the

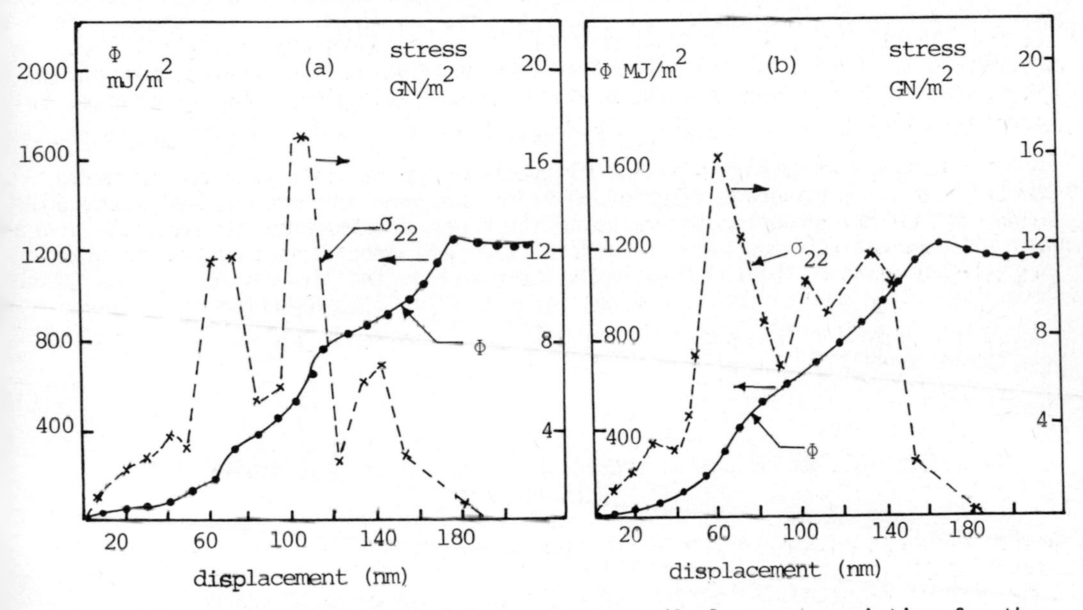

Fig. (3) - Interfacial potential Φ and stress-displacement variation for the: a) θ = 36.9 boundary; and b) 28.1° boundary in copper

data for Φ have been spline-fitted so as to minimize error in the stress calculation. The variation of stress for each of the boundaries is shown in Figures 3a and 3b.

The calculated parameters from the results of the simulation are given in Table 1. λ_o and E_b were calculated from equations (9) and (10) respectively after fitting Φ to the cubic equation (8) for small $\delta \cdot \sigma_a^N$ was calculated on the basis of equation (14) assuming the friction factor q=β and a mean grain diameter d = 2 x 10^{-5} m corresponding to the value for the fine grain Cu-Bi specimens tested by Hondros and McLean [4]. σ_{mb}, f(∞), f_b and γ were directly from Figures 3a and 3b.

TABLE 1 - SIMULATED GRAIN BOUNDARY FRACTURE PARAMETERS IN COPPER

θ	λ_o (nm)	E_b GN/m^2	σ_a^N MN/m^2	σ_{mb} GN/m^2	$f(\infty)$ J/m^2	f_b J/m^2	γ J/m^2	S	R_o	R_o^*
36.9	280	32.4	20.6	17.0	2.31	1.04	1.27	0.112	2.21	0.91
28.1	282	27.4	19.2	16.5	2.23	0.98	1.25	0.108	2.24	0.92

In order to test whether the pure boundaries will cleave or not, the critical fracture equation (18) was examined, assuming, in line with Rice [6], that the terms in [---] in equations (19) for R_o and S are approximately equal to 1, and $\zeta_o = 2$. It is worth noting that assuming $\zeta_o = 1$ leads to $R_o = 1.11$ and 1.12 for the $\theta = 36.9$ and 28.1 grain boundaries respectively. Thus, the present model shows that the pure copper boundaries should not cleave since $R_o > R_o^*$ in each case. However, while the case $\zeta_o = 2$ predicts high ductility, the case $\zeta_o = 1$ predicts a very narrow margin of ductility in agreement with the previous observations [4,7]. The average value of 20 MN/m^2 obtained for the nucleation stress is well within the experimental value of 120 MN/m^2 obtained for the fracture stress of the Cu-Bi alloy [4]. This shows that the process of crack nucleation is an easier process than propagation which provides the limiting condition. Assuming the experimental fracture stress of 120 MN/m^2, the Griffith crack length (2c) as per equation (25) is about 4×10^{-6} m, i.e., about d/5 where d is the mean grain diameter.

CONCLUSIONS

The following conclusions have emanated from the present study:

1. The process of intergranular fracture can be modelled reasonably well as one involving crack nucleation and propagation.

2. Utilization of the stress-concentrating effect due to the double slip dislocation band to model crack nucleation; and the consideration of the propagation of the nucleated crack in terms of blunting by nucleated dislocations, give conditions for nucleation and propagation which depend critically upon the reversible work of fracture.

3. The work of fracture, γ, can be determined within the framework of two limiting processes: one at constant potential and the other at constant concentration of segregant in the interface.

4. No experimental technique is available to determine the work of fracture at ordinary temperatures. Hence, recourse is made to theoretical evaluation.

5. An atomic model which evaluates the bond energies of the atoms in the boundary provides a useful framework for the determination of γ. However, such model depends critically upon the availability of interatomic potential to describe the various atomic interactions.

6. Application of the atomistic model to the analysis of pure boundaries in copper leads to prediction in agreement with experimental observations.

7. A combination of the present atomistic model with the analytical model of the variation of γ with segregant may be the logical path to the theoretical formulation of intergranular fracture.

REFERENCES

[1] Hopkins, B. E. and Tipler, H. R., J. Iron Steel Ins., Vol. 177, p. 110, 1954; ibid, Vol. 188, p. 218, 1959.

[2] McMahon, C. J., Temper Embrittlement in Steel, STP No. 407, ASTM, p. 127, 1968.

[3] Rees, W. P., Hopkins, B. E. and Tipler, H. R., J. Iron Steel Ins., Vol. 169, p. 157, 1951.

[4] Hondros, E. D. and McLean, D., Phil. Mag., Vol. 29, p. 771, 1974.

[5] Powell, B. D. and Mykura, H., Acta Met., Vol. 21, p. 1151, 1973.

[6] Rice, J. R., Effect of Hydrogen on Behavior of Materials, A. W. Thompson and I. M. Berstein, eds., AIME, p. 455, 1976.

[7] Mason, D. D., Phil. Mag., Vol. 39, p. 455, 1979.

[8] Bamiro, O. A. and Provan, J. W., Res. Mechanica Letters, Vol. 1, p. 125, 1981.

[9] Cottrell, A. H., Trans. AIME, p. 192, 1958.

[10] Stroh, A. N., Advances in Physics, Vol. 6, p. 418, 1957.

[11] Smith, E. and Barnby, J. T., Met. Sci. J., Vol. 1, p. 56, 1967.

[12] Kittle, C., Introduction to Solid State Physics, Wiley, New York, 3rd Ed., p. 184, 1976.

[13] Prinz, P., Karnthaler, H. P. and Kirchner, H. O. K., Acta Met., Vol. 29, p. 1029, 1981.

[14] Evans, J. T., Phil. Mag., Vol. 29, p. 1095, 1974.

[15] Eshelby, J. D., Frank, F. C. and Nabbarro, F. R. N., Phil. Mag., Vol. 42, p. 351, 1951.

[16] Rice, J. R. and Thomson, R., Phil. Mag., 29, p. 73, 1974.

[17] Hirth, J. and Loethe, J., Theory of Dislocations, McGraw-Hill, p. 212, 1968.

[18] Bamiro, O. A., Res. Mechanica Letters, in press.

[19] Seah, M. P., Surface Sci., Vol. 53, p. 168, 1975.

[20] Asaro, R. J., Phil. Trans. R. Soc. Lond., Vol. A295, p. 151, 1980.

[21] Hirth, J. P., Phil. Trans. R. Soc. Lond., Vol. A295, p. 139, 1980.

[22] Hirth, J. P. and Rice, J. R., Met. Trans., ASME, Vol. 11A, p. 1501, 1980.

[23] Biggs, J. W., Thermodynamics, Dover, New York, p. 219, 1961.

[24] Barenblatt, G. I., J. Appl. Math. Mech., Vol. 23, p. 622, 1959.

[25] Provan, J. W. and Bamiro, O. A., Acta. Met., Vol. 25, p. 309, 1977.

[26] Hashimoto, M., Ishida, Y., Yamamoto, R. and Doyama, M., Acta. Met., Vol. 29, p. 617, 1981.

[27] Cotteril, R. M. J. and Doyama, M. J., Argonne National Laboratory Report, Chicago, 1965.

ACKNOWLEDGEMENT

The NSERC/CIDA Canadian Fellowship is acknowledged. The ever useful and inspirational discussions with Professor J. W. Provan are also appreciated. The computing services at the McGill University, Canada and University of Ibadan, Nigeria, are acknowledged. Thanks to Mr. E. C. Ihedigbo for typing this paper.

STATISTICAL MECHANICS OF EARLY GROWTH OF FATIGUE CRACKS

W. J. Pardee, W. L. Morris, B. N. Cox and B. D. Hughes*

Rockwell International Science Center
Thousand Oaks, California 91360

ABSTRACT

This paper presents a mechanistic theory of the statistics of crack initiation and early growth in a polycrystalline alloy subjected to uniform, fully reversed cyclic loads. The physical basis, mathematical form, and numerical solution of the theory are described, and experimental tests are proposed. For brevity and clarity, the simplified version presented here describes surface cracks propagating with one tip permanently blocked.

INTRODUCTION

Fatigue crack initiation and early growth in metals are well-known to be macroscopically stochastic. Even when specimens are machined from a single heat, their macroscopic residual stresses eliminated, and are fatigued under identical loading conditions, the scatter in lifetime, or in time-to-(engineering) crack initiation is large [1,2]. Similar scatter is observed when the growth rate of short cracks is plotted versus ΔK, the stress intensity range. A typical example [3] is shown in Figure 1. This scatter occurs because the growth rate of a short crack depends on other variables in addition to ΔK, and some of those variables are material properties that change rapidly on a microscopic scale as the crack grows.

The problem addressed in this paper is the calculation of such distributions of crack growth rate or of time-to-crack-initiation from reasonably measurable distributions of microscopic properties. Though this theory has some formal similarity to McCartney's [4], there is here no assumption of a statistically varying (local) failure criterion. Here, cracks initiate and grow according to deterministic laws with the parameters of those laws varying from grain to grain. Models are used to relate crack initiation and growth to local properties such as grain size, shape, and orientation. The theory does not attempt to predict the failure of any particular specimen, but, rather, for example, the number of fatigue cycles after which the first 1% of the specimens have failed. Such information about the scatter in lifetime is often of greater practical importance than the average lifetime.

---

*Present address, Department of Chemical Engineering, University of Minnesota, Minneapolis, Minnesota 55455.

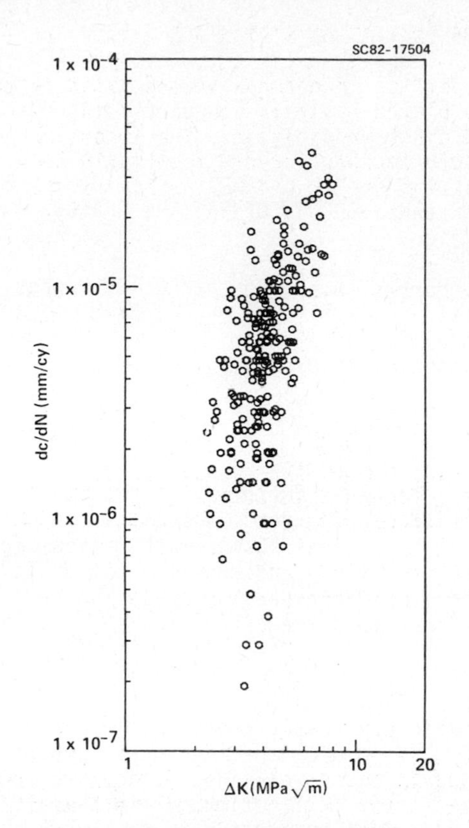

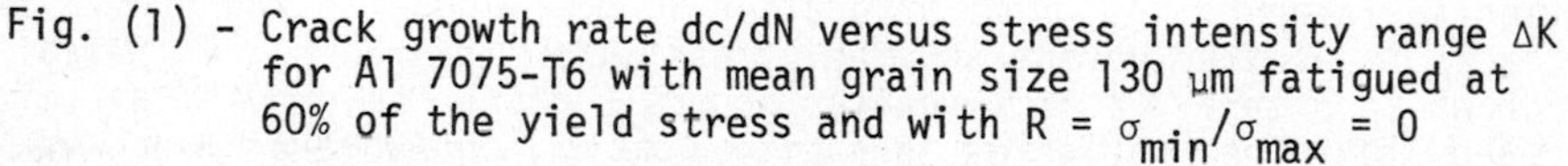

Fig. (1) - Crack growth rate dc/dN versus stress intensity range ΔK for Al 7075-T6 with mean grain size 130 μm fatigued at 60% of the yield stress and with $R = \sigma_{min}/\sigma_{max} = 0$

The analytic theory presented here contains essentially the same microscopic models [5,6] as computer simulations previously published [7]. Computer simulation offers greater modeling flexibility, but is computationally slower. Because the physical assumptions of the analytic theory are more succinctly and precisely exhibited, and because the predicted distributions of the analytic theory are continuous functions of, e.g., crack length, it is possible to define more sensitive statistical tests against experimental measurements. For brevity and clarity, this paper presents a simplified version of the theory. This version describes only one tip of each crack. This choice is made for pedagogic reasons, but it physically corresponds to situations where, e.g., there is a local reflection symmetry about the point of crack initiation, or where growth of one tip of each crack is permanently blocked by a phase boundary. The paper first introduces the microscopic phenomena that must be modeled to obtain a locally deterministic theory, then defines the essential local microstructural variables and the statistical densities. The differential equations and boundary conditions are then derived to describe the evolution of the distribution of cracks, and a geometric picture of their solution is presented. Applications to the interpretation of crack growth data are described. The theory discussed here is really a theory of early growth, but the extension to include initiation is indicated.

A MICROSCOPIC PICTURE OF INITIATION STATISTICS

The most important physical phenomena incorporated in the following theory are related to the role played by grain boundaries in restricting surface micro-plastic deformation and crack propagation. The importance of grain boundaries in interrupting the growth of short cracks is clearly seen in Figure 2, which is a histogram of crack tips per fractional number of grains traversed. The

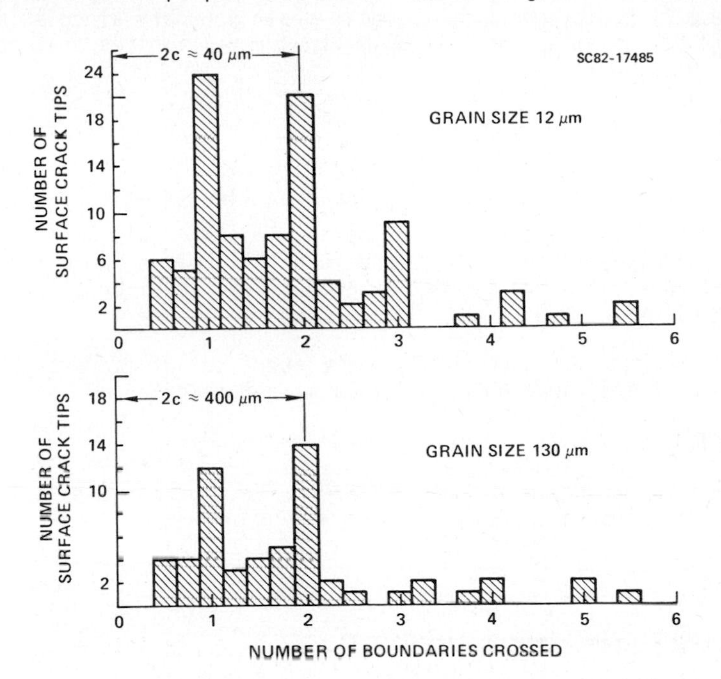

Fig. (2) - Number of crack tips per grain fraction traversed for Al 7075-T6 with, respectively, 12 μm and 130 μm mean grain size. Specimens fatigued at 60% of the yield stress in tension-tension (R=0) loading

data [7] are for Al 7075-T6 subjected to tension-tension flexural loading at 60% of the yield stress. A crack tip that progressed from a fractured constituent particle in the middle of one grain across one boundary to stop at the next grain boundary is plotted at a normalized length of 2. If it crossed that second boundary and 40% of the distance to the third, it would be plotted at 2.4. The peaks at one, two, and three grain diameters exhibit the large fraction of crack tips whose growth has been interrupted by a grain boundary.

For short cracks, plastic deformation at the surface extends from a crack tip only to the next grain boundary [8]. When the growth of a crack tip is interrupted by a grain boundary, continued opening and closing of the crack gradually generates plastic deformation in the next grain. This accumulating plastic deformation is the incubation of a new plastic zone which must precede renewed crack growth. Let β denote the accumulated plastic deformation at the crack tip, normalized so that $\beta=1$ is that of a mature plastic zone. That is, the crack resumes growth when β=1. The rate at which β increases with fatigue depends on crack length and on properties of the grain such as orientation and size. The model of Morris, James and Buck [5] relates this deformation rate $d\beta/dN$, denoted here by v_β,

$$d\beta/dN = v_\beta \tag{1}$$

to the maximum slip distance D to the next grain boundary along a line at 45° to the principle stress axis. Specifically, that model implies that in elastically loaded aluminum alloys

$$v_\beta = b(2c)^{1/2} D(\tau-\tau_o)^2\exp(-\theta N) \tag{2}$$

where b is a material parameter, 2c is the crack length at the surface, τ_o is a friction stress resisting dislocation motion, τ is the maximum effective applied shear stress parallel to the surface and in a [111] plane, N is the number of fatigue cycles, and θ is a coefficient related to the level of internal hydrogen for aluminum alloys.

A moving crack tip is governed by a different set of microstructural variables. The geometrical length 2c increases at a rate given by

$$v_c = dc/dN = A(\Delta K_{eff})^2 \tag{3}$$

where A is a material parameter and the effective stress intensity range ΔK_{eff} is corrected for closure stress σ_{cc}

$$\Delta K_{eff} = \alpha(2c)^{1/2} (\sigma_a-\sigma_{cc}) \tag{4}$$

where σ_a is the amplitude of the alternating stress and α is a geometrical factor of order unity. It has been shown [9] empirically that the closure stress is approximately governed by the distance z, Figure 3, between the (moving) tip and the next grain boundary. In fact, it is approximately true in Al 2219-T851 that

$$\sigma_{cc} = 0.8\ \sigma_a\ z/(2c) \tag{5}$$

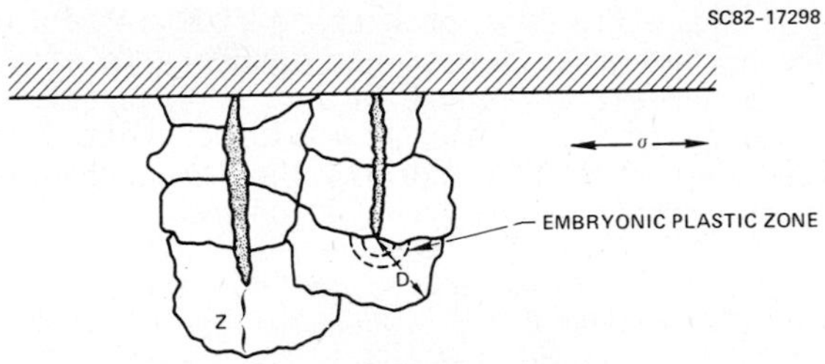

SC82-17298

EMBRYONIC PLASTIC ZONE

σ

D

z

Fig. (3) - Illustration of geometrical parameters for a type one (left) and a type two crack

These relationships are depicted in Figure 3. All cracks are assumed to have their upper tip permanently blocked. The cracks with their other tip stopped at a boundary are called type zero (zero velocity), and those with one free end are called type one. The general theory requires three statistical densities, one each for cracks with zero, one, or two propagating tips, and includes transitions among the three types. The microstructural variables specified provide a deterministic prediction of a crack's state after successive loading cycles. Thus, with the above model for v_β, knowledge of N, c, β, D, and τ is sufficient to predict the cycles required to liberate a type zero crack. The appropriate statistical density can be defined by requiring that after N cycles, the average number of cracks per unit area with length between 2c and 2(c+dc), accumulated deformation between β and β+dβ, and with the next grain having maximum 45° slip distance between D and D+dD, and surface shear stress between τ and τ+dτ be given by

$$f_o(N,c,\beta,D,\tau)dcd\beta dDd\tau \tag{6}$$

An equivalent definition is as the derivative

$$f_o = \frac{\partial^4 F_o}{\partial c \partial \beta \partial D \partial \tau} \tag{7}$$

where F_o is the (cumulative) distribution function, the average number of cracks per unit area with length less than c, plastic deformation less than β, etc.

The expected number of type one cracks per unit area is related to a similarly defined statistical density f_1, a function of N, c, and z. If F_1 (N,c,z) is the expected number of cracks per unit area with length less than 2c and distance to the next grain boundary less than z, one has

$$f_1 = \frac{\partial^2 F_1}{\partial c \partial z} \tag{8}$$

The average crack velocity and the distribution in crack velocity at fixed stress intensity are quantities of considerable engineering importance. For fixed maximum applied stress, σ_a, (cyclic, fully reversed loading), the dependence on stress intensity is obtained from the elimination of crack length via the relation

$$c = (\Delta K/(\alpha\sigma_a))^2 \tag{9}$$

The average velocity $\nu(N,\Delta K)$ is given by

$$\nu(N,\Delta K) = \rho(N,c)^{-1} \int dz\, v_c(c,z)\, f_1(N,c,z) \tag{10}$$

where the normalization ρ is given by

$$\rho(N,c) = \int dz\, f_1(N,c,z) + \int d\beta d\tau dD\, f_0(N,c,\beta,D,\tau) \tag{11}$$

and equation (9) is used to replace c by ΔK. The probability $\psi(N,v,\Delta K)dv$ that the velocity of a crack with stress intensity range ΔK has value in the interval (v,v+dv) is obtained from

$$\psi(N,v,\Delta K) = \rho(N,c)^{-1} \int dz\delta(v-v_c(c,z))f_1(N,c,z) \quad (12)$$

or

$$\psi(N,v,\Delta K) = f_1(N,c,z_o)/[\rho(N,c)|dv_c/dz|] \quad (13)$$

where z_o is the root of $v-v_c(c,z_o) = 0$. The probability Ψ that the crack velocity is less than v is given by

$$\Psi(N,v,\Delta K) = \int_0^v dv'\psi(N,v',\Delta K) \quad (14)$$

This cumulative distribution can be used to generate percentile velocity curves giving, for example, a curve that is the upper bound velocity for 90% of the cracks.

THE KINETIC THEORY OF CRACKS [9]

Let us suppose that the loading frequency f_o (cycles/sec) is constant and use "time" (= N/f_o) as equivalent to "number of fatigue cycles" as we have tacitly taken "crack velocity" to be equivalent to "crack growth rate".

The equations governing the time evolution of the average crack densities f_0 and f_1 are most easily obtained by considering a small region of phase space at time N, following it to time N+ΔN, and keeping track of the processes that add or or subtract cracks. This is illustrated in Figure 4 in the c,z plane for type one

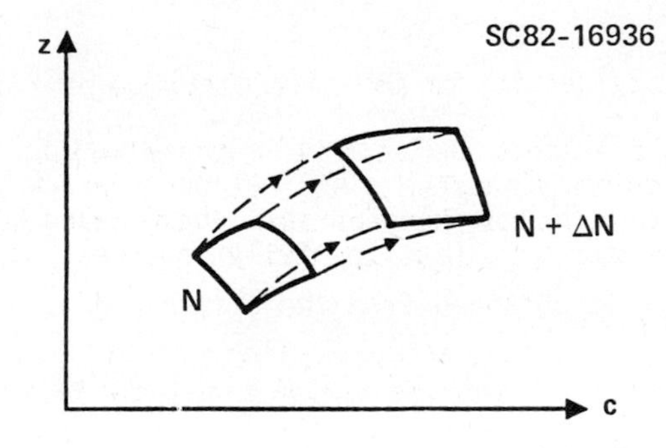

Fig. (4) - Small region of type one phase space evolving in time

cracks. The initial region is bounded by the curves $c_1(\beta) < c_2(\beta)$ and $z_1(c) < z_2(c)$. Each point $\underset{\sim}{p}_o = (c_o,z_o)$ in this region moves along a streamline defined by the rate equation

$$\frac{d\underset{\sim}{p}}{dN} = \underset{\sim}{v}^{(o)} = (v_c,v_z) \quad (15)$$

and the initial conditions

$$p(N) = p_o \tag{16}$$

The only possible sources and sinks of cracks are initiation of new cracks, escape of cracks from a grain boundary (a transfer from $f_0 \rightarrow f_1$), and blockage of a propagating (f_1) crack by a grain boundary (a transfer $f_1 \rightarrow f_0$). New cracks are initiated without a plastic zone, so they are described by a boundary contribution to the f_o distribution at $\beta = 0$. The number of blocked (f_0) cracks that escape to join this region of the type one phase space in the interval $(N, N+\Delta N)$ is given by

$$\Delta n_{0\rightarrow 1} = \int_{N}^{N+\Delta N} dN' \int_{c_1(N',z)}^{c_2(N',z)} dc' \int_{z_1(N',c)}^{z_2(N',c)} dz' \int dDd\tau \, J_\beta(N,c,D,z) G(z'|D) \tag{17}$$

where the flux J_β of type zero cracks across the $\beta=1$ boundary is given by

$$J_\beta(N,c,D,\tau) = v_\beta(D,\tau) f_0(N,c,1,D,\tau) \tag{18}$$

The conditional probability density $G(z\ D)$ is defined by the property that the quantity P,

$$P = G(z|D)dz \tag{19}$$

is the probability that the maximum direct distance z_{MAX} to the next grain boundary is between z, z+dz given that the maximum 45° slip distance was D. G is thus a measure of grain shape. It is related to the joint probability $G(z,D)$ by

$$G(z|D) = G(z,D)/\Phi(D) \tag{20}$$

where Φ is the probability density of 45° slip distances.

Thus, no new cracks are added to the type 1 distribution, and the transfer of type 0 to type 1 is described by equation (17). The loss of type 1 to type 0 occurs as cracks arrive at grain boundaries, $z \rightarrow 0$. Thus, the number Δn_1 of cracks in the region bounded by c_1, c_2, z_1, z_2 at time N,

$$\Delta n_1(N) = \int dcdz f_1(N,c,z) \tag{21}$$

is equal to the number $\Delta n_1'$ in the image region minus the transfers from type 0,

$$\Delta n_1(N) + \Delta n_{0\rightarrow 1} = \Delta n_1'(N+\Delta N) \tag{22}$$

The volume magnification factor is given by the Jacobian,

$$\frac{\partial(c',z')}{\partial(c,z)} = 1 + \Delta N \nabla \cdot v^{(1)} + O(\Delta N^2) \tag{23}$$

The resulting equation for f_1 as $\Delta N \rightarrow 0$ is, from equation (22),

$$\frac{\partial f_1}{\partial N} + \nabla \cdot (\underset{\sim}{v}^{(1)} f_1) = \int dDd\tau G(z|D) v_\beta(D,\tau) f_0(N,c,1,D,z) \quad (24)$$

The divergence operator in equations (23) and (24) refers, of course, to the c, z variables.

Because the crack length is constant in type 0 phase space

$$v_c^{(0)} = 0 \quad (25)$$

the time evolution of the area is very simple. Cracks are initiated and gained from f_1 at β=0 and lost to f_1 at β=1. Thus, for a point interior to the phase space, the analogue of the foregoing argument produces the equation

$$\frac{\partial f_0}{\partial N} + v_\beta \frac{\partial f_0}{\partial \beta} = 0 \quad (26)$$

Let the rate dn_I/dN at which cracks are initiated in the intervals, (c,c+dc), (D,D+dD), (τ,τ+dτ) be given in terms of a density $s_I(N,c,D,\tau)$ by

$$\frac{dn_I}{dN} = s_I(N,c,D,\tau)dcdDd\tau \quad (27)$$

The rate at which type 1 cracks stop to join the type 0 distribution is given by

$$\frac{dn_{1\to 0}}{dN} = - v_z(c,0) f_1(N,c,0)\Phi(D)H(\tau) \quad (28)$$

where H is the probability density of resolved shear stress in a [111] plane parallel to the surface, and Φ is the probability density introduced in equation (20). The sum of these two contributions if the flux across the β=0 boundary,

$$v_\beta(D,\tau) f_0(N,c,0,D,\tau) = s_I(N,c,D,\tau) - v_z(c,0)\Phi(D)H(\tau) f_1(N,c,0) \quad (29)$$

The differential equations (24) and (26) and the boundary condition equation (29) are sufficient to calculate f_0 and f_1 given their initial values and the source function s_I. The model is completed by specifying the dynamic models v_β, v_c and v_z and the microstructural distributions Φ, H, G. The model for v_β has been given and v_c is determined by equations (3)-(5) to be

$$v_c = A\sigma_a c(1-\alpha z/(2c))^2 \Theta(2c-\alpha z) \quad (30)$$

where the material parameters A and α have been redefined and the function Θ is one when its argument is positive and zero otherwise. The remaining velocity v_z is simply given by

$$v_z = - 2v_c \quad (31)$$

A very simple and useful result can be obtained for the density ρ defined by equation (10). Calculating the time derivative and using the equations of motion and the boundary condition one easily obtains

$$\frac{\partial\rho}{\partial N} + \frac{\partial}{\partial c}(\nu\rho) = \int dDd\tau s_I(N,c,D,\tau) \tag{32}$$

where $\nu(N,\Delta K)$ is the average velocity defined by equation (9). This is the evolution equation postulated by McCartney [4], but here, it and the average velocity ν are derived from a microscopic theory. In some cases, equation (29) is a numerically convenient basis for obtaining N by calculating ρ at different times.

SOLUTION BY METHOD OF CHARACTERISTICS

The arguments leading to the basic differential equations (24) and (26) were based on following a small region of phase space as its boundaries evolved in time along the natural crack trajectories or characteristic curves defined by equations (15) and (16) for type one phase space and by

$$\frac{d\tilde{r}}{dN} = (v_c^{(0)}, v_\beta) = (0, v_\beta) \tag{33}$$

and

$$\tilde{r}(N_o;\tilde{r}_0) = r_0 \tag{34}$$

for type zero phase space. The solution of equations (24) and (26) is based on essentially the same procedure. The function f_1 evolves very simply along a characteristic curve. Let the value of f_1 along a particular characteristic curve $\tilde{p}(N;\tilde{p}_o)$ be denoted by u_1,

$$u_1(N;\tilde{p}_o) = f_1(N,p_i(N;\tilde{p}_o)) \tag{35}$$

This family of functions u_1 is indexed by the label $\tilde{p}_o$ of the characteristic curve $\tilde{p}(N;\tilde{p}_o)$ and satisfy the ordinary differential equation

$$\frac{du_1}{dN}(N;\tilde{p}_o) = -\nabla \cdot v^{(1)}(c',z')u_1(N;\tilde{p}_o) + \int dDd\tau G(z'|D)v_\beta(D,\tau)f_0(N,c',1,D,\tau) \tag{36}$$

where c' and z' are defined by

$$(c',z') = p_i(N;\tilde{P}_o) \tag{37}$$

The geometrical interpretation of equation (36) is simple; the first term is the volume dilation factor describing the spreading of cracks in phase space, and the second term is the rate cracks are added from the type zero distribution.

The type one characteristic curves in the c-z plane are illustrated in Figure 4. The velocities v_c and v_z both vanish above the line

$$z = 2c/\alpha \tag{38}$$

In the region above this line, equation (24) has the directly integrable form

$$\frac{\partial f_1}{\partial N} = \int dDd\tau G(z|D)v_\beta(D,\tau)f_0(N,c,1,D,\tau) \tag{39}$$

This is a region where, although the cracks have a mature plastic zone, they are frozen in place by a large closure stress. The differential equation is continuous across this boundary, and so, therefore, is f_1.

In the region below the line equation (38), the differential equation for u_1 has the form

$$\frac{du_1}{dN} = Y(N,u_1,\underset{\sim}{p}(N)) \tag{40}$$

with an implicit dependence on f_0. This equation has a standard form and can be integrated by any of a number of techniques. The distribution of interest is $f_1(N,c,z)$, and it is obtained from its values at N-ΔN by finding the preimage of (c,z); that is, the point $\underset{\sim}{p}_o$ for which

$$\underset{\sim}{p}_o(N;\underset{\sim}{p}_o) = (c,z) \tag{41}$$

and

$$\underset{\sim}{p}(N-\Delta N;\underset{\sim}{p}_o) = \underset{\sim}{p}_o \tag{42}$$

Then f_1 is obtained from

$$f_1(N,c,z) = u_1(N;\underset{\sim}{p}_o) \tag{43}$$

and the right side of equation (43) is obtained by integrating equation (40) subject to the initial condition

$$u_1(N-\Delta N;\underset{\sim}{p}_o) = f_1(N-\Delta N;\underset{\sim}{p}_o) \tag{44}$$

The characteristic curves for f_0 in the cβ plane are simply straight lines perpendicular to the c axis. The simple time dependence of equation (2),

$$v_\beta = v_\beta(0)\exp(-\theta N) \tag{45}$$

permits analytic solution of equation (26) in terms of the initial and boundary values of f_0. For any particular β, if the time N satisfies

$$N < \theta^{-1} \ell n(1-\theta\beta/v_\beta(0))^{-1} \tag{46}$$

then the trajectory passing through β at time N passed through the point β_0, $0<\beta_o<1$, at N=0, with β_o given by

$$\beta_0 = \beta - v_\beta(0)(1-e^{-\theta N})/\theta \tag{47}$$

Thus, for N satisfying equation (46),

$$f_0(N,c,\beta) = f_0(0,c,\beta_0) \tag{48}$$

For N not satisfying equation (46), the characteristic curve passing through β at N passed through $\beta=0$ at a time N_0 given by

$$N_0 = \theta^{-1}\ \ell n(e^{-\theta N}+\theta\beta/v_\beta(0)) \tag{49}$$

and the solution to equation (26) is given by

$$f_0(N,c,\beta,D,\tau) = s_I(N_0,c,D,\tau) - \Phi(D)H(\tau)f_1(N_0,c,0)/v_\beta(N_0,c,D,\tau) \tag{50}$$

APPLICATION TO EXPERIMENT

This model and its generalization to two ended cracks can be used to predict the results of several kinds of experiment, including time to engineering crack initiation. It is desirable to test the early growth models separately from the initiation model s_I. An experimental technique for producing a well-characterized distribution of genuine fatigue cracks has been used with considerable success [6] to study the statistics of early growth. It is not necessary or even desirable that this initial distribution be sharply peaked in length, and that is a significant experimental simplification.

After creating a set of cracks with lengths $c_i(N_o)$ at time N_o, the experimenter observes the lengths of these cracks at subsequent times N_i. The simplest test is of the theoretical cumulative distributions F_0 and F_1. Let the average number of cracks per unit area with lengths less than 2c be denoted by F(N,c),

$$F(N,c) = \int_0^c dc'\rho(N,c') \tag{51}$$

and the total number of cracks per unit area by Q(N),

$$Q(N) = \lim_{c\to\infty} F(N,c) \tag{52}$$

The experimentally observed fraction of cracks with lengths less than 2c and N cycles should converge to F(N,c)/Q(N), and these two functions of c (theoretical and experimental) can be compared for each observation time N_i. The cumulative distributions facilitate comparison by avoiding the necessarily arbitrary choice of bin size needed to define an experimental density, and by smoothing statistical scatter.

The distribution in crack velocity is also very interesting, and provides additional information. The experimenter measures a velocity averaged over time. That time interval should be long enough to see a measurable change in crack length, but short enough that the local environment of the crack tip does not change much. It would be desirable to measure the length at two closely spaced times, subtract the lengths to obtain a velocity, then let the tip propagate through several grains before making another velocity measurement. This procedure makes subsequent measurements on a single crack reasonably independent. The empirical density corresponding to Ψ (equation (14)) suffers from an ambiguity in the definition of the ΔK bin size, so it is preferable to calculate the experimental fraction of cracks with velocity less than v and stress intensity range less than ΔK and compare that to

$$\int_0^{\Delta K} dK' \Psi(N,v,K') \tag{53}$$

One of the most interesting applications of the theory is to the prediction of the probability $P(N,c^*)$ that a crack longer than some length $2c^*$ exists after N fatigue cycles. This is zero if $Q(N) = 0$, so let us assume $Q(N) > 0$. Let us consider a unit area large enough that $Q(N) \gg 1$, and first consider an area A large enough that the expected number of cracks $\nu = QA$ is large. Then the probability, P, that at least one crack has reached length $2c^*$ is given by

$$P = 1 - q^{\nu} \tag{54}$$

where q is the probability than any one of the ν cracks is shorter than $2c^*$. This is just given by F/Q, or

$$P = 1 - (F/Q)^{\nu} = 1 - [1-(Q-F)/Q]^{QA} \tag{55}$$

and since $Q \gg 1$,

$$P = 1 - e^{-(Q-F)A} = 1 - e^{-RA} \tag{56}$$

with

$$R(N,c) = Q(N) - F(N,c) = \int_c^{\infty} dc' \rho(N,c') \tag{57}$$

Thus, $R(N,c^*)$ is the average number of cracks per unit area with lengths greater than or equal to $2c^*$. The result (equation (56)) is also valid when QA is not large, as can be seen by considering an ensemble of n little areas A such that $nQA \gg 1$. The probability that none of the little regions has such a crack is q^n, but because $nQA \gg 1$, this must be given by

$$q^n = e^{-nRA} \tag{58}$$

Experimental measurements of time-to-(engineering)-crack initiation are often displayed by plotting $\phi(N,c^*)$ versus N where $\phi(N,c^*)\Delta N$ is the fraction of samples that had no crack longer than $2c^*$ at $N-\Delta N$, but have at least one after N cycles. The theoretical expression for ϕ is obtained by differentiation of equation (56) with respect to N,

$$\phi(N,c^*) = \nu(N,c^*)\rho(N,c^*)A\exp(-R(N,c^*)A) \tag{59}$$

where equation (32) has been used.

SUMMARY

A mechanistic theory of the statistics of early growth of short fatigue cracks has been presented. The simple version presented here was chosen for pedagogic value, but has physical realizations in a system with a large phase boundary blocking one tip of all cracks permanently. The theory captures much of the physics of its far more complicated generalization to cracks with two tips. The form of the experimental tests and the probabilistic predictions given here are nearly identical to those appropriate to the more general theory.

ACKNOWLEDGEMENT

This work has been supported by DARPA under Contract MDA-903-80-C-0641. The authors are grateful for several conversations with Ken Fertig and Gene Meyer.

REFERENCES

[1] Sinclar, G. M. and Dolan, T. J., Transactions of the ASME, July 1953.

[2] Bloomer, N. T. and Roylance, T. F., The Aeronautical Quarterly, XVI, p. 307, 1965.

[3] Zurek, A. K., James, M. R. and Morris, W. L., submitted to Met. Trans.

[4] McCartney, L. N., Int. J. Fract., 15, p. 477, 1979.

[5] Morris, W. L., James, M. R. and Buck, O., Met. Trans. A, 12A, p. 57, 1981.

[6] Morris, W. L. and James, M. R., Proceedings of ASTM Conference on Quantitative Measurement of Fatigue Damage, Dearborn, Mich., 1982 (in press).

[7] Morris, W. L., James, M. R. and Buck, O., Eng. Fract. Mech., 13, p. 213 1980.

[8] Morris, W. L., Met. Trans. A, 11A, p. 1117, 1980.

[9] The kinetic theory presented here is a specific realization of a general formulation first suggested to the authors by J. M. Richardson (unpublished).

THE EFFECT OF HOMO- AND HETEROGENEOUS MECHANISMS COUPLING ON MICROCRACK NUCLEATION IN METALS

J. Krzemiński

Polish Academy of Sciences
Warsaw 00-049, Poland

ABSTRACT

A hypothesis of vacancy-microcrack formation in strained metals is considerably extended by coupling of homo- and heterogeneous mechanisms in the global process of nucleation. The first one is a spatial lattice diffusion of vacancies which allows to form a homogeneous vacancy cluster (a microvoid, assumed as spherical) within the grain. The second one consists actually of two different simultaneous mechanisms which are responsible for generation of a heterogeneous vacancy aggregation (a cap-shaped microcrack) on the grain boundary. Joining of vacancies from the grain space to the arising cluster (spatial diffusion) as well as joining of already absorbed on the grain boundary vacancies (surface diffusion) is taken into account. The global microcrack nucleation rate results from coupling of the aforesaid mechanisms and is not a simple sum of the corresponding rates calculated for independent mechanisms. The considerations are based on the continuum analysis of the process kinetics.

INTRODUCTION

In a series of papers [1-5], a hypothesis of vacancy-microcrack origination in strained metals was proposed. The hypothesis was based on the qualitative similarity between the Griffith theory of brittle fracture and the process of phase transition. This similarity suggested the idea that a microcrack may be initiated in an analogous manner as a nucleus of a new phase, composed of a set of vacancies situated at neighboring lattice points. First, using the discrete analysis of the classical theory of phase transitions, the homogeneous nucleation rate of microcracks in a uniaxially strained single crystal was derived [1,2]. Then the theory was extended to polycrystalline materials and the heterogeneous nucleation of microcracks at the grain boundaries was described. In the latter problem, two different analytical treatments of the process were distinguished. The first one permitted introduction of macroscopic parameters, and hence a thermodynamic model of the system [3]. On the other hand, for very small nucleus sizes, application of the bulk variables could be seriously questionable and the use of microscopic quantities (and so of an atomistic model) was required. In this case, the considerations were carried out on the basis of statistical mechanics [4]. Next, the continuum approach to the kinetics of the homogeneous process of microcrack nucleation in single crystals was presented [5]. The continuum analysis gave the same nucleation rate equation as that obtained by the discrete method, but the unknown non-equilibrium density of vacancy clusters was also found.

In spite of the large number of papers on nucleation theory, very little attention has been paid to what happens when both the homogeneous and heterogeneous mechanisms operate simultaneously [6-10]. In the present paper, an attempt is made to take into consideration the coupling of these two mechanisms in the global process of vacancy-microcrack formation in uniaxially strained metals. The total microcrack nucleation rate resulting from coupling of the aforesaid mechanism is not a simple sum of the corresponding rates calculated for independent mechanisms. Macroscopic thermodynamic properties are ascribed to the subcritical and critical clusters, and the considerations are based on the isothermal continuum analysis of the process kinetics and on a recent contribution to the theory of nonhomogeneous phase transitions proposed by Ziabicki [10]. Moreover, it is assumed that the dislocation density is negligibly small everywhere except on the grain boundaries.

QUALITATIVE DESCRIPTION OF THE PHENOMENON

Let us consider an internal region of a polycrystalline sample subject to tension and kept in a uniaxial state of strain. The observed region contains at least one grain (with its boundary) which together with all separate vacancies (phase I), dispersed inside the grain space or adsorbed on the grain boundary, as well as all types of vacancy aggregations (phase II) - constitutes our system. Within the grain, isotropic clusters (assumed as spherical) are formed according to the homogeneous mechanism and under the classical assumption that particular aggregates can grow or disintegrate only by successive associations or separations of single vacancies. At the same time, on the grain boundary, heterogeneous cap-shaped islands are created by two different simultaneously operating mechanisms: by direct addition of individual vacancies from the grain space to the already existing clusters and by successive joining of separate advacancies (adsorbed vacancies) via the surface diffusion along the boundary.

The smallest clusters of both types with equal probability of growth or disintegration are called the critical clusters or microcrack nuclei.

When the linear strain of the analysed crystal sample is equal to zero, the system considered is in the thermodynamic equilibrium. As the strain increases, the system becomes supersaturated with vacancies and aggregates of different sizes are generated. When the degree of supersaturation is high enough, and correspondingly the strain, critical aggregates are also nucleated.

BASIC PRELIMINARY FORMULAE

Below some fundamental formulae are listed. They will be used in the next sections.

Due to the strain, the activation energy of the spatial vacancy motion in the direction of the crystal elongation is reduced. The decrease of this energy (ΔU) was derived from simple energy considerations [1]. Subtracting the potentia energy of two neighboring lattice elements of an unstrained crystal from that of an elongated one, it was found that

$$\Delta U(\varepsilon) = aR_o^{-m} \{1 - (1+\varepsilon)^{-m} - \frac{m}{n} [1 - (1+\varepsilon)^{-n}]\} \tag{1}$$

where a, m, n are positive material constants, ε denotes the linear strain of the crystal, and R_o is the lattice constant.

The average velocity of a vacancy within the grains of the strained crystal can be approximately expressed in the form

$$v = \frac{R_o}{\tau} = \nu R_o \exp[- \frac{(U_m - \Delta U)}{kT}] \tag{2}$$

where τ is the lifetime of a vacancy at a lattice point, $\nu \approx 10^{13}$ sec^{-1} is the atomic vibrational frequency, k is Boltzmann's constant, T is the absolute temperature and U_m is the activation energy of motion of vacancies.

The average initial (ε=0) volume concentration of vacancies is given by the known formula

$$c_o = N \exp(- \frac{U_f}{kT}) \tag{3}$$

where N is the total number of atomic sites per unit volume, and U_f is the activation energy of formation of a vacancy.

As a consequence of the crystal extension, the energy of vacancy formation is also reduced. Assuming that this energy decrease is ΔU, the actual volume vacancy concentration reads

$$c = N \exp[- \frac{(U_f - \Delta U)}{kT}] = c_o \exp(\frac{\Delta U}{kT}) \tag{4}$$

It was proposed [11] that the magnitude of the formation energy decrease amounts to $\Delta U_1 = (\kappa\bar{\varepsilon})^2 E R_o^3$, where κ is the concentration coefficient, $\bar{\varepsilon}$ denotes the mean linear strain, and E is Young's modulus. The quantity ΔU_1 represents the elastic energy in that volume element in which a vacancy is formed. The difference between ΔU and ΔU_1 is, in general, rather small. However, the use of ΔU_1 requires not only the continuum approach but also the assumption of a constitutive equation. For this reason, we choose to work with ΔU as the reduction quantity of the formation energy.

From equation (4), we receive the ratio of the vacancy concentrations after and before the crystal is strained as a measure of the supersaturation of vacancies inside the grains

$$S_{ho} = \frac{c}{c_o} = \exp(\frac{\Delta U}{kT}) \tag{5}$$

Within the grain space the mean number of vacancies per unit time condensing on a unit area (the incidence, or capture, rate of vacancies) may be written

$$P_c = p_g c v \tag{6}$$

where p_g is the probability that a vacancy will jump in the desirable direction, i.e., towards an aggregate. For simple cubic lattice, p_g is simply 1/6, so that

$$P_c = \frac{1}{6} cv \tag{7}$$

For isotropic continuum model of the crystal, P_c can be evaluated in the form [12]

$$P_c = \frac{c}{\tau} \int_0^{R_o} \frac{2\pi R_o(R_o-x)}{4\pi R_o^2} dx = \frac{1}{4} cv \tag{8}$$

Here, $1/\tau$ is the jump frequency of vacancies, and the integrand is the ratio of the surface area of a spherical segment to the surface area of the whole sphere; it represents the probability that a vacancy placed at a distance x from the aggregate surface will jump towards the aggregate. The jump distance is equal to the lattice constant R_o, and vacancies located farther than R_o have zero probability to reach the aggregate with one hop.

Substituting equations (2) and (4) into equation (6), one obtains for the incidence rate the final relation

$$P_c = p_g \nu R_o c_o \exp[- \frac{(U_m-2\Delta U)}{kT}] \tag{9}$$

A similar notion is needed to describe the heterogeneous nucleation on the grain boundary. This is the rate of impingement of advacancies on a line of unit length on the grain boundary surface (called also the capture rate). This quantity can be determined from kinetic considerations. When advacancies diffuse over the boundary with a mean velocity v_{sd}, N(1) is the number of single advacancies per unit area of the boundary, and p_b is the probability factor that an advacancy will jump in the right direction, then

$$\omega_c = p_b N(1) v_{sd} \tag{10}$$

Assuming a square lattice on the surface of the boundary, p_b is simply 1/4, i.e.,

$$\omega_c = \frac{1}{4} N(1) v_{sd} \tag{11}$$

For continuum model of the crystal, ω_c may be calculated similarly as for the three-dimensional case

$$\omega_c = \frac{N(1)}{\tau_{sd}} \int_0^d \frac{2d \arccos \frac{x}{d}}{2\pi d} dx \tag{12}$$

where τ_{sd} is the surface diffusion lifetime of an advacancy at a lattice site, and the integrand is the ratio of the arc length of a circular segment to the circumference of the whole circle; it represents the probability that an advacancy located at a distance x from the periphery of the base circle of the cap-shaped aggregate will jump toward the aggregate. The jump distance is equal to the interatomic spacing d on the boundary surface, and advacancies at distances larger than d have zero probability to be incorporated into the aggregate with one hop.

The average migration velocity of a single advacancy over the grain boundary surface is approximately

$$v_{sd} = \nu d \exp\left(-\frac{U_{sd}}{kT}\right) \tag{13}$$

In the above formula, the vibrational frequency, parallel to the surface, is assumed equal to that of regular atoms in the three-dimensional lattice ($\simeq 10^{13}$ sec^{-1}), U_{sd} is the activation energy for surface autodiffusion.

Substituting equation (13) into equation (10), we obtain

$$\omega_c = p_b \nu d N(1) \exp\left(-\frac{U_{sd}}{kT}\right) \tag{14}$$

HOMOGENEOUS NUCLEATION OF MICROCRACKS WITHIN THE GRAINS

In a supersaturated system, it is always possible to define the equilibrium aggregate distribution function, $A^o(i)$, when there are still no perceptible traces of the new phase nuclei. This function is given, for a temperature T, by the well-known stationary Boltzmann distribution for a diluted system [12-14]

$$A^o(i) = A^o(1)\exp\left[-\frac{\Delta F_g(i) - kT\ln\frac{N}{A^o(1)}}{kT}\right] = N\exp\left[-\frac{\Delta F_g(i)}{kT}\right] \tag{15}$$

Here, $\Delta F_g(i)$ is the free energy of formation of an aggregate of i vacancies from isolated vacancies in an isothermal reversible process ignoring the entropy of mixing; it is the sum of two terms, of which the first one is always positive and represents the work spent in forming the surface of the aggregate, and the second one is the work gained in forming the new volume. If nucleation is thermodynamically admissible ($S_{ho} > 1$) the second term is always negative, otherwise no phase transition is possible. Taking into account the minus sign for the second term, we have for spherical isotropic clusters

$$\Delta F_g(i) = O(i)\alpha - V(i)\,\frac{kT}{V(1)}\,\ell n\,\frac{c}{c_o} = 4\pi r^2\alpha - \frac{4}{3}\,\pi r^3\,\frac{kT}{V_v}\,\ell n S_{ho}\,,\ i \geq 2$$ (16

where O(i), V(i) and r = r(i) are surface area, volume and radius of a spherical i-mer, respectively, α is the surface tension of the crystal, $V_v = V(1)$ is the volume of a vacancy.

The numerator of the exponent in the first part of equation (15)

$$\tilde{\Delta}F_g(i) = \Delta F_g(i) - kT\ell n\,\frac{N}{A^o(1)}$$ (17

presents the cluster formation energy completed with the configurational entropy term $- kT\ell n(N/A^o(1))$ [15], under the assumption that $N >> \Sigma A^o(i)$, i.e., that the number of all the aggregates is small compared with the total number of atomic sites. Since no energy is needed to form an aggregate of one vacancy from single vacancies, the total free energy of formation $\tilde{\Delta}F_g$ should be zero for i=1. Thus

$$\tilde{\Delta}F_g(1) = \Delta F_g(1) - kT\ell n\,\frac{N}{A^o(1)} = 0$$ (18

and we have

$$\Delta F_g(1) = U_f = kT\ell n\,\frac{N}{A^o(1)}$$ (19

Because $A^o(1) = c$, this is in accordance with equation (3). From equations (17) and (19), it follows that

$$\tilde{\Delta}F_g(i) = \Delta F_g(i) - \Delta F_g(1)$$ (20

Utilizing the relations

$$i = \frac{V(i)}{V(1)} = \frac{4\pi r^3}{3V_v}$$ (21

$$O(i) = (36\pi)^{1/3}\, V_v^{2/3}\, i^{2/3}$$ (21

we can express ΔF_g as a function of the variable i

$$\Delta F_g(i) = \begin{cases} (36\pi)^{1/3}\, V_v^{2/3}\, \alpha i^{2/3} - ikT\ell n S_{ho} & \text{for } i \geq 2 \quad (22 \\ kT\ell n\,\dfrac{N}{c_o} & \text{for } i = 1 \quad (22 \end{cases}$$

When $S_{ho} > 1$, the first derivative of ΔF_g equals zero

$$\frac{\partial \Delta F_g(i)}{\partial i} = \frac{2}{3}(36\pi)^{1/3} V_v^{2/3} \alpha i^{-1/3} - kT\ell n S_{ho} = 0 \qquad (23)$$

at a point $i = i_g^*$

$$i_g^* = \frac{32\pi\alpha^3 V_v^2}{3(kT\ell n S_{ho})^3} \qquad (24)$$

Since the second derivative of ΔF_g is negative at that point

$$\left[\frac{\partial^2 \Delta F_g(i)}{\partial i^2}\right]_{i=i_g^*} = -\frac{2}{9}(36\pi)^{1/3} V_v^{2/3} \alpha i_g^{*-4/3} < 0 \qquad (25)$$

the function ΔF_g reaches a maximum at the critical cluster of i_g^* vacancies

$$\Delta F_g(i_g^*) = \Delta F_g^* = \frac{16\pi\alpha^3 V_v^2}{3[\Delta U(\varepsilon)]^2} \qquad (26)$$

Let us now consider an arbitrary internal subregion of a grain of the crystal sample subject to uniaxial strain. When, under the increasing strain, the supersaturation of vacancies in the considered system attains a sufficiently high value, a certain number of critical aggregates (microcrack nuclei) is formed. Let the number of nuclei formed per unit volume and unit time, at a given temperature T and strain ε, be $J_g(i,t)$. It is called the homogeneous nucleation rate, and results from a flux which flows through all the aggregate classes from smaller to larger aggregates. Its intensity is equal to the difference between the numbers of aggregates that pass to the next higher class by addition of single vacancies and those that return to the lower class by separation of single vacancies. The nucleation rate $J_g(i,t)$ can, therefore, be written

$$J_g(i,t) = A(i-1,t)P_c O(i-1) - A(i,t)P_e(i)O(i), \quad i \geq 2 \qquad (27)$$

where $A(i,t)$ is the non-equilibrium density of aggregates of i vacancies at a time t, and $P_e(i)$ is the emission rate of vacancies per unit area of the surface of an aggregate of size i.

Under equilibrium conditions, there is no phase transition and the net flow of aggregates must vanish. Consequently, $J_g(i,t) = 0$ and $A(i,t) = A^o(i)$, which allows to eliminate $P_e(i)$ from equation (27) and yields

$$J_g(i,t) = P_c O(i-1)\left[A(i-1,t) - A(i,t)\frac{A^o(i-1)}{A^o(i)}\right] \qquad (28)$$

Now, in equation (28), we assume that $O(i-1) \approx O(i)$, and we expand the functions $A(i-1,t)$ and $A^o(i-1)$ into Taylor series near the point i, linearizing the expansions. This leads to

$$J_g(i,t) = P_c O(i)[\frac{A(i,t)}{A^o(i)} \frac{\partial A^o(i)}{\partial i} - \frac{\partial A(i,t)}{\partial i}] \tag{29}$$

Differentiating the function (15) and substituting the result into equation (29), we find

$$J_g(i,t) = -P_c O(i)[\frac{A(i,t)}{kT} \frac{\partial \Delta F_g(i)}{\partial i} + \frac{\partial A(i,t)}{\partial i}] \tag{30}$$

The functions $A(i,t)$ and $J_g(i,t)$ must satisfy, for each ε, the equation of continuity which for the considered sourceless process is reduced to the form

$$\frac{\partial A(i,t)}{\partial t} + \frac{\partial J_g(i,t)}{\partial i} = 0 \tag{31}$$

Inserting the flux, equation (30), into equation (31), we receive

$$\frac{\partial A(i,t)}{\partial t} - P_c \frac{\partial}{\partial i} \{O(i)[\frac{A(i,t)}{kT} \frac{\partial \Delta F_g(i)}{\partial i} + \frac{\partial A(i,t)}{\partial i}]\} \tag{32}$$

Assuming that the process achieves a steady state (the time delay to attain this steady state is usually negligibly short [12,16]), we put $\partial A(i,t)/\partial t = 0$ in equation (32), and the problem is reduced to the following linear differential equation for the distribution function $A(i)$

$$O(i) \frac{\partial A(i)}{\partial i} + \frac{O(i)}{kT} \frac{\partial \Delta F_g(i)}{\partial i} A(i) = C \tag{33}$$

where C is a function which does not depend on i.

The general solution of equation (33) reads

$$A(i) = \exp[\frac{\Delta F_g(1)-\Delta F_g(i)}{kT}]\{C_1 + C\exp[-\frac{\Delta F_g(1)}{kT}] \int_1^i \frac{1}{O(i)} \exp[\frac{\Delta F_g(i)}{kT}]di\} \tag{34}$$

where C_1 is a constant, and $\Delta F_g(1)$ is given by equation (19). We then set the boundary conditions for the function $A(i)$:

For $i = 1$, $A(1) = c$ - the actual vacancy concentration (35

For $i = \infty$, $\lim_{i\to\infty} A(i) = 0$ (35

Taking into account the following limit

$$\lim_{i\to\infty} \frac{\Delta F_g(i)}{kT} = -\infty$$

the conditions (35a,35b) imposed on the function (34) allow to specify the constants C and C_1 and lead to the particular solution of equation (33) for the non-equilibrium stationary distribution of aggregates

$$A(i) = NS_{ho}\exp[-\frac{\Delta F_g(i)}{kT}]\left\{1 - \frac{\int_1^i \frac{1}{O(i)}\exp[\frac{\Delta F_g(i)}{kT}]di}{\int_1^\infty \frac{1}{O(i)}\exp[\frac{\Delta F_g(i)}{kT}]di}\right\} \quad (36)$$

Substituting equation (36) into the flux equation (30), we determine the steady state nucleation rate

$$J_g(i,t) = J_g = \frac{P_c NS_{ho}}{\int_1^\infty [O(i)]^{-1}\exp[\frac{\Delta F_g(i)}{kT}]di} = \text{const.} \quad (37)$$

To estimate the integral in the denominator of equation (37), we follow a simplification suggested by Frenkel [14]. We namely expand the function $\Delta F_g(i)$ into Taylor series at the point of its maximum i_g^*, neglecting all the terms above the third one, i.e.,

$$\Delta F_g(i) \approx \Delta F_g^* + [\frac{\partial \Delta F_g(i)}{\partial i}]_{i_g^*}(i-i_g^*) + \frac{1}{2}[\frac{\partial^2 \Delta F_g(i)}{\partial i^2}]_{i_g^*}(i-i_g^*)^2 \quad (38)$$

Taking advantage of equations (23) and (25) and replacing O(i), given by equation (21b), by its value for $i = i_g^*$ (in view of a very sharp maximum of the integrand $\exp[\Delta F_g(i)/kT]$ at this point), we get an integral related to the Gauss probability integral which can be simply evaluated to give

$$\int_1^\infty [O(i)]^{-1}\exp[\frac{\Delta F_g(i)}{kT}]di \approx \frac{1}{2V_v}\sqrt{\frac{kT}{\alpha}}\exp(\frac{\Delta F_g^*}{kT}) \quad (39)$$

Substituting the integral (39) into equation (37), we obtain the final form of the homogeneous stationary nucleation rate of vacancy-microcracks within monocrystalline grains of the strained crystal

$$J_g = 2NS_{ho}P_c V_v \sqrt{\frac{\alpha}{kT}}\exp(-\frac{\Delta F_g^*}{kT}) \quad (40)$$

Since the entropy of mixing is included in the total formation energy of an aggregate $\tilde{\Delta F}_g(i)$, equation (40) differs in the preexponential factor from the analogical equation derived earlier [1,5].

Coming back to the non-equilibrium distribution of vacancy clusters, we insert the integral (39) into equation (36) arriving at the final expression for A(i)

$$A(i) = NS_{ho}\exp[-\frac{\Delta F_g(i)}{kT}]\{1 - 2V_v\sqrt{\frac{\alpha}{kT}}\exp(-\frac{\Delta F_g^*}{kT})\int_1^i \frac{1}{O(i)}\exp[\frac{\Delta F_g(i)}{kT}]di\} \quad (41)$$

The appearance of the three-dimensional flux of vacancies in the homogeneous as well as in the heterogeneous nucleation kinetics makes both the processes coupled rather than linearly superposed. However, rough quantitative estimates show that the effect of coupling on the homogeneous nucleation mechanism is very small and has been neglected here. It is taken into consideration in the heterogeneous nucleation (see Section: Coupling of the Two Mechanisms in the Heterogeneous Nucleation).

HETEROGENEOUS MICROCRACK NUCLEATION DUE TO THE SOLE ADDITION OF VACANCIES FROM THE GRAIN SPACE

Beside the homogeneous clusters formed within the grain, heterogeneous vacancy aggregation in the shape of spherical segments are originated at the grain boundary. Two mechanisms are involved in the formation of each heterogeneous cluster: direct impingement of single vacancies from the grain on the surface of the cluster, and joining of separate advacancies to the periphery of the cluster due to the surface diffusion. Both mechanisms are coupled. However, in thi section, and the next one, we will treat them formally as uncoupled assuming, in the first place, that the process involving addition of vacancies from the grain space is the only active mechanism.

Let us fix our attention on a segment of the grain boundary transverse to th direction of the crystal elongation. We assume that the cap-shaped vacancy islands lie on a two-dimensional square lattice of lattice parameter d, and that t equilibrium distribution of aggregate sizes $N^o(i)$ is determined, for a given temperature, by the formula

$$N^o(i) = N^o(1)\exp[-\frac{\Delta F_{bg}(i) - kT\ln\frac{N_o}{N^o(1)}}{kT}] = N_o\exp[-\frac{\Delta F_{bg}(i)}{kT}] \quad (4$$

Here, N_o is the density of discrete adsorption sites per unit area of the grain boundary ($N_o \approx 1/d^2 \approx 10^{15}$ cm^{-2}), the mixing entropy term $- kT\ln(N_o/N^o(1))$ [17] is included and it is assumed that $N_o >> N^o(1) \approx \Sigma N^o(i)$, which means that the fraction of the unit area of the grain boundary surface covered by the aggregate of various classes is negligibly small; $\Delta F_{bg}(i)$ is the free energy of formation of a single cluster of i vacancies in an isolated reversible process involving solely the growth of cluster by acquisition of vacancies from the grain space (

curvature effect), and with the configurational entropy disregarded. $\Delta F_{bg}(i)$ is postulated in the form

$$\Delta F_{bg}(i) = W(i) - \overline{V}(i)\,\frac{kT}{V_v}\,\ln\frac{c}{c_o},\quad i \geq 2 \tag{43}$$

where W(i) is the work spent in forming the surface of the cluster, and the second term represents the work gained in forming the new volume; $\overline{V}(i)$ is the volume of an i-mer of the shape of a spherical segment.

The total free energy of formation, appearing in the exponent of the first part of equation (42)

$$\tilde{\Delta}F_{bg}(i) = \Delta F_{bg}(i) - kT\ln\frac{N_o}{N^o(1)} \tag{44}$$

should vanish for i = 1. Consequently, it follows that

$$\Delta F_{bg}(1) = kT\ln\frac{N_o}{N^o(1)} \tag{45}$$

and

$$\tilde{\Delta}F_{bg}(i) = \Delta F_{bg}(i) - \Delta F_{bg}(1) \tag{46}$$

Taking advantage of the geometry of the aggregate and introducing formally the notion of the equilibrium contact angle θ, which makes the aggregate with the grain boundary, we can evaluate the first term in equation (43). To this end, we accept the Young's formula in which θ is determined by the relevant specific interfacial free energies at adjacent surfaces

$$\alpha_{bg} = \alpha_{ba} + \alpha\cos\theta \tag{47}$$

where the subscripts b, g and a refer to the grain boundary, grain space and the aggregate, respectively, and α is the surface free energy of the crystal. If S(i) and S'(i) denote the surface area of the cap of the aggregate and the area of its base, and r is the radius of curvature of the aggregate, we can write, using equation (47)

$$W(i) = S(i)\alpha + S'(i)(\alpha_{ba}-\alpha_{bg}) = 4\pi r^2\alpha\Phi(\theta) \tag{48}$$

where

$$\Phi = \frac{1}{4}\,(2 - 3\cos\theta + \cos^3\theta) \tag{49}$$

is the ratio of the volume of a spherical segment to the volume of the entire sphere.

Making use of the relation

$$i = \frac{\overline{V}(i)}{V_v} = \frac{4\pi r^3}{3V_v}\,\Phi(\theta) \qquad (50$$

in equations (48) and (43), we obtain ΔF_{bg} as a function of i

$$\Delta F_{bg}(i) = \begin{cases} (36\pi)^{1/3}\, V_v^{2/3}\, \Phi^{1/3}\, \alpha i^{2/3} - ikT\ln S_{ho} & \text{for } i \geq 2 \qquad (51 \\ kT\ln \dfrac{N_o}{N^o(1)} & \text{for } i = 1 \qquad (51 \end{cases}$$

It is worth noting that for $i \geq 2$, $\Delta F_{bg}(i)$ differs from $\Delta F_g(i)$ in the first term only (cf., equation (22a)), in view of the change of the cluster shape.

For supersaturation, $S_{ho} > 1$, we have

$$\frac{\partial \Delta F_{bg}(i)}{\partial i} = \frac{2}{3}\,(36\pi)^{1/3}\, V_v^{2/3}\, \Phi^{1/3}\, \alpha i^{-1/3} - kT\ln S_{ho} = 0 \qquad (52$$

at a point

$$i^*_{bg} = \frac{32\pi\alpha^3 V_v^2}{3(kT\ln S_{ho})^3}\,\Phi \qquad (53$$

Since

$$\left[\frac{\partial^2 \Delta F_{bg}(i)}{\partial i^2}\right]_{i=i^*_{bg}} = -\frac{2}{9}\,(36\pi)^{1/3}\, V_v^{2/3}\, \Phi^{1/3}\, \alpha i_{bg}^{*-4/3} < 0 \qquad (5$$

$\Delta F_{bg}(i)$ exhibits a maximum at the critical aggregate consisting of i^*_{bg} vacancies

$$\Delta F_{bg}(i^*_{bg}) = \Delta F^*_{bg} = \frac{16\pi\alpha^3 V_v^2}{3(\Delta U)^2}\,\Phi \qquad (5$$

Similarly, as in the homogeneous case, we consider now the net flow of heter geneous void embryos, denoting it by $J_{bg}(i,t)$. Since the growth process is limited to the sole addition of vacancies from the grain space, the nucleation rate equation reads

$$J_{bg}(i,t) = N(i-1,t)P_c S(i-1) - N(i,t)P_e(i)S(i), \; i \geq 2 \qquad (5$$

where $N(i,t)$ is the non-equilibrium surface density of heterogeneous clusters of size i at a time t, and $S(i)$ being the surface area of the cluster cap is, according to equation (50), given by the relation

$$S(i) = 2\pi r^2(1-\cos\theta) = \left(\frac{9\pi}{2}\right)^{1/3} V_v^{2/3}\ \Phi^{-2/3}\ (1-\cos\theta) i^{2/3} \tag{57}$$

If the system is in a state of equilibrium, there is no nucleation so that $J_{bg}(i,t) = 0$ and $N(i,t) = N^o(i)$ in equation (56). This enables us to eliminate $P_e(i)$ from equation (56). Next, the functions $N(i-1,t)$ and $N^o(i-1)$ are expanded into Taylor series in the neighborhood of the point i, and all the terms above the second one are disregarded. Moreover, we put $O(i-1) \approx O(i)$, and after differentiating the function (42), we substitute the derivative $\partial N^o(i)/\partial i$ into the transformed flux equation, getting

$$J_{bg}(i,t) = -P_c S(i)\left[\frac{N(i,t)}{kT}\frac{\partial \Delta F_{bg}(i)}{\partial i} + \frac{\partial N(i,t)}{\partial i}\right] \tag{58}$$

By substituting equation (58) into the equation of continuity, and assuming that a steady state is attained, i.e., $\partial N(i,t)/\partial t = 0$ (or $J_{bg}(i,t) = \text{const.}$), we receive a linear differential equation for the function $N(i)$. Finally, using the same procedure as in the Section, Homogeneous Nucleation of Microcracks within the Grains, we obtain the isolated heterogeneous cap surface-growth-controlled nucleation rate of microcracks on the grain boundaries

$$J_{bg} = N_o S_{het} P_c V_v \sqrt{\frac{\alpha}{kT}}\ (1-\cos\theta)\Phi^{-1/2} \exp\left(-\frac{\Delta F^*_{bg}}{kT}\right) \tag{59}$$

where

$$S_{het} = \frac{N(1)}{N^o(1)} \tag{60}$$

is defined as the supersaturation of advacancies at the grain boundaries (S_{het} is calculated in the next section, cf., equation (66).

In a similar way, we find the non-equilibrium distribution function $N(i)$

$$N(i) = N_o S_{het} \exp\left[-\frac{\Delta F_{bg}(i)}{kT}\right]\left\{1 - V_v \sqrt{\frac{\alpha}{kT}}\ (1-\cos\theta)\Phi^{-1/2} \exp\left(-\frac{\Delta F^*_{bg}}{kT}\right) \times \right.$$
$$\left. \times \int_1^i \frac{1}{S(i)} \exp\left[\frac{\Delta F_{bg}(i)}{kT}\right] di\right\} \tag{61}$$

HETEROGENEOUS NUCLEATION BY THE ISOLATED SURFACE DIFFUSION MECHANISM

Consider again a segment of the grain boundary perpendicular to the direction of the crystal elongation. We assume now that the problem of heterogeneous nucleation of vacancy microcracks at the boundary is uncoupled and depends only on a surface diffusion mechanism.

Under the conditions of adsorption equilibrium at the grain boundary surface, the incidence flux of vacancies will equal the desorption flux. This can be written approximately

$$P_c(1-f_b) = \frac{N(1)}{\tau_{des}}(1-f_g) \tag{62}$$

where $f_b = N(1)/N_o$, so that $(1-f_b)$ is the fraction of a unit area of the boundary surface not occupied by advacancies, $f_g = c/N$ is the fractional concentration of vacancies within the grain,

$$\tau_{des} = \frac{1}{\nu}\exp\left(\frac{U_{des}-\Delta U}{kT}\right) \tag{63}$$

is the mean lifetime of an advacancy on the grain boundary before desorption, and U_{des} is the activation energy for desorption of an advacancy from the grain boundary surface which is also lowered by $\Delta U(\varepsilon)$, similarly as in the case of the activational energy of vacancy motion. U_{des} is connected with the activation energy for adsorption, $U_{ad} = U_m$, through the heat of adsorption E_a

$$E_a = U_{des} - U_m \tag{64}$$

Solving equation (62) for N(1) we obtain

$$N(1) = \frac{P_c}{\frac{1}{\tau_{des}}\left(1 - \frac{c}{N}\right) + \frac{P_c}{N_o}} \tag{65}$$

The surface supersaturation of advacancies for the heterogeneous process was defined in the Section, Heterogeneous Microcrack Nucleation Due to the Sole Addition of Vacancies from the Grain Space, equation (60), as the ratio $S_{het} = N(1)/N^o(1)$. Consequently, writing equation (65) for the actual state ($\varepsilon>0$) and for the initial one ($\varepsilon=0$), and using equations (5), (9), (63) and (64), we find

$$S_{het} = \frac{N(1)}{N^o(1)} = S_{ho}H(\varepsilon) \tag{66}$$

where

$$H(\varepsilon) = \frac{\exp(-\frac{E_a}{kT})(N-c_o) + \frac{p_g N R_o}{N_o} c_o}{\exp(-\frac{E_a}{kT})(N-c) + \frac{p_g N R_o}{N_o} c} \tag{67}$$

We notice that for $\varepsilon=0$, $\Delta U(\varepsilon) = 0$ and $c = c_o$ so that $H(\varepsilon) = 1$ and $S_{het} = S_{ho} = 1$.

In order to determine the final form of the impingement rate of advacancies on a line of unit length on the boundary surface (ω_c), we substitute N(1) from equation (65) into equation (14), getting

$$\omega_c = \frac{p_b p_g \nu R_o N c_o d}{\exp(-\frac{E_a}{kT})(N-c) + \frac{p_g N R_o}{N_o} c} \exp(\frac{\Delta U - U_{sd}}{kT}) \tag{68}$$

The equilibrium distribution of vacancy clusters $N^o(i)$ is now assumed in the form

$$N^o(i) = N^o(1)\exp[-\frac{\Delta F_{bs}(i) - kT\ln\frac{N_o}{N^o(1)}}{kT}] = N_o \exp[-\frac{\tilde{\Delta}F_{bs}(i)}{kT}] \tag{69}$$

where the sum $\tilde{\Delta}F_{bs}(i) = \Delta F_{bs}(i) - kT\ln(N_o/N^o(1))$ is the total energy of formation of a cluster of i advacancies in a process involving only the surface diffusion mechanism; $\Delta F_{bs}(i)$ is a size-dependent term, and $-kT\ln(N_o/N^o(1))$ [17] is the correction for the free energy of distributing the clusters on the N_o available adsorption sites (configurational entropy term). Since $\tilde{\Delta}F_{bs}(1)$ must be zero, $\Delta F_{bs}(1) = kT\ln(N_o/N^o(1))$, and for cap-shaped clusters, the size-dependent term $\Delta F_{bs}(i)$ can be postulated in the form

$$\Delta F_{bs}(i) = \begin{cases} (36\pi)^{1/3} V_v^{2/3} \Phi^{1/3} \alpha i^{2/3} - ikT\ln S_{het} & \text{for } i \geq 2 \quad (70a) \\ kT\ln\frac{N_o}{N^o(1)} & \text{for } i = 1 \quad (70b) \end{cases}$$

Because the growth of clusters is governed now by the heterogeneous supersaturation S_{het}, we see that for $i \geq 2$, $\Delta F_{bs}(i)$ differs from $\Delta F_{bg}(i)$ in the second term only, cf., equation (51a).

Similarly, as ΔF_g and ΔF_{bg}, the function ΔF_{bs} has a maximum

$$\Delta F_{bs}(i^*_{bs}) = \Delta F^*_{bs} = \frac{16\pi\alpha^3 V_v^2}{3(kT\ln S_{het})^2} \Phi \tag{71}$$

at the critical cluster size

$$i^*_{bs} = \frac{32\pi\alpha^3 V_v^2}{3(kT\ell nS_{het})^3}\,\Phi \tag{72}$$

Passing now to the kinetics we denote here the isolated flux of clusters by $J_{bs}(i,t)$. Since in this case the growth mechanism is controlled only by the surface diffusion of advacancies, the nucleation rate equation reads

$$J_{bs}(i,t) = N(i-1,t)\omega_c\ell(i-1) - N(i,t)\omega_e(i)\ell(i),\ i \geq 2 \tag{73}$$

where $\omega_e(i)$ is the emission rate of advacancies from the unit length of the periphery of the base of a cap-shaped i-mer, and $\ell(i)$ is the length of this periphery which, by using the relation (50), can be expressed as

$$\ell(i) = 2\pi r\sin\theta = (6\pi^2)^{1/3}\ V_v^{1/3}\ \Phi^{-1/3}\ \sin\theta i^{1/3} \tag{74}$$

We now follow again the lines of Section, Heterogeneous Microcrack Nucleation Due to the Sole Addition of Vacancies from the Grain Space, bearing in mind that here instead of $S(i)$, P_c, and $\Delta F_{bg}(i)$, we have to use the quantities $\ell(i)$, ω_c and $\Delta F_{bs}(i)$, respectively. Consequently, we first eliminate $\omega_e(i)$ from equation (73) by putting $J_{bs} = 0$ (equilibrium condition), and next using the linearized expansions of the functions $N(i-1,t)$ and $N^o(i-1)$ and the approximation $\ell(i-1) \approx \ell(i)$, as well as the derivative $\partial N^o(i)/\partial i$ (from equation (69)), we are led to the flux equation in the form

$$J_{bs}(i,t) = -\omega_c\ell(i)\left[\frac{N(i,t)}{kT}\frac{\partial\Delta F_{bs}(i)}{\partial i} + \frac{\partial N(i,t)}{\partial i}\right] \tag{75}$$

Substituting J_{bs} to the continuity equation and solving the resulting differential equation, under the conditions $\partial N(i,t)/\partial t = 0$ and $N(\infty) = 0$, we arrive at the non-equilibrium surface density of the clusters on the grain boundary

$$N(i) = N_o S_{het}\exp\left[-\frac{\Delta F_{bs}(i)}{kT}\right]\left\{1 - \frac{1}{2}\sqrt{\frac{kT}{\alpha\Phi}}\,\ell n(S_{het})\sin\theta\exp\left(-\frac{\Delta F^*_{bs}}{kT}\right) \times \right.$$
$$\left. \times \int_1^i \frac{1}{\ell(i)}\exp\left[\frac{\Delta F_{bs}(i)}{kT}\right]di\right\} \tag{76}$$

Inserting equation (76) in equation (77) delivers finally the separated hetero geneous peripheral-growth-determined nucleation rate of microcracks in a steady state

$$J_{bs} = \frac{1}{2} N_o S_{het} \ell n(S_{het}) \omega_c \sqrt{\frac{kT}{\alpha\Phi}} \sin\theta \exp(-\frac{\Delta F^*_{bs}}{kT}) \tag{77}$$

COUPLING OF THE TWO MECHANISMS IN THE HETEROGENEOUS NUCLEATION

We are now in a position to find the total coupled flux of heterogeneous clusters on the grain boundary. However, in the presence of both the surface and peripheral-growth mechanisms, we do not know the coupled formation energy of clusters and can not postulate the equilibrium distribution function $N^o(i)$. Therefore, we start at once from the kinetics and the flux equation.

Denoting the coupled heterogeneous nucleation rate on the grain boundary by $J_{gs}(i,t)$, we can write the net flow of cap-shaped vacancy clusters in the form

$$J_{gs}(i,t) = J_{bg}(i,t) + J_{bs}(i,t) = N(i-1,t)P_c S(i-1) - N(i,t)P_e(i)S(i)$$
$$+ N(i-1,t)\omega_c \ell(i-1) - N(i,t)\omega_e(i)\ell(i), \ i \geq 2 \tag{78}$$

Using the equilibrium conditions in which both the component fluxes J_{bg} and J_{bs} disappear, we can eliminate $P_e(i)$ and $\omega_e(i)$ from equation (78). Further treatment follows the same procedure as in the last two sections so that we can take advantage of equations (58) and (75). Thus

$$J_{gs}(i,t) = -\{[P_c S(i) + \omega_c \ell(i)]\frac{\partial N(i,t)}{\partial i} + [\frac{P_c S(i)}{kT}\frac{\partial \Delta F_{bg}(i)}{\partial i}$$
$$+ \frac{\omega_c \ell(i)}{kT}\frac{\partial \Delta F_{bs}(i)}{\partial i}]N(i,t)\} \tag{79}$$

Next, the equation of continuity which for the case considered is of the shape

$$\frac{\partial N(i,t)}{\partial i} + \frac{\partial}{\partial i}[J_{bg}(i,t) + J_{bs}(i,t)] = 0 \tag{80}$$

must be solved under the steady state condition $\partial N(i,t)/\partial t = 0$. This yields the differential equation for the stationary non-equilibrium distribution function $N(i)$

$$J_{gs}(i,t) = (P_c S + \omega_c \ell(i)\frac{\partial N(i)}{\partial i} + \frac{1}{kT}(P_c S \frac{\partial \Delta F_{bg}}{\partial i} + \omega_c \ell \frac{\partial \Delta F_{bs}}{\partial i})N(i) = C \tag{81}$$

with a constant C. The general solution of equation (81) has the form

$$N(i) = \exp\left(- \frac{1}{kT} \int_1^i \frac{P_c S \frac{\partial \Delta F_{bg}}{\partial i} + \omega_c \ell \frac{\partial \Delta F_{bs}}{\partial i}}{P_c S + \omega_c \ell} di\right) \times$$

$$\times \left[N(1) + C \int_1^i \frac{1}{P_c S + \omega_c \ell} \exp\left(\frac{1}{kT} \int_1^i \frac{P_c S \frac{\partial \Delta F_{bg}}{\partial i} + \omega_c \ell \frac{\partial \Delta F_{bs}}{\partial i}}{P_c S + \omega_c \ell} di\right)\right] \quad (82)$$

where the integral in the exponent can be evaluated to give

$$\tilde{\Delta}F_{gs}(i) = \int_1^i \frac{P_c S \frac{\partial \Delta F_{bg}}{\partial i} + \omega_c \ell \frac{\partial \Delta F_{bs}}{\partial i}}{P_c S + \omega_c \ell} di = \{\Delta F_{bg}(i) - \frac{3kT\ell nH}{D^3} \times$$

$$\times [\frac{D^2}{2} i^{2/3} - Di^{1/3} + \ell n(Di^{1/3} + 1)]\} - \{\Delta F_{bg}(1)$$

$$- \frac{3kT\ell nH}{D^3} [\frac{D^2}{2} - D + \ell n(D+1)]\} \quad (83)$$

Here,

$$D = \left(\frac{3}{4\pi}\right)^{1/3} \frac{P_c}{\omega_c} V_v^{1/3} \Phi^{-1/3} \frac{1-\cos\theta}{\sin\theta} \quad (84)$$

$\Delta F_{bg}(i)$, $\Delta F_{bg}(1)$ and $H(\varepsilon)$ are given by equations (51a,51b) and (67), respectively.

In equilibrium, $J_{gs}(i,t) = 0$, $N(i) = N^o(i)$ and equation (81) becomes a homogeneous equation (C=0) solution of which yields the equilibrium distribution of clusters

$$N^o(i) = N^o(1)\exp[- \frac{\tilde{\Delta}F_{gs}(i)}{kT}] \quad (85)$$

where $\tilde{\Delta}F_{gs}$, given by equation (83), constitutes the total combined energy of formation of an i-mer in the coupled heterogeneous process involving both the surface and peripheral mechanisms of growth. $\tilde{\Delta}F_{gs}$ is clearly a sum of two terms

$$\tilde{\Delta}F_{gs}(i) = \Delta F_{gs}(i) - \Delta F_{gs}(1) \quad (86)$$

where the first term is a size-dependent part of the formation energy

$$\Delta F_{gs}(i) = \Delta F_{bg}(i) - \frac{3kT\ell nH}{D^3} [\frac{D^2}{2} i^{2/3} - Di^{1/3} + \ell n(Di^{1/3} + 1)] \quad (87)$$

and the second one

$$\Delta F_{gs}(1) = \Delta F_{bg}(1) - \frac{3kT\ell nH}{D^3}\left[\frac{D^2}{2} - D + \ell n(D+1)\right] \tag{88}$$

is the mixing entropy contribution ($-\Delta F_{bg}(1) = -kT\ell n(N_o/N^o(1))$), and the effect of coupling. Similarly, as in Sections - Basic Preliminary Formulae; Homogeneous Nucleation of Microcracks within the Grains; and Heterogeneous Microcrack Nucleation Due to the Sole Addition of Vacancies from the Grain Space - the total combined free energy of formation of the monomer itself is evidently zero, $\tilde{\Delta}F(1) = 0$.

In view of equations (85), (83) and (87), the equilibrium surface density of vacancy clusters finally reads

$$N^o(i) = N_o \exp\{-\frac{3\ell nH}{D^3}\left[\frac{D^2}{2} - D + \ell n(D+1)\right]\}\exp\left[-\frac{\Delta F_{gs}(i)}{kT}\right] \tag{89}$$

It can be shown that the function $\Delta F_{gs}(i)$ has a maximum at the critical cluster of i^*_{gs} vacancies. Maximizing ΔF_{gs} with respect to i we find

$$i^*_{gs} = (8i^*_{bs})^{-1} D^{-3}\{i^{*1/3}_{bg} i^{*1/3}_{bs} D - i^{*1/3}_{bg} + [(i^{*1/3}_{bg} i^{*1/3}_{bs} D - i^{*1/3}_{bg})^2 + 4i^{*1/3}_{bg} i^{*2/3}_{bs} D]^{1/2}\}^3 \tag{90}$$

and

$$\Delta F_{gs}(i^*_{gs}) = \Delta F^*_{gs} = (36\pi)^{1/3} V_v^{2/3} \Phi^{1/3} \alpha i^{*2/3}_{gs} - i^*_{gs}\Delta U - \frac{3kT\ell nH}{D^3} \times \left[\frac{D^2}{2} i^{*2/3}_{gs} - Di^{*1/3}_{gs} + \ell n(Di^{*1/3}_{gs} + 1)\right] \tag{91}$$

Coming back to the non-equilibrium distribution of clusters, equation (82), we can write it now, in view of the integral (83) and equation (86), in the form

$$N(i) = \exp\left[-\frac{\Delta F_{gs}(i)-\Delta F_{gs}(1)}{kT}\right]\{N(1) + C\int_1^i \frac{1}{P_c S+\omega_c \ell} \times \exp\left[\frac{\Delta F_{gs}(i)-\Delta F_{gs}(1)}{kT}\right]di\} \tag{92}$$

Addition of the boundary condition $N(\infty) = 0$ enables us to determine the constant C and gives the particular solution for the function N(i)

$$N(i) = N_o S_{het} \exp\{- \frac{3\ell nH}{D^3} [\frac{D^2}{2} - D + \ell n(D+1)] - \frac{\Delta F_{gs}(i)}{kT}\} K(i) \tag{93}$$

where

$$K(i) = 1 - \frac{\int_1^i \frac{1}{P_c S + \omega_c \ell} \exp[\frac{\Delta F_{gs}(i)}{kT}] di}{\int_1^\infty \frac{1}{P_c S + \omega_c \ell} \exp[\frac{\Delta F_{gs}(i)}{kT}] di} \tag{94}$$

Equation (94) is then substituted into equation (79), delivering the total steady state nucleation rate

$$J_{gs}(i,t) = J_{gs} = \frac{N_o S_{het} \exp\{- \frac{3\ell nH}{D^3} [\frac{D^2}{2} - D + \ell n(D+1)]\}}{\int_1^\infty \frac{1}{P_c S + \omega_c \ell} \exp[\frac{\Delta F_{gs}(i)}{kT}] di} \tag{95}$$

The integral appearing in the denominator of equations (94) and (95) can be approximated by the method shown in Section, Homogeneous Nucleation of Microcracks within the Grains. This leads us to the final form of the coupled heterogeneous nucleation rate of microcracks on the grain boundaries

$$J_{gs} = (6\pi kT)^{-1/2} N_o S_{het} \{P_c[(\frac{9\pi}{2})^{1/3} V_v^{2/3} \Phi^{-2/3} (1-\cos\theta) i_{gs}^{*1/3}] + \omega_c[(6\pi^2)^{1/3} V_v^{1/3} \Phi^{-1/3} \sin\theta]\}[i_{bg}^{*1/3} i_{gs}^{*-2/3} \Delta U - \frac{kTD\ell nH}{(Di_{gs}^{*1/3} + 1)^2}]^{1/2} \times$$
$$\times \exp\{- \frac{3\ell nH}{D^3} [\frac{D^2}{2} - D + \ell n(D+1)] - \frac{\Delta F_{gs}^*}{kT}\} \tag{96}$$

Similarly, we can obtain the distribution function N(i).

GLOBAL STEADY STATE NUCLEATION RATE

Consider finally the entire grain of our observed internal region of the strained polycrystal. When all the fluxes are active in the formation of various classes of both homo- and heterogeneous clusters, we can write a balance equation for the steady state global nucleation rate J_o

$$J_o V_g = J_g V_g + J_{gs} S_g \tag{97}$$

where V_g and S_g are volume and surface area of the grain, respectively, the homogeneous flux is given by equation (40) and the coupled heterogeneous one by equation (96).

In equation (97), the product J_oV_g means the sum of all the microcracks formed in the grain per unit time irrespectively of their shape or type. Consequently, it includes the homogeneous clusters generated within the grain space as well as the heterogeneous ones nucleated on the grain boundary.

From equation (97), J_o can be easily found

$$J_o = J_g + J_{gs}\frac{S_g}{V_g} \tag{98}$$

CONVENTIONAL MICROFRACTURE CRITERION

In the analysis of the course of fracture, three main stages of the process can be distinguished. The first one is the microfracture stage in which stable microcracks of the size of atomic order are formed. In the second stage, their further growth (or perhaps joining) to macroscopic dimensions of the order of a Griffith crack ($\approx 1\mu$) takes place. Finally, the third one is the macrofracture stage in which the macrocracks (called simply cracks) propagate further leading immediately to the actual fracture. While for the first stage, the present hypothesis can be applied, for the second one, another theory should be introduced. The last cracking process (third stage) can follow the existing continuum hypotheses in which, however, the pre-existence of the crack is assumed a priori. One of such theories is the Griffith theory of brittle fracture in which the dependence of the potential energy of the sample with a crack on the initial length of the crack ℓ is very similar to the relation between the energy of formation of a new phase i-mer in a super-saturated system and the size of the i-mer. Consequently, from the energy point of view, there is a close analogy between the micro- and macrofracture mechanisms.

From the above, it follows that individual stages of the fracture process are separated by two energy thresholds, each corresponding to suitable fracture mechanism. They may be called the microfracture and the macrofracture thresholds, respectively. The former is determined by the critical size i^* of a microcrack (or by the corresponding strain ε^*), and the latter by the critical crack length ℓ_{cr} (or by the corresponding critical strain ε_{cr}).

Now, a formal convention concerning a critical value of the global nucleation rate, J_o^*, should be introduced to get a conceivably invariant measure of approaching the strain after which a crack of Griffith length can be originated. This means that the microfracture threshold is reached and the microfracture stage is terminated. At this strain, stable microvoids, ready for further spontaneous growth, are generated. However, in most known phase transformation processes, the period of time during which the actual number of the new phase nuclei is formed is short and difficult to determine precisely by either measurement or calculation [16]. Consequently, the factual nucleation frequency can be established only theoretically. Fortunately, the nucleation rate is so sensitive to the supersaturation that one may specify a critical supersaturation, S^*, below which nucleation rate is negligible ($J \approx 0$) and above which it is very high ($J \approx \infty$). For these reasons, the critical supersaturation should correspond to the rate of nucleation having only some perceptible value, and it is generally agreed that meaningful data on homogeneous as well as heterogeneous nucleation in vapors and liquids are reported as supersaturations S^* corresponding to $J \sim 1$ nucleus per sec and cm^3 or cm^2. For metals, such an agreement cannot be accepted. To establish a possibly invariant microfracture criterion for metals, it seems that it should be connected with the dimensions of the grains. If we now assume that the supersaturation increases sufficiently slowly so that the induction period to establish a steady state of the system $\tau_1 << 1$ sec [12,16,25], the following conventional microfracture criterion for uniaxially strained metals may be proposed: The global nucleation rate τ_o^* corresponding to one arbitrary nucleus per second in the smallest grain is critical.

Thus, assuming the grain as a sphere with the diameter δ_{min} and volume V_{gmin}, we have

$$J_o^* = \frac{1}{V_{gmin}} = \frac{6}{\pi\delta_{min}^3} \frac{\text{nuclei}}{\text{cm}^3\text{sec.}} \tag{99}$$

For this number of nuclei we should find a critical strain ε_o^*. Similar procedure can be formally applied also to isolated homogeneous and heterogeneous nucleations to find the corresponding critical strains ε_g^*, ε_{bg}^*, ε_{bs}^* and ε_{gs}^*. For each separated heterogeneous nucleation on the grain boundary, we have to use $1/S_{gmin}$ instead of $1/V_{gmin}$ in equation (99).

CONCLUSION AND FINAL REMARKS

The vacancy hypothesis of microfracture of strained metals has been essentially extended by coupling of homo- and heterogeneous mechanisms in the process of global nucleation.

In view of the idealized model of the metal, disregarding the effect of other kinds of defects, some formally introduced physical quantities, and many unknown or uncertain data, the vacancy mechanism of microfracture of strained metals has mainly theoretical character. Moreover, it is not quite clear whether the actual supersaturations caused by the strain in real metals are not too low to nucleate a vacancy-microcrack. In the present theory, the critical global nucleation rate requires supersaturations ranging from a rather impossibly high value of 10^8 to around 3.5. They are dropping very rapidly with rising temperature so that beginning from 800°K they do not exceed 50. On the other hand, it was reported [18-21] that the highest supersaturation at which voids are produced amounts to merely 1.01. Such a low supersaturation was concluded from diffusion experiments without any external tension applied. It is of interest to note here that in the paper [22], a critical vacancy supersaturation for pore formation during the interdiffusion of copper and zinc was found to be about 1.5, which is contrary to the above statements. Much higher supersaturations at which voids are generated appear also in irradiated metals [12,13] where, however, the presence of interstitials cannot be neglected because of the competing effects of interstitials and vacancies on the void formation.

In spite of the doubts about the sufficiency of vacancy supersaturation, a number of papers [19,23,24] is in agreement with the suggestion that stable void nuclei are induced by the tensile stress. Moreover, the vacancy hypothesis explains to some degree the beginning of the process of microfracture and the nature of origination of submicroscopic cracks which can initiate the formation of a Griffith crack. Furthermore, it seems that the considerations presented here deliver, beside the paper [10], some contribution to the general theory of non-homogeneous phase transitions. Finally, from this theory, some important quantitative conclusions can be derived. Firstly, it can be shown that the effect of the homogeneous flux J_g on the global nucleation rate J_o is negligibly small and can be disregarded. Thus the global nucleation rate depends almost exclusively on the coupled heterogeneous one. This result supports earlier statements that the vacancy clusters observed are nucleated heterogeneously [18,19,25]. Secondly, it comes out that in the coupled heterogeneous nucleation, the contribution of the cap surface-controlled mechanism (J_{bg}) depends strongly on temperature and metal properties, and for high temperatures can be much greater than it was estimated in [6].

The formation of a microcrack nucleus results here from the application of an elastic strain to the metal and in the first (microscopic) stage is a reversible process. On unloading the vacancy supersaturation decreases the nuclei cease to be stable and disintegrate and the metal returns to the original state. When the unloading process is very rapid, the healing of microcracks may proceed slower than their growth. On the other hand, once the growing nucleus has reached a macroscopic size, its further growth is irreversible; otherwise, no crack could exist in free of load bodies.

ACKNOWLEDGEMENT

The author wishes to express his appreciation to Professor A. Ziabicki for helpful discussions and suggestions which have been incorporated in this paper.

REFERENCES

[1] Krzemiński, J., Arch. Mech. Stos., 21, 3, p. 215, 1969.

[2] Krzemiński, J., Arch. Mech. Stos., 21, 4, p. 429, 1969.

[3] Krzemiński, J., IFTR Reports, 44, 1974.

[4] Krzemiński, J., Arch. Mech., 25, 6, p. 903, 1973.

[5] Krzemiński, J., ZAMM, 56, p. T122, 1976.

[6] Pound, G. M., Simnad, M. T. and Yang, L., J. Chem. Phys., 22, p. 1215, 1954.

[7] Chakraverty, B. K., J. Phys. Chem. Solids, 28, p. 2413, 1967.

[8] Gretz, R. D., Surface Sci., 6, p. 468, 1967.

[9] Sigsbee, R. A., J. Crystal Growth, 13/14, p. 135, 1972.

[10] Ziabicki, A., IFTR Reports, 4, 1977.

[11] Pines, B. J., J. Techn. Phys. (in Russian), Vol. XXV, 8, p. 1399, 1955.

[12] Katz, J. L. and Wiedersich, H., J. Chem. Phys., 55, p. 1414, 1971.

[13] Russell, K. C., Acta Met., 19, p. 753, 1971.

[14] Frenkel, J., Kinetic Theory of Liquids, Dover Publications, Inc., New York, p. 376, 1955.

[15] Weertman, J. and Weertman, J. R., Elementary Dislocation Theory, Macmillan, London, 1967. (Polish translation, p. 109, PWN, Warsaw, 1969).

[16] Hirth, J. P. and Pound, G. M., Condensation and Evaporation, Pergamon Press, Oxford, p. 22, 1963.

[17] Lothe, J. and Pound, G. M., J. Chem. Phys., 36, p. 2080, 1962.

[18] Balluffi, R. W., Acta Met., 2, p. 194, 1954.

[19] Balluffi, R. W. and Seigle, L. L., Acta Met., 3, p. 170, 1955.

[20] Brinkman, J. A., Acta Met., 3, p. 140, 1955.

[21] Machlin, E. S., Trans. AIME, J. of Met., 206, p. 106, 1956.

[22] Resnick, R. and Seigle, L., Trans. AIME, 209, p. 87, 1957.

[23] Greenwood, J. N., Miller, D. R. and Suiter, J. W., Acta Met., 2, p. 250, 1954.

[24] Hull, D. and Rimmer, D. E., Phil. Mag., 4, p. 673, 1959.

[25] Davis, T. L. and Hirth, J. P., J. Appl. Phys., 37, p. 2112, 1966.

SECTION III
SLIP-BANDS AND CRACK INITIATION

A MODEL OF HIGH-CYCLE FATIGUE-CRACK INITIATION AT GRAIN BOUNDARIES BY PERSISTENT SLIP BANDS

H. Mughrabi

Max-Planck-Institut für Metallforschung, Institut für Physik,
7000 Stuttgart 80, Federal Republic of Germany

ABSTRACT

Recent observations on copper polycrystals fatigued in the high-cycle range have shown that crack initiation occurs not only in emerging persistent slip bands (PSBs) but also at grain boundaries (GBs) at sites where PSBs impinge. A dislocation model of the latter type of PSB-GB crack initiation is formulated. The model is based on a recent description of the growth of extrusions at emerging PSBs in single crystals. It is assumed that PSB-GB crack initiation results from the fact that PSBs in constrained grains exert a stress on the GBs because the latter impede the growth of extrusions. From a comparison with experimental data it is concluded that the proposed model provides a satisfactory semi-quantitative description of PSB-GB fatigue crack initiation.

INTRODUCTION

Fatigue crack initiation in wavy-slip single-phase materials is generally believed to occur mainly transgranularly by slip-band cracking in persistent slip bands (PSBs) at low amplitudes (high-cycle fatigue) and intergranularly at high amplitudes (low-cycle fatigue), cf. the review by Laird and Duquette [1]. Recent work on copper polycrystals fatigued at constant plastic strain amplitudes at room temperature has, however, shown that crack initiation at grain boundaries (GBs) occurs fairly commonly also in the high-cycle range and is induced by impinging PSBs [2-5]. In the following a dislocation model of this type of PSB-GB crack initiation will be presented which is an extension of a recent microscopic model of PSBs by Essmann, Goesele and Mughrabi [6] which will be referred to as the EGM-model subsequently.

PERSISTENT SLIP BANDS AND GROWTH OF EXTRUSIONS IN SINGLE CRYSTALS

The PSB-model of EGM refers to the so-called wall or ladder structure of PSBs which is characteristic of fatigued single-phase wavy-slip materials. It is based on a description of steady-state cyclic deformation in PSBs in terms of a dynamic equilibrium between the generation and annihilation of dislocations. In particular, it is shown that the combined effect of dislocation glide and mutual annihilation of close unlike edge dislocations on glide planes less than the annihilation distance y_e apart (in copper, $y_e \approx 1.6$ nm) has two

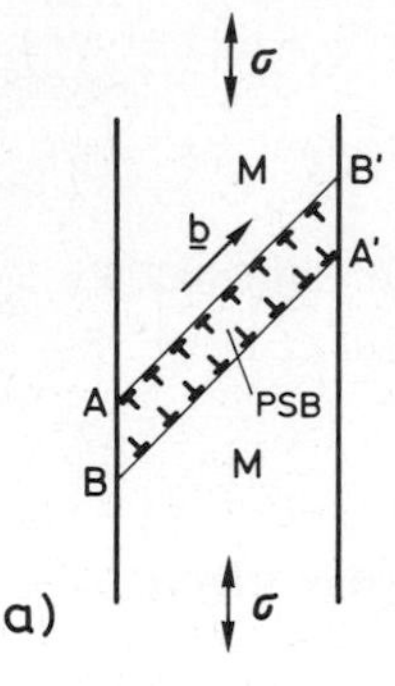

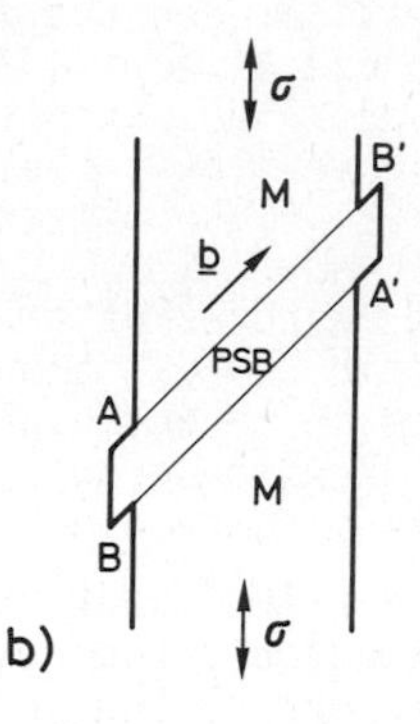

(a) PSB-matrix interface-dislocation multipole-layers

(b) Extrusions formed by emergence of PSB-matrix interface dislocations

Fig. (1) - Formation of extrusions by emergence of PSB-matrix interface dislocations (schematically) σ: axial stress, M: matrix

consequences: (1) Point defects which have predominantly vacancy character accumulate, and (2) Edge dislocations are deposited at the PSB-matrix interfaces as shown schematically in Figure 1(a) for a single crystal. The mean atomic concentration of vacancy-type defects prevailing in the PSBs in cyclic saturation is governed by their rate of formation by edge-dislocation annihilation and by their rate of annealing-out at gliding dislocations and is given by

$$\bar{C}_v^{sat} \approx y_e^2 \, \bar{\rho}_e^{sat} \, b/2y_p , \tag{1}$$

where $\bar{\rho}_e^{sat}$ is the mean density of edge dislocations in the PSBs, b the modulus of the Burgers vector and y_p the capture distance within which a vacancy is swept up by a passing dislocation. With the values $y_e \approx 1.6$ nm and $\bar{\rho}_e^{sat} \approx 5 \times 10^{14}\ m^{-2}$ which are typical for copper and assuming $y_p \approx b$, one obtains $\bar{C}_v^{sat} \approx 6 \times 10^{-4}$. The PSB-matrix interface dislocations superimpose on the high "background density" of edge-dislocation dipoles constituting the dipolar PSB-walls which have been omitted from Figure 1(a) for the sake of clarity. They form an interstitial-type dislocation-multipole layer in such a way that the number of atoms contained in the extra atomic planes is equal to the number of vacancies present in the PSB under the condition that the vacancy-type defects formed cannot migrate from the PSB into the matrix. Hence the number m of interface dislocation-dipoles constituting an interface multipole-layer is simply given by

$$mb = \bar{C}_v^{sat} \, D \approx 6 \times 10^{-4} \, D, \tag{2}$$

where D denotes the diameter of the single crystal measured in the direction of b. Under the combined action of the applied stress and of the repulsive internal stress between the interface dislocations, the latter can emerge stepwise at A and A' during the tensile phases and at B and B' during the compressive phases. Thus extrusions are formed on both sides of the PSB, as indicated in Figure 1(b). The maximum heights e of these extrusions will thus be simply mb/2 (in the direction of b).

The basic features of the EGM-model with regard to the formation of extrusions and the general shape of the surface profiles at emerging PSBs have been confirmed in recent experimental work on fatigued copper single crystals. The rate of extrusion growth has been investigated on crystals ($D \approx 0.6$ cm) fatigued at room temperature at resolved shear strain amplitudes γ_{pl} between 5×10^{-4} and 2×10^{-3} and at a frequency of 5 Hz. These observations [4,5] show that extrusion growth is very rapid ($\approx$ 10 nm/cycle) immediately after the PSBs have been formed and becomes progressively slower subsequently in accord with the EGM-model, without, however, reaching a true saturation [4]. This behavior which reflects that the vacancy-type defect concentration lags systematically behind the attainment of steady-state conditions leads to extrusion heights which are typically, after 10^5 cycles, $\approx$ 4 times larger than the value given by equation (2). The extrusion height e, normalized with respect to the specimen diameter D, as a function of the age of the PSBs, ΔN (in number of cycles), can be represented approximately (for $\Delta N > 1000$) by

$$\frac{e}{D} = \frac{mb}{2D} = A\,(\Delta N)^n \qquad (3)$$

The constants A and n are found to be 3.3×10^{-5} and 0.33, respectively.

PERSISTENT SLIP BANDS AND PSB-GB CRACK INITIATION IN POLYCRYSTALS

As outlined by EGM, similar considerations apply to PSBs in polycrystals. In this case D represents the grain diameter measured in the direction of b. With regard to PSB-GB crack initiation the behavior in surface grains is of particular concern. If b has a large component perpendicular to the surface, then the operation of the PSBs and the formation of extrusions are almost unconstrained. On the other hand, if b lies in the surface, the operation of PSBs will be severely constrained at the GBs and the suppression of extrusion growth can cause PSB-GB crack initiation. In the latter case two relevant processes must be considered which are indicated schematically in Figure 2 for the tensile phase: (1) The homogeneous localized shearing of the PSB lamella (Figure 2(a)), and (2) The piling-up of the PSB-matrix interface dislocations against the GBs (Figure 2(b)). The former effect is indicated in dashed lines in Figure 2(a) for the hypothetical case that the displacements at the GBs be unconstrained. More realistically, the rigidity of the surrounding material will only permit rather local displacements at the GBs, as sketched on the right of Figure 2(a) for the region A'B'. Thus transverse tensile and compressive stresses are provoked at the GB-sites A' and B', respectively. This effect is partly counteracted by the simultaneous piling-up of the interface dislocations at A' (and A). A very crude consideration suggests that the latter effect will become dominant, if $mb > 0.5s$, where s is the displacement due to the homogeneous shearing of the PSB, cf. Figure 2(a). In a PSB of thickness h with a local plastic shear strain amplitude $\gamma_{pl,PSB}$, $s = h \cdot \gamma_{pl,PSB}$.

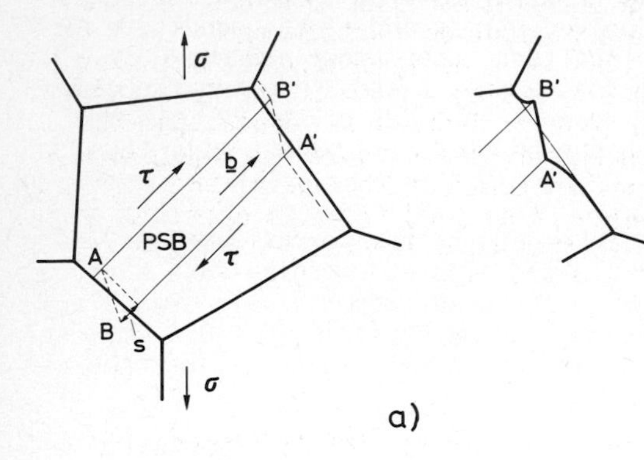

(a) Shearing of PSB lamella

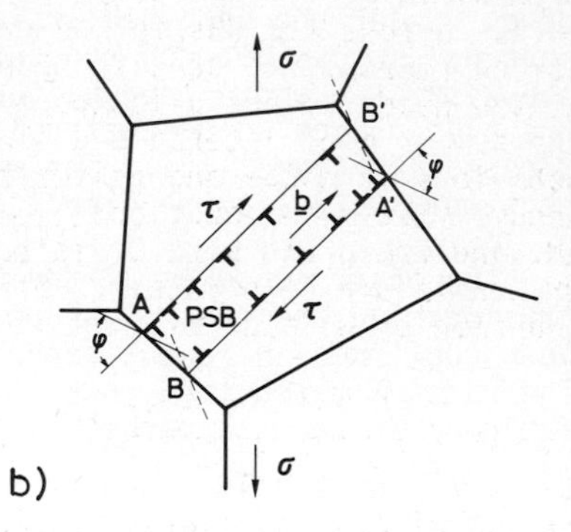

(b) Piling-up of PSB-matrix interface dislocations against GBs.

Fig. (2) - Interaction of PSB with grain boundaries.
τ: shear stress.

With the aid of equation (2), the condition $mb > 0.5s$ can be written as

$$D > 0.5\ h \cdot \gamma_{pl,PSB}/\bar{C}_v^{sat} \quad . \tag{4}$$

Equation (4) indicates that with values typical for copper, i.e. $h \approx 1\ \mu m$ and $\gamma_{pl,PSB} \approx 7.5 \times 10^{-3}$, the dislocation pile-up effect is dominant for $D \gtrsim 6\ \mu m$, i.e. for most grain sizes of practical interest.

According to Stroh [7], cf. also Smith and Barnby [8], the tensile stresses occurring at the head of a dislocation pile-up are largest across a plane making an angle $\phi = 70.5°$ with the pile-up. These planes are indicated by faint lines in Figure 2(b). (The faint dashed lines denote the corresponding planes in the compression phase). Thus, if the GBs lie approximately along these planes of large transverse tensile stress, then Zener-type GB-cracking can occur, if Stroh's crack-initiation criterion

$$m\tau > 3\pi^2\alpha_{eff}/8b \tag{5}$$

is fulfilled. Here τ denotes the acting shear stress and α_{eff} is an effective surface energy.

The shear stress τ acting in PSBs in copper single crystals <u>and</u> in copper polycrystals in the low-amplitude range is $\tau \approx \tau_{PSB} \approx 28$ MPa [3]. This stress is much larger than that necessary to separate the two dislocation groups which constitute the PSB-matrix interface dislocation multipole and which are spaced $h \approx 1\ \mu m$ apart. Most other dislocations in the PSBs are narrow edge-dislocation dipoles with well-screened elastic strain fields. It therefore appears justified to consider the long-range stress effect of the PSB with respect to PSB-GB cracking as due to the independent piling-up of the two interface edge-dislocation groups in opposing directions against the GBs and to apply equation

(5) to these pile-ups. Compared to the large tensile stresses that are required to act across the GBs for GB-cracking, the contribution from the applied stress (during the tensile phase) is small and will be ignored. The stresses at the heads of the pile-ups increase with increasing age ΔN of the PSBs, as the number m of interface dislocations per pile-up increases. Since the PSB-behaviors in mono- and polycrystals are similar in the high-cycle range, we assume that m in equation (5) can be expressed in terms of the age of the PSBs, ΔN, and the grain size D via equation (3) which is derived from the extrusion growth in single crystals. Stroh's crack initiation criterion (equation can then be written as

$$\Delta N > \left(\frac{5.6 \times 10^4 \ \alpha_{eff}}{\tau_{PSB} \cdot D}\right)^{1/n} . \quad (6)$$

In the high-cycle fatigue range, ΔN, as given by equation (6) is expected to represent a major portion of the fatigue life-time N_f. Equation (6) predicts that ΔN is smaller for large than for small grain sizes. Hereby it is of course assumed that the PSBs extend across the entire grains which may not be true for very large grain sizes. In this latter case D should be taken as the effective extension of the PSBs in the direction of $\underline{b}$. We wish to point out that equation (6) presumably underestimates ΔN by approximately a factor of 2, since it has been observed that the action of PSBs in polycrystals (with $D \approx 25$ μm) is constrained, rendering the value of the local plastic shear strain amplitude in PSBs, $\gamma_{pl,PSB}$, only about half as large as in single crystals [4]. This effect will be ignored in the following.

For "brittle-type" PSB-GB cracking, equations (5) and (6) describe both the condition for crack initiation and propagation, cf. [7] and [8]. In this case we write for the effective surface energy

$$\alpha_{eff} = \alpha_s - \frac{1}{2}\alpha_{gb} , \quad (7)$$

where α_s and α_{gb} denote the surface energy and the GB-energy, respectively. With the values $\alpha_s \approx 1.65$ J/m^2 and $\alpha_{gb} \approx 0.32\ \alpha_s \approx 0.53$ J/m^2 for copper, cf. [9], $\alpha_{eff} \approx 1.4$ J/m^2. In the case of some plastic deformation at the PSB-GB crack tip, equation (6) describes merely the condition for PSB-GB crack nulea-tion and additional processes such as, e.g., environmental interaction are required in order to allow the PSB-GB cracks to propagate. For this more general situation we write [5]

$$\alpha_{eff} = \alpha_s - \frac{1}{2}\alpha_{gb} + \alpha_p - \alpha_{ads} , \quad (8)$$

where a specific plastic-deformation energy α_p and a possible lowering of α_{eff} by α_{ads}, due to gas adsorption or chemisorption [10], are taken into account.

COMPARISON WITH EXPERIMENTAL OBSERVATIONS

In the range of plastic strain amplitudes just above the strain fatigue limit the PSB-behavior can be described using the data for PSBs in single crystals, cf. [3-5], i.e., τ_{PSB} = 28 MPa. The high-cycle fatigue life-time data of Wang [4] on copper polycrystals with D ≈ 25 μm (in air and at room temperature) can be described by the Coffin-Manson relation

$$\Delta\varepsilon_{pl} \cdot N_f^{0.62} \approx 0.94 \quad , \tag{9}$$

where $\Delta\varepsilon_{pl}$ is the axial plastic strain range. In the immediate vicinity of the fatigue limit, i.e., at a value $\Delta\varepsilon_{pl} \approx 5 \times 10^{-5}$ above which PSB-formation sets in [3,4], equation (6) yields $N_f \approx 7.8 \times 10^6$. From equation (6) we obtain $\Delta N \approx N_f$, if we use a value $\alpha_{eff} \approx 2.5\ J/m^2$ which is only larger by a factor of ≈ 2 than the value obtained from equation (7). Considering also equation (8), this result suggests that PSB-GB crack initiation in copper in air is possibly a quasi-brittle process with little plastic blunting and is enhanced by environmental interaction (oxidation). Some support for this possibility can be derived from the work of Coffin [11] who showed for a number of materials that in vacuum fatigue-cracking occurs largely only transgranularly. In addition the detailed observations of PSB-GB cracks indicate that crack initiation and propagation may be additionally facilitated by the co-operative action of groups of PSBs, by the gradual widening of PSBs and possibly also by deformation-induced vacancy migration and coalescence at GBs and by GB-sliding [5].

At plastic strain ranges higher than $\Delta\varepsilon_{pl} \approx 5 \times 10^{-5}$ the volume fraction occupied in individual grains by PSBs has been observed to increase strongly and at $\Delta\varepsilon_{pl} = 2 \times 10^{-3}$ the PSBs occupy about 80% of the grain volumes [3,4]. For this reason we attribute the decrease in fatigue life-times above $\Delta\varepsilon_{pl} \approx 5 \times 10^{-5}$ and up to $\Delta\varepsilon_{pl} \approx 2 \times 10^{-3}$ to the increasing co-operative action of larger numbers of PSBs both in individual grains and in adjacent grains at the GBs.

In the above considerations the competing process of fatigue crack initiation in emerging PSBs (slip-band cracking) has not been taken into account for the sake of simplicity. Moreover, no quantitative model of slip-band cracking exists to-date. Recent investigations indicate that slip-band cracking may be less significant in polycrystals than in single crystals because of the much smaller extrusion heights [5], cf. equation (2). Hence only surface roughening by random slip at emerging PSBs remains as the major effective mechanism of slip-band cracking [4,5] in polycrystals. Observations of fracture surfaces of copper polycrystals fatigued at low $\Delta\varepsilon_{pl}$ reveal a mixed-mode fracture and suggest indeed that intergranular cracking is as important as slip-band cracking [4,5].

CONCLUSIONS

A model of PSB-GB fatigue crack initiation in wavy-slip materials fatigued in the high-cycle range has been presented and has been shown to be in fair agreement with recent experimental observations. The model is based directly on detailed evaluations of the observed behavior of PSBs and of extrusion growth in fatigued copper single crystals. In this respect the model differs considerably from a recent related model of PSB-GB fatigue crack initiation by Tanaka and Mura [12]. The latter model is based entirely on micromechanical considerations which have been worked out in considerable detail and the reader is referred to the original paper for a comparison with the present model. Finally, it should be pointed out that, with an appropriate definition of D, both models appear to be equally applicable also to PSB-induced fatigue-cracking at interfaces other than GBs, such as, for example, at twin boundaries or at inclusion-matrix interfaces.

ACKNOWLEDGMENT

The author is very grateful to Drs. K. Differt, U. Essmann. U. Goesele and R. Wang for many fruitful discussions.

REFERENCES

[1] Laird, C. and Duquette, D. J., "Mechanisms of fatigue crack nucleation", in Proceedings of International Conference on Corrosion Fatigue, Storrs, Conn., 1971, National Association of Corrosion Engineers, Houston, Texas, NACE-2, p. 88, 1972.

[2] Figueroa, J. and Laird, C., personal communication 1979.

[3] Mughrabi, H. and Wang, R., "Cyclic strain localization and fatigue crack initiation in persistent slip bands in face-centered cubic metals and single-phase alloys", in Proceedings of the International Symposium "Defects and Fracture", Tuczno, Poland, Oct. 1980, G. C. Sih and H. Zorski, Eds., Martinus Nijhoff Publishers, The Hague, Boston, London, p. 15, 1982.

[4] Wang, R., "Untersuchungen der mikroskopischen Vorgänge bei der Wechselverformung von Kupfereinkristallen und -vielkristallen", Doctorate Thesis, Stuttgart University, 1982.

[5] Mughrabi, H., Wang, R., Differt, K. and Essmann, U., "Fatigue crack-initiation by cyclic slip irreversibilities", presented at the International Conference on Quantitative Measurement of Fatigue Damage, May 10-11, 1982, Dearborn, Michigan, to appear in ASTM STP.

[6] Essmann, U., Goesele, U. and Mughrabi, H., "A model of extrusions and intrusions in fatigued metals, Part I:", Phil. Mag. A, 44, 405, 1981.

[7] Stroh, A. N., "A theory of the fracture of metals", Adv. in Physics, 6, 418, 1957.

[8] Smith, E. and Barnby, J. T., "Crack nucleation in crystalline solids", Met. Sci. J., 1, 56, 1967.

[9] Tyson, W. R. and Miller, W. A., "Surface free energies of solid metals: estimation from liquid surface tension measurements", Surface Science, 62, 267, 1977.

[10] Duquette, D. J. and Gell, M., "The effect of environment on the mechanism of stage I fatigue fracture", Met. Trans. 1, 1325, 1971.

[11] Coffin, L. F., "Fatigue at high temperature - prediction and interpretation", Proc. Instn. Mech. Engrs., 188, 109, 1974.

[12] Tanaka, K. and Mura, T., "A dislocation model for fatigue crack initiation", J. of Appl. Mech., 103, 97, 1981.

SHAPE AND STRUCTURE OF PERSISTENT SLIP BANDS IN IRON CARBON ALLOYS

K. Pohl

Schweißtechnische Lehr- und Versuchsanstalt Mannheim GmbH

P. Mayr

Stifung Institut für Härterei-Technik Bremen

and

E. Macherauch

Institut für Werkstoffkunde I, Universität Karlsruhe

INTRODUCTION

Already in 1903 Ewing and Humfrey [1] reported, that fracture of their fatigued Swedish iron samples started preferentially at slip bands, which developed at the specimen surface during fatigue loading. In spite of these early observations only some years ago, major advance has been made in understanding the most important microstructural details of these persistent slip bands (PSB) in fcc single and polycrystals with high stacking fault energies [2-10].

Much more restricted and partly controversial information is available about the occurrence of PSB's in materials with bcc structure. So, in an TEM study on dislocation arrangements in carburized α-iron it is pointed out, that PSB's develop as straight, relatively dislocation free channels [7,11]. In an other investigation on iron polycrystals it is argued, that in the very surface grains PSB's with a ladder-like structure as in fcc metals should develop. More recently, this suggestion has been confirmed by the results of an investigation on a structural steel with 0,1 wt % C which shows that at room temperature the ladder-like dislocation structure is not restricted to the surface ferrite grains only, but can also develop in the bulk as an intermediate dislocation arrangement [12].

In the present paper it will be shown, that in iron carbon alloys depending on the carbon content and the test temperature three different types of slip bands can develop at the surface of fatigue loaded samples.

A. Experimental Details and Materials

The stress controlled single step tests have been carried out in a Schenck type servohydraulic testing equipment at zero mean stress with a 5Hz triangular wave function. An integrated controlled atmosphere (Argon) furnace enabled experiments at elevated temperatures.

During the tests the plastic strain amplitude $\varepsilon_{a,p}$ (Figure 1) has been measured using an on line data acquisition system. The plots $\varepsilon_{a,p}$ vs number

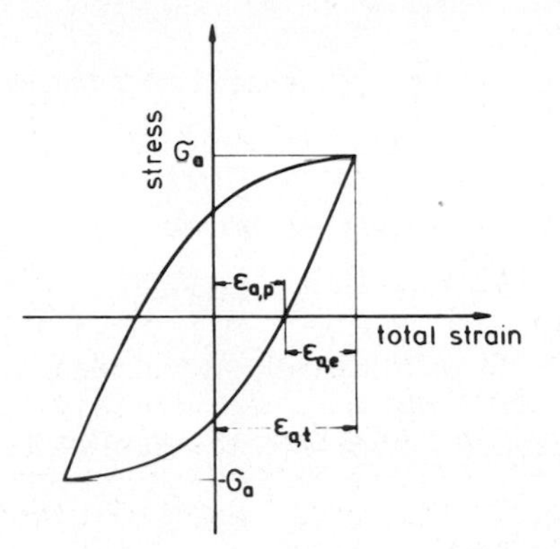

Fig. (1) - Stress-strain hysteresis loop

of cycles N for the different imposed stress amplitudes reveal the cyclic deformation curves by which the changes in the mechanical behavior of the material during fatigue loading can be characterized. To get some information about the microstructural processes during the lifetime of the specimens, extensive microscopical studies have been performed using light-, scanning- and transmission microscopy. The preparation of the transmission electron microscopy foils has been done in a commercial electrolytic double jet apparatus. The TEM used, was a Siemens Elmiskop 102, the SEM a Cambridge Mark 2.

In the investigation a high purity iron (Vacufer, Vakuumschmelze Hanau) and a commercial structural steel with 0,1 wt % carbon (German Grade Ck 10) have been used, with the chemical composition as follows:

α-Fe (Al 0,0005; As 0,0002; C 0,005; Co 0,0002; Cr 0,01; Cu 0,0002; Fe 99,96; Mn 0,0001; Mo 0,01; Ni 0,01; P 0,0005; Pb 0,0001; S 0,0015; Si 0,0001; Zn 0,0002)

Ck 10 (C 0,12; Cr 0,07; Cu 0,05; Fe 93,88; Mn 0,54; Mo 0,01; Ni 0,05; P 0,028; S 0,026; Si 0,28).

Shape and dimensions of the specimen used are shown in Figure 2.

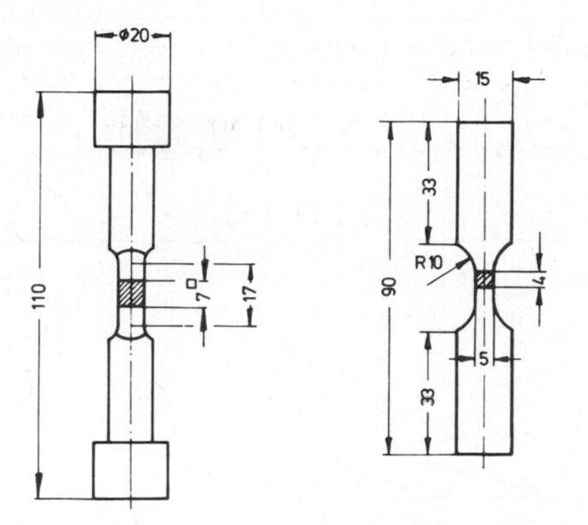

Fig. (2) - Specimen

After shape cutting, all Ck 10 specimens have been normalized 30 minutes at 1250 K in an argon atmosphere and were subsequently cooled down in the furnace to room temperature. The α-Fe samples were annealed at 873 K in an hydrogen atmosphere for 45 minutes and slowly furnace-cooled to room temperature. Two opposite surfaces in the gauge length of each specimen have been carefully electropolished to make microscopical studies possible.

Some relevant mechanical properties of the heat treated materials are listed in Table 1. The tensile test data are related to a strain rate $\dot{\varepsilon} = 2 \cdot 10^{-4}\ s^{-1}$.

TABLE 1 - MECHANICAL PROPERTIES

	Grain size µm	Yield strength (N/mm²) R_{eH}	R_{eL}	Tensile strength R_m (N/mm²)	Total elongation %	Reduction of Area %
α-Fe	54	163	130	206	107	2,8
Ck 10	30	260	235	377	28	2

B. Results

1. Mechanical Measurements

The cyclic deformation curves (CDC) of a material can be successfully used to recognize changes in the operative microstructural processes during fatigue loading. As can be seen from the room temperature CDC's of the pure iron samples compiled in Figure 3, a cyclic softening process is active at the beginning of the test if stress amplitudes lower than the yield stress R_{eL} are imposed. The amount and length of the cyclic softening is strongly dependent on the stress amplitude, as well as the subsequent observed cyclic

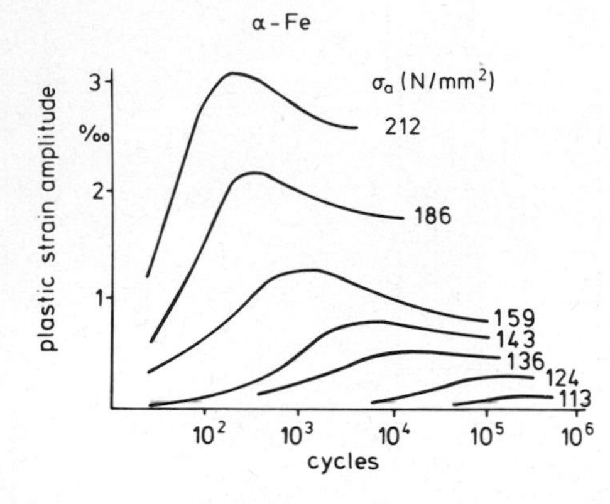

Fig. (3) - Cyclic deformation curves of α-iron at room temperature

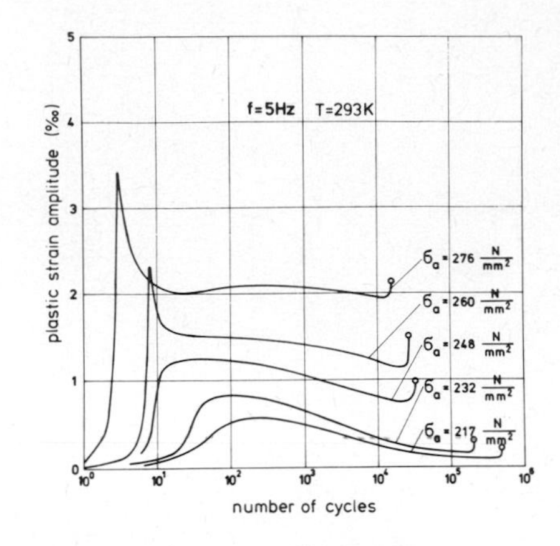

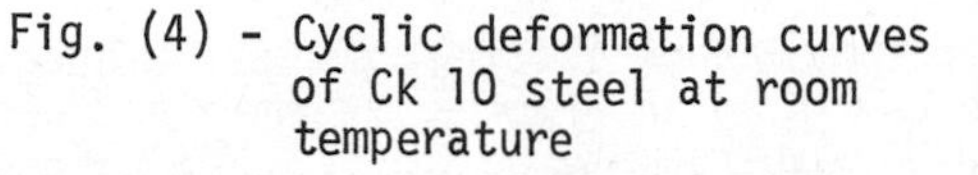

Fig. (4) - Cyclic deformation curves of Ck 10 steel at room temperature

hardening. A quite similar behavior show the CDC's of Ck 10 at room temperature (Figure 4). The obvious differences to the α-iron are the pronounced softening peaks for $\sigma_a > 232$ N/mm² which can be followed partly by a weak secondary softening period.

Completely different CDC's result if elevated test temperatures are used, as can be seen from the data in Figure 5 representing the material response at the temperature T = 420 K.

Only the smallest amplitude used ($\sigma_a = 217$ N/mm²) shows within the first 10 cycles a softening behavior followed by an exceptional strong cyclic hardening yielding plastic strain amplitudes in an order of magnitude close to the resolution of the strain measurement system. An increase in stress amplitude results in cyclic hardening from the beginning which at a sufficient number of cycles changes into a pronounced cyclic softening. At the highest amplitude used, a secondary hardening period is observed, therefore it is obvious, that the preceding softening is not associated with propagating fatigue cracks.

As a possible basis to compare the different materials under the various test conditions identical numbers of cycles have been chosen. If a mean lifetime $N_f \sim 5 \cdot 10^4$ is taken, the CDC's compiled in Figure 6 have to be considered, which are associated with the following microstructures.

2. Microscopical Observations

Using light microscopy on the surface of pure iron samples with increasing number of cycles, a very characteristic wavy topography is observed.

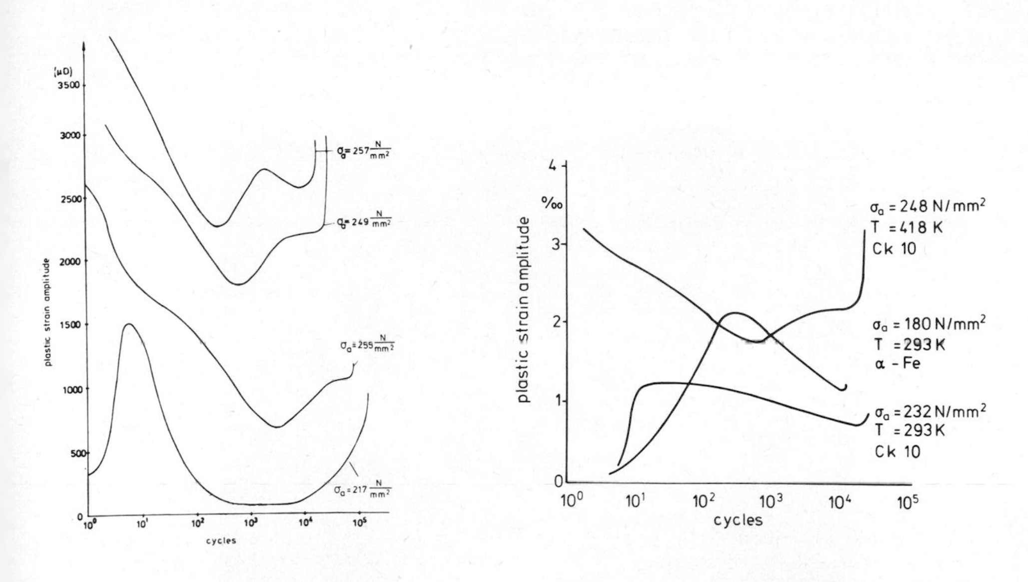

Fig. (5) - Cyclic deformation curves of Ck 10 steel at 420 K

Fig. (6) - Cyclic deformation curves for α-iron and Ck 10 samples at different test conditions yielding approximately identical cycles to fracture

As a typical example Figure 7 shows a micrograph taken at $N = 1.4 \cdot 10^4$. At higher resolution it can be seen, that the wavy profile contains at first perpendicular oriented parallel markings which develop later on into the pronounced

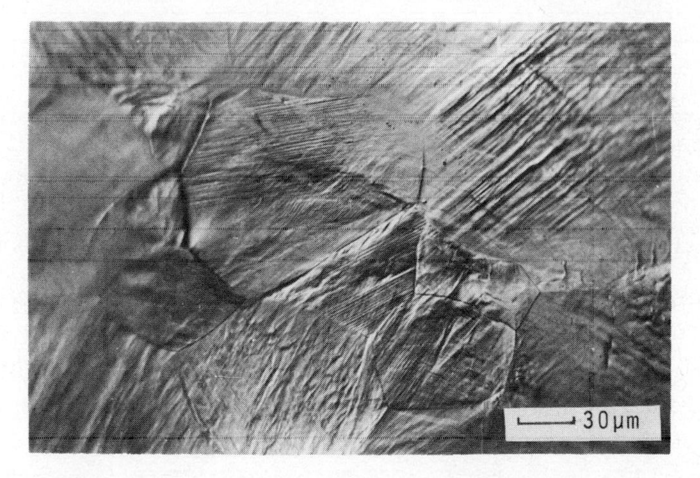

Fig. (7) - Surface topography of α-iron at $N = 1.4 \cdot 10^4$

lamellar structure shown in Figure 8. The evaluation of several micrographs taken from samples which have been fatigued with different stress amplitudes, revealed nearly identical width of the lamellae of about ~ 0,5 μm.

Fig. (8) - Lamellar PSB;s at the surface of α-iron

Carefully prepared TEM specimen taken below the lamellae near the surface show a typical cell structure with cell diameters in the order of 10 μm, as can be seen from Figure 9 representing the microstructure at $N = 3 \cdot 10^3$ and the stress amplitude σ_a = 180 N/mm^2.

Pronounced differences in the surface topography of pure iron and structural steel samples can be identified. Relatively early, i.e. $N \sim 0{,}01\ N_f$, in favorable oriented surface ferrite grains of Ck 10 slip bands are developed which are interspaced by practically undeformed material. Already at the magnification of a light microscope (Figure 10) it can be recognized, that the slip bands are associated with a characteristic extrusion-intrusion topography, which is more clearly shown in the SEM-micrograph in Figure 11. Detailed studies revealed, that the overall slip band density in the ferrite surface grains increases with the number of cycles. But due to the fact, that slip band formation starts at different times in each grain a very inhomogenous slip band density in a specimen is the consequence.

In the ferrite of the bulk material at $N \sim N_f$ the dislocations are predominantly arranged in a cell structure (Figure 12). The mean cell diameter at the stress amplitude σ_a = 248 N/mm^2 is d ~ 1 μm, which is ten times smaller than the cell size in the pure iron specimen. This is remarkable in the sense, that the CDC of Ck 10 exhibits over the relevant period of

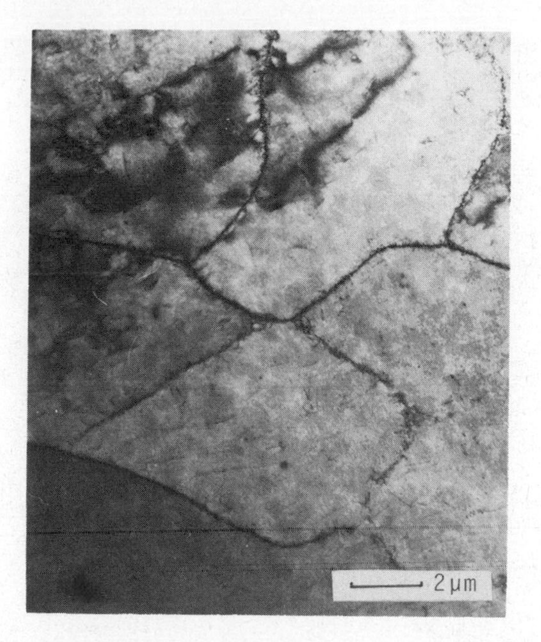

Fig. (9) - Cell structure in the bulk of α-iron samples at $N - 3 \cdot 10^3$

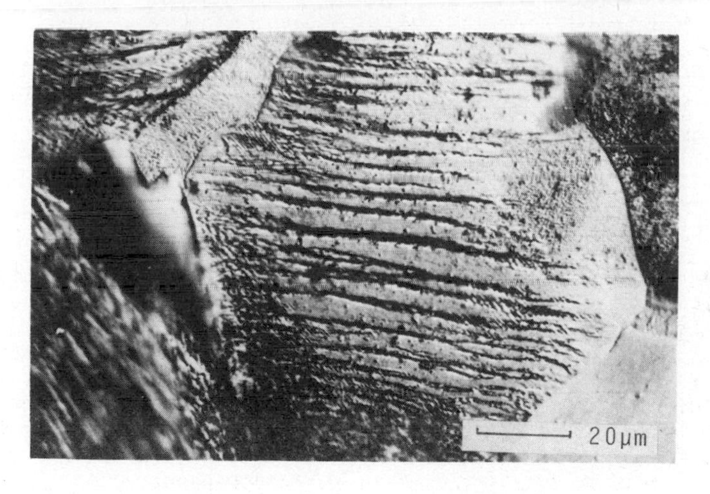

Fig. (10) - PSB's in a surface ferrite grain of Ck 10

specimen life significantly smaller plastic strain amplitudes (Figure 6).

TEM studies in the perior 0, 1 $N_f \leq N \leq N_f$ reveal in some part of the bulk ferrite grains ladder-like dislocation arrangements which are typical for PSB's

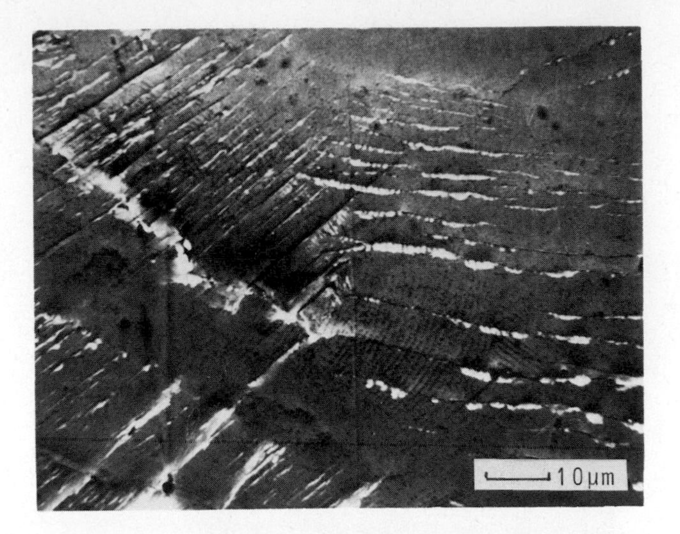

Fig. (11) - Extrusions and intrusions in PSB's of Ck 10 at room temperature

in surface grains. A characteristic example is shown in Figure 13 which has been taken at N - 10^4 cycles, close to the center of the specimen.

Fig. (12) - Cell structure in bulk ferrite grains

Ck 10 specimens from elevated temperature tests exhibit a surface structure which is quite different from the observations at room temperature. For example in Figure 14 it can clearly be seen, that at 418 K in the surface ferrite grains very narrow slip bands in a high density develop. At higher

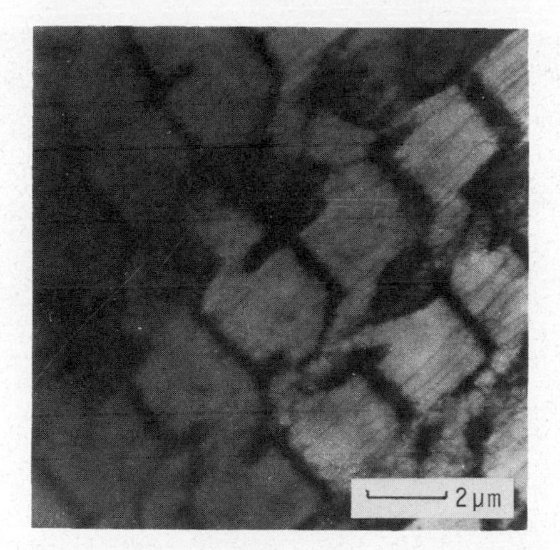

Fig. (13) - PSB's in bulk ferrite grains of Ck 10

magnification additional differences to the room temperature observations are discernible. So the SEM micrograph in Figure 15 reveals much more pronounced extrusions and intrusions, which are already present at $N = 5 \cdot 10^3$, that is 10% of the total life. Also the dislocation arrangement in the interior differs markedly from the room temperature observations. The prominent feature of the elevated temperature bulk dislocation structure are straight, narrow channels (Figure 16) with a low dislocation density in the interior and interfaces to the matrix with significantly increased dislocation density. Furthermore channels with different directions are observed in mostly all grains which can intersect each other. The ferrite grains which exhibit this typical channel structure increase with the number of cycles during the test.

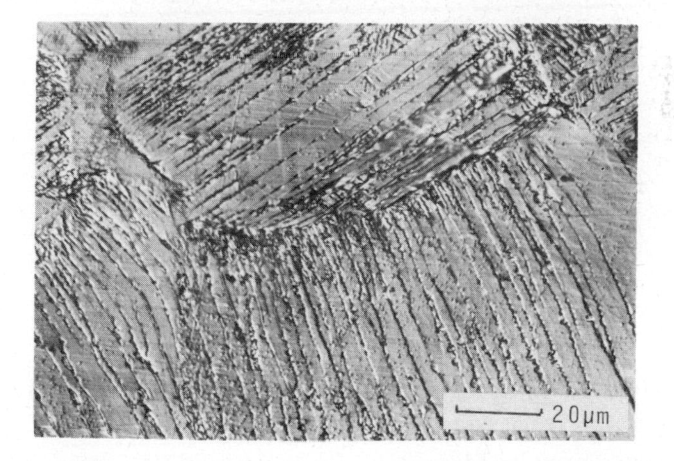

Fig. (14) - Surface topography of Ck 10 fatigued at 418 K

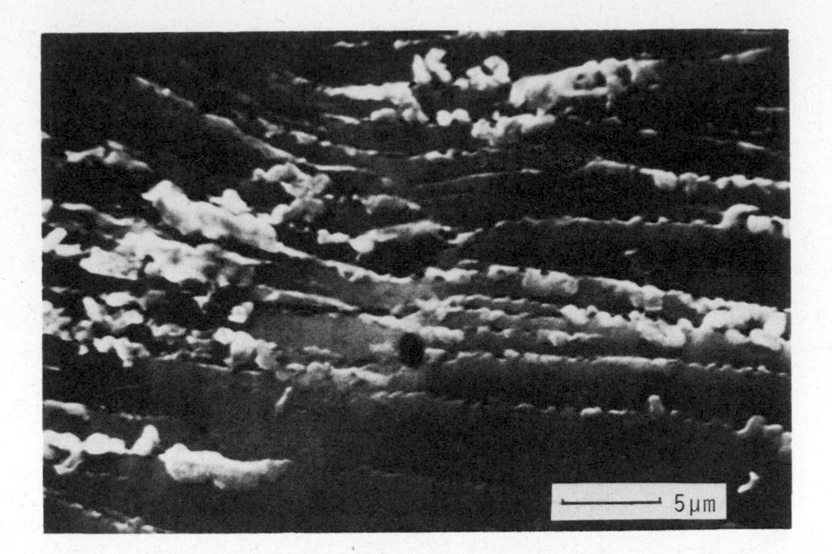

Fig. (15) - Extrusions and intrusions in PSB's at 418 K

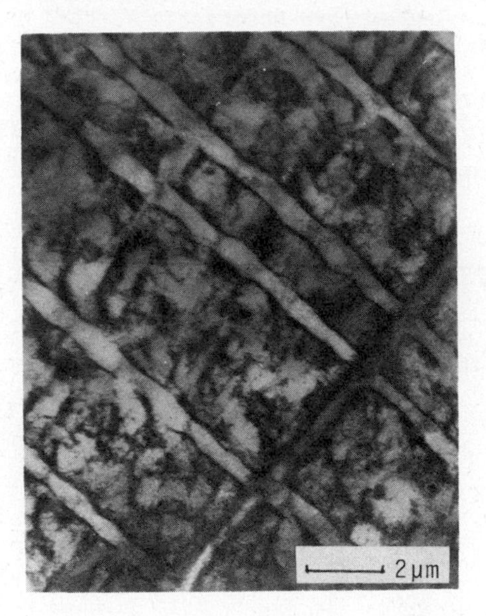

Fig. (16) - Channel structure in the interior of Ck 10
fatigued at 418 K

DISCUSSION

The pronounced wavy shaped roughening of the whole specimen surface which has been observed after a sufficient number of cycles on the investigated pure iron and Ck 10 specimen at room temperature seems to be a general feature of carbon steels with a mainly ferrite matrix. This can be concluded with respect to additional results on hypoeutectoid steels [13]. The basic mechanism of this deformation pattern is assumed to be associated with the asymmetric slip of screw dislocations in the bcc ferrite [14].

Only in the pure iron at medium and low amplitudes the above mentioned long range wavy surface roughening is superimposed by a well developed lamellar structure. It is interesting, that the form of the lamellae is most clearly defined near the fatigue limit and gets indistinct at high amplitudes. Therefore at low amplitudes the fatigue cracks initiate predominantly as a consequence of the associated local stress fields around the sharp notch. Now the question is, whether the lamellae are the result of slip processes in the surface grains analogous to tensile deformation (sometimes referred in the literature as T-bands) or did they show the typical subsurface dislocation arrangements as PSB's. All TEM micrographs taken during this investigation from the bulk as well as near surface regions gave no information about the existence of PSB i.e. a ladder-like dislocation arrangement in accordance with theoretical considerations [7], which postulate that in pure bcc materials (via. mainly in absence of interstitial carbon) PSB's cannot be formed because of the differences in screw and edge dislocation mobility. Not in line with this consideration are the observed pronounced PSB's in the surface ferrite grains of the structural steel. Due to the very low solubility of carbon at room temperature which is estimated to be roughly 1 ppm one would expect, that the behavior of the ferrite in carbon steel coincides with the pure iron investigated. Therefore the conclusion has to be drawn, that other interstitials as for instance nitrogen or substitional atoms are responsible for the observed differences. The ladder-like PSB structure in the bulk grains of Ck 10 represents an intermediate structure. The analysis of a large number of TEM micrographs shows, that at first in individual grains due to plastic deformation a tangle like dislocation arrangement is formed. This structure is followed later on by multipole veins. Further cycling will result in an increasing diameter of the veins up to the point at which local instabilities in the veins occur and the ladder-like structure of the "interior" PSB's develops. Continuing the fatigue loading the whole grain fills up with PSB's. Until now the deformation process could be accomplished only by single slip whereas the subsequent transition to the cell structure requires necessarily multiple slip processes. It must be pointed out, that the foregoing substructure development will occur only in such grains which are subjected to a sufficiently high shear stress amplitude. In a material with randomly oriented grains, the PSB structure therefore occurs only in a certain part of the bulk ferrite grains and also at different times. Therefore the strain softening which is associated with the PSB development is not reflected to the CDC.

The occurrence of the elevated temperature PSB's on the other hand is accompanied with significant changes in the CDC. This results from their quite different structure and origin compared with the room temperature PSB's. The formation of elevated temperature PSB's presupposes dynamic strain aging processes. It is suggested, that the strong cyclic work hardening which is associated with dynamic strain aging results from the high density of randomly distributed dislocations. Therefore the material cannot meet the deformation requirements by homogeneous slip. Obviously, local concentration of slip occurs and leads to the formation of very narrow channels with a low dislocation density in which glide dislocations can move easily over long distances with velocities high enough, that interaction with moving interstitials are suppressed. A further characteristic difference to room temperature PSB's is their formation in different slip systems so that frequently intersections of channel-PSB's are observed.

Summarizing the results, one has to proceed, that three different types of PSB's can form in the ferrite of bcc iron. They are quite different in shape and dislocation structure as is indicated in the schematic Figure 17, but are responsible in the same way for the formation of pronounced surface slip markings from which predominately fatigue cracks initiate.

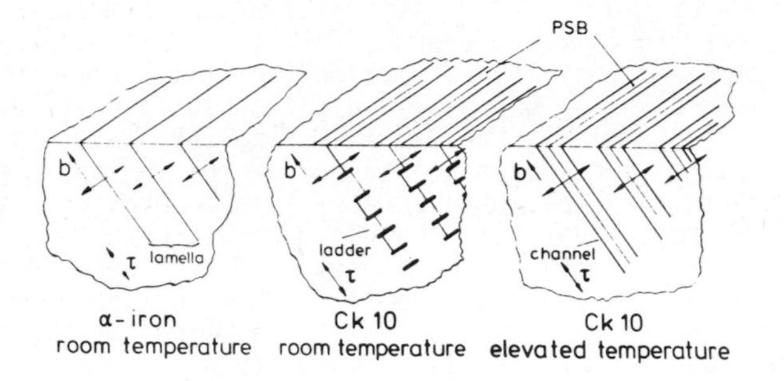

Fig. (17) - PSB development in different materials at different test conditions

REFERENCES

[1] Ewing, J. A.,and Humfrey, J. C. W., "The Fracture of Metals Under Repeated Alternations of Stress", Phil. Trans. A 200, p. 241, 1903.

[2] Woods, P. J., "Low Amplitude of Copper and Copper- 5 at. % Aluminum Single Crystals", Phil. Mag., 28, pp. 155-191, 1973.

[3] Basinski, S. J., Basinski, Z. S., and Howie, A., "Early Stages of Fatigue in Copper Single Crystals", Phil. Mag., 19, No. 161, pp. 899-924, 1969.

[4] Lukas, P., Klesnil, M. and Krejci, J., "Dislocations and Persistent Slip Bands in Copper Single Crystals Fatigued at Low Stress Amplitude", Phys. Stat. Sol., 27, pp. 545-558, 1968.

[5] Antonopoulos, G. J. and Winter, A. T., "Weak Beam Study of Dislocation Structures", Phil. Mag., 33, pp. 87-95, 1976.

[6] Mughrabi, H., "Warum ermüden Metalle?", Umschau 77, Heft 3, pp. 80-81, 1977.

[7] Mughrabi, H., Ackermann, F. and Herz, K., "Persistent Slip Bands in Fatigued Face Centered and Body Centered Cubic Metals", ASTM-STP 675, pp. 69-105, 1979.

[8] Laufer, E. E., and Roberts, W. N., "Dislocation Structures in Fatigued Copper Single Crystals", Phil. Mag., 10, No. 107, pp. 883-885, 1964.

[9] Laufer, E. E., and Roberts, W. N., "Dislocation and Persistent Slip Bands in Fatigued Copper", Phil. Mag., Vol. 14, No. 127, pp. 56-77, 1966.

[10] Mecke, K., "TEM-Investigations of the Cyclic Stress-Strain Behaviour and the Formation of Persistent Slip Bands in Fatigued Single Crystals of Nickel Using Changing Amplitude Tests", Phys. Stat. Sol. (a) 25, K, pp. 93-98, 1974.

[11] Mughrabi, H., Herz, K. and Stark, X., "Cyclic Deformation and Fatigue Behaviour of α-iron Mono- and Polycrystals", Int. Journ. Fracture 17, pp. 193-220, 1981.

[12] Pohl, K., Mayr, P., and Macherauch, E., "Persistent Slip Bands in the Interior of a Fatigued Low Carbon Steel", Scripta Met. 14, pp. 1167-1169, 1980.

[13] Mayr, P., "Habilitationsschrift Universität Karlsruhe", 1978.

[14] Hughrabi, H., Proc. of NATO Advanced Study Institute on Surface Effects in Crystal Plasticity, Noordhoff, Leyden, pp. 479-485, 1977.

ESTIMATION OF CRACK INITIATION IN PLAIN CARBON STEELS BY THERMOMETRIC METHODS

H. Harig and M. Weber

Universität Essen
Federal Republic of Germany

ABSTRACT

Temperature changes during deformation of metals mainly occur due to the thermoelastic effect and to moving dislocations. It is pointed out that a cyclic yield stress can be determined by thermometric methods corresponding to formation of slipbands. Estimations of endurances seem to be possible.

INTRODUCTION

It is known that for constant stress amplitudes greater than the fatigue limit, there are four consecutive stages in a material's behavior during cyclic stressing. In the first stage, materials will change their mechanical properties due to microstructural processes, no cracks will occur by stressing. During the second stage, microcracks form near the surface of a specimen. At least one of these microcracks extends to a macrocrack and propagates through the specimen during stage three. Stable crack propagation will lead to unstable crack growth ıd at last to complete fracture of the test piece in the fourth stage [1].

For the characterization of material's behavior and for the calculation of endurance limits especially of structural parts, it will be necessary to separate the above-mentioned stages.

This presentation will mainly be concerned with estimation of crack initiation by determination of cyclic yield stress. Measurements of load and deformation and, furthermore, microscopic investigations were done besides temperature measurements. Changes of temperature appear due to energy transformation during cycling. Some advantages of thermometric methods are high sensitivity, local or areal measurement, simplicity of measuring device and feasibility of non-contact gauging.

ENERGY CHANGE DURING DEFORMATION

During elastic deformation of polycrystalline metals, changes in volume occur so that analogous to the gas laws temperature changes ΔT arise due to changes in stress $\Delta\sigma$:

$$\Delta T = - \frac{T_o \cdot \alpha_1}{c \cdot \rho} \cdot \Delta\sigma$$

where: T_o = start temperature (K)

α_1 = coefficient of linear thermal expansion

c = specific heat

ρ = density

In the case of nonadiabatic condition and due to inelastic mechanisms, there will appear some dissipation of energy. The resulting amount of mean temperature rise can be assumed to be small and constant all over the stressed volume.

During plastic deformation, moving dislocations increase atomic oscillations regardless of sign of external stress. Thus, most of the work done during deformation W is changed into heat while only a little part increases internal energy. Rise of temperature in a definite test volume depends on produced heat Q, mass, specific heat and heat transfer in the contact surface. It should be stated that energy dissipation during plastic deformation may be localized microscopically due to employed load and microstructural conditions.

Monotonic deformations lead to temperature changes according to Figure 1. For plain carbon steel (0.4%C, room-temperature, adiabatic conditions), we get

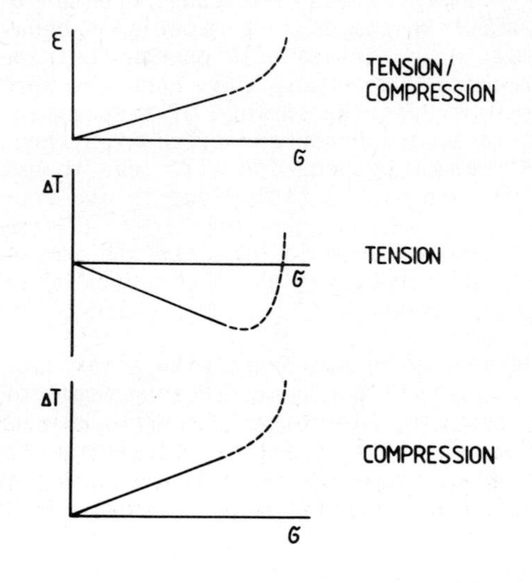

Fig. (1) - Monotonic stress strain curve and changes of temperature (schematically)

a constant of proportionality in the elastic region of about:

$$\Delta T/\Delta\sigma = - 0.00091 \quad K \cdot mm^2 \cdot N^{-1}$$

Energy dissipated during cyclic deformation may be estimated by determination of the area enclosed by stress strain hysteresis loop. Exact determinations can be realized only in a stabilized state of material, i.e., when the loops remain closed. In this case, it can be written:

area of hysteresis A = work done during deformation W = produced heat Q

On the assumption that the amount of heat transferred is constant, changes of mean temperature ΔT in a stressed volume may be accepted proportional to changes of area of hysteresis loop until macro crack propagation starts and deformation becomes strongly heterogeneous. It is well-known that highly localized deformation processes for instance resulting in the formation of some few slipbands only within particular grains can be determined normally neither by deformation measurements nor by temperature measurements. We assume that these processes are distributed statistically over the stressed volume. Finally, it should be noted that temperatures can be related exactly to plastic strain amplitudes only with regard to constant stress amplitudes.

THERMOMETRIC METHODS

Fundamentally, all common temperature measuring devices like thermo-couples, resistance thermometers, infrared thermometers, infrared cameras or liquid crystals may be used. Their selection should be done due to the thermal, timely and geometric resolution intended and due to the accessibility of the measuring point.

Special measuring devices like flow calorimeter may be necessary under certain test conditions for instance when using a cooling system in order to minimize heating of the specimen. Best experience was made with thermo-couples iron-constantan. The two separated thermo wires were only pressed on the surface of the specimen in order to minimize influence on material's properties. Infrared thermometers were used in rotating bending tests.

CHANGES OF TEMPERATURE DURING FATIGUE TESTS

As mentioned above, cyclic deformations will lead to cyclic temperature changes which are proportional to stress amplitudes if strain is macroscopic elastic, if the strain rate is not too high, and if the test condition is nearly adiabatic. Deviations from this proportionality will indicate the beginning of plasticity.

Common fatigue tests, however, are carried out at frequencies, which do not allow to measure changes in temperature due to elastic deformations. Thus, the temperature of the specimen will remain nearly constant until plastic deformation occurs. Figure 2 shows typically cyclic temperature curves at different stress amplitudes. These curves are similar to the well-known cyclic strain curves; see [1].

In this diagram, the rise of temperature indicates cyclic softening, while the decrease of temperature results from cyclic hardening. The sensitivity of the thermometric method used is better than that of deformation measurement at the frequency of 37 Hz. Specimens which reach $5 \cdot 10^6$ cycles show temperature changes of about 10 K, while the corresponding plastic strain is in the magnitude of 10^{-4}.

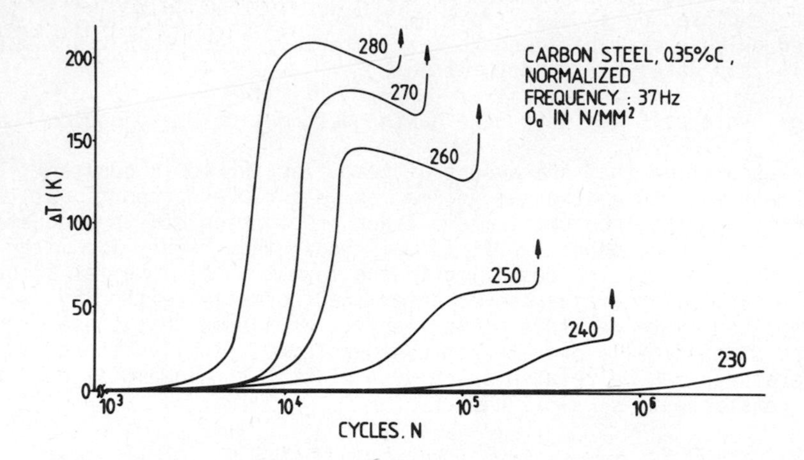

Fig. (2) - Cyclic temperature curves, constant amplitude tests, tension-compression

Figure 3 points out changes of temperature resulting from special mechanisms during cyclic deformation. Stress amplitude σ_{a1} will lead to only very small

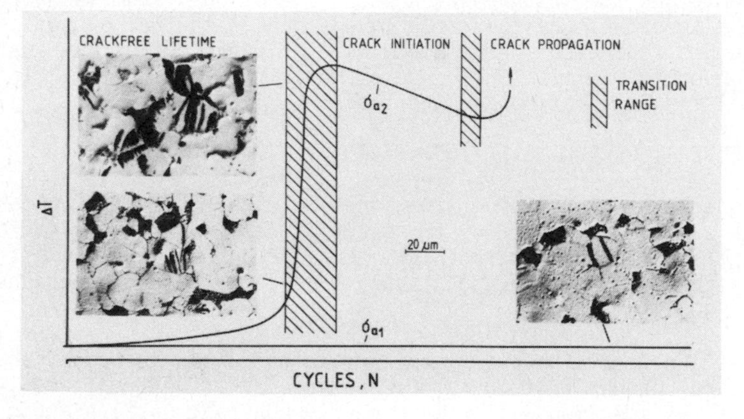

Fig. (3) - Cyclic temperature curves, normalized steel, $\sigma_a < R_{el}$

temperature rises. First, flat deformation marks could be observed on the specimen's electropolished surface with the aid of a light microscope after 10^5 cycles.

Stress amplitudes like σ_{a2} lead to formation of isolated slipbands after only a few cycles. Slipband formation is indicated by an increase of temperature. Furthermore, the number of slipbands increases with the number of cycles, some of them become more and more sharply limited. It might be possible to correlate certain ranges of the cyclic temperature curve registered at a definite stress amplitude with the above marked fatigue stages.

Rapid estimations of stress amplitudes which form slipbands can be made with cyclic progressively-increasing load tests. Figure 4 points out the results of

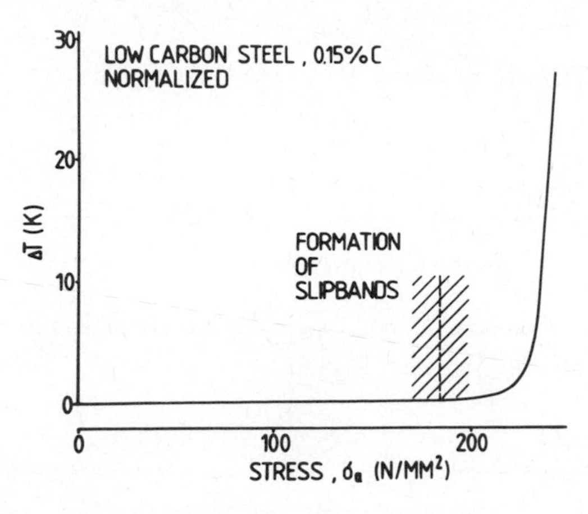

Fig. (4) - Cyclic stress temperature curve, stepwise load increase

an experiment where load was increased stepwise. Progressive temperature rise obviously is caused by formation of some slipbands. According to general behavior of polycrystalline materials, only a range of first slipband formation can be defined instead of a well-marked stress level. Moreover, this range is influenced a little by the actual structure of surface layer and by the applied test conditions. It may be defined more exactly only by means of statistics.

ESTIMATION OF ENDURANCE LIMIT

Cyclic stress strain curves indicate material's behavior due to certain numbers of cycles. Conventions are necessary how to establish these curves. Strain may be determined for instance, when half of cycles to failure was reached, but there is no physical sense for this procedure. Thus, correlations with a certain failure criterion can be made only empirically.

As shown in Figure 5 schematically, a stress temperature curve can be determined, Figure 5b, analogous to the well-known stress strain curve, Figure 5a. Until to the marked cyclic yield stress S_y (in our experiments at deviations of 0.2K from linearity of $\Delta T/\sigma$-curve), only small rises of temperature with increased stress can be observed, above S_y progressive temperature rises are significant. According to a definite sensitivity of the temperature measuring device, the cyclic yield stress S_y may be correlated to slipband formation of certain density.

Figures 5c and 5d point out schematically that the cyclic yield stress may be determined in the above-mentioned test procedure with increasing load [2]. Coordination of values S_y resulting from constant amplitude tests and from increasing load tests depends on sensitivity of measuring device and rate of load increase. This is pointed out in analogy to problems with evaluations of monotonic test procedures.

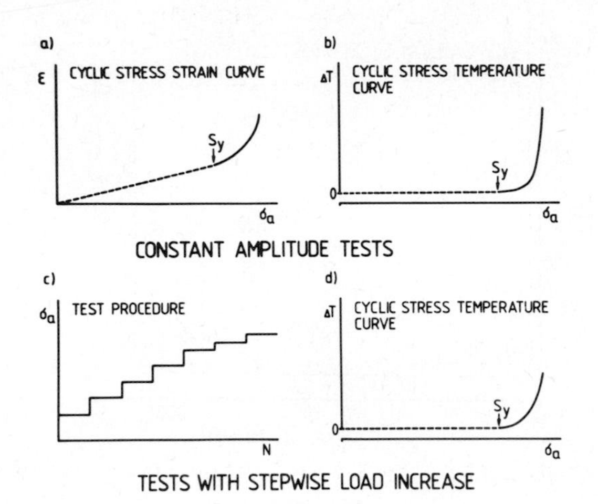

Fig. (5) - Evaluation of the cyclic yield stress (schematically)

Surface investigations suppose the thermometrically determined cyclic yield stress to be a boundary for initiation of microcracks. Temperature rises above S_y occur due to plastic deformation in a few small areas of surface grains. They must be much higher in these regions than the recorded values. Thus, the high correlation of the cyclic yield stress - which doesn't scatter very much - with the fatigue limit of low probability of failure may be understood. Figure 6 points out this fact schematically. Experimental verification has been published for different steels and heat treatments [3-5].

It must be emphasized that a high correlation between the cyclic yield stress and the low stress limit of the Wöhler transition range can be pointed out only if constant amplitude tests are carried out on the basis of statistical planning and evaluation [6]. Additionally, this is valid on the assumption that cracks initiate at slipbands [7].

Figure 7 once more shows results of temperature measurements during constant amplitude tests. For stress amplitudes, $\sigma_a < S_y = 200$ N/mm^2, only small rises of temperature can be measured. It is obvious, however, that all specimens cycled with amplitudes within the transition range of the Wöhler diagram ($\sigma_a > S_y$) get warm. Those specimens which break before reaching the ultimate number of cycles show specific rises of temperature in a very early stage of life [8].

According to the material's behavior, mentioned above, estimations of endurance in the range of low strain fatigue may be possible in an early stage. In Figure 8 are shown temperature curves for five specimens of the same heat treatment from constant amplitude tests at one stress level. It can be determined

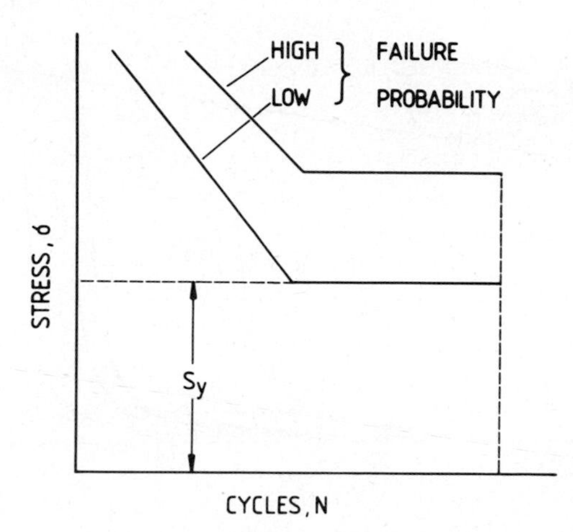

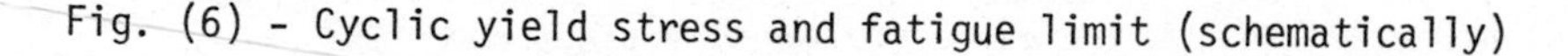
Fig. (6) - Cyclic yield stress and fatigue limit (schematically)

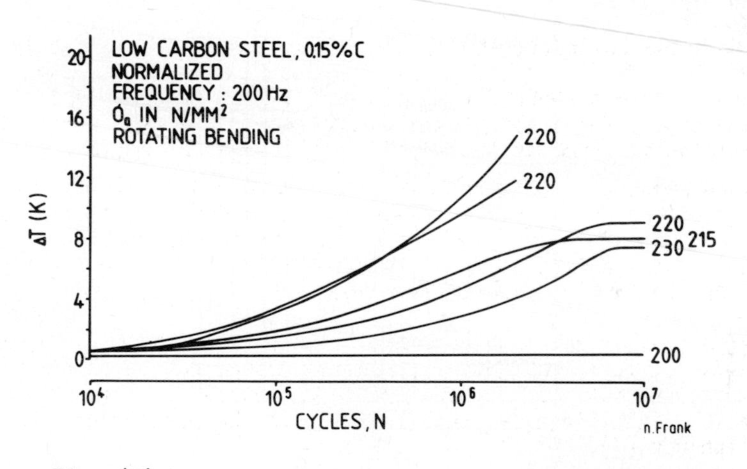

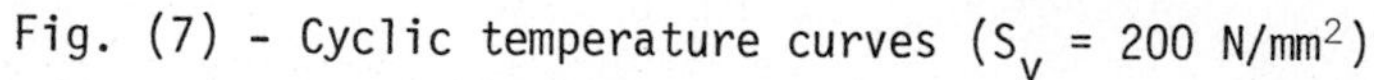
Fig. (7) - Cyclic temperature curves (S_y = 200 N/mm²)

numbers of cycles to obtain a definite temperature rise $N_{\Delta T_x}$, for instance ΔT = 4K. The numbers of cycles can be plotted together with those of other stress amplitudes in relation to their cycles to failure N_f or to formation of macrocracks. The latter may be understood better. Well-known difficulties in determination of crack sizes justify the first procedure. As shown in Figure 9, all values fit fairly well in the scattering band. It can be written

$$N_f = C \cdot N_{\Delta T_x}^{k}$$

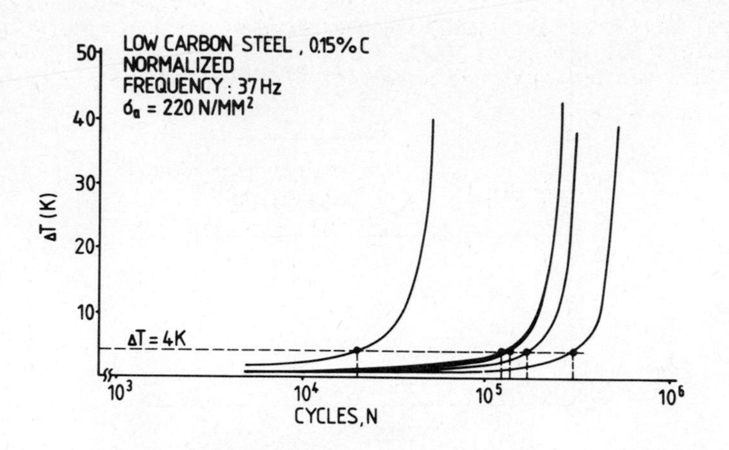

Fig. (8) - Cyclic temperature curves, determination of cycles of temperature change ΔT = 4K, 5 specimens

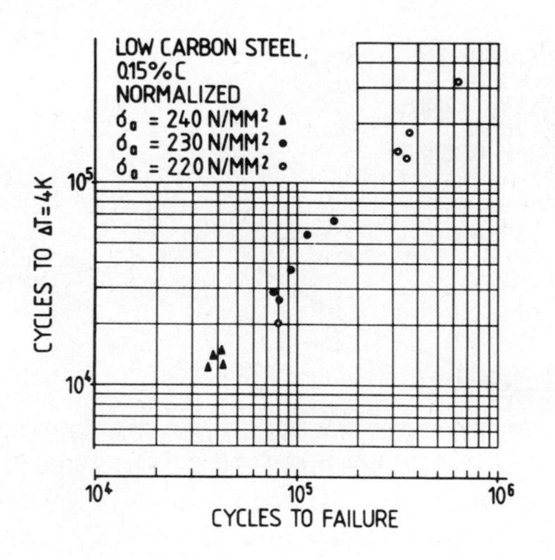

Fig. (9) - Estimation of endurances by determination of temperature changes

Constants C and k are introduced for certain structures and test conditions.

As shown in Figure 2, temperature measurements may mark the start of macroscopic crack propagation too. Experimental assurances, however, are still necessary especially regarding different alloys and heat treatments.

Influences of mean stress on the cyclic yield stress are shown in Figure 10. The approximate equality of yield stress in tension and compression range for normalized steel may be accentuated. In the range of compression, the cyclic yield stress points out the stage of fatigue better than a certain criterion of flaw can do. This can be explained with restrained formation and growth of macrocracks.

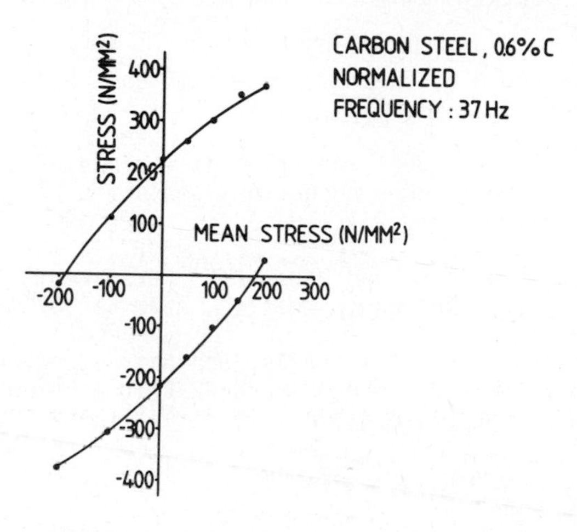

Fig. (10) - Smith-diagram on the relation of cyclic yield stress

A last example may point out that crack initiation can be determined thermometrically also with specimens, which are surface hardened. Cyclic yield stresses were determined in rotating bending tests. For nitrided carbon steel (normalized condition, 0.15%C, nitrided 580 °C/2h) the value of cyclic yield stress determined thermometrically was about 15% below the lower limit of the transition range in Wöhler's diagram.

Metallographic investigations showed slipband formation in the transition zone from the surface layer to the matrix when local stress amplitude exceeds yield stress. TEM observations pointed out that slipbands were dislocation-poor channels with walls of high dislocation density. Discrepancy of cyclic yield stress and fatigue limit according to the ultimate number of cycles $N_u = 10^7$ can also be explained with restrained formation and propagation of macro-cracks [8].

REFERENCES

[1] Macherauch, E. and Mayr, P., "Some basic principles of the fatigue behavior of plain carbon steels", in: Proceedings of the International Symposium on Low-Cycle Fatigue Strength and Elasto-Plastic Behavior of Materials, Stuttgart, pp. 129-168, 1979.

[2] Harig, H. and Weber, M., "Lebensdaueruntersuchungen an unlegierten Kohlenstoffstählen bei Zug-Druck-Beanspruchung", Materialprüfung 24, No. 5, pp. 169-172, 1982.

[3] Klages, H., "Untersuchungen über den Einfluß variierter Wärmebehandlungszustände auf die Dauerfestigkeit der Vergütungsstähle C 35 und 34 CrMo4 durch thermometrische Kurzzeitprüfverfahren und vergleichende Wöhler-Versuche", Dr.-Ing. Dissertation, TU Berlin, 1979.

[4] Dengel, D. and Harig, H., "Estimation of the fatigue limit by progressively-increasing load tests", Fatigue of Engineering Materials and Structures, 3, pp. 113-128, 1980.

[5] Harig, H. and Frank, E., "Thermometry - a method of short term fatigue testing", Wire 31, pp. 227-231, 1981.

[6] Dengel, D., "Die arc sin $\sqrt{P}$-Transformation - ein einfaches Verfahren zur graphischen und rechnerischen Auswertung geplanter Wöhlerversuche", Z. f. Werkstofftechnik 6, pp. 253-261, 1975.

[7] Mughrabi, H., Ackermann, F. and Herz, K., "Persistent slip bands in face-centered and body-centered cubic metals", ASTM STP 675, pp. 69-105, 1979.

[8] Frank, E., "Untersuchungen zum Einfluß bon Nitrierbehandlungen auf das Ermüdungsverhalten der Stähle Ck 15 und 42 CrMo 4 bei niedrigen Spannungsamplituden", Dr.-Ing. Dissertation, TU Berlin, 1982.

INHOMOGENEOUS WORK-SOFTENING DURING CYCLIC LOADING OF SAE 4140 IN DIFFERENT HEAT TREATED STATES

D. Eifler and E. Macherauch

Institut für Werkstoffkunde I, Universität Karlsruhe (TH)
Kaiserstr. 12, D-7500 Karlsruhe 1, Federal Republic of Germany

ABSTRACT

Stress-controlled push-pull tests have been performed to investigate the influence of microstructure on the fatigue behavior of normalized as well as different quenched and tempered specimens of SAE 4140. In the initial stage of cyclic loading, normalized specimens exhibit quasi elastic behavior followed by inhomogeneous work-softening effects. After reaching a maximum plastic strain amplitude, a work-hardening period follows with homogeneous plastic strains. On the other hand, quenched and tempered specimens show quite a different behavior in the crack free period of fatigue life. Independent of the tempering treatment chosen (T_t > 450°C) in all cases after a quasi elastic initial period work-softening occurs until crack initiation. Photoelastic investigations reveal that work-softening appears extremely localized. The areas of the specimen's softening first continuously accumulate plastic strain during further life. Thus, plastic strains are particularly concentrated on small fatigue zones. Very often, cyclic deformation processes seem to be mainly restricted to one or two fatigue zones. In these zones, cracks initiate, grow and finally lead to failure of the specimen.

INTRODUCTION

The cyclic deformation behavior of normalized plain carbon steels is well documented [1-3]. On the other side, the fatigue properties of quenched and tempered steels which are of special importance for technical applications are mainly assessed by the aid of classical S-N curves [4-7] and some strain controlled low cycle fatigue tests [8,9]. Almost no investigations are known concerning the cyclic deformation processes in the macro free stage of life. Therefore, this paper presents some results of stress controlled uniaxial push-pull fatigue tests with normalized as well as quenched and tempered SAE 4140 steel specimens. Besides mechanical measurements, photoelastic investigations, localized strain gage measurements and microhardness determinations have been carried out. The main aim of the experiments was to obtain detailed information about the deformation processes during the cyclic softening period of quenched and tempered steels.

MATERIAL AND HEAT TREATMENT

The material used for the investigation was a low alloyed SAE 4140 steel (German grade 42CrMo4) with the following chemical composition (in wt. %): 0.44 C; 0.32 Si; 0.73 Mn; 0.014 P; 0.028 S; 1.11 Cr; 0.22 Mo and 0.11 Ni. Besides the normalized state, four quenched and different tempered states have been investigated. The heat treatment parameters chosen are summarized in Table 1. Characteristic properties of the mentioned materials states determined in

TABLE 1 - MONOTONIC TENSILE PROPERTIES AS A FUNCTION OF THE HEAT TREATMENT

Heat Treatment	R_{eS} (N/mm^2)	$R_{p\,0.2}$ (N/mm^2)	R_m (N/mm^2)	A [%]	HV 0.1
Normalized 3h/930°C	400	-	740	19.4	230
Austenized at 850°C, oil quenched and tempered:					
2h/450°C	-	1315	1375	7.3	405
4h/570°C	-	1000	1120	12.0	350
2h/650°C	875	-	940	15.0	305
2h/730°C	680	-	735	27.5	208

monotonic tensile tests with a strain rate $\dot{\varepsilon} = 2 \cdot 10^{-4}\ s^{-1}$ are also given. Shape and dimensions of the specimens used in cyclic deformation experiments are shown in Figure 1a. The photoelastic investigations required special specimens (Figure 1b).

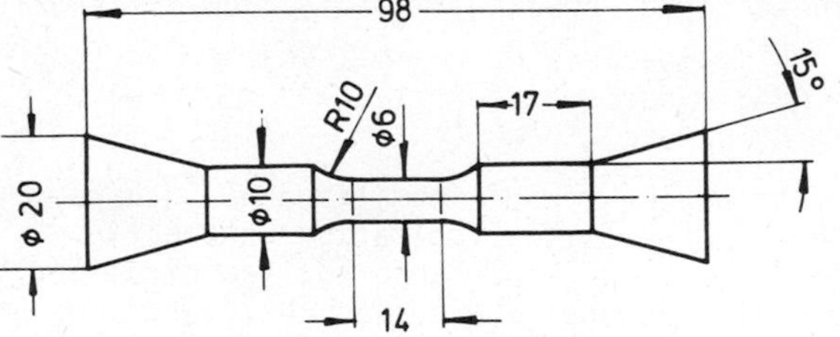

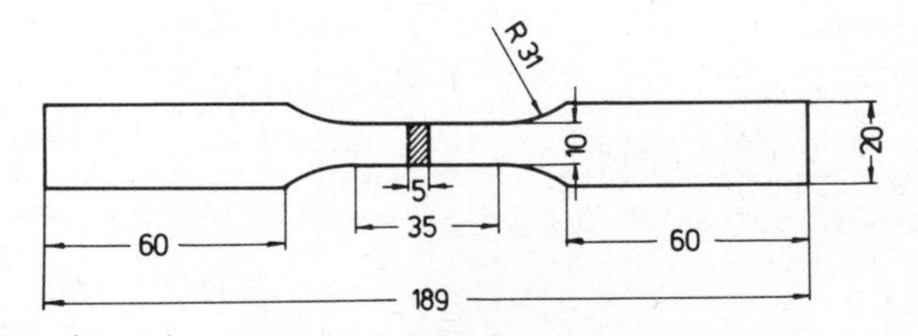

Fig. (1a,b) - Shape and dimensions of the specimens

EXPERIMENTAL PROCEDURE

Push-pull fatigue tests were performed under laboratory conditions at room temperature with a 83 KN Schenck servohydraulic testing machine using fully reversed triangular load waves. The test frequency was 5 Hz. Some of the photoelastic experiments demanded a reduced frequency of 0.5 Hz. Strain measurements were conducted with a capacitive extensometer fixed parallel to the gage length of the specimens. For calibration purposes and for experiments to determine localized inhomogeneous strain distributions, strain gages were used.

During cyclic loading, the hysteresis loops were recorded with an on-line data acquisition system at preselected numbers of cycles. A typical (idealized) hysteresis loop with characteristic features is shown in Figure 2. From these data, the nominal stress amplitudes σ_a and the plastic strain amplitudes $\varepsilon_{a,p}$ (strain at zero load) were calculated.

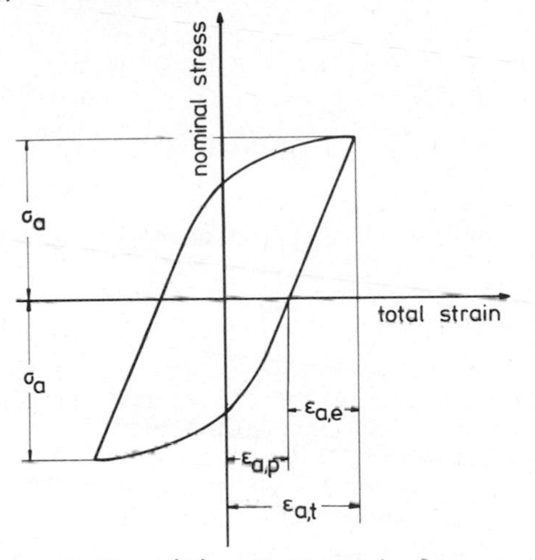

Fig. (2) - Hysteresis loop

The relation between plastic strain amplitude and logarithm of the number of cycles is called cyclic deformation curve and corresponds to the microstructural changes in the structure during cyclic loading. The loading induced strain distributions in the surface layers of flat specimens can be investigated with a photoelastic method. For that purpose, suitable specimens, Figure 1b, were covered with thin foils of araldit which change their refractive index proportionally to the difference of the occurring main strains. The strain distribution can be valued by means of the resulting coloured interference patterns (isochromatic lines).

After mechanical testing, the surface and bulk structural features of the specimens were examined by microhardness measurements and transmission electron microscopic (TEM) observations. The foils for TEM investigations of the heat treated as well as fatigued materials were prepared from parts of the specimens cut from the gage length.

EXPERIMENTAL RESULTS

A. Cyclic Deformation Behavior

Characteristic cyclic deformation curves ($\varepsilon_{a,p}$ versus lg N curves) for the normalized materials state are shown in Figure 3. During stress controlled

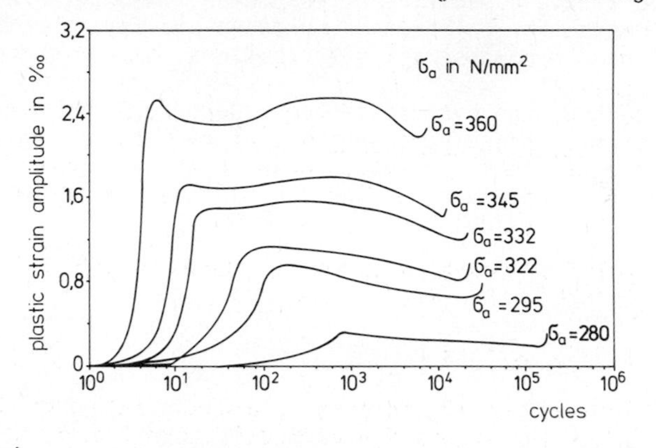

Fig. (3) - Cyclic deformation curves of normalized SAE 4140

push-pull tests with R = -1, the plastic strain amplitudes in a systematic manner change with the stress amplitudes applied ($\sigma_a < \sigma_y$). According to the integral strain measurement over the whole gage length of the specimen, in the initial stage of cyclic loading, always a quasi linear elastic behavior is found. This stage is followed by a period of increasing plastic strain amplitudes due to cyclically induced softening effects in the material. The primary incubation period shortens and the extent of cyclic softening grows simultaneously with increasing stress amplitude σ_a. After reaching a maximum of softening, the materials behavior is different under large respectively small amplitudes. For $\sigma_a \lesssim 325$ N/mm^2, the material work-hardens and the plastic strain amplitudes decrease. For $\sigma_a \gtrsim 325$ N/mm^2, after a period of decreasing or constant strain amplitude, a secondary work-softening peak occurs. At the end of the specimen's life, the plastic strain amplitudes increase again, however fictitious, as a consequence of changes in the compliance of the specimens due to crack propagation. In Figure 4, characteristic cyclic deformation curves of quenched and tempered specimens are shown. In the crack-free life, a fatigue behavior quite different from that in normalized specimens is exhibited, Figure 3. At the beginning of cyclic loading, in all cases, a quasi linear elastic deformation is observed with $\varepsilon_{a,p}$ values smaller 10 to 50 $\cdot$ 10^{-6}. Increasing incubation periods are shown with decreasing stress amplitudes. After passing the end of the incubation period, the plastic strain amplitudes increase and a continuous transition occurs from work-softening to crack propagation. The other quenched and tempered materials states, see Table 1, behave in a similar way as the one just discussed [10]. All quenched and tempered specimens investigated macroscopically showed cyclic softening effects. However, as can be seen from the cyclic deformation curves in Figure 4 at fatigue life > 10^4 develop only small plastic strain amplitudes of about 1% or less.

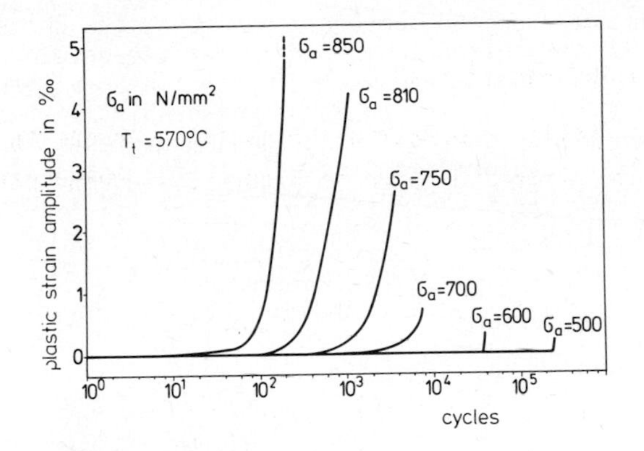

Fig. (4) - Cyclic deformation curves of quenched and tempered (4h/570°C) SAE 4140

B. Inhomogeneous Deformations

The initial increase of the integral plastic strain amplitudes of the normalized material after the incubation period is connected with inhomogeneous deformation processes. Figure 5 shows a typical result pointing at this phenomenon. The data are valid for a stress amplitude σ_a = 322 N/mm². Besides,

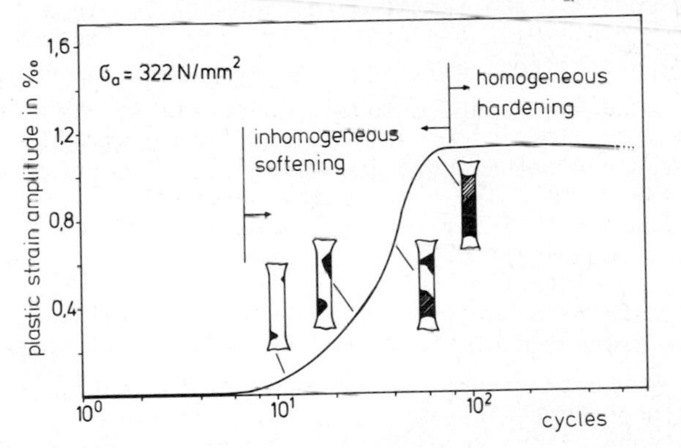

Fig. (5) - Inhomogeneous strain distribution during cyclic softening of a normalized specimen

the cyclic deformation curve, the photographically recorded strain distributions along the gage length of the specimen are plotted as hatched areas indicating different states of cyclic softening. Usually the inhomogeneous cyclic plastic deformation of the specimens starts in the transition areas between grip ends and gage length and afterwards propagates with increasing number of cycles completely over the gage length. In the following work-hardening stage, homogeneous elastic plastic deformations occur. Loading conditions yielding endurances of $N_f < 10^5$ always cause plastic deformations in the whole deformable volume of the specimen.

In the case of quenched and tempered specimens, the photoelastic observation technique applied very often indicated an earlier development of plastically deformed areas than could be detected by the integral strain measurement over the whole gage length. However, integral measurably plastic strain amplitudes usually developed within the following loading cycles. Continued cycling increases the local plastic deformations as well as the integral plastic strain amplitudes. Figure 6 presents a characteristic result for a quenched and tem-

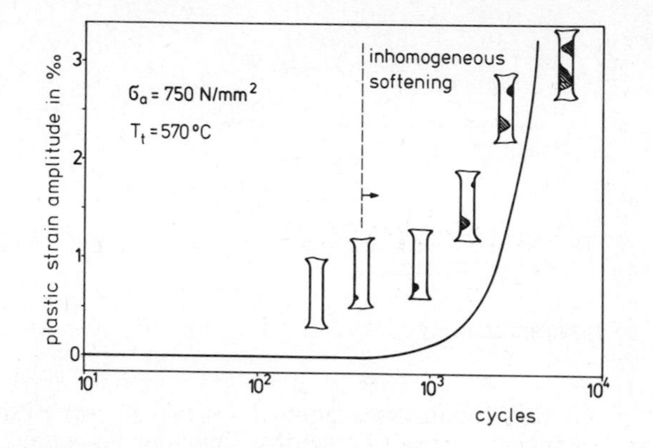

Fig. (6) - Inhomogeneous strain distribution during cyclic softening of a quenched and tempered specimen

pered specimen loaded with stress amplitudes σ_a = 750 N/mm^2. As in the case of normalized specimens also cyclic softening often starts at the transition from the grip ends of the specimens to the gage length. Subsequently, the plastically deformed areas grow with increasing cycles and - dependent on the stress amplitude applied - finally more or less of the specimen is included. The experiments establish the important result that the areas of the specimen in which the first photoelastically identified plastic deformations occur, during continued cycling permanently accumulate further plastic strains. Thus, nearly all plastic strains are concentrated on small fatigue zones. As a consequence, these zones later on act as sources of fatigue cracks which finally lead to failure of the specimen.

Plastic strain amplitudes in the fatigue zones cannot be detected correctly with integral measurements since the measurements with the capacitive extensometer are related to the whole gage length. In order to quantify the extremely localized plastic strains, some specimens covered with photoelastic foils have been cycled until the first plastic deformation was visible. Afterwards, the foils were detached and three strain gages (S.G.) were glued to the specimen. The first strain gage was fixed in an area photoelastically identified as plastically deformed and the second one just beside the first. The third strain gage was applied in such an area of the specimen where up to the interruption of the test only pure elastic deformation was observed. Additionally, an integral strain measurement with the capacitive system was carried out. The typical results of such an experiment are presented in Figure 7. The measured specimen showed the first plastic deformation zone after $5 \cdot 10^2$ cycles. Thereafter, further cycling yields to the plotted values. Strain gage I attached in the center of the fatigue zone measures the largest plastic strain amplitudes until fracture. Strain gage

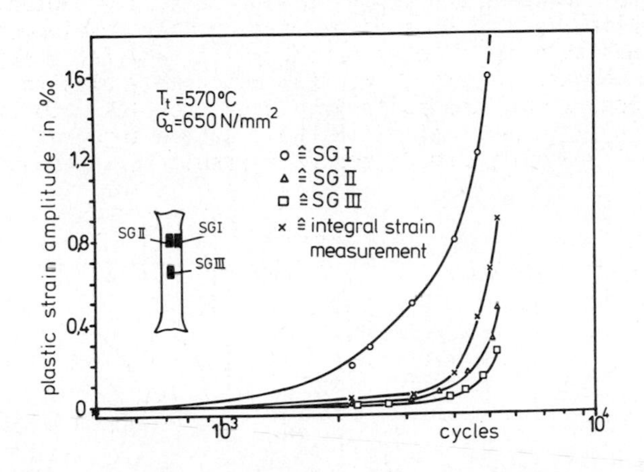

Fig. (7) - Inhomogeneous strain distribution of quenched and tempered SAE 4140

II measures at first only elastic deformations. However, with succeeding softening and the according increasing size of the fatigue zone also plastic strain amplitudes were recorded. Strain gage III shows only small plastic strain amplitudes towards the end of the specimen's life. As can be seen from the curves presented in Figure 7, the integral measurements are 2-8 times smaller than the strain amplitudes indicated from the strain gage I in the center of the fatigue zone.

Microhardness measurements in and around the above mentioned fatigue zones could prove the statement that the deformation processes observed are really connected with cyclic softening. Therefore, hardness measurements have been performed with a load of 100 p at specimens with well developed fatigue zones. A typical result of such measurements is given in Figure 8. Lines of identical hardness were determined in the plastically deformed areas of the specimen which on their part were indicated before with the photoelastic method. It emerges that the hardness values increase continuously from the points of the first plastic deformation (e.g., edge of the specimen) to the middle of the specimen and pass into the initial hardness of the uncycled material.

Additional experiments with photoelastic foils attached to the front and the back of quenched and tempered specimens revealed that the localized cyclic softening processes spread through the thickness of the specimens. It was proved that plastic deformations develop on both sides of the specimen at nearly the same point and at nearly the same time. Figure 9 shows the results of a corresponding experiment realized by the aid of photoelastic foils and microhardness measurements. This particular result undoubtedly demonstrates that the fatigue zones observed can occur in all parts of the bulk of the specimen and are not necessarily combined with edge effects.

The relation appearing between the relative amount of the plastically deformed volume of the specimens and the logarithm of the number of cycles is shown in Figure 10. A_{pl} is the plastically deformed area, A_o the complete gage

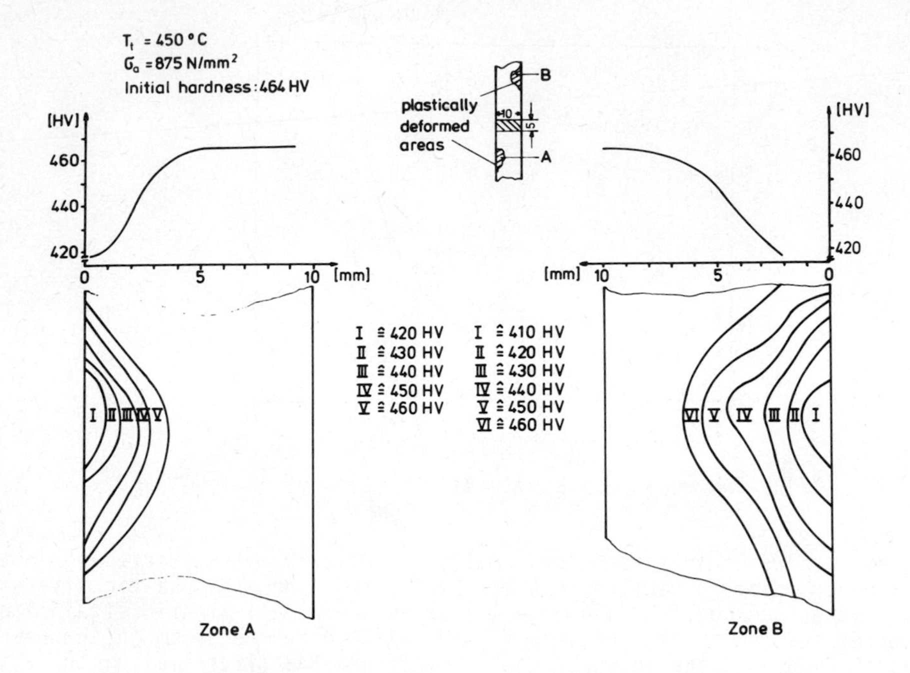

Fig. (8) - Lines of identical hardness in the fatigue zones of quenched and tempered SAE 4140

area. As expected, the A_{pl}/A_o versus lg N curves which show the relative growth of the plastically deformed materials' volume with increasing cycles are similar to the cyclic deformation curves of corresponding stress amplitudes. Figure 11 is summarizing the influence of the materials state of the quenched and tempered specimens on the formation of the inhomogeneously strained volume of the specimen under cyclic loading with different stress amplitudes. The data for tempering temperatures of 450°, 570° and 650°C show the situation at the number of cycles at crack initiation. Obviously, plastic strains develop earlier and more extended in the higher tempered materials state. Thus, the inhomogeneities of the cyclic plastic deformation increase with decreasing tempering temperature.

DISCUSSION

The first sections of cyclic deformation curves of the normalized materials state (see Figure 3) under loadings $\sigma_a < \sigma_y$ are characterized by an incubation period with quasi elastic behavior. This is a consequence of a very small density of mobile dislocations in the normalized state. The following softening effects strongly depend on the stress amplitudes applied and are associated with the development and propagation of fatigue-Luedersbands [2]. During this period, inhomogeneous plastic strain distributions and locally different dislocation densities are observed. After reaching a maximum plastic strain amplitude, homogeneous work-hardening predominates work-softening at lower stress amplitudes. The degree of changing the plastic strain amplitude under a given load depends on

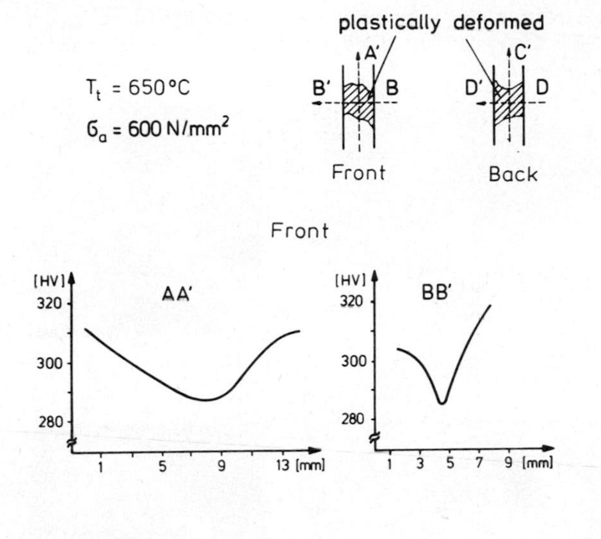

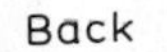

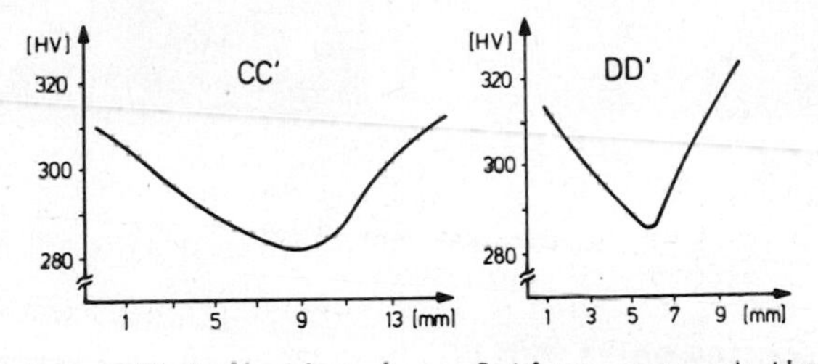

Fig. (9) - Hardness distribution in a fatigue zone at the front and the back of a quenched and tempered specimen

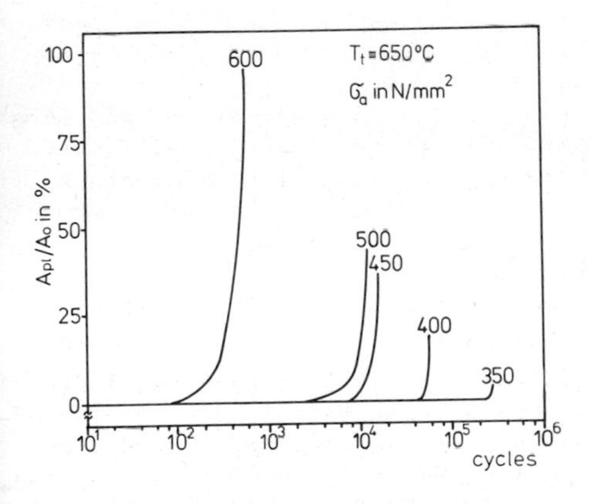

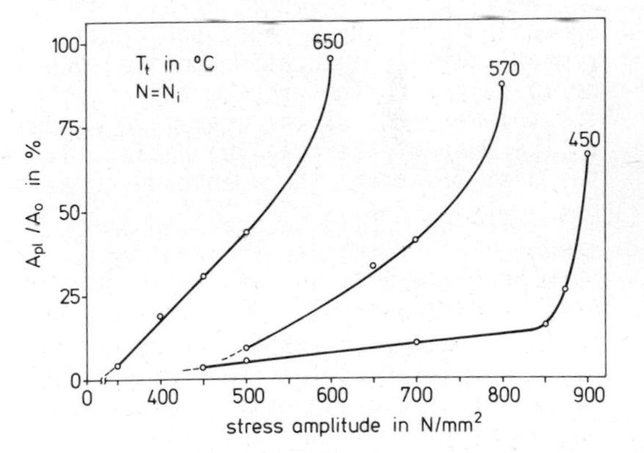

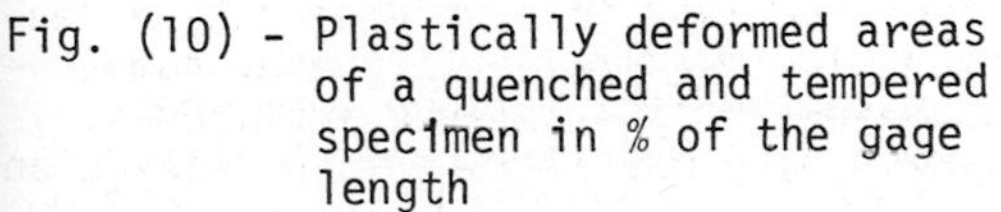
Fig. (10) - Plastically deformed areas of a quenched and tempered specimen in % of the gage length

Fig. (11) - Plastically deformed areas of quenched and tempered specimens for $N = N_i$ in % of the gage area the dislocation structure developing in the ferrite grains. An example is shown in Figure 12. At N_1 dislocation bundles appear. At N_2 besides a bundle struc-

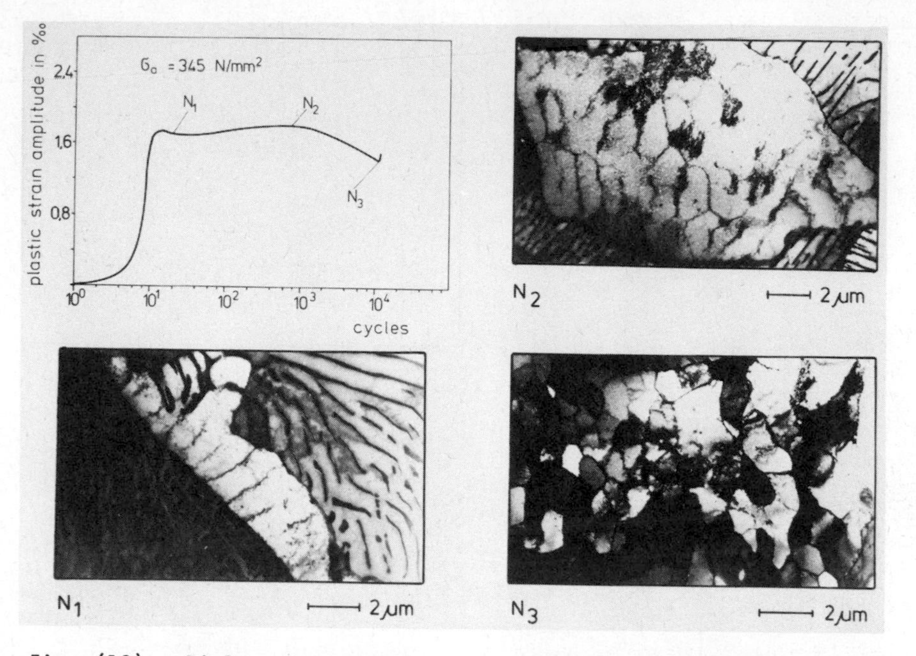

Fig. (12) - Dislocation structures in ferrite grains of normalized SAE 4140 at different fatigue states

ture the first dislocation cells can be observed. In the life between N_2 and N_3 a cell structure is formed.

Whereas normalized material exhibits inhomogeneous cyclic deformation processes only during a short period of life quenched and tempered materials after a quasi elastic incubation period show extremely inhomogeneous cyclic softening up to failure. The processes in both materials states are quite different. In the volume parts of the normalized steel passed through by fatigue-Luedersbands an increase of the original dislocation density occurs. On the other hand, after the heat treatment the quenched and tempered states exhibit a dislocation density of ~ 10^{11} cm^{-2} [11]. Localized softening is believed to occur by dislocation rearrangements resulting from the repeated interaction of oscillating dislocations with structural inhomogeneities. The new configurations offer less resistance to plastic deformation. Thus, with an increasing number of cycles the total dislocation density decreases and the mean free path of movable dislocations increases In favorable areas these effects promote the localized growth of the plastic strain amplitude and easily deformable areas (fatigue zones) are developed. At the boundaries of the fatigue zones the dislocation density reaches nearly the value of the uncycled structure. Careful investigations have demonstrated that these fatigue zones are softer than the surrounding matrix material (see Figures 8 and 9). A minimum hardness appears in the middle of the fatigue zone where the point of maximum cyclic softening is assumed. Furthermore, a transition of the hardness values to the initial hardness of the structure is observed at the bounds of the fatigue zone.

Depending on the materials state during cyclic loading, either the dislocation density increases in the prior martensite boundaries or cell-like structures are formed in areas with a greater distance between the carbides. That can be seen from the TEM-micrographs shown in Figure 13. The upper part of the figure

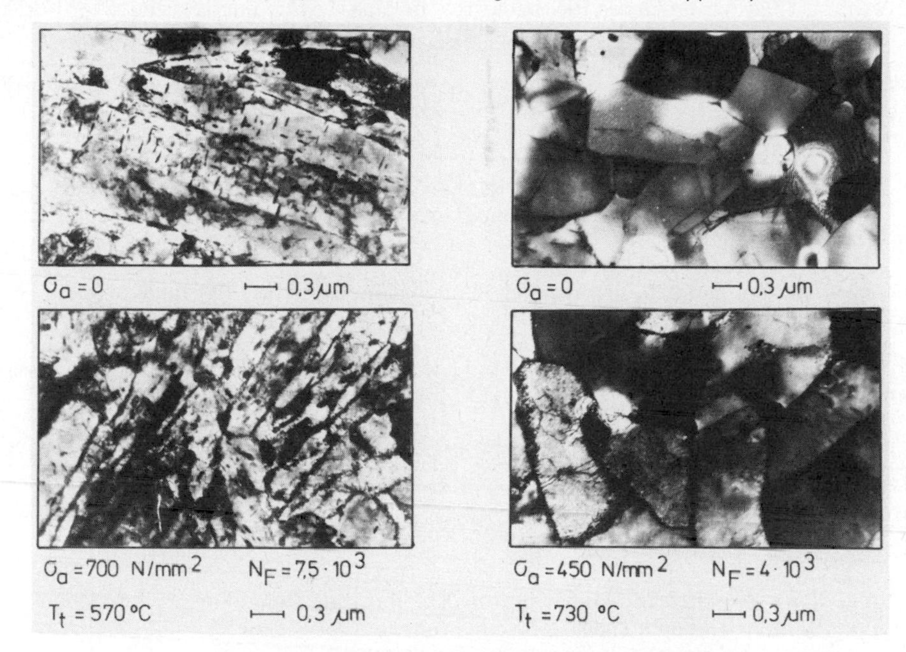

Fig. (13) - Bulk dislocation structures of uncycled and cycled quenched and tempered SAE 4140

presents uncycled materials states, the lower part materials states after ~ 5 $\cdot 10^3$ cycles. On the left hand side (T_t = 570°C), the increase of the dislocation density at the prior martensite boundaries is well documented. On the right hand side (T_t = 730°C) the development of dislocation cells is clearly shown.

SUMMARY

The cyclic deformation behavior of normalized steel SAE 4140 with loadings $\sigma_a < \sigma_y$ is characterized by softening and hardening phenomena. Cyclic softening processes are combined with inhomogeneous plastic deformations and the propagation of a fatigue-Luedersband. The following hardening process occurs homogeneously.

At stress amplitudes $\sigma_a < \sigma_y$, the different quenched and tempered conditions (T_t > 450°C) work-soften continuously until crack initiation. Plastic deformation processes occur extremely inhomogeneous and are concentrated on small fatigue zones. The extent of inhomogeneity in the plastic deformation state of the specimens was found to be strongly dependent on the stress amplitude applied. For a given stress amplitude the relative amount of the plastically deformed volume of the specimen increases with increasing tempering temperature. This is to be expected since the microstructure of the quenched and tempered specimens change considerably with the tempering condition.

REFERENCES

[1] Pilo, D., "Cyclic deformation behavior of unalloyed normalized steel with 0.1 to 1.02 wt.-% C" (in German), Ph.D. Thesis, Univ. Karlsruhe, 1979.

[2] Pilo, D., Reik, W., Mayr, P. and Macherauch, E., "Inhomogeneous deformation processes in the incipient crack free fatigue stage of unalloyed steels" (in German), Arch. Eisenhüttenwes. 48, p. 575, 1977.

[3] Pilo, D., Reik, W., Mayr, P. and Macherauch, E., "Cyclic induced creep of a plain carbon steel at room temperature", Fatigue of Eng. Mat. and Struc., Vol. 1, p. 287, 1979.

[4] Frankel, H. E., Benett, J. A. and Pennington, W. A., "Fatigue properties of high strength steels", Trans. of the ASM, Vol. 52, 1960.

[5] Borik, F., Chapman, R. D. and Jominy, W. E., "The effect of percent tempered martensite on endurance limit", Trans. of the ASM, Vol. 50, 1958.

[6] Sato, T. and Tomomoto, K., "Effects of microstructure on the fatigue property of high carbon steel", Trans. ISIJ, Vol. 20, p. 821, 1980.

[7] Thomson, R. F., "Fatigue behavior of high-carbon high hardness steels", Trans. of the ASM, Vol. 56, 1963.

[8] Landgraf, R. W., "Cyclic deformation and fatigue behavior of hardened steels", T.& A.M. Report No. 320, Illinois, 1968.

[9] Thielen, P. N., "Cyclic stress strain relations and strain-controlled fatigue of 4140 steel", Acta Met., Vol. 24, p. 1, 1976.

[10] Eifler, D., Mayr, P. and Macherauch, E., "Cyclic stress-strain behavior of SAE 4140 after different heat treatments", Stahl u. Eisen 101, p. 131, 1981.

[11] Malik, L. and Lund, J. A., "A study of strengthening mechanisms in tempered martensite from a medium carbon steel", Metallurg. Trans., Vol. 3, p. 1403, 1972.

FATIGUE CRACK INITIATION IN IRON

C. V. Cooper and M. E. Fine

Northwestern University
Evanston, Illinois 60201

ABSTRACT

Fatigue crack initiation in iron of 99.93$^+$ w/o purity has been carefully characterized at constant plastic strain amplitudes varying from 10^{-4} to 10^{-2}. Fatigue microcracks initiate transgranularly at all amplitudes; the initial microcracks which form are less than 1 μm in length, as determined by transmission electron microscopy of replicas. At high plastic strain amplitudes, however, grain boundary initiation of cracks was also observed, and the fraction of grain boundary segments containing cracks was measured as a function of plastic strain amplitude. The material compliance as determined at strain reversal from maximum tension has been established as an indication of fatigue damage and very early microcrack initiation. Finally, the Coffin-Manson relation for microcrack initiation has been evaluated for iron and deviations from the relation discussed.

INTRODUCTION

The results of a careful, thorough investigation into the nature of and mechanisms by which fatigue microcracks initiate in pure polycrystalline iron are presented for a broad range of plastic strain amplitudes. Fatigue crack initiation was first documented in 1903 by Ewing and Humfrey [1], who studied Swedish iron cycled in rotating bending. They reported that a few, small slip lines formed during the initial stages of cycling and served as preferential sites for fatigue crack initiation. It is now commonly accepted that, for a wide variety of metals and alloys cycled at low to intermediate amplitudes, persistent slip bands form initially, broaden and deepen to form intrusions and extrusions while increasing in density across the surface of the material, and finally grow or coalesce to form microcracks. This process is reported for virtually all FCC pure metals and alloys, many BCC alloys, and some HCP metals. As the initiation of slip band microcracks is highly sensitive to shear strains, they will always form first within those crystals exhibiting slip bands oriented for the highest resolved shear strains. However, very little is known about the exact processes and mechanisms leading to crack initiation and the characteristics of the freshly initiated fatigue cracks. Changes in mechanical properties associated with crack initiation have not previously been established.

In this study, fatigue crack initiation has been correlated with changes in mechanical properties which occur during the course of cyclic straining. Inasmuch as the initiation process is a local phenomenon, discrete in occurrence, changes in first order bulk properties such as maximum tensile or compressive stresses will likely be insensitive to the onset of microcracking. Higher order bulk elastic properties, i.e., $\partial\sigma/\partial\varepsilon$ and $\partial^2\sigma/\partial\varepsilon\partial N$, where σ is the stress, ε is the strain and N is the number of cycles, are expected to be more sensitive to microcrack initiation.

Thus, the most promising parameter to monitor is expected to be the slope of the stress-strain curve on unloading from maximum tension, since microcracks opened by maximum tension would begin to close immediately upon strain reversal.

EXPERIMENTAL PROCEDURES

Rather pure, unalloyed iron was vacuum melted and hot-rolled by the Inland Steel Company Research Laboratory to ¼ inch nominal thickness plate; the chemical composition of this heat is given in Table 1. Smooth samples having gage dimensions of 8 mm long and 6 mm by 5 mm cross-section were milled from the as-received plate such that the straining axis was coincident with the rolling direction. Following machine grinding to produce flat, parallel surfaces, all samples were annealed for one hour at 850°C at a vacuum of 5×10^{-6} torr, resulting in equiaxed grains having an average diameter of 100 μm. Hand polishing with silicon carbide to a 17 μm finish was followed by chemical polishing for 90 seconds duration in a solution of 95 v/o H_2O_2 and 5 v/o HF, removing approximately 0.1 mm of surface. TEM examination of replicas taken from the as-polished surface revealed a smooth, unpitted, unetched surface at 30,000 times magnification. Completely reversed, cyclic plastic strains (R' = -1) resulting in fatigue damage and consequent microcrack initiation were imposed using a closed-loop, servo-controlled, electrohydraulic MTS machine. All fatigue experiments were conducted in ambient laboratory air of approximately 35% relative humidity. Throughout this study, a constant total strain rate, $\dot{\varepsilon}_{tot}$, of 10^{-3} sec^{-1} was maintained. Cyclic stress-strain responses were recorded with the use of an X-Y recorder, from which compliance upon strain reversal from maximum tension, i.e., $(\partial\sigma/\partial\varepsilon)_{\sigma^+_{max}} \equiv E_T$, the ratio of $\sigma^+_{max}/\sigma^-_{max}$, and other mechanical properties were determined. Direct determination of fatigue damage and microcrack initiation and growth was accomplished by replicating the specimen surface at pre-determined intervals during the course of cyclic straining. Prepared using the conventional two-stage technique, these replicas were shadowed with platinum-paladium and

TABLE 1 - ANALYZED CHEMICAL COMPOSITION OF IRON

Element	Amount in wt.%
Carbon	0.017
Nitrogen	0.002
Oxygen	0.016
Aluminum	0.008
Manganese	0.010
Sulphur	0.003
Phosphorus	0.008
Iron	Balance

coated with carbon, both in an evacuated bell jar at a vacuum of approximately 5×10^{-5} torr. In situ examination of the specimen surface with the use of a 750 times magnification Olympus metallurgical microscope with a long working distance objective lens was used to determine N_i (opt.), the number of cycles required to produce the first discernable microcrack using this technique. Such microcracks were characteristically 15 to 20 μm in length; using optical microscopy, shorter microcracks could not be distinguished from slip bands or other types of fatigue damage occurring on the surface.

RESULTS AND DISCUSSION

A. Microstructural Changes Leading to Fatigue Crack Initiation

Microcrack initiation in the iron for this study occurred by two modes: slip band (transgranular) initiation and grain boundary (intergranular) initiation. Throughout the entire range of plastic strain amplitudes investigated ($1 \times 10^{-4} \leq \Delta\varepsilon_p/2 \leq 7.5 \times 10^{-3}$), microcrack initiation occurred by the slip band mode. In general, the process of slip band initiation may be described by the following scenario: cyclic straining produces slip bands, the density of which is a function of strain amplitude. With continued cycling, intrusions and/or extrusions form, and microcracks less than 1 μm long initiate on the intrusions. Figures 1 and 2 are typical evidence of early fatigue damage leading to trans-

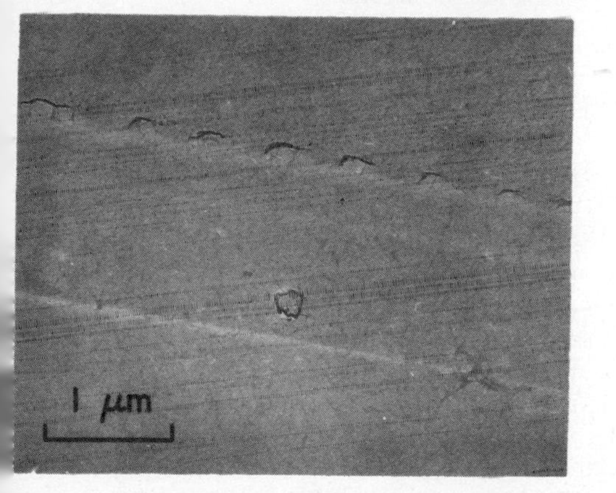

Fig. (1) - Periodically spaced intrusions and parallel slip bands after 9000 cycles at $\Delta\varepsilon_p/2 = 1 \times 10^{-4}$

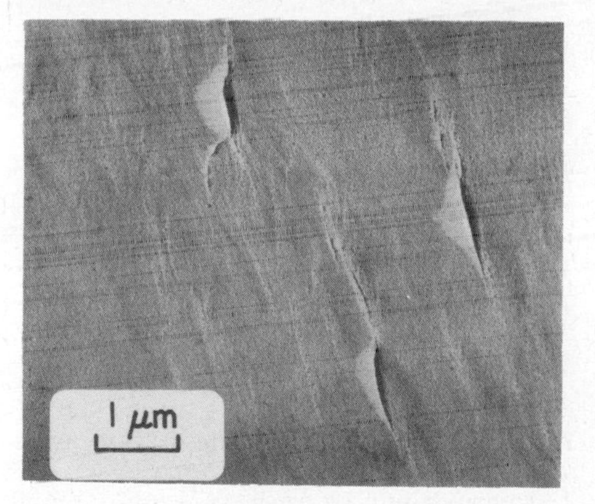

Fig. (2) - Transgranular microcracks approximately 1 μm long. N = 26,000; $\Delta\varepsilon_p/2 = 1 \times 10^{-4}$

granular microcrack initiation. Figure 1 shows two parallel slip bands on which small, periodically spaced intrusions have formed after the application of 9000 cycles at a plastic strain amplitude of 1×10^{-4}. After 26,500 cycles at this amplitude, Figure 2, these intrusions have grown to form bona fide microcracks approximately 1 μm long. It is evident from these photomicrographs that the initial microcracks are very small and discontinuous. Fatigue cracks which have initiated along slip bands do not extend, initially, across entire grains, as modeled by some recent theories [2]. Furthermore, these microcracks initiate in the central regions of the grains; therefore, a dislocation pileup phenomenon at a grain boundary seems to be an unlikely origin for this mode of fatigue crack initiation.

Figure 3 shows a TEM micrograph of the first evidence of grain boundary damage after 1500 cycles at a plastic strain amplitude of 1×10^{-3}. It is estimated from this photomicrograph that the depth of the microcrack is approximately 500Å. As in the initial stages of transgranular microcrack initiation, the microcracks along the grain boundary are discontinuous; here the length of the first observed microcracks is estimated to be 0.5 μm. Fatigue damage in the interior of the grain adjacent to the damaged grain boundary segment can be described as a rumpled surface with intrusion-like microcracks on the peaks of the rumples.

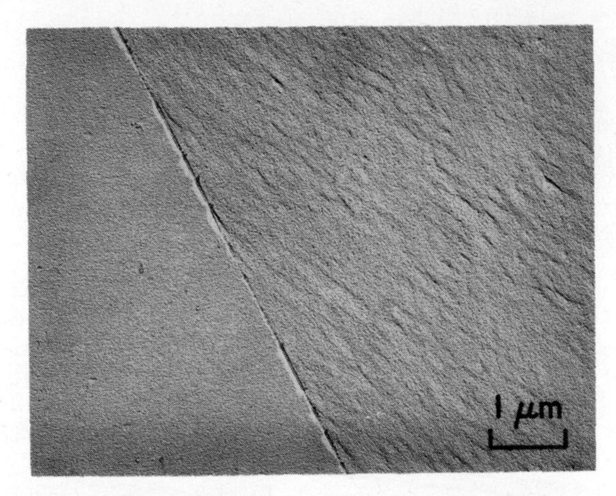

Fig. (3) - Grain boundary damage after 1500 cycles at $\Delta\varepsilon_p/2 = 1 \times 10^{-3}$

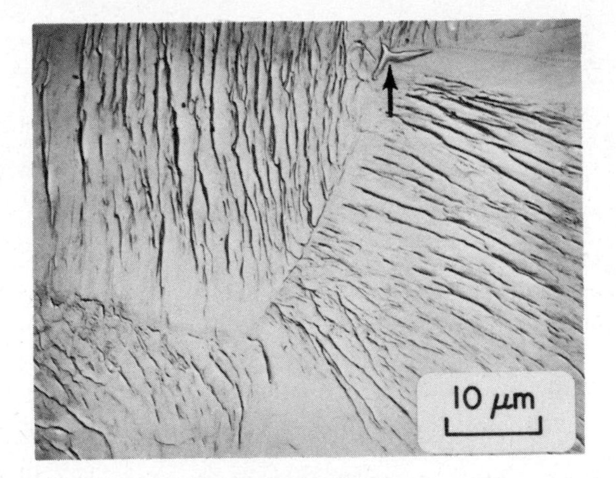

Fig. (4) - Densely-packed, transgranular microcracks and damage to a grain boundary triple-point. N = 500; $\Delta\varepsilon_p/2 = 5 \times 10^{-3}$

Figure 4 shows surface damage which is characteristic of the higher plastic strain amplitudes. Taken after 500 cycles at a plastic strain amplitude of 5×10^{-3}, this photomicrograph indicates a rather dense packing of transgranular microcracks as well as damage to a grain boundary triple point, indicated by the arrow. The sequence of photomicrographs from Figure 1 to Figure 3 is intended to provide qualitative evidence of the increasing tendency toward the formation of grain boundary microcracks with increasing plastic strain amplitude; this tendency will be quantified in the following section.

B. Quantitative Description of Grain Boundary Cracking

Presented in Figure 5 is the fraction of grain boundary segments containing microcracks, FB, as a function of plastic strain amplitude. As stated previously, the tendency for grain boundary failure increases markedly with increasing strain amplitude. This fraction was determined at each plastic strain amplitude by direct TEM examination of surface replicas. For each amplitude evaluated, a minimum of fifty grain boundary segments were examined at the termination of the

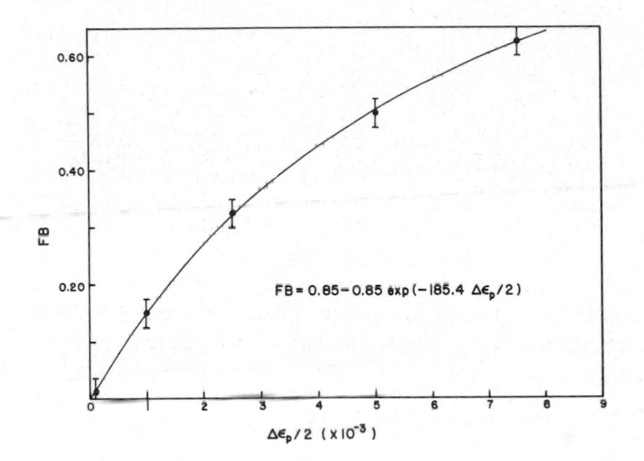

Fig. (5) - Fraction of grain boundary segments with microcracks versus plastic strain amplitude

experiment. Experiment termination was based on the fulfillment of either of the following criteria: (1) The optical detection, at 750 times magnification, of one or more cracks a minimum of a grain diameter in length, or (2) a measurable change in the ratio of $\sigma^{+}_{max}/\sigma^{-}_{max}$ or graphical evidence of macrocrack closure as depicted schematically in Figure 6. The second criterion was applied only when the "fatal" crack initiated outside the area observable by the optical microscope. In general, the percentage of grain boundary segments containing cracks follows the empirical expression

$$FB = FB_{max} - FB_{max} \exp(-k\Delta\varepsilon_p/2) \qquad (1)$$

where FB is the fraction of grain boundary segments containing cracks, FB_{max} is the maximum fraction of segments broken at the maximum imposable plastic strain amplitude, k is an empirically determined constant, and $\Delta\varepsilon_p/2$ is the plastic strain amplitude. One experimental boundary condition is met at $\Delta\varepsilon_p/2 = 0$, since

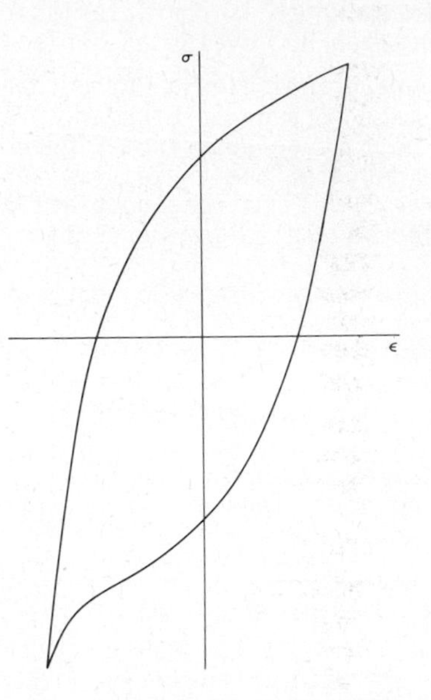

Fig. (6) - Schematic representation of the hysteresis loop and inflection point as evidence of macrocracking

the model insists that grain boundaries maintain their integrity under purely elastic deformation. The second boundary condition can be assessed by the fraction of segments containing cracks after monotonic tensile fracture (assuming no changes in the mode of failure), since this condition establishes a maximum plastic strain amplitude. The value for FB_{max} resulting in the best fit of empirical data was determined in an iterative manner, and the resulting equation is

$$FB = 0.85 - 0.85 \exp [- (185.4)(\Delta\varepsilon_p/2)] \tag{2}$$

The implication in equation (2) is that, with increasing plastic strain amplitude the fraction of grain boundary segments with cracks asymptotically approaches 85% assuming a constant failure mode.

The numerical value of "k" in equations (1) and (2) is an indication of the relative rapidity with which the fraction of damaged grain boundary segments saturates with increasing plastic strain amplitude. Its physical interpretation provides a quantitative measure of the capability of the material to accommodate cyclic plasticity of low amplitude without grain boundary failure. Conversely, the numerical value of FB_{max} in equations (1) and (2) should be interpreted as a description of material behavior at high plastic strain amplitudes, since it establishes the asymptote. As the strength or load carrying capability of the grain boundaries increases, the value of the asymptote, FB_{max}, decreases. Although it it suggested that both FB_{max} and k in equations (1) and (2) are materials parameters, these values are likely to be affected by grain size, grain texture, degree of prior cold work, and state of impurity segregation.

C. Compliance Changes as Evidence of Microcrack Initiation

Previously, other researchers have sought to find changes in mechanical properties which occur during the course of cyclic straining or stressing as an indication of fatigue crack initiation [3-6]. The criteria applied include a decrease in the ratio of $\sigma^+_{max}/\sigma^-_{max}$ [3], a drop, by a specified percentage, in the maximum tensile load (under strain control) [4,6], and the manifestation of an inflection point during the compressive portion of the hysteresis loop [5], as depicted schematically in Figure 6. Changes in mechanical properties of this order (here, defined as first order changes) are insensitive to the onset of microcrack initiation. As shown in the present study, the application of these criteria allows only for the detection of fatigue macrocracks which have eliminated a significant portion of the gage cross-section. As an example, at a plastic strain amplitude of 5 x 10^{-3}, first order changes in mechanical properties as cited above were not manifested until a crack on a microscopically unobserved surface had eliminated an estimated 25% of the load-bearing, cross-sectional area.

E_T, defined as the slope of the stress-strain curve immediately upon strain reversal (also defined as $\partial\sigma/\partial\varepsilon$), is considered, for the purposes of this study, to represent a second order change in mechanical properties. The change in this slope with cycling, $\partial^2\sigma/\partial\varepsilon\partial N$, as mentioned in the introduction, was thought to be the mechanical property having the greatest degree of promise for microcrack detection. A compliance measurement such as this should be little affected by changes in cyclic plasticity properties due to accumulated plastic strain prior to microcrack initiation. Presented in Figure 7 is a schematic

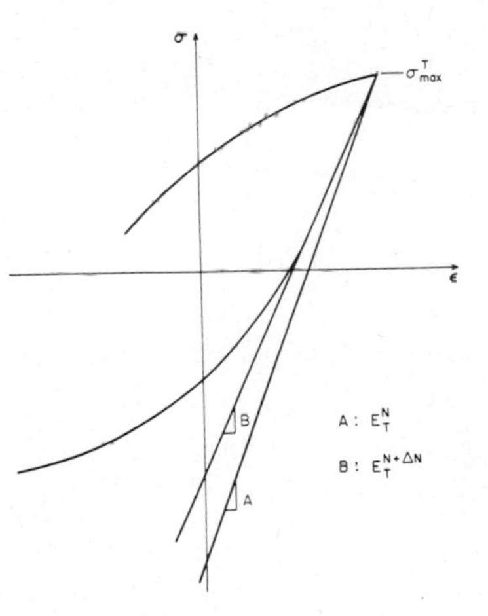

Fig. (7) - Schematic representation of E_T during the Nth and $N+\Delta N^{th}$ cycle depicting changes due to microcrack formation and/or propagation

representation of the decrease in E_T with increasing cycle number; the slope of the linear portion is determined graphically. Here E_T^N is evaluated after the

N^{th} cycle and $E_T^{N+\Delta N}$ after some increment in cycling, N+ΔN. The decrease in E_T, it is proposed, is due exclusively to the loss in the load-bearing, cross-sectional area of the sample due to the initiation and very early propagation of microcracks. The length and depth of the microcracks at the various cycling intervals and plastic strain amplitudes were determined by TEM replica examination [7].

Figures 8, 9 and 10 show E_T as a function of ε_{pa} for plastic strain amplitudes of 1×10^{-3}, 5×10^{-3} and 7.5×10^{-3}, respectively. Here, E_T is

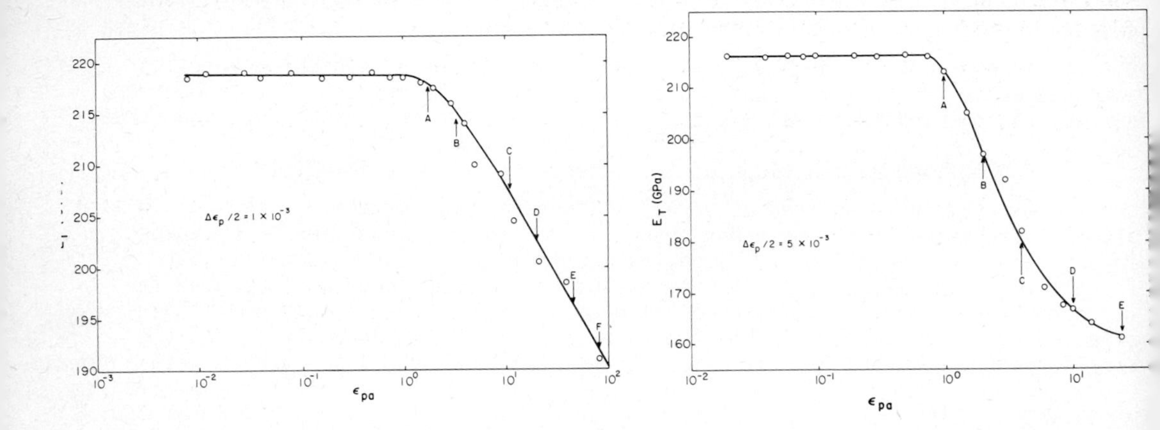

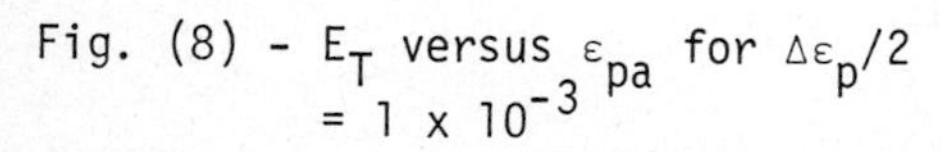
Fig. (8) - E_T versus ε_{pa} for $\Delta\varepsilon_p/2 = 1 \times 10^{-3}$

Fig. (9) - E_T versus ε_{pa} for $\Delta\varepsilon_p/2 = 5 \times 10^{-3}$

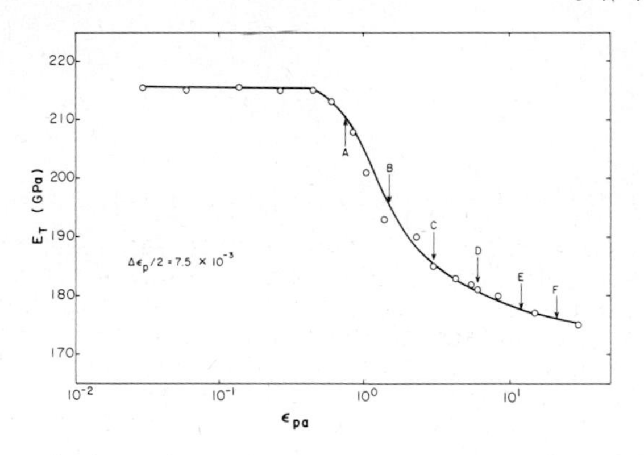

Fig. (10) - E_T versus ε_{pa} for $\Delta\varepsilon_p/2 = 7.5 \times 10^{-3}$

as previously defined, and ε_{pa} is the accumulated plastic strain, integrally related to N, the cycle number, through the equation

$$\varepsilon_{pa} = 4N(\frac{\Delta\varepsilon_p}{2}) \quad (3)$$

As can be seen readily, Figures 8, 9 and 10 may be characterized by a period during which E_T remains constant (initial stages of cycling), followed by a monotonic decrease in E_T with increasing cumulative plastic strain. The lettered points on these figures correspond to various degrees of fatigue damage and extent of microcracking, as detailed in Table 2 for plastic strain amplitudes of 1×10^{-3}, 5×10^{-3} and 7.5×10^{-3}, respectively. In every case, the values in this table are the results from TEM examination of replicas taken at the various stages indicated by the letters in Figures 8, 9 and 10.

TABLE 2 - CRACK LENGTHS FOR THREE PLASTIC STRAIN AMPLITUDES

$\Delta\varepsilon_p/2$	Point	Crack Length, μm
1×10^{-3}	A	2
	B	5
	C	8
	D	25
	E	46
	F	56
5×10^{-3}	A	3
	B	6
	C	16
	D	33
	E	50
7.5×10^{-3}	A	1.3
	B	5
	C	15
	D	31
	E	40
	F	52

The application of the present technique using changes in E_T as an indication of microcracking results in the reliable detection of fatigue damage in iron and is capable of resolving microcracks as small as 5 μm long. The explanation and justification for the success of this technique are simple. Microcracks which form on the surface of the gage section have a maximum crack tip opening displacement at maximum stress and plastic strain in tension. Upon strain reversal, the microcracks tend to close immediately, decreasing the slope of the stress-strain curve; with increases in the extent of microcracking, E_T decreases.

While small changes in the modulus of elasticity of a few percent due to prior cold work have been reported for many materials [8-10], the results presented here are not attributed to this phenomenon on the basis of several arguments. First, E_T during the "crack-free" portion of the histogram remains constant. If changes in E_T were to be ascribed, in part, to changes due to plastic deformation, such changes would be manifested before such a large accumulation of plastic strain. Second, throughout the entire range of plastic strain amplitudes and accumulated plastic strains, the slopes of the stress-strain curve immediately upon strain reversal from maximum compression remain constant. Changes due to the accumulation of plastic strain would certainly be expected in compression as well as in tension. Finally and perhaps most importantly, increases, not decreases, in the modulus of elasticity were reported for iron with increasing prior cold work due to a reduction in the magneto-mechanical hysteretic strain [11]. Thus, the decrease in E_T is clearly not a consequence of changes in the deformation state of the uncracked material.

The trend established by the three data sets presented in Figures 8, 9 and 10 is consistent and conclusive; changes in E_T constitute a reliable technique in the detection of fatigue microcrack initiation in iron.

D. The Coffin-Manson Relation Applied to Microcrack Initiation in Iron

The Coffin-Manson relation [12,13] empirically relates the number of reversals to failure, $2N_f$, to the plastic strain amplitude, $\Delta\varepsilon_p/2$. The general form of the equation is

$$\frac{\Delta\varepsilon_p}{2} = \varepsilon_f' \, (2N_f)^c \tag{4}$$

where ε_f' and c are empirically-determined material constants. As originally reported, the relation was intended to be a general description of material behavior in the high-amplitude, low-cycle fatigue regime. More recently, Lukas et al [14] extended the relationship to include the low-amplitude, high-cycle region for a wide variety of materials including iron-based alloys, a copper-zinc binary alloy and aluminum. Since initiation, microcrack growth, and macrocrack growth are highly different processes, the Coffin-Manson relation must represent summation of a number of "Coffin-Manson type" relations.

In this study, the Coffin-Manson relation was evaluated for fatigue microcrack initiation in iron. The form of the equation was accordingly modified:

$$\frac{\Delta\varepsilon_p}{2} = \varepsilon_i' \, (2N_i)^{c_i} \tag{5}$$

In the above equation, ε_i' and c_i are still empirically determined constants, now based on data for microcrack initiation rather than total life, and N_i is the number of cycles required to initiate a microcrack of a given size.

Figure 11 is a graph of the plastic strain amplitude versus the number of reversals to fatigue crack initiation. As indicated in the figure, two criteria were applied, resulting in independent assessments of the number of reversals necessary for fatigue crack initiation. On the left hand side of the diagram, the criterion for fatigue crack initiation corresponds to the inflection

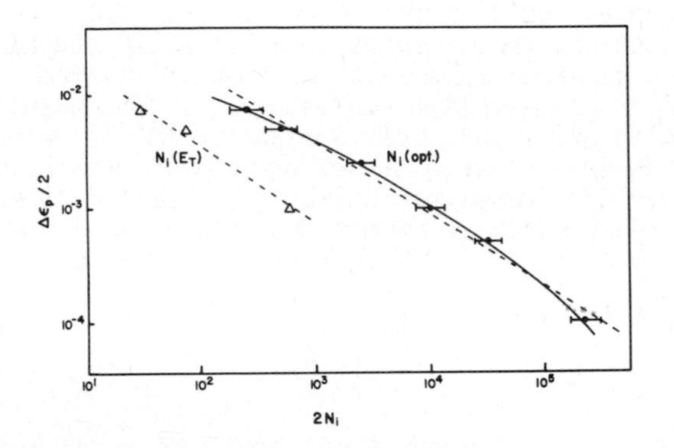

Fig. (11) - Modified Coffin-Manson diagram with plastic strain amplitude versus number of reversals to initiation. Open triangles represent data from the inflection points of Figures 8, 9 and 10, and solid circles represent data from 750 times magnification, optical microscopy

points of the E_T-ε_{pa} curves presented in Figures 8, 9 and 10. On the right hand side of the diagram, optical detection of microcrack initiation at 750 times magnification is the criterion plotted versus plastic strain amplitude. The dotted lines represent a least-squares linear fit of the data, and the solid curve represents a smooth curve fitting of the data. It can be seen that there is significant deviation from the least squares linear fit for the data corresponding to optical detection of initiation. At both low and high plastic strain amplitudes, the number of cycles indicated by the linear fit overestimates the actual number of cycles required. Many deviations from the Coffin-Manson relation for cycles to failure as well as cycles to a drop in σ^+_{max} have been reported in the literature [15,6], so these results for microcrack initiation are not surprising.

SUMMARY AND CONCLUSIONS

1. Transgranular microcrack initiation occurred along slip bands throughout the range of plastic strain amplitudes investigated ($1 \times 10^{-4} \leq \Delta\varepsilon_p/2 \leq 7.5 \times 10^{-3}$). Microcracks as small as 1 μm were documented, and these were preceded by small intrusions.

2. Microcrack initiation along grain boundary segments was not observed in the low range of plastic strain amplitudes, but the probability of grain boundary segment cracking increased rapidly with plastic strain amplitude. Such cracks as small as 0.5 μm were observed.

3. The cyclic change in the effective elastic unloading modulus, $\partial^2\sigma/\partial\varepsilon\partial N$, is very sensitive to early microcracking and has been demonstrated to reliably detect microcracks 5 μm in length or less.

4. The Coffin-Manson relation applied to microcrack initiation in iron showed appreciable curvature, suggesting that it is not an accurate description of microcrack initiation in iron.

ACKNOWLEDGEMENT

This research was sponsored by the National Science Foundation, Grant No. DMR-7926179 under the Industry/University Cooperative program. The authors appreciate much helpful discussion with Professor Y. W. Chung and Debassis Majumdar of Northwestern University and with Drs. Shrikant Bhat and Nassos Laziridis of Inland Steel Company Research Laboratories, who also furnished the iron. Use of the Central Facilities of Northwestern University's Materials Research Center, supported under the NSF-MRL program (Grant No. DMR79-23573), is gratefully acknowledged.

REFERENCES

[1] Ewing, J. A. and Humfrey, J. C., Phil. Trans., Vol. A200, p. 241, 1903.

[2] Tanaka, K. and Mura, T., J. Appl. Mech., Vol. 48, p. 97, 1981.

[3] Santner, J. S. and Fine, M. E., Scripta Met., Vol. 11, p. 159, 1977.

[4] ASTM Committee on Fatigue (E-9), Standard E606-80, ASTM Annual Book of Standards, Part 10, p. 694, 1980.

[5] Quesnel, D. J. and Meshii, M., Proceedings of the Second Int. Conf. on Mechanical Behavior of Materials, p. 764, 1976.

[6] Thielen, P. N., Fine, M. E. and Fournelle, R. A., Acta Met., Vol. 24, p. 1, 1976.

[7] Murr, L. E., "Electron optical applications in materials science", New York, McGraw-Hill Book Co., pp. 170-175, 1970.

[8] Fine, M. E. and Kennedy, N. T., J. Metals, Vol. 4, p. 151, 1952.

[9] Bradfield, G. and Pursey, H., Phil. Mag., Vol. 44, p. 437, 1953.

[10] Smith, A. D. N., Phil. Mag., Vol. 44, p. 453, 1953.

[11] Fine, M. E. in Tegart, W. J. McG., "Elements of mechanical metallurgy", The MacMillan Co., New York, p. 95, 1966.

[12] Coffin, L. F., Trans. ASME, Vol. 76, p. 931, 1954.

[13] Manson, S. S., NACA Technical Note 2933, 1954.

[14] Lukas, P., Klesnil, M. and Polak, J., Mater. Sci. Eng., Vol. 15, p. 239, 1974.

[15] Tomkins, B., Phil. Mag., Vol. 23, p. 687, 1971.

THE CYCLIC RESPONSE AND STRAIN LIFE BEHAVIOUR OF POLYCRYSTALLINE COPPER AND α-BRASS

N. Marchand, J.-P. Bailon and J. I. Dickson

Ecole Polytechnique
Montréal, Québec, Canada H3C 3A7

ABSTRACT

The cyclic-hardening behaviour, the cyclic stress-strain (CSS) response and the strain life behaviour were studied for copper and α-brass of two grain sizes at constant total strain amplitudes. The saturated stress for copper was grain size independent at low amplitudes, where a lower plateau was observed in the CSS curves. This plateau is rationalized in terms of the strong influence of the control parameter on the cyclic behaviour at low amplitudes. The cyclic response of α-brass was grain size dependent. An intermediate plateau in the CSS curve was obtained for the coarser grain size. The different stages in the CSS curves were reflected in the Manson-Coffin plots.

INTRODUCTION

Much of our fundamental knowledge about fatigue mechanisms stem from careful studies on single crystals of f.c.c. metals, e.g., [1-5]. It has been shown that the plateau in the CSS curve is associated with the formation of persistent slip bands (PSB's) within a less deformable matrix and with the volume occupied by PSB's increasing with increasing amplitude.

The cyclic deformation of polycrystals is less well understood. Dipolar walls of edge dislocations similar to those found in the ladder-like structures of PSB's have been observed [6-11]. However, since the grain boundaries act as barriers to PSB's, the resulting constraint may influence the CSS behaviour [1,2]. Whether a true plateau exists in the CSS curve of polycrystals has been a matter of controversy [12]. Pedersen et al [12] indicated that, although evidence may exist for a plateau in the case of coarse grain, texture-free copper [13,14], the CSS curve displays a positive slope for $\Delta\varepsilon_p > 3 \times 10^{-4}$. Lukas and co-workers [15,16] and Figueroa et al [8], have shown that the shape of the CSS curve depends on the testing variable controlled. For solution-treated Cu-4% Ti, Sinning and Haasen [17] observed a plateau in the CSS of polycrystals but not for monocrystals. They explained their results in terms of the annihilation distance of screw dislocations, which has been proposed [18] as the relevant parameter to characterize the cyclic slip mode. These previous studies suggest that the presence of the wall substructure might not be solely responsible for the existence of a plateau in the CSS curve for polycrystals. The present study is part of an investigation of the influence of grain size and cyclic plastic strain amplitude on the fatigue behaviour of copper and α-brass [19]. In the present paper, only the cyclic response and associated fatigue lives are considered.

EXPERIMENTAL PROCEDURE

The copper employed contained the following impurities in weight ppm: 420 Oxygen, 100 S, 20 Fe, 9 Pb, 9 Zn, <10 Sn and <5 As. The α-brass contained 29.9% Zn and the following impurities in weight ppm: 40 Fe, 110 S, 47 Pb, <10 Sn, <10 O, <5 As. The cast copper ingot was reduced 21% by hot-rolling at approximately 850°C followed by a water quench. It was further reduced 80% by cold rolling and annealed in argon either for 3 hrs at 315°C or for 1 hr at 805°C to obtain average grain sizes of 12 ±2μm, or 120 ±5μm respectively, as determined by an intercept method. The finer grain size is typical of the lower limit found in the literature for pure Cu [20,21]. The α-brass was reduced 35% by cold-working annealed 1 hr at 600°C and further reduced 60% by cold-working. It was then annealed 3 hrs at 420°C or 1 hr at 620°C to obtain average grain sizes d of 12 ±2μm or 120 ±10μm, respectively.

Cylindrical specimens of 6.35 mm diameter and 15.5 mm gauge length were machined, with final passes removing less than 0.13 mm. The specimens were polished longitudinally with 600 grit emery paper and chemically polished to remove a surface layer of at least 125μm. The low cycle fatigue test were performed at room temperature on a computer controlled servohydraulic testing system using Wood's Metal grips. Fully reversed pull-push strain cycling was employed at constant total strain amplitude $\Delta\varepsilon_t$ of ±0.00075-0.015 as measured with a clip-on axial extensometer. To prevent knife-edge crack initiation methyl-acrylate glue was applied near both shoulders. Only fatigue lives for fracture in the middle of the gauge length were considered. A symmetric sawtooth wave form was employed with the first half-cycle applied in tension. The product of the frequency and total strain was kept constant at 0.00125 s^{-1} for the fine-grained and 0.00250 s^{-1} for the coarse-grained materials. The hysteresis loops were recorded by computer at selected intervals. The cyclic stress amplitudes ($\Delta\sigma$) were obtained by averaging successive peak tensile and compressive stresses. The plastic strain ($\Delta\varepsilon_p$) was measured at zero mean stress and taken as the mid-width of the hysteresis loop.

RESULTS

Figures 1 and 2 show for copper and α-brass respectively, the variation in $\Delta\sigma$ with N, the number of cycles. The curves plotted in Figure 1 for low $\Delta\varepsilon_t$ are those which show the greatest difference in $\Delta\sigma$ in order to prevent these curves for copper from running into each other. Figures 3 and 4 display the corresponding variation in $\Delta\varepsilon_p$. The CSS curves are plotted on a linear scale in Figures 5 and 6 and on a semi-logarithmic scale in Figure 7. The curves are generated from data taken at mid-life, which for copper corresponded to stabilized $\Delta\sigma$ and $\Delta\varepsilon_p$ values. Figures 8 and 9 display the $\Delta\varepsilon_t$ and $\Delta\varepsilon_p$ (mid-life values) versus fatigue life curves. The Manson-Coffin exponents for the different strain amplitude range are given in Table 1.

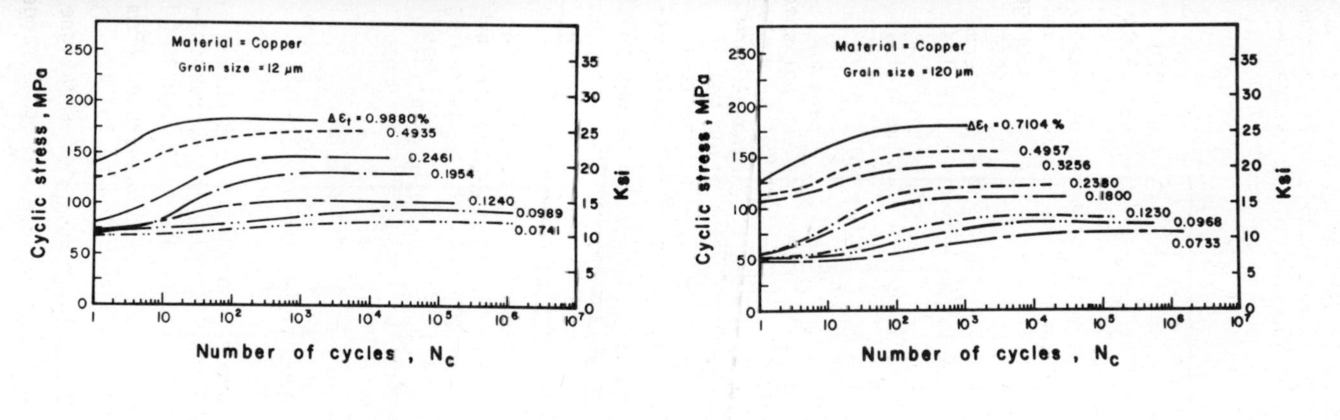

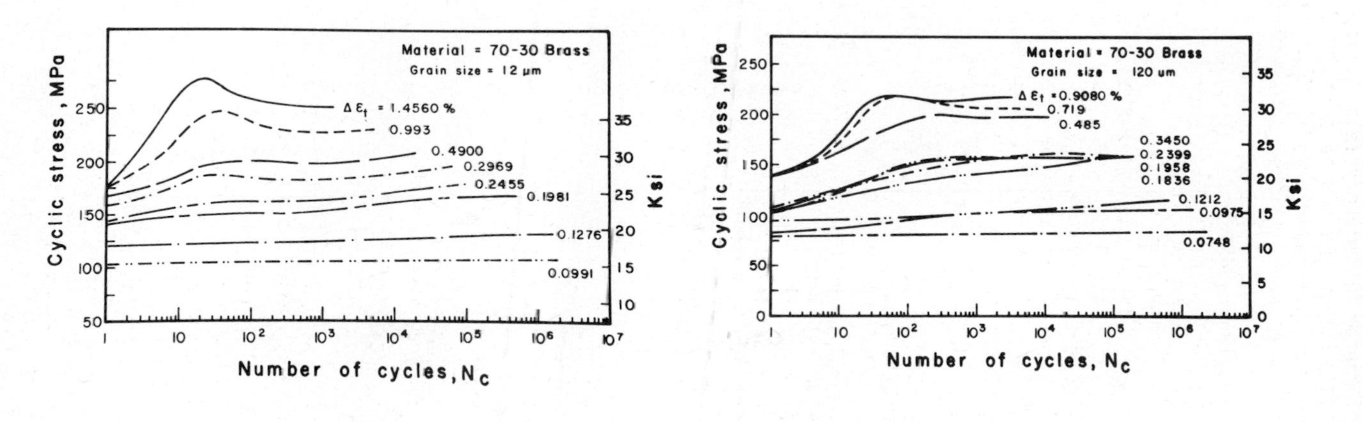

Fig. (1) - Cyclic stress-log N curves for copper a) d = 12 μm, b) d = 120 μm

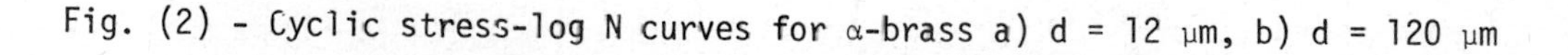

Fig. (2) - Cyclic stress-log N curves for α-brass a) d = 12 μm, b) d = 120 μm

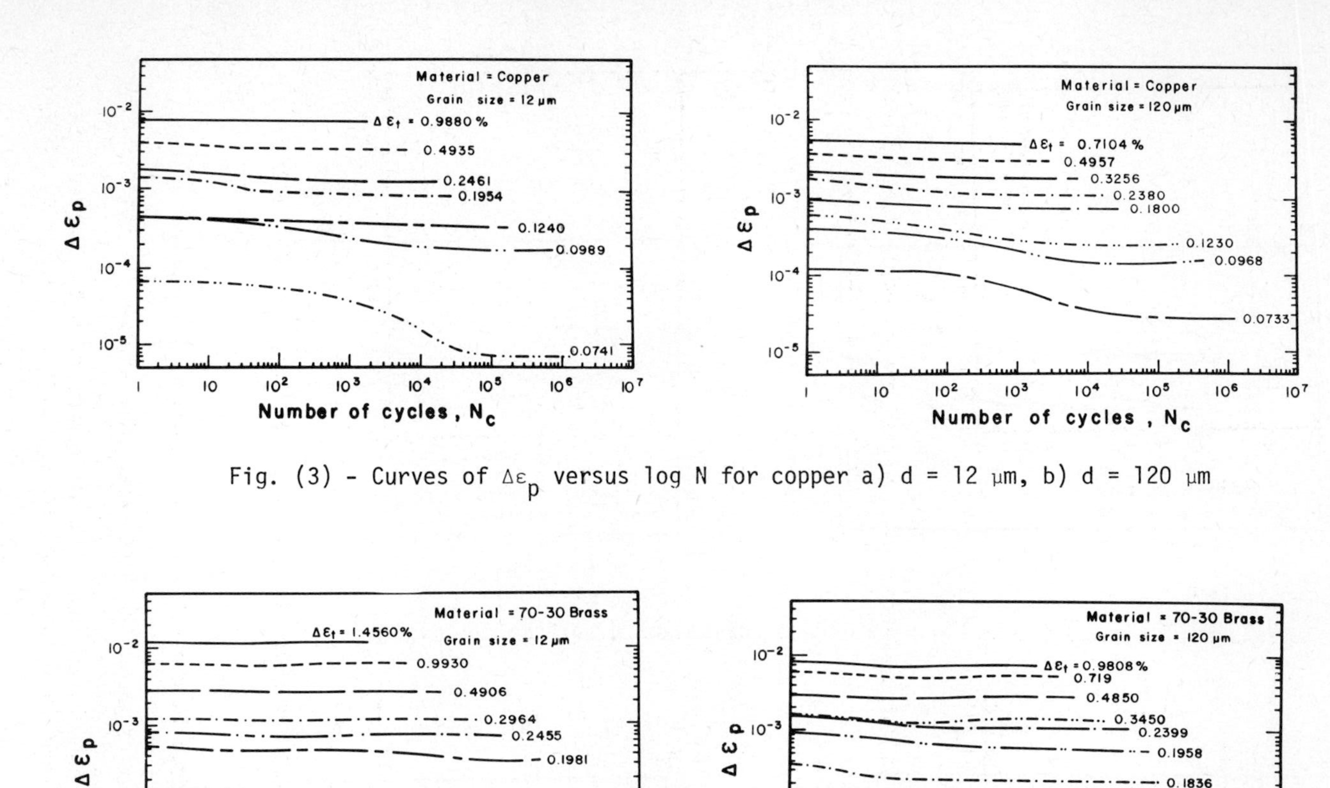

Fig. (3) - Curves of $\Delta\varepsilon_p$ versus log N for copper a) d = 12 μm, b) d = 120 μm

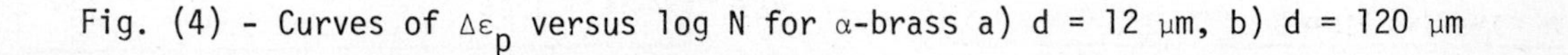

Fig. (4) - Curves of $\Delta\varepsilon_p$ versus log N for α-brass a) d = 12 μm, b) d = 120 μm

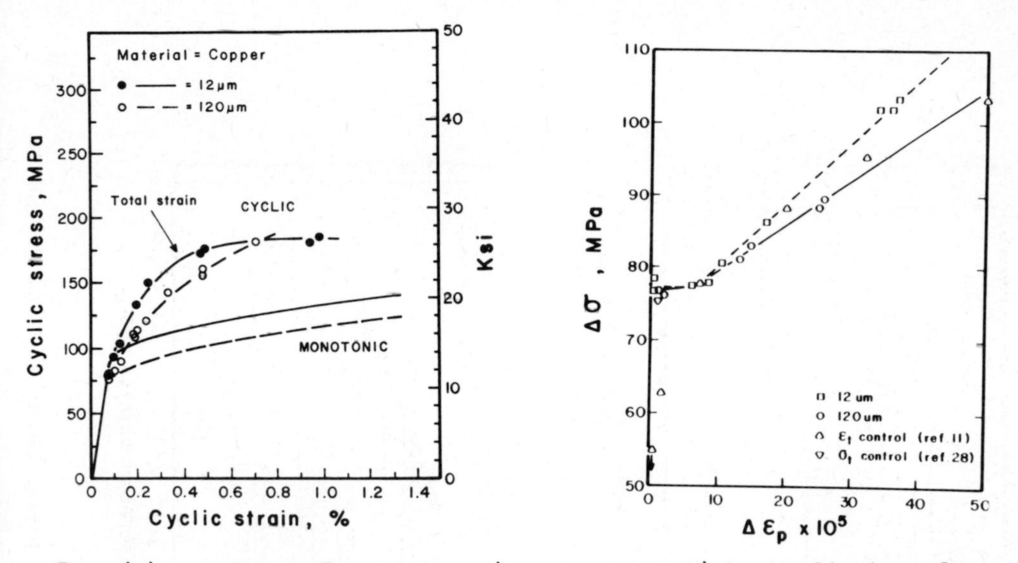

Fig. (5) - CSS curves for copper a) full curves, b) low amplitudes only

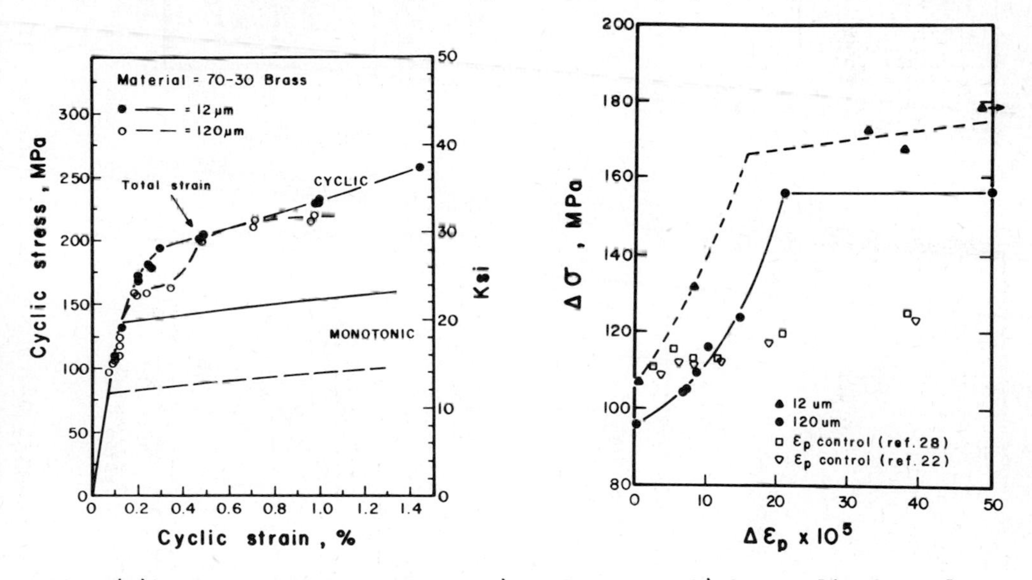

Fig. (6) - CSS curves for α-brass a) full curves, b) low amplitudes only

DISCUSSION

A. The Lower Plateau in the CSS Curve for Copper

For copper polycrystals, the appearance of a plateau in the CSS curves is indicated both on the basis of semi-logarithmic, Figure 7a, and linear plots, Figure 5b. In Figures 3a and 3b, an important gap can be seen for samples cycled at $\Delta\varepsilon_t$ = 0.00074 and 0.000984 which have similar saturated $\Delta\sigma$ values, although the

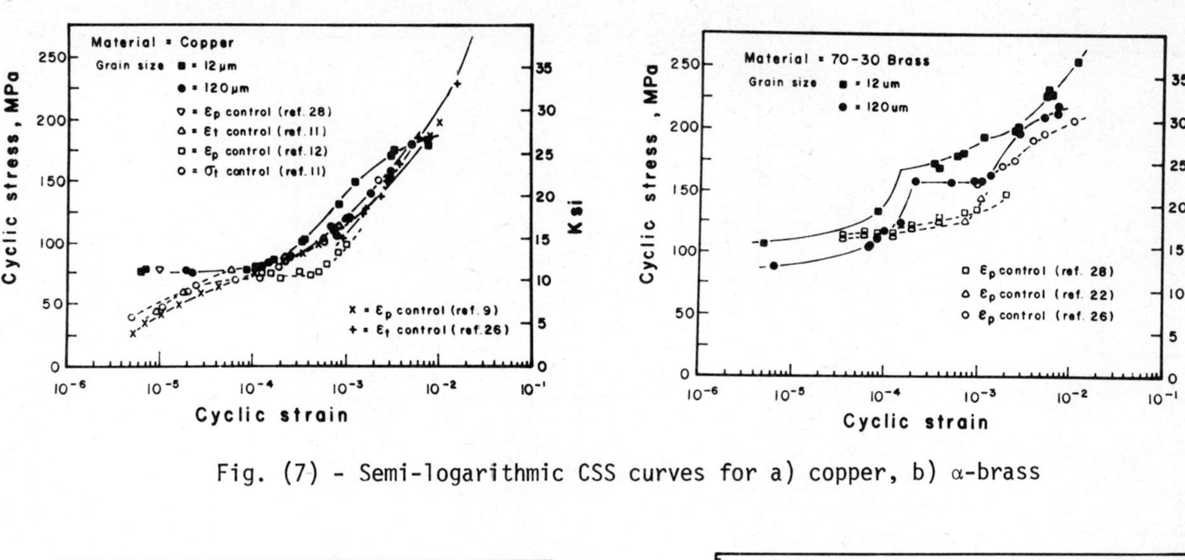

Fig. (7) - Semi-logarithmic CSS curves for a) copper, b) α-brass

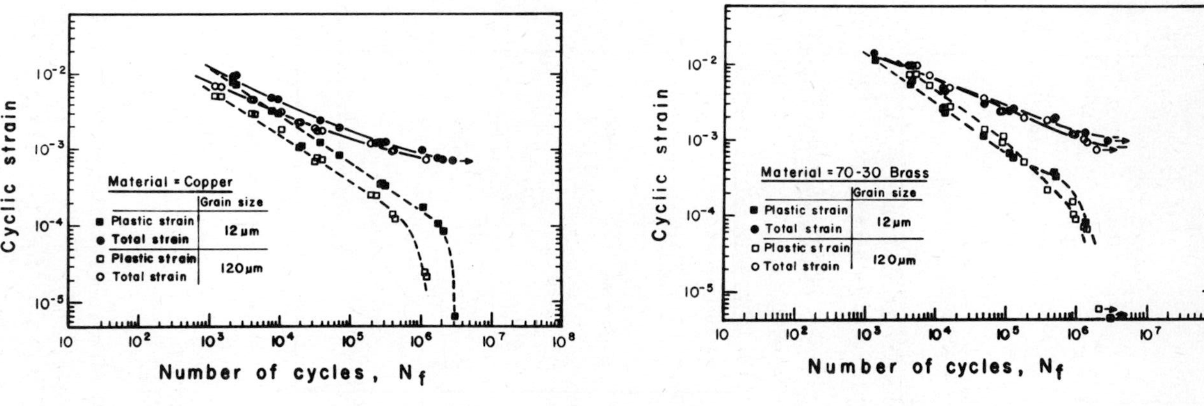

Fig. (8) - Manson-Coffin plots for copper

Fig. (9) - Manson-Coffin plots for α-brass

TABLE 1 - MANSON-COFFIN EXPONENTS (β)

Material	Grain Size	Plastic Strain Range ($\Delta\varepsilon_p$)		
		$\Delta\varepsilon_p < 2 \times 10^{-4}$	$10^{-4} < \Delta\varepsilon_p < 10^{-3}$	$\Delta\varepsilon_p > 10^{-3}$
copper	12 μm	not calculated (sharp knee)	$\beta = -0.63$	$\beta = -0.63$
	120 μm	not calculated (sharp knee)	$\beta = -0.61$	$\beta = -0.61$
α-brass	12 μm	not calculated	not calculated	$\beta = -0.67$
	120 μm	not calculated (sharp knee)	$\beta = -0.96$	$\beta = -0.65$

hardening period is much longer at the lower $\Delta\varepsilon_t$. As already mentioned, the curves plotted in Figure 1 for low $\Delta\varepsilon_t$ were those with the greatest differences in $\Delta\sigma$. The spreading effect observed, for which the saturated value of $\Delta\varepsilon_p$ increases disproportionately with an increase in the corresponding value of $\Delta\varepsilon_t$ is expected for a plateau behaviour. Figures 10 and 11 compare the cyclic hardening

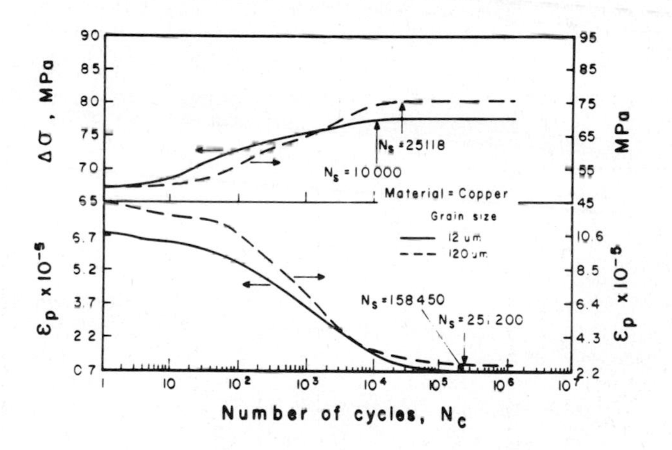

Fig. (10) - Cyclic hardening curves for copper, $\Delta\varepsilon_t \simeq 0.00098$

behaviour as observed by the variation in $\Delta\sigma$ and $\Delta\varepsilon_p$ for two of the lower values of $\Delta\varepsilon_t$. For such small amplitudes, the stress hardening period is much shorter than the strain hardening period, with this difference increasing as $\Delta\varepsilon_t$ decreases. For example, for d = 120 μm, the hardening periods differ by approximately 225,000 and 12,000 cycles for $\Delta\varepsilon_t$ values of 0.000733 and 0.000968. This difference in hardening period is consistent with a plateau behaviour and would also be consistent with a dislocation substructure formed towards the end of the stress harden-

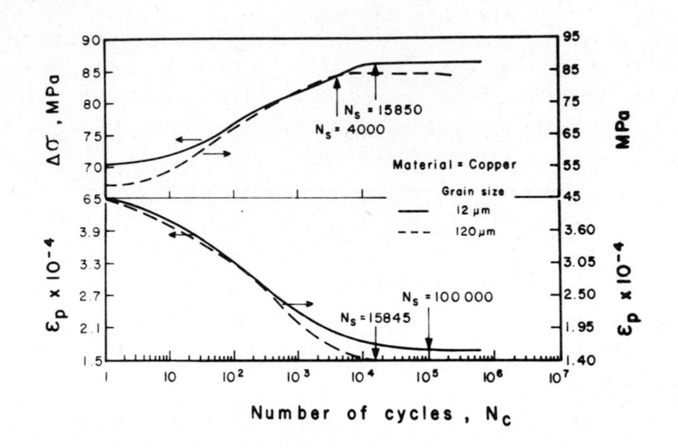

Fig. (11) - Cyclic hardening curves for copper, $\Delta\varepsilon_t$ = 0.00075

ing period, which after the saturated stress has been reached, gradually spreads through the sample volume.

Previous studies, however, have indicated that the testing procedure strongly influences the shape of the CSS curve. As well, Mughrabi did not obtain a plateau for $\Delta\varepsilon_p$ control for a grain size of 25 μm [9]. In $\Delta\varepsilon_t$ and $\Delta\sigma$ controlled tests at low amplitudes, there is a considerable relative difference between the values of $\Delta\varepsilon_p$ in the early cycles and the saturated values. Since recovery mechanisms are sluggish at these low amplitudes, strain history effects could be important in $\Delta\varepsilon_t$ control and more so in $\Delta\sigma$ control. Such effects can also influence in total plastic strain ($\Delta\varepsilon_a + \Delta\varepsilon_p$) control tests, as a result of the strong variation in $\Delta\varepsilon_a$ during tests at low amplitudes. Because of the differing influence of the strain history as well as the differences in the nature of the tests, the relative importance of the $\Delta\varepsilon_a$ contribution would also depend on the parameter controlled.

In $\Delta\varepsilon_t$ controlled tests, Figueroa et al [8] for a 350 μm grain size did not obtain a plateau. In the present study, the 12 μm grain size gives a clearer and possibly a longer plateau than the 120 μm grain size. It appears that in $\Delta\varepsilon_t$ control, the tendency for a plateau increases as d decreases.

From previous transmission electron microscopy observations on coarse grained copper polycrystals [6,8] and taking into account the decrease in $\Delta\varepsilon_p$ that occurs in tests at low $\Delta\varepsilon_t$, it can be assumed that for the plateau region, the microstructure consists of patches or veins of dipole loops, probably containing some regularly spaced walls and possibly loose arrangement of dipoles and debris [8]. It is the dislocation substructure in the vicinity of the grain boundaries, however, that is of interest concerning the influence of grain size on the plateau behaviour. Less is known about this substructure. The fraction of the volume occupied by this substructure, however, should increase as the grain size decreases. As well, this substructure is probably particularly important in causing strain history effects. Its presence probably is also directly or indirectly responsible for increasing the relative importance of the $\Delta\varepsilon_a$ contribution as the grain size decreases, Figure 12, in which $\Delta\varepsilon_a$ is taken as $(\Delta\sigma/E) - \Delta\varepsilon_p$. Thus, the clearer tendency for a plateau for the finer grain

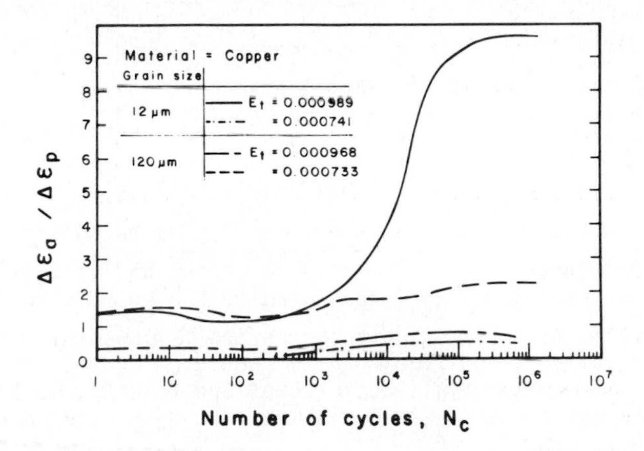

Fig. (12) - $\Delta\varepsilon_a/\Delta\varepsilon_p$ versus N for copper at low amplitudes

sizes is consistent with this plateau being associated with the manner in which the cyclic tests are controlled. It appears probable that in addition to texture differences [24], effects due to differences in control parameter are important in explaining the discrepancies [12] between the different results on polycrystalline copper. It is interesting to note that above $\Delta\varepsilon_p \simeq 0.0002$, the CSS curves in the literature appear to be similar for grain sizes $\geq$25 μm, while the CSS curve for the very fine (12 μm) grain size is displaced to the left, Figure 7a.

B. Cyclic Hardening Behaviour of 70-30 Brass

The cyclic hardening response of α-brass as well as the initial stresses obtained for the lower amplitudes show a more important variation with $\Delta\varepsilon_t$ compared to copper. At higher amplitudes, the cyclic hardening curves pass through a maximum, which is more pronounced for the finer grain size. In Cu-4% Ti, Sinning and Haasen [17] observed a similar maximum. In other single-phased metals, such a maximum has been related to pronounced changes in dislocation substructures [8,25]. In particular, in commercial-purity zirconium and titanium [25], which initially behave as planar slip materials, this maximum is related to the onset of cross-slip which permits mutual annihilation of screw dislocations. It appears probable that the maxima in Figure 2 are also associated with a change in dislocation substructure; however, transmission electron microscopy observations are required to identify the nature of this change. The pronounced maximum obtained at $\Delta\varepsilon_t \simeq 0.001$ for the finer grain size may be related to the greater amount of multiple slip giving rise to a harder dislocation substructure during the initial cycling, which acts as a stronger driving force for the occurrence of recovery mechanisms such as cross-slip and for a transition to softer dislocation substructures.

At low amplitudes, the $\Delta\sigma$-log N curves when plotted on a sufficiently-extended abscissa show a slow but steady semi-logarithmic hardening rate with little tendency to saturate. At these amplitudes, the dislocation should be arranged in planar arrays, which are likely to undergo reversible slip responsible for promoting long fatigue lives [26]. The ratio $\Delta\varepsilon_a/\Delta\varepsilon_p$ increases when $\Delta\varepsilon_t$ decreases and for example varies from 0.10 to 1.0 for $\Delta\varepsilon_t$ = 0.000975 to 0.000748 for the coarse-grained α-brass. The low hardening rate suggests that the dislocation substructures change little during the cycling but that there is a slow increase in dislocation density.

At intermediate amplitudes ($0.002 < \Delta\varepsilon_t < 0.0035$), the $\Delta\sigma$-log N behaviour changes gradually with increasing $\Delta\varepsilon_t$ from continuous hardening with little tendency to saturate to hardening followed by softening and/or quasi-saturation. The results of Hatanaka and Yamada [27] for relatively high values of $\Delta\varepsilon_p$, show a similar change from continuous hardening to hardening followed by quasi-saturation. With their heat treatment, their grain size should be comparable to our coarse-grained material. For lower amplitudes, Lukas and Polak [16] at 80 Hz found much longer hardening stages (>85% N_f) and claimed that saturation was obtained in constant $\Delta\varepsilon_p$ tests but not in $\Delta\sigma$ control. The gradual change from slow continuous cyclic hardening to rapid hardening-softening-quasi-saturation is consistent with an increasing importance of multiple slip with increasing $\Delta\varepsilon_t$. Multiple slip should promote rapid hardening which can result in dislocation substructures which become unstable. Recovery mechanisms, probably cross-slip, become activated and softening results.

For the intermediate strain amplitudes, the $\Delta\sigma$ values merge together while this behaviour is not observed in the $\Delta\varepsilon_p$-log N curves. The result is a plateau in both the linear and semi-logarithmic plots of the CSS curves, Figures 6b and 7b. For the 12 μm grain size, the linear plot is consistent with a quasi-plateau in the same $\Delta\varepsilon_p$ range and a small positive slope is obtained. The results indicate that the presence of the plateau is a true material response, which is strongly influenced by the grain size. As previously mentioned, Sinning and Haasen [17] observed a similar plateau for approximately the same strain range in Cu-4% Ti of similar grain size (110 μm). The shape of their CSS curve moreover is similar to that presently obtained for the 120 μm grain size except in the high strain range $\Delta\varepsilon_p > 10^{-3}$, where they did not observe a tendency for $\Delta\sigma$ to saturate. They explained the absence of the plateau in Cu-4% Ti single crystals by grain boundaries being required to give rise to the stress concentrations and dislocation densities necessary to provoke cross-slip and to cause persistent slip band-like structures. They observed surface markings within slip bands, which they pointed out appeared similar to the ladder-like substructure. As well, they observed that prominent slip bands in adjoining grains often tended to coincide at grain boundaries and explained the plateau by the spread of PSB-like substructures across grain boundaries. Our preliminary optical microscopy observations suggest increased amount of cross-slip above the plateau but do not suggest a strong tendency for prominent slip bands to coincide across grain boundaries in the plateau region. If the pla teau is assumed to be associated with PSB-like substructures and assuming that the grain boundary regions act largely to accommodate the strain differences between adjoining grains, the volume available for spreading of these structures within grains can be considered to be less in the finer grains. As well, finer grains tend to promote slip homogenization and multiple slip which should disfavor slip localization, to reduce the extent of the plateau region and to increase the slope of the CSS curve. The result would be a quasi-plateau. As pointed out previously [17], further information about dislocation substructures produced during the cyclic deformation of materials with similar tendencies for planar slip is required before these tentative explanations can be verified. Although there are significant differences in the slip modes of the two materials and in the $\Delta\varepsilon_t$ values involved, fine-grain copper appears to favor a lengthening rather than a shortening of the plateau obtained at low amplitudes in $\Delta\varepsilon_t$ control, which further suggests that the plateau in copper is strongly influenced by the test procedure.

C. Strain Life Behaviour

The Manson-Coffin plots for copper show that a smaller grain size promotes a longer fatigue life. The $\Delta\varepsilon_t$ versus N_f curves, however, indicate no influence of α on fatigue lives at low $\Delta\varepsilon_t$ and an increasing beneficial effect of fine grains as $\Delta\varepsilon_t$ increases. This is similar to the tendency observed by Thompson and Backofen [29] in stress controlled tests. This similarity can also be related with our CSS curves which show a $\Delta\sigma$ plateau at $\simeq$75 MPa, independent of grain size. The difference with the $\Delta\varepsilon_p$ versus N_f curve point out the less important influence of $\Delta\varepsilon_a$ and the elastic strain amplitude in promoting fracture. The influence of grain size on N_f at a given $\Delta\varepsilon_p$ indicates differences in the effectiveness of the plastic strain amplitude in causing strain localization which leads to earlier failure. As expected, a finer grain size promotes strain homogenization and a longer fatigue life. The apparent discrepancy between our results and those of Lukas and co-workers [15,16] appears largely related to our measurement of $\Delta\varepsilon_p$ as the half-width of the hysteresis loops rather than as the total plastic strain amplitude, $\Delta\varepsilon_p + \Delta\varepsilon_a$. Adding $\Delta\varepsilon_a$ to the present $\Delta\varepsilon_p$, values would remove the sharp knee in the Manson-Coffin plot. As well, it is at the lowest amplitudes employed that $\Delta\varepsilon_p$ varies most strongly during a test performed under $\Delta\varepsilon_t$ control and that the early cycles at higher $\Delta\varepsilon_t$ can be expected to have some influence.

The $\Delta\varepsilon_t$ versus N_f curves for α-brass, Figure 9, indicate a grain size effect at low amplitudes but little influence at high amplitudes. Thompson and Backofen [29] observed a similar trend, although they noted some grain size effect throughout their $\Delta\sigma$ range. For high amplitudes ($\Delta\varepsilon_p > 10^{-3}$), it is interesting to note that the coarse-grain α-brass show longer lives than the fine-grain material for a constant $\Delta\varepsilon_p$ value. The strain-life curve of Hatanaka and Yamada [27] also lies above that for our 12 μm grain size. Further study is required to explain this result and to determine its relationship to the influence of d on the CSS and cyclic hardening behaviour.

Since the fatigue life is determined primarily by the cyclic plastic strain which is related to the cyclic stress through the CSS curve, it is logical to check for inter-relationships between this curve and the Manson-Coffin plot. It is of particular interest to note that each stage of the CSS curve is mirrored by a stage in the $\Delta\varepsilon_p$ - life curve, as indicated by changes in the value of the Manson-Coffin exponent β. For example, the plateau observed in the CSS curve for copper corresponds to the knee in the Manson-Coffin plots. The coarse grain α-brass results provides a more stringent check for correspondence between the stages in the two curves. For the high strain amplitude range, the Manson-Coffin exponent, (-0.65), is similar to the value (-0.67) obtained by Hatanaka and Yamada [27] for $\Delta\varepsilon_p > 10^{-3}$. For amplitudes corresponding to the intermediate plateau region of the CSS curve, $\beta = -0.96$ is obtained compared to the value of -0.91 obtained by Klesnil and Lukas [15] for a similar amplitude, although their results did not indicate a plateau. For lower strain amplitudes, the present results suggest a tendency to a lower value of β. Since the CSS curve is related to the different dislocation mechanisms that operate to accommodate the plastic strain and the Manson-Coffin plot is related to the effectiveness of these mechanisms to cause damage leading to fracture, the correspondence between the two curves is not surprising.

It has previously been suggested [30] that the fatigue life may be independent of $\Delta\varepsilon_p$ in the plateau region. This is not the case for the coarse-grained α-brass, in agreement with previous results [31] on copper single crystals. The change in the Manson-Coffin exponent, however, indicates that the effectiveness of $\Delta\varepsilon_p$ to promote failure decreases more slowly with decreasing $\Delta\varepsilon_p$ in the plateau region than at higher amplitudes.

CONCLUSIONS

From the present study, it can be concluded that in the materials studied, different stages in the CSS curves are reflected by different stages in the Manson-Coffin plots. The latter plots are grain-size dependent except possibly for high amplitudes in α-brass. The plateau obtained in the CSS curve for copper is relate to the testing mode employed, which causes strain history effects and which influences the relative importance of the reversible strain amplitudes $\Delta\varepsilon_a$ obtained. The intermediate plateau for α-brass of 120 μm grain size, however, should be obtainable in $\Delta\varepsilon_p$ controlled tests.

ACKNOWLEDGEMENTS

Support from the Natural Sciences and Engineering Research Council of Canada and from the Ministry of Education of Quebec (FCAC program) is acknowledged.

REFERENCES

[1] Mughrabi, H., Proc. 5th Int. Conf. on the Strength of Metals and Alloys, Aachen, West Germany, August 1979, edited by P. Haasen, V. Gerold and G. Kostorz, Pergamon Press, pp. 1615-1638, 1979.

[2] Winter, A. T., Phil. Mag., Vol. 30, p. 719, 1974.

[3] Mughrabi, H., Mater. Sci. Eng., Vol. 33, p. 207, 1978.

[4] Basinski, Z. S., Korbel, A. S. and Basinski, S. J., Acta Met., Vol. 28, p. 191, 1980.

[5] Grosskreutz, J. C. and Mughrabi, H., in Constitutive Equations in Plasticity, A. S. Argon, ed., MIT Press, Cambridge, Mass., pp. 251-326, 1975.

[6] Winter, A. T., Pedersen, O. B. and Rasmussen, K. V., Acta Met., Vol. 29, p. 735, 1981.

[7] Winter, A. T., Acta Met., Vol. 28, p. 963, 1980.

[8] Figueroa, J. C., Bhat, S. P., DelaVeaux, R., Murzenski, S. and Laird, C., Acta Met., Vol. 29, p. 1667, 1981.

[9] Mughrabi, H. and Wang, R., Proc. of the 2nd Riso Int. Sym. on Met. and Matls. Sci., September 14-18, 1981, N. Hansen, A. Horsewell, T. Leffers and H. Lilholt, eds., pp. 87-98, 1981.

[10] Charsley, P., Mater. Sci. Eng., Vol. 47, p. 181, 1981.

[11] Pohl, K., Mayr, P. and Macherauch, E., Scripta Met., Vol. 14, p. 1167, 1980.

[12] Pedersen, O. B., Rasmussen, K. V. and Winter, A. T., Acta Met., Vol. 30, p. 57, 1982.

[13] Rasmussen, K. V. and Pedersen, O. B., Acta Met., Vol. 28, p. 1467, 1980.

[14] Kettunen, P. and Tiainen, T., Scand. J. Met., Vol. 10, p. 253, 1981.

[15] Lukas, P. and Klesnil, M., Mater. Sci. Eng., Vol. 11, p. 345, 1973.

[16] Lukas, P. and Polak, J., in "Work Hardening in Tension and Fatigue", A. W. Thompson, ed., AIME, pp. 177-205, 1977.

[17] Sinning, H. R. and Haasen, P., Z. Metall., Vol. 72, p. 807, 1981.

[18] Mughrabi, H. and Wang, R., in Proc. Int. Symp. on Defects and Fracture, Tuczno, Poland, G. C. Sih and H. Zorski, eds., Martinus-Nijhoff Pub., 1980.

[19] Marchand, N., M.Sc. Thesis, Ecole Polytechnique de Montreal, 1982.

[20] Thompson, A. W. and Backofen, W. A., Met. Trans. Vol. 2, p. 2004, 1971.

[21] Johnston, T. L. and Feltner, C. E., Met. Trans., Vol. 1, pp. 1161-1167, 1970.

[22] Franciosi, P., Berveiller, M. and Zaoui, A., Proc. 5th Int. Conf. on the Strength of Metals and Alloys, Aachen, West Germany, August 1979, edited by P. Haasen, V. Gerold and G. Kostorz, Pergamon Press, pp. 23-28, 1979.

[23] Kuhlmann-Wilsdorf, D., Mater. Sci. Eng., Vol. 39, p. 231, 1979.

[24] Burke, M. A. and Davies, G. J., Proc. 5th Int. Conf. on the Strength of Metals and Alloys, Aachen, West Germany, August 1979, edited by P. Haasen, V. Gerold and G. Kostorz, Pergamon Press, pp. 1181-1185, 1979.

[25] Handfield, L. and Dickson, J. I., this symposium.

[26] Saxena, A. and Antolovich, S. D., Met. Trans., Vol. 6A, p. 1809, 1975.

[27] Hatanaka, K. and Yamada, T., Bull. of JSME, Vol. 24, p. 613, 1981.

[28] Lukas, P. and Klesnil, M., Phys. Stat. Sol., Vol. 37, p. 833, 1970.

[29] Thompson, A. W. and Backofen, K. A., Acta Met., Vol. 19, p. 597, 1971.

[30] Laird, C., Finney, J. M. and Kuhlmann-Wilsdorf, D., Mater. Sci. Eng., Vol. 50, p. 127, 1981.

[31] Cheng, A. S. and Laird, C., Mater. Sci. Eng., Vol. 51, p. 55, 1981.

SECTION IV
MATERIAL DAMAGE AND FRACTURE

PREDICTION OF DAMAGE SITES AHEAD OF A MOVING HEAT SOURCE

M. Matczyński* and G. C. Sih

Lehigh University
Bethlehem, Pennsylvania 18015

ABSTRACT

When an intense moving energy source impinges on a solid, the local stress and strain state are disturbed and cause distortion and dilatation of the material. The degree of disturbance depends on the speed and intensity of the moving source and material properties. Examined is the possibility of a threshold condition such that the material may damage by a combination of yielding and fracture.

The problem is analyzed by application of the linear theory of thermoelasticity. Galilean transformation is used to eliminate the time dependence in the heat conduction equation such that the problem becomes one of steady state as the observer is made to move with the energy source. Calculated is the strain energy density function, dW/dV, for examining possible sites of material damage by yielding and fracture. This is achieved by determining the locations of the relative maximum, $(dW/dV)_{max}$, and relative minimum, $(dW/dV)_{min}$, of the strain energy density function. The onset of yielding and fracture are assumed to occur when $(dW/dV)_{max}$ and $(dW/dV)_{min}$ reach their respective critical values. Yielding is predicted to occur behind the moving heat source while fracture initiates off to the sides. These results are displayed graphically. The influence of additional applied mechanical stress is also discussed.

INTRODUCTION

Generally speaking, material can fail in a variety of modes depending on the nature and magnitude of loading, geometry and size of the component and microstructure of the material. These combinations can yield failure ranging from brittle to plastic collapse. The construction of a unified theory is difficult because of the enormous number of varying parameters that could affect the material behavior leading to failure.

From the viewpoint of continuum mechanics, a solid undergoes volume and shape change when load is applied. For a linear elastic material, these two changes

---

*On leave from the Institute of Fundamental Technological Research, Polish Academy of Sciences, Warsaw, Poland.

can be expressed in terms of the strain energy density component $(dW/dV)_v$ associated with dilatation and $(dW/dV)_d$ associated with distortion. The sum is the strain energy density function dW/dV which can be computed from the area under the true stress and strain curve [1]. It is also known [2] that in the neighborhood of each point within the continuum, there prevails a relative maximum, $(dW/dV)_{max}$, and a relative minimum, $(dW/dV)_{min}$, of the strain energy density function. This is referenced with respect to the coordinates that locate the position of an element near the point. These stationary values of dW/dV naturally lead to a general failure criterion [3] that can predict the locations of the onset of yielding and fracture.

This work is concerned with possible material damage due to a moving heat source. The intensity and speed of the source are incorporated into the thermoelastic stress and failure analysis for evaluating the dilatational and distortional component of the strain energy density function. Analyzed in detail is the redistribution of the local stress and energy field. These disturbances could attribute to the instability of the system if additional mechanical loads are also present.

TIME DEPENDENT TEMPERATURES AND THERMOELASTIC STRESSES FOR A MOVING HEAT SOURCE

Consider the problem of a heat source with magnitude Q_0 moving at a constant speed v along the x_1-axis as shown in Figure 1. The dimensions of the plate are

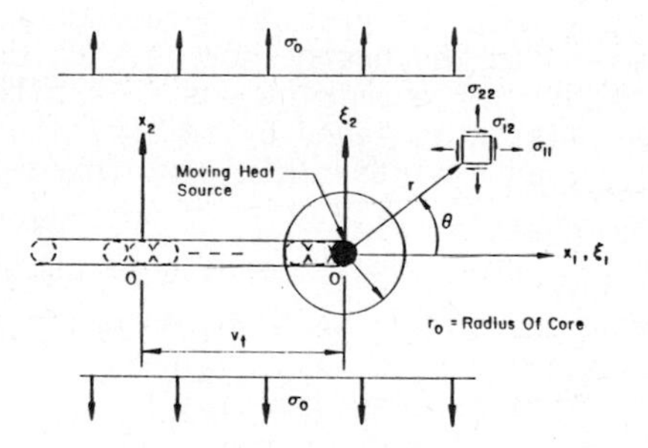

Fig. (1) - A moving heat source in a plate

assumed to be large enough such that the edges are sufficiently far away from the source and will not affect the local temperature field $T(x_i,t)$ and stress state $\sigma_{ij}(x_i,t)$ where i,j = 1,2 and t stands for the time.

As $x_i \to \infty$, the heat flow or temperature gradients $\partial T/\partial x_i$ and stresses σ_{ij} are taken as zero. Refer to [4] for a more detailed discussion of this problem.

Heat Conduction. With reference to the stationary coordinates x_i (i = 1,2) in Figure 1, the differential equation governing $T(x_i,t)$ is given by [5]

$$\kappa\nabla^2 T(x_i,t) - \dot{T}(x_i,t) = Q(x_i,t),\ i = 1,2 \tag{1}$$

in which $\nabla^2 = \partial^2/\partial x_1^2 + \partial^2/\partial x_2^2$ is the Laplacian operator in two dimensions and $\dot{T} = \partial T/\partial t$. The thermal diffusivity is $\kappa = k/\rho C$ with k being the thermal conductivity coefficient, ρ the mass density and C the specific heat.

Now, introduce the coordinate axes ξ_i ($i = 1,2$) such that they move with the heat source at the same speed v:

$$x_1 = \xi_1 + vt,\ x_2 = \xi_2 \tag{2}$$

Applying equation (2) and letting $Q(x_i,t)$ be a point source with intensity Q_0, equation (1) can be transformed to read as

$$\kappa\nabla_1^2 T(\xi_i) + 2c\kappa[T(\xi_i)]_{,1} = -\,Q_0\delta(\xi_1)\delta(\xi_2) \tag{3}$$

where $2c = v/\kappa$ and $\delta(\xi_i)$ is the Dirac delta function. The Laplacian operator in the moving coordinate system is $\nabla_1^2 = \partial^2/\partial\xi_1^2 + \partial^2/\partial\xi_2^2$ and $T_{,i} = \partial T/\partial\xi_i$. It can be shown that [5]

$$T(\xi_i) = \frac{Q_0}{2\pi\kappa}\exp(-c\xi_1)K_0(cr) \tag{4}$$

In equation (4), $K_n(cr)$ is the modified Bessel function of the second kind of order $n=0$ with argument cr where $r = (\xi_1^2+\xi_2^2)^{1/2}$.

Thermoelasticity. Once the temperature distribution is known, the thermal stresses can be found by making use of the displacement potential $\Phi(x_i,t)$ for the steady state case, [6]:

$$u_i(x_i,t) = \Phi_{,i} \tag{5}$$

where $\Phi_{,i} = \partial\Phi/\partial x_i$. The rectangular Cartesian components of the displacement vector are denoted by $u_i(x_i,t)$ ($i = 1,2$). For plane strain, $u_3 = 0$. Note that $\Phi(x_i,t)$ satisfies the equation

$$\nabla^2\Phi = \alpha\beta T \quad \text{(plane strain)} \tag{6}$$

in which α is the coefficient of thermal expansion and

$$\beta = \frac{1+\nu}{1-\nu} \tag{7}$$

with ν being the Poisson's ratio. The stress-displacement relations give

$$\sigma_{ij}(x_i,t) = 2\mu(\Phi_{,ij} - \delta_{ij}\nabla^2\Phi),\ i,j = 1,2 \tag{8}$$

with δ_{ij} being the Kronecker delta. The shear modulus of elasticity is μ and the transverse normal stress component $\sigma_{33}(x_i,t)$ takes the form

$$\sigma_{33}(x_i,t) = -\,2\mu\nabla^2\Phi \tag{9}$$

Making use of equation (3), equation (6) may be expressed as

$$\nabla_1^2\Phi,_1 = -\frac{\alpha\beta}{2c}\nabla_1^2(T+\Phi_o) \tag{10}$$

such that

$$\Phi_o(\xi_i) = \frac{Q_o}{2\pi\kappa}\log r \tag{11}$$

and remembering that $\delta(\xi_1)\delta(\xi_2) = (\log r)/2\pi$. A particular solution of equation (10) is

$$\Phi,_1 = -\frac{\alpha\beta}{2c}(T+\Phi_o) \tag{12}$$

which yields zero stress state at infinity. Defining the constant

$$N = \frac{\mu\alpha\beta Q_o}{2\pi\kappa} \tag{13}$$

the transient thermal stresses expressed in terms of ξ_i $(i = 1,2)$ are given by

$$\begin{aligned}\sigma_{11}(\xi_i) &= -N\{[K_o(cr) - \frac{\xi_1}{r}K_1(cr)]\exp(-c\xi_1) + \frac{\xi_1}{cr^2}\}\\ \sigma_{22}(\xi_i) &= -N\{[K_o(cr) + \frac{\xi_1}{r}K_1(cr)]\exp(-c\xi_1) - \frac{\xi_1}{cr^2}\}\\ \sigma_{12}(\xi_i) &= N[K_1(cr)\exp(-c\xi_1) - \frac{1}{cr}]\cdot\frac{\xi_2}{r}\end{aligned} \tag{14}$$

while $\sigma_{33}(\xi_i)$ is

$$\sigma_{33}(\xi_i) = -2NK_o(cr)\exp(-c\xi_1) \tag{15}$$

This completes the stress solution of the problem which is a prerequisite to the failure analysis.

STRAIN ENERGY DENSITY CRITERION

In order to be more specific, the strain energy density criterion [2,3] in its general form will be stated. It should be mentioned that

$$\frac{dW}{dV} = \int_0^{\varepsilon_{ij}} \sigma_{ij}d\varepsilon_{ij} \tag{16}$$

applies to all materials, elastic or elastic-plastic while equation (16) reduces to $\sigma_{ij}\varepsilon_{ij}/2$ only for the linear elastic material. The basic assumptions are:

(1) Yielding and fracture are assumed to coincide with *locations* of maximum of the local maximum and minimum of the strain energy density function $(dW/dV)_{max}$ and $(dW/dV)_{min}$, respectively.

(2) Yielding and fracture are assumed to occur when the maximum of $(dW/dV)_{max}$ and $(dW/dV)_{min}$ reach their respective critical values.

(3) The amount of incremental growth $r_1, r_2, \ldots, r_j, \ldots, r_c$ is governed by

$$\left(\frac{dW}{dV}\right)_c = \frac{S_1}{r_1} = \frac{S_2}{r_2} = \ldots = \frac{S_j}{r_j} = \ldots = \frac{S_c}{r_c} = \text{const.} \tag{17}$$

if the process of yielding and/or fracture leads to global instability*, i.e.,

$$\begin{aligned} &S_1 < S_2 < \ldots < S_j < \ldots < S_c \\ &r_1 < r_2 < \ldots < r_j < \ldots < r_c \end{aligned} \tag{18}$$

The criterion stated above is extremely general and applies to media with or without initial defects. In this work, the effect of loading rates will not be considered and hence Condition (3) is not needed but is given for the sake of completeness. The rate of energy transferred to the system will only depend on the velocity of the heat source.

Strain Energy Density Expression. The strain energy density function for a linear elastic material subjected to the plane strain condition can be computed from

$$\frac{dW}{dV} = \frac{1}{4\mu}\left[(1-\nu)(\sigma_{11}+\sigma_{22})^2 - 2(\sigma_{11}\sigma_{22}-\sigma_{12}^2)\right] + \mu(1+\nu)(\alpha T)^2 \tag{19}$$

Substitution of equations (14) into (19) yields

$$\frac{dW}{dV} = \frac{N^2}{2\mu}\left\{\frac{3-5\nu}{1+\nu}\left[K_o(cr)\exp(-c\xi_1)\right]^2 + \left[K_1(cr)\exp(-c\xi_1) - \frac{1}{cr}\right]^2\right\} \tag{20}$$

which can be computed numerically once the constants in equation (20) are known.

Dilatational and Distortional Components. It is well-known in the linear theory of elasticity that dW/dV in equation (19) may be divided into two components as follows:

---

*If the loading is such that yielding and/or fracture are arrested, then the ratio S_c/r_c in equation (18) should be replaced by S_o/r_o with

$$S_1 > S_2 > \ldots > S_j > \ldots > S_o$$

$$r_1 > r_2 > \ldots > r_j > \ldots > r_o$$

$$\frac{dW}{dV} = (\frac{dW}{dV})_v + (\frac{dW}{dV})_d \tag{21}$$

The dilatational component $(dW/dV)_v$ can be written as

$$(\frac{dW}{dV})_v = \frac{(1+\nu)(1-2\nu)}{12\mu}(\sigma_{11}+\sigma_{22}-2\mu\alpha T)^2 \tag{22}$$

and the distortional component $(dW/dV)_d$ is given by

$$(\frac{dW}{dV})_d = \frac{1}{6\mu}[(1-\nu+\nu^2)(\sigma_{11}+\sigma_{22})^2 + 3(\sigma_{12}^2-\sigma_{11}\sigma_{22})] + \frac{1+\nu}{3}[(1-2\nu)(\sigma_{11}+\sigma_{22})$$
$$+ 2\mu(1+\nu)\alpha T]\alpha T \tag{23}$$

The stress component σ_{33} is equal to $\nu(\sigma_{11}+\sigma_{22}) - 2\mu(1+\nu)\alpha T$ as the condition of plane strain has been assumed. Equations (14) can be put into equations (22) and (23) to obtain the numerical results.

Superposition of Applied Mechanical Stress. Suppose that a uniform stress of magnitude σ_o is also applied to the plate in the ξ_2 direction as shown in Figure 1, i.e., $\sigma_{22}(\infty) = \sigma_o$ = const. Since the problem is linear, the superposition principle prevails and equations (14) are altered only by adding σ_o to the expression for $\sigma_{22}(\xi_i)$. The strain energy density expression in equation (20) will be modified to

$$\frac{dW}{dV} = \frac{N^2}{2\mu}\{\frac{3-5\nu}{1+\nu}[K_o(cr)\exp(-c\xi_1)]^2 + [K_1(cr)\exp(-c\xi_1) - \frac{1}{cr}]^2\}$$
$$- \frac{N\sigma_o}{2\mu}\{[(1-2\nu)K_o(cr) + \frac{\xi_1}{r}K_1(cr)]\exp(-c\xi_1) - \frac{\xi_1}{cr^2}\} + \frac{(1-\nu)\sigma_o^2}{4\mu} \tag{24}$$

Note that the sign in the second term will change depending on whether σ_o is tensile or compressive.

DISCUSSION OF NUMERICAL RESULTS

Let the plate material in Figure 1 be made of 2024-T3 aluminum with the following properties:

$$\alpha = 13 \times 10^6 \text{ m/m/°F} \qquad \nu = 0.33$$
$$k = 1.903 \times 10^2 \text{ w/m°K} \qquad \mu = 2.748 \times 10^4 \text{ MPa} \tag{25}$$
$$C = 9.623 \times 10^2 \text{ J/Kg°K} \qquad \rho = 2.770 \times 10^3 \text{ kg/m}$$

From equation (24), it is seen that dW/dV for a given material will vary as a function of the parameters r, v and σ_o. The speed of the heat source will be set at v = 1.0 cm/sec while the values of $r = 10^{-3}$, 10^{-2} and 10^{-1} cm and $\sigma_o/N = -0.1$, 0.1 and 1.0 will be used for calculating dW/dV at the different angles θ around the heat source, Figure 1.

Intensity of Strain Energy Density. The material elements near the heat source tend to dilate and distort more than those far away. This can be seen by calculating the relative maximum and minimum values of dW/dV with reference to θ for $r = 10^{-3}$, 10^{-2} and 10^{-1} cm. The applied mechanical stress σ_o is fixed at a value of $\sigma_o = -0.1$ N corresponding to an uniaxial compression. Table 1 gives the

TABLE 1 - STATIONARY VALUES OF dW/dV FOR v = 1.0 cm/sec AND $\sigma_o/N = -0.1$

Radial distance r(cm)	$\frac{2\mu}{N^2}(\frac{dW}{dV})_{max}$	$\frac{2\mu}{N^2}(\frac{dW}{dV})^{max}_{v/d}$	$\frac{2\mu}{N^2}(\frac{dW}{dV})_{min}$	$\frac{2\mu}{N^2}(\frac{dW}{dV})^{min}_{v/d}$
10^{-3}	56.507 (θ = 180°)	1.975	55.533 (θ = 88°)	2.073
10^{-2}	27.582 (θ = 180°)	1.895	26.312 (θ = 79°)	2.076
10^{-1}	9.866 (θ = 180°)	1.686	7.737 (θ = 61°)	1.871

results for $(dW/dV)_{max}$ and $(dW/dV)_{min}$ corresponding to locations of possible initiation of yielding and fracture. The quantity $(dW/dV)^{max}_{v/d}$ represents the ratio of the dilatational to distortional strain energy density component for $(dW/dV)_{max}$ and $(dW/dV)^{min}_{v/d}$ for $(dW/dV)_{min}$. They all decrease in magnitude as r is increased from 10^{-3} to 10^{-1} cm. The results are displayed graphically in Figures 2 to 4. For $r = 10^{-3}$ cm, Figure 2(a) shows that $2\mu(dW/dV)/N^2$ at $\theta = 0°$ is 56.361 and decreases to a minimum of 55.533 at $\theta = 88°$. It then increases to a maximum value of 56.507 at $\theta = 180°$. For sufficiently large σ_o and N, yielding is predicted to initiate behind the heat source at $\theta = 180°$ where $(dW/dV)_{max}$ occurs and fracture at $\theta = 88°$ coincides with $(dW/dV)_{min}$. The variations of $(dW/dV)_v$ and $(dW/dV)_d$ with θ for $r = 10^{-3}$ cm are shown in Figures 2(b) and 2(c). Note that $(dW/dV)^{min}_{v/d}$ is always larger than $(dW/dV)^{max}_{v/d}$ regardless of r, Table 1. This can also be seen from the curves plotted in Figures 3 and 4. The general trends are the same as those in Figures 2.

Effect of Applied Mechanical Stress. The additional application of a mechanical load will affect the local energy state in a complex fashion depending on the sign of σ_o. Let attention be focused at the radial distance $r = 10^{-2}$ cm while the heat source speed is kept at v = 1.0 cm/sec. The values of $(dW/dV)_{max}$ and $(dW/dV)_{min}$ for $\sigma_o/N = -0.1$, 0.0, 0.1 and 1.0 are summarized in Table 2. As σ_o/N is increased from a state of compressive stress to tensile stress passing through

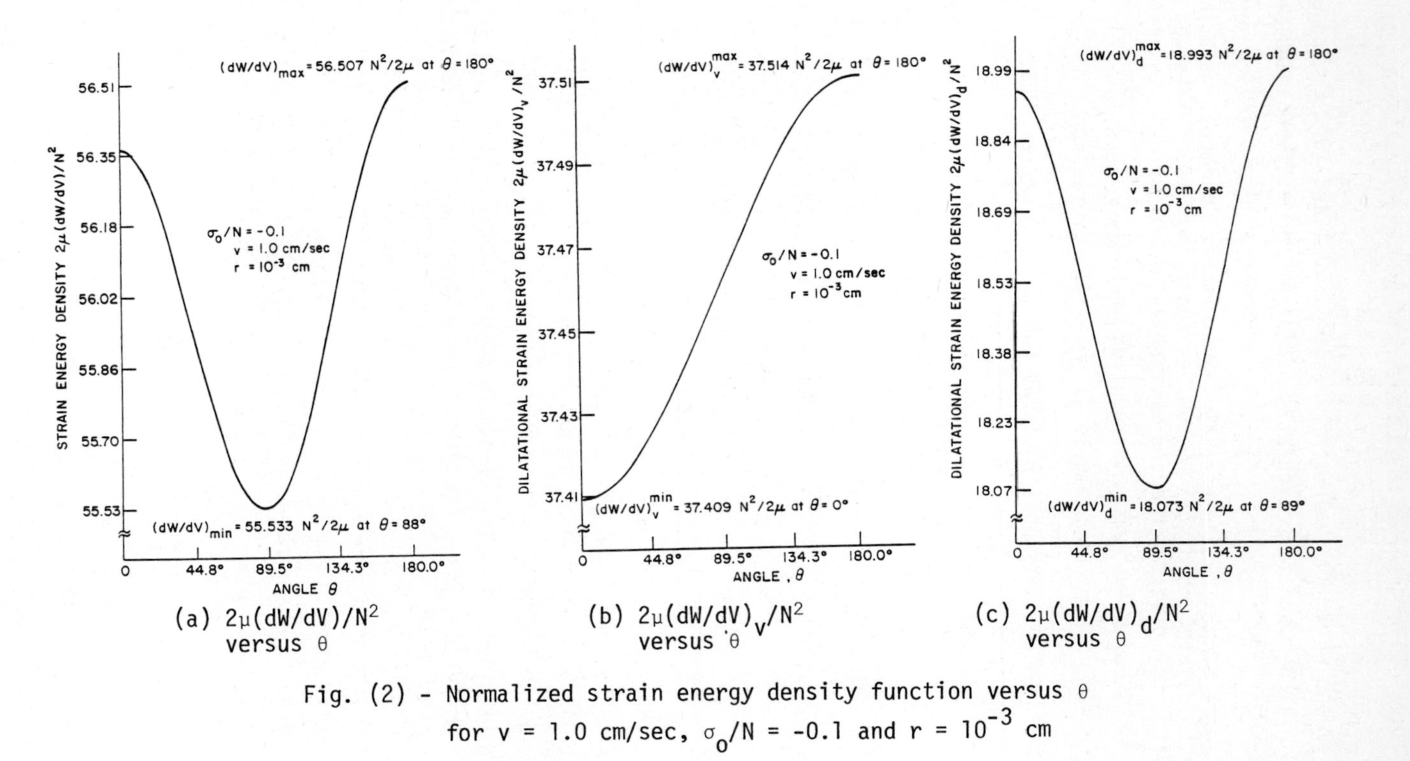

(a) $2\mu(dW/dV)/N^2$ versus θ

(b) $2\mu(dW/dV)_v/N^2$ versus θ

(c) $2\mu(dW/dV)_d/N^2$ versus θ

Fig. (2) - Normalized strain energy density function versus θ for $v = 1.0$ cm/sec, $\sigma_o/N = -0.1$ and $r = 10^{-3}$ cm

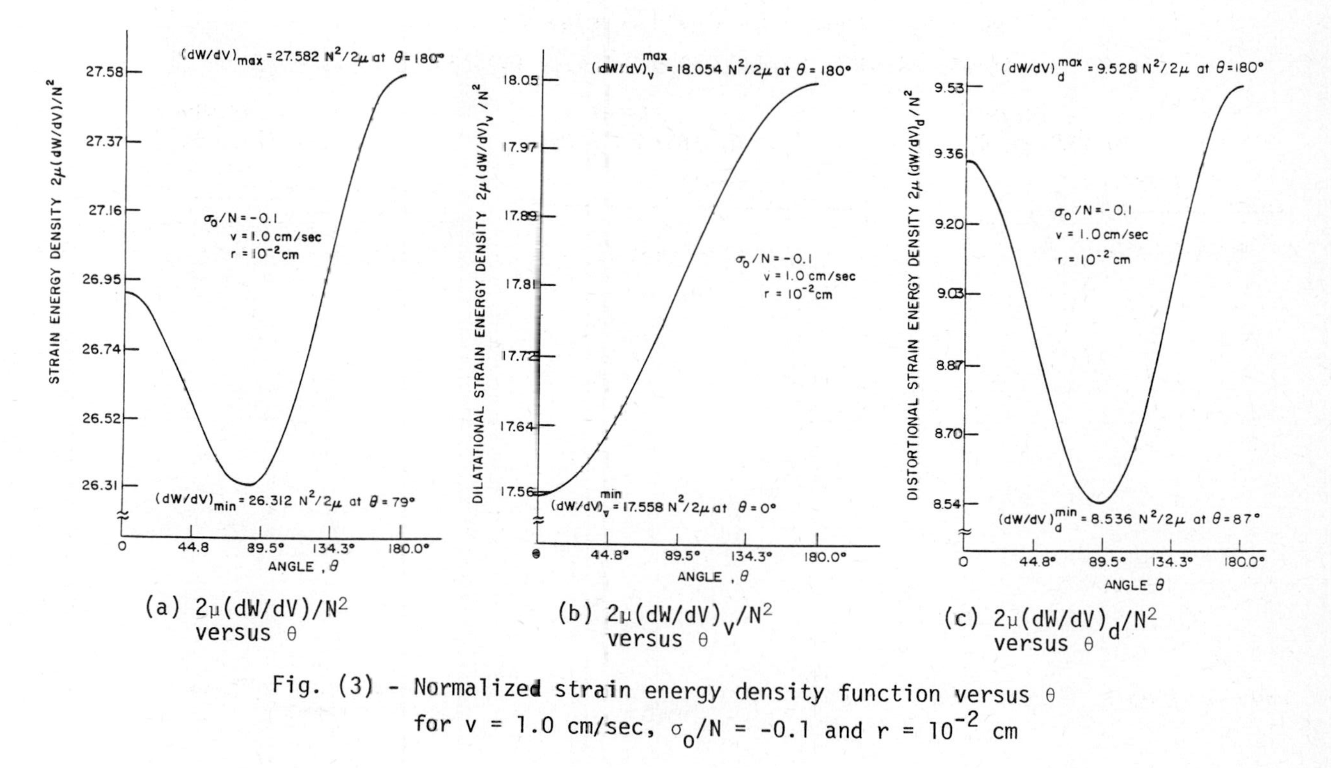

(a) $2\mu(dW/dV)/N^2$ versus θ

(b) $2\mu(dW/dV)_v/N^2$ versus θ

(c) $2\mu(dW/dV)_d/N^2$ versus θ

Fig. (3) - Normalized strain energy density function versus θ for $v = 1.0$ cm/sec, $\sigma_o/N = -0.1$ and $r = 10^{-2}$ cm

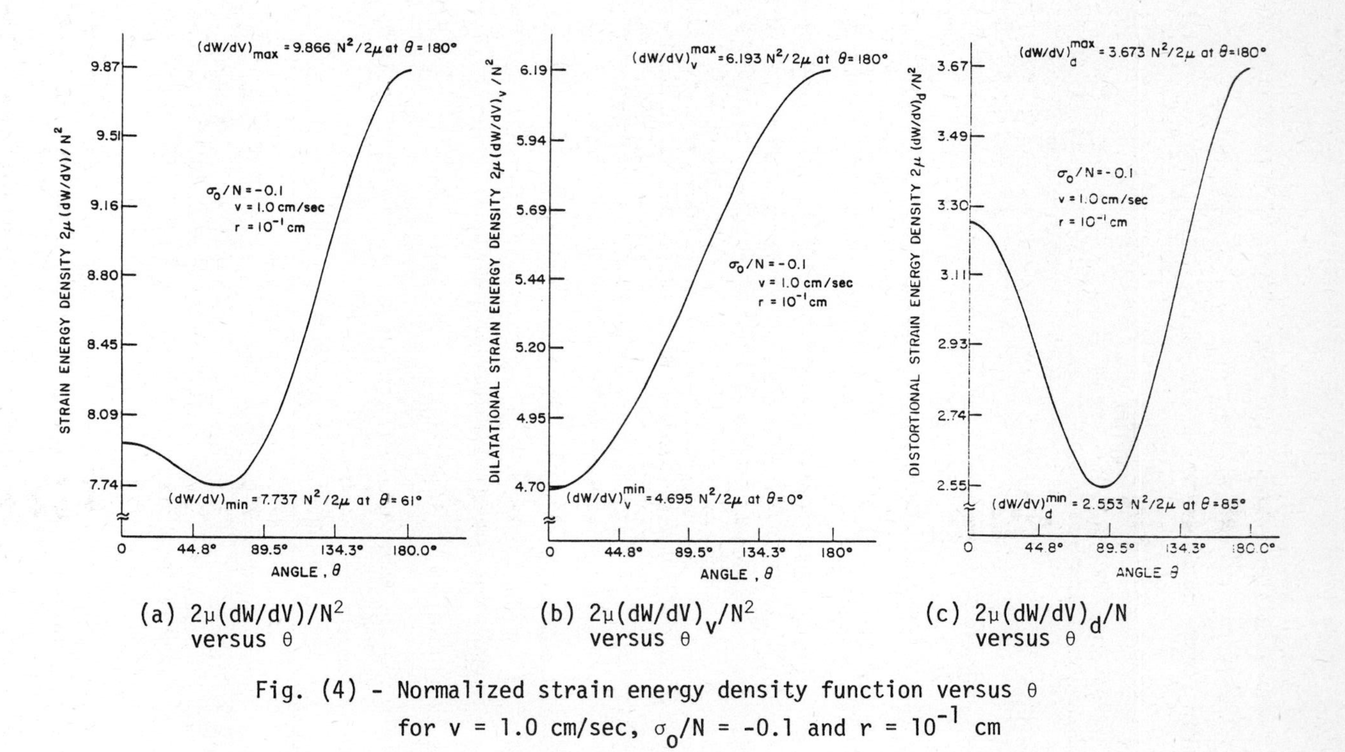

(a) $2\mu(dW/dV)/N^2$ versus θ (b) $2\mu(dW/dV)_v/N^2$ versus θ (c) $2\mu(dW/dV)_d/N$ versus θ

Fig. (4) - Normalized strain energy density function versus θ for $v = 1.0$ cm/sec, $\sigma_o/N = -0.1$ and $r = 10^{-1}$ cm

zero, Table 2 shows that both $(dW/dV)_{max}$ and $(dW/dV)_{min}$ decrease monotonically. Using $\sigma_o/N = 0.0$ as the reference state, the energy produced by compressive mechanical loading tends to add to that by the heat source while the opposite is ob-

TABLE 2 - EFFECT OF APPLIED MECHANICAL STRESS ON THE STATIONARY VALUES OF dW/dV FOR $v = 1.0$ cm/sec AND $r = 10^{-2}$ cm

Stress ratio σ_o/N	$\frac{2\mu}{N^2}(\frac{dW}{dV})_{max}$	$\frac{2\mu}{N^2}(\frac{dW}{dV})^{max}_{v/d}$	$\frac{2\mu}{N^2}(\frac{dW}{dV})_{min}$	$\frac{2\mu}{N^2}(\frac{dW}{dV})^{min}_{v/d}$
-0.1	27.582 ($\theta = 180°$)	1.895	26.312 ($\theta = 79°$)	2.076
0.0	27.503 ($\theta = 180°$)	1.841	26.140 ($\theta = 81°$)	2.038
0.1	27.431 ($\theta = 180°$)	1.787	25.973 ($\theta = 82°$)	1.998
1.0	27.084 ($\theta = 180°$)	1.354	24.764 ($\theta = 86°$)	1.625

served for tensile mechanical loading. Behind the heat source at $\theta = 180°$, the distortional component $(dW/dV)_d$ increases while the dilatational component $(dW/dV)_v$ decreases with increasing mechanical loading. This results in a reduction of $(dW/dV)^{max}_{v/d} = [(dW/dV)_v/(dW/dV)_d]_{max}$ corresponding to an increase in yielding as σ_o/N is raised. The possible sites of fracture initiation are assumed to coincide with the locations of $(dW/dV)_{min}$ which occurred at $\theta = \pm 79°$ to $\pm 86°$ as shown in Table 2. A maximum of $(dW/dV)^{min}_{v/d} = [(dW/dV)_v/(dW/dV)_d]_{min}$ is observed at $\sigma_o/N = 0.0$ when mechanical loading is absent.

CONCLUDING REMARKS

The energy state near a moving heat source decreases in magnitude with distance and is a function of its traveling speed. Analyzed in this work is only the case of $v = 1.0$ cm/sec although numerical results for other values of v can also be obtained to show that less energy would be transferred to the system if the velocity of the heat source is increased.

The angular variations of the strain energy density function around the heat source provide information on possible sites of failure by yielding and fracture. Figures 5 illustrate this pictorially for $\sigma_o/N = 0.1$. As mentioned earlier, the intensities of dW/dV decrease as a function of the distance from the heat source. Figures 5(a) to 5(c) show this reduction for r varying from 10^{-3} to 10^{-1} cm. The material elements behind the heat source corresponding to locations of $(dW/dV)_{max}$ at $\xi_1 = -r$ are proned to yielding. Possible sites of fracture initiation are assumed to correspond with locations of $(dW/dV)_{min}$. They occur at different positions of $\theta = \pm 85°$ to $\pm 89°$ depending on the value of r.

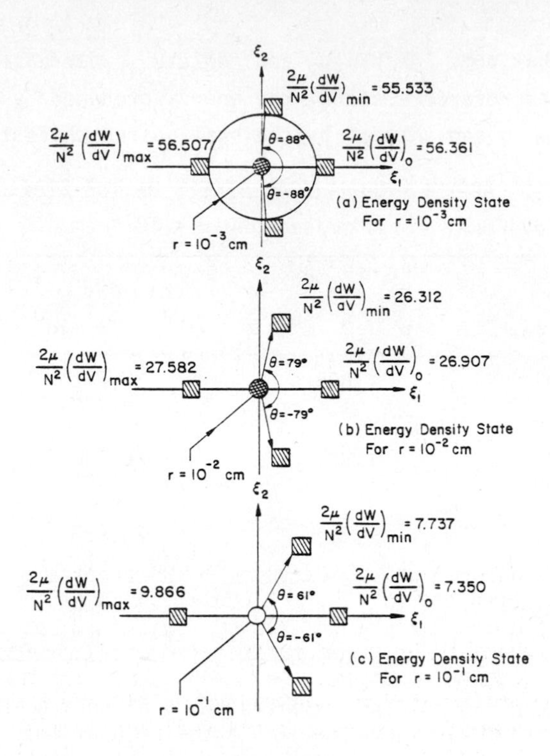

Fig. (5) - Strain energy density field around heat source for $v = 1.0$ cm/sec and $\sigma_o/N = 0.1$

Mechanical loading can increase or decrease the energy state near a moving heat source depending on whether the loading is compressive or tensile. A uniaxial compressive stress is found to be more damaging as it tends to increase the strain energy density level around the source. This causes more dilatation as well as distortion of the local material elements.

REFERENCES

[1] Gillemot, L. F., "Criterion of crack initiation and spreading", Journal of Engineering Fracture Mechanics, Vol. 8, pp. 239-253, 1976.

[2] Sih, G. C., "A Special Theory of Crack Propagation", Mechanics of Fracture, Vol. 1: Methods of Analysis and Solutions of Crack Problems, edited by G. C. Sih, Noordhoff International Publishers, The Netherlands, pp. XXI-XLV, 1973.

[3] Sih, G. C., "Experimental Fracture Mechanics: Strain Energy Density Criterion", Mechanics of Fracture, Vol. 7: Experimental Evaluation of Stress Concentration and Intensity Factors, edited by G. C. Sih, Martinus Nijhoff Publishers, The Netherlands, pp. XVII-LVI, 1981.

[4] Nowacki, W., Thermoelasticity, Pergamon Press, PWN, 1962.

[5] Carslaw, H. S. and Jaeger, J. C., Conduction of Heat in Solids, Oxford University Press, 1959.

[6] Goodier, J. N., "On the integration of the thermoelastic equations", Phil. Mag., Vol. 23, pp. 1017-1032, 1937.

A PSEUDO-LINEAR ANALYSIS OF YIELDING AND CRACK GROWTH: STRAIN ENERGY DENSITY CRITERION

G. C. Sih and P. Matic

Lehigh University
Bethlehem, Pennsylvania 18015

ABSTRACT

A typical phenomenon of ductile fracture is the process of crack initiation, slow growth and final termination. This process, being sensitive to the rate of loading, results in global nonlinearity of the load and displacement as the material may be damaged locally by yielding and/or fracture. The progressive damage of material being a path dependent process requires the stress and failure analysis be performed in tandem for each increment of loading. This is accomplished by a finite element procedure in conjunction with the strain energy density criterion. The configuration of a through crack under a rising load is analyzed where yielding of the elements near the crack are assumed to coincide with $\Delta W/\Delta V$ reaching some critical value $(\Delta W/\Delta V)_y$. This is translated into a permanent change of local stiffness for each increment of loading. Further damage of the material will be distinguished by the local value of $(\Delta W/\Delta V)$. The nonuniform segment of crack growth is assumed to occur along the path of $\Delta W/\Delta V$ being a relative minimum regardless of whether the material separates elastically or plastically and is assumed to be governed by the condition $S_1/r_1 = S_2/r_2 = --- = S_c/r_c = (\Delta W/\Delta V)_c^*$, where $(\Delta W/\Delta V)_c^*$ is the relative strain energy density. The strain energy density of virgin undamaged material is $(\Delta W/\Delta V)_c$. The pseudo-linear stress and failure analysis lead to a nonlinear response of load and displacement for the cracked medium. It is shown that for materials with a relatively low value of critical strain energy density more damage is inflicted due to crack growth or fracture than yielding while the reverse holds for materials with a relatively high value of critical strain energy density.

The proposed method of analyzing crystalline materials damaged by yielding and fracture is unique in that the same criterion is applied to yielding and fracture. This is unlike all previous works where yielding usually follows the von Mises condition while fracture is assumed to be governed by some other conditions such as maximum normal strain, critical strain energy density, etc.

INTRODUCTION

Ductile fracture refers to material damage by yielding and creation of macroscopic free surface. It is a process that involves subcritical crack growth prior to global instability and nonlinear load-displacement response. This is a manifestation of the fact that material is being damaged at both the microscopic and macroscopic scale level [1]. The precise nature of this behavior is dictated by the component geometry, loading rate, temperature and metallurgical structure of the material. Therefore, uniaxial tensile test data alone are not sufficient for predicting structure failure unless a consistent analytical procedure can be developed to predict failure of larger size structural components from small specimen data. Such a capability has not yet been established because the theories of deformable continuum mechanics cannot consistently account for irreversibility due to material damage.

Although the theory of linear elastic fracture mechanics (LEFM) has had success for the prevention of brittle fracture [2], not all structures exhibit such a behavior. It is desirable to allow a material to dissipate energy by plastic deformation so that the available energy to drive a potentially dangerous macrocrack could be reduced. Previous works on elastic-plastic fracture mechanics [3-5] have been concerned with plastic deformation around a crack for each increment of growth. The von Mises yield condition is usually invoked in the incremental theory of plasticity while a separate criterion is assumed for crack growth. Since yielding and fracture belong to the same unique phenomenon of material damage, their interplay should not be addressed separately by two unrelated criteria. This shortcoming has prevailed mainly because of the lack of a general failure criterion. Traditionally, different parameters are assumed to denote the thresholds of yielding and fracture.

The introduction of the strain energy density criterion [6,7] represents a departure from the classical notion of yielding and fracture. The proportion of energy associated with volume change (dilatation) and shape change (distortion) within each material element is assumed to govern the final failure mode. The time rate of change of volume and surface of material elements can be related to the relative minima and maxima of the strain energy density function $\Delta W/\Delta V$. Each point within a system can be checked for the initiation of yielding and/or fracture by evaluating whether $(\Delta W/\Delta V)_{max}$ and/or $(\Delta W/\Delta V)_{min}$ have exceeded their respective critical values. This criterion provides a consistent way of evaluating material damaged simultaneously by yielding and fracture.

FAILURE CRITERION IN YIELDING AND FRACTURE

The foundation of continuum mechanics relies on the assumption that the material properties for each constituent of a system can be obtained from simple tests while the global response of the system can be predicted from the governing equations combining the effects of geometry, loading, etc. Such an approach is known to yield inaccurate results if the energy dissipated by material damage in the form of yielding and fracture is a significant portion of the total energy within the system. Hence, the specimens for collecting material property data must be designed and loaded such that the load-displacement response is representative of the local situations within the system where damage is occurring. The fact that ductile fracture remains as one of the least understood subjects in fracture mechanics is because of the inability to translate elastic-plastic small specimen data to predict the ductile behavior of larger size structural components. The proportion of energy dissipated by yielding and fracture is obviously a function of the specimen size, loading rate, etc.

Uniaxial Tensile Test. The uniaxial tensile test is historically the standard method of obtaining the elastic modulus E and Poisson's ratio ν for a linear response of the material. When the response becomes nonlinear, additional parameters are required to describe the material behavior. The strain hardening coefficients, for example, are used to characterize the plastic response of metals. The wide disparity in the nonlinear response range for different test conditions suggest that material damage leading to final fracture are dependent on the geometry, specimen size and loading rate. Keeping this in mind, the true stress and true strain curve in Figure 1 may be used to quantitatively assess the local material behavior provided that appropriate changes are made when the

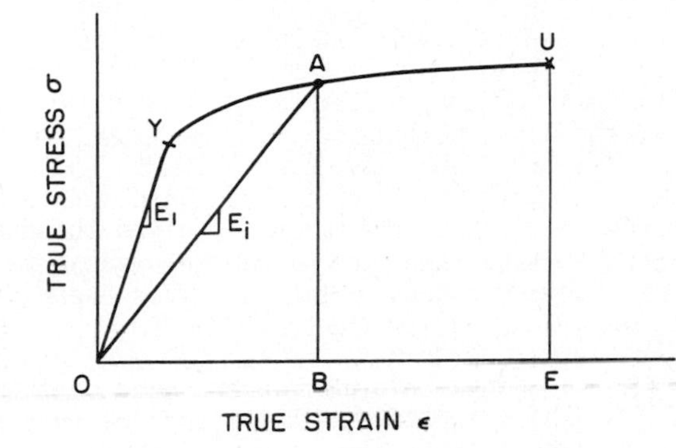

Fig. (1) - True stress-true strain curve

test conditions deviate significantly from those within the system. Traditionally, the point Y corresponds to yielding with a stress σ_y and U represents final fracture.

Strain Energy Density. The area under the true stress and true strain curve, say OYAB, can be computed as

$$\left(\frac{\Delta W}{\Delta V}\right)_{OYAB} = \int_0^{\varepsilon_B} \sigma_{ij} d\varepsilon_{ij} \tag{1}$$

such that

$$\sigma = \begin{cases} E\varepsilon \ , & \sigma < \sigma_y \\ A\varepsilon^n, & \sigma \geq \sigma_y \end{cases} \tag{2}$$

with A and n being constants that characterize the uniaxial tensile behavior of the material. In the nonlinear range for $\varepsilon_B > \varepsilon_Y$:

$$\left(\frac{\Delta W}{\Delta V}\right)_{OYAB} = \int_0^{\varepsilon_Y} \sigma d\varepsilon + \int_{\varepsilon_Y}^{\varepsilon_B} \sigma d\varepsilon = \frac{1}{2}\frac{\sigma_Y^2}{E} + \frac{A}{n+1}\left(\varepsilon_B^{n+1} - \varepsilon_Y^{n+1}\right) \tag{3}$$

Consider the area bounded by OYA that represents the energy dissipated due to material damage and the portion OAB that is recoverable. Hence

$$\left(\frac{\Delta W}{\Delta V}\right)_{OYAB} = \left(\frac{\Delta W}{\Delta V}\right)_{OYA} + \left(\frac{\Delta W}{\Delta V}\right)_{OAB} = \left(\frac{\Delta W}{\Delta V}\right)_{dam.} + \left(\frac{\Delta W}{\Delta V}\right)_{elas.} \quad (4)$$

The critical strain energy density function $(\Delta W/\Delta V)_c$ corresponds to final fracture of the material at point U. This can be computed from the total area:

$$\left(\frac{\Delta W}{\Delta V}\right)_c = \left(\frac{\Delta W}{\Delta V}\right)_{dam.} + \left(\frac{\Delta W}{\Delta V}\right)_c^* \quad (5)$$

in which

$$\left(\frac{\Delta W}{\Delta V}\right)_c^* = \left(\frac{\Delta W}{\Delta V}\right)_{elas.} + \left(\frac{\Delta W}{\Delta V}\right)^* \quad (6)$$

and $(\Delta W/\Delta V)^*$ is simply the area BAUE. Equation (6) gives the energy density required in addition to that already used in the damage process to produce final local separation of material in the body. Note that for energies below yield, $(\Delta W/\Delta V)_c^* = (\Delta W/\Delta V)_c$.

Failure Criterion. Based on the information provided by the uniaxial tensile specimen, a consistent criterion for yielding and fracture can be stated in terms of the strain energy density near the potential failure sites. When yielding occurs, i.e.,

$$\frac{\Delta W}{\Delta V} > \left(\frac{\Delta W}{\Delta V}\right)_y \quad (7)$$

material is being damaged permanently. This damage is translated into a local reduction in the elastic modulus of the material,

$$E_i < E_1 \quad (8)$$

where E_1 is the modulus of the virgin material. Conceptually, yielding is a continuous process so that the local modulus of the damaged material will take on a spectrum of decreasing values.

In the neighborhood of a potential failure site such as a macrocrack tip, the strain energy density field can be written in the form

$$\frac{\Delta W}{\Delta V} = \frac{S}{r} \quad (9)$$

where the strain energy density factor S is a function of the angular coordinate θ and r is the radial distance measured from the crack tip.

The strain energy density theory [6,7] may be stated in terms of three fundamental hypotheses:

(I) Yielding and fracture are assumed to coincide with the locations of relative maximum $(\Delta W/\Delta V)_{max}$ and minimum $(\Delta W/\Delta V)_{min}$ of the strain energy density function.

(II) Yielding and fracture are assumed to occur when $(\Delta W/\Delta V)_{max}$ and $(\Delta W/\Delta V)_{min}$ reach their respective critical values.

(III) The amount of incremental growth in yielding and fracture, say, $r_1, r_2, ---, r_j, ---, r_c$ is governed by

$$\left(\frac{\Delta W}{\Delta V}\right)_c^* = \frac{S_1}{r_1} = \frac{S_2}{r_2} = --- = \frac{S_j}{r_j} = --- = \frac{S_c}{r_c} = \text{const.} \tag{10}$$

Global instability* of the system corresponds to S and r reaching their critical values S_c and r_c.

METHODS OF ANALYSIS AND RESULTS

Assumed in this work is that the behavior of each material element follows a linear elastic stress-strain response up to yielding beyond which point permanent damage will be done producing a change in the local modulus, equation (8). This leads to a nonuniform distribution of material moduli near the macrocrack and to a nonlinear global response of the system. Therefore, it suffices to carry out a series of linear elastic stress calculations.

Finite Element. The finite element method provides a practical means of evaluating the material damage process described earlier. A center cracked panel shown in Figure 2 is subjected to uniform tension. It will be loaded incremental-

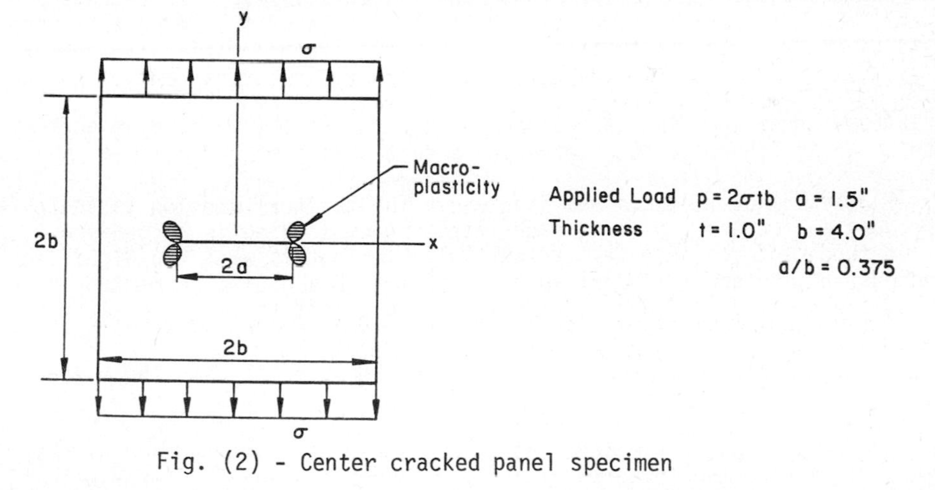

Fig. (2) - Center cracked panel specimen

*If yielding and fracture come to a rest, then S_c/r_c in equation (10) should be replaced by S_o/r_o corresponding to the arrest values of S and r.

ly to generate a series of global load-displacement curves. The system is discretized into 77 elements with cubic displacement shape functions. Refer to Figure 3. All stress analyses was performed by the Axisymmetric/Planar Elastic Structural (APES) finite element program [8].

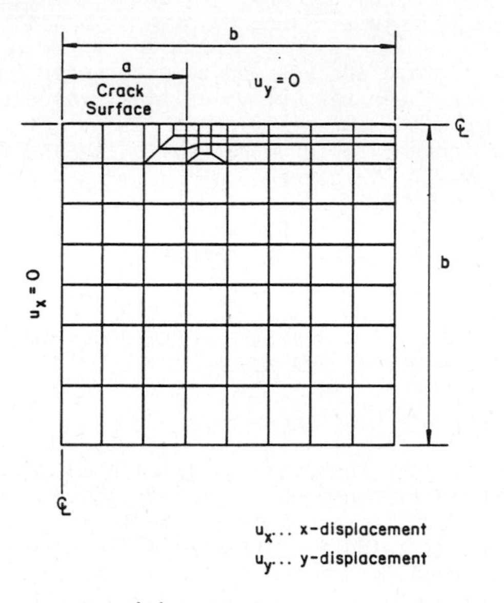

Fig. (3) - Finite element grid

Materials. At each load, the local damage is assessed by determining those finite elements with average strain energy density levels that correspond to the threshold value. Twenty-four strain energy density levels were translated into twenty-four decreasing values of elastic moduli based on the uniaxial tensile data of the selected material. Considered are two materials with $(\Delta W/\Delta V)_c$ = 64.2 in-lb/in^3 being relatively brittle and $(\Delta W/\Delta V)_c$ = 466.7 in-lb/in^3 being relatively ductile. The general characteristics of the local displacement curves are shown in Figure 4. E_{25} corresponds to the largest reduction in modulus.

Proportion of Yielding and Fracture. The tacit assumption is that yielding precedes macrocrack growth at each increment of loading in accordance with the three fundamental hypotheses stated earlier. In this way, the global nonlinear response of the cracked panel in terms of load P and displacement δ is obtained. This is displayed in Figure 5. It is apparent that the material with the lower toughness value of $(\Delta W/\Delta V)_c$ produces an approximately 60% greater derivation from the initial linear response line than the material with the higher $(\Delta W/\Delta V)_c$ value.

The proportion of yielding and fracture can be separated and displayed on the load-displacement diagram by applying the yield and fracture criteria separately, each in the absence of the other. For the more brittle material with $(\Delta W/\Delta V)_c$ = 64.2 in-lb/in^3, Figure 6 shows that the response due to fracture is more predominant as compared to yielding. The opposite trend is seen in Figure

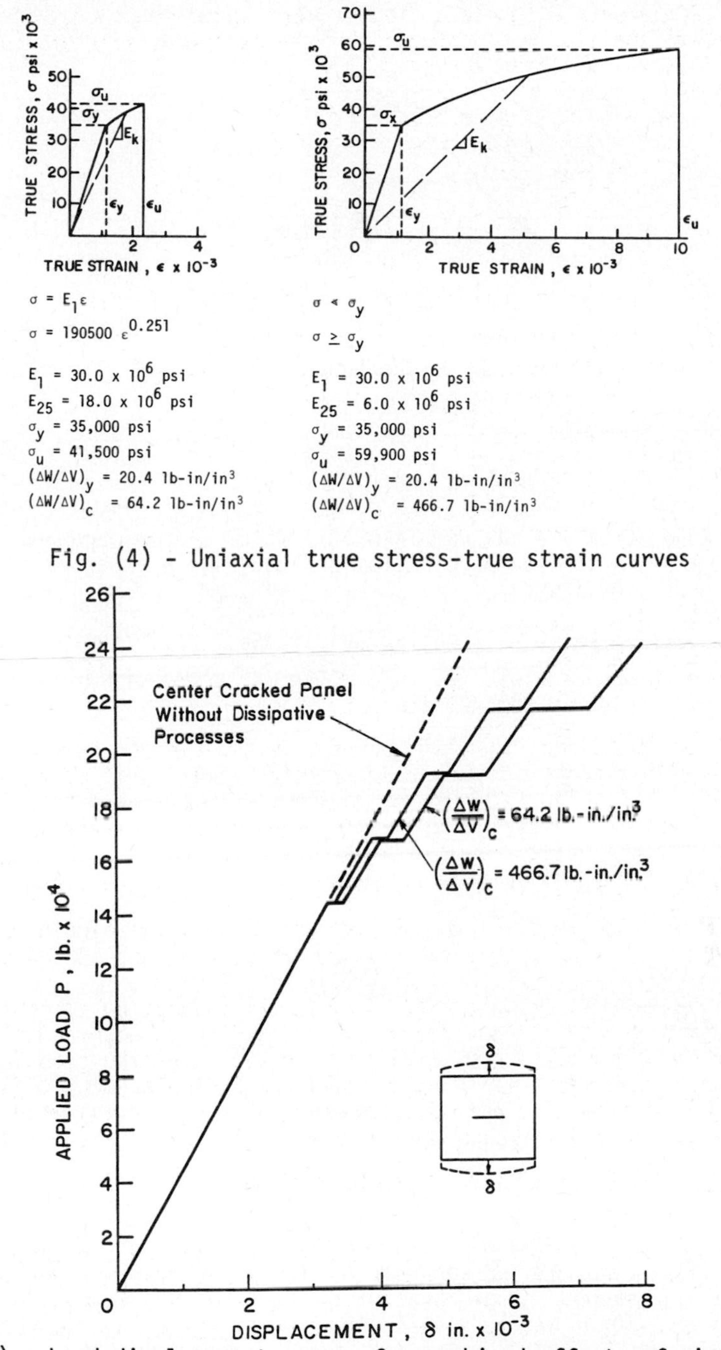

Fig. (4) - Uniaxial true stress-true strain curves

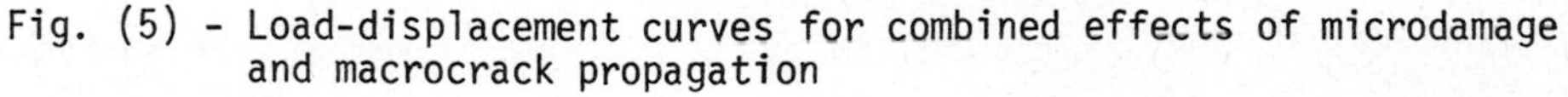

Fig. (5) - Load-displacement curves for combined effects of microdamage and macrocrack propagation

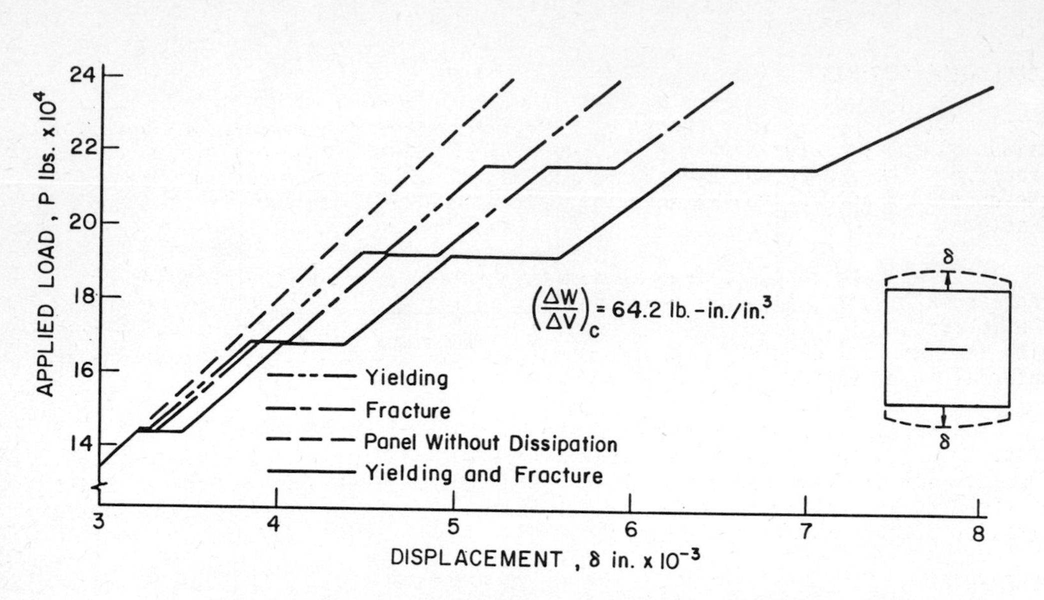

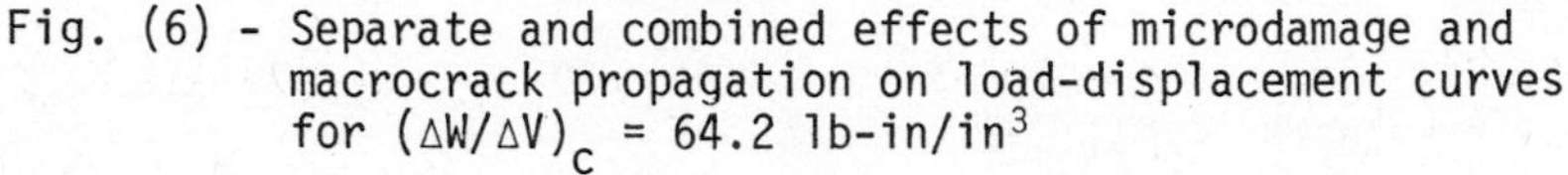
Fig. (6) - Separate and combined effects of microdamage and macrocrack propagation on load-displacement curves for $(\Delta W/\Delta V)_c$ = 64.2 lb-in/in^3

7 for the more ductile material with $(\Delta W/\Delta V)_c$ = 466.7 in-lb/in^3. In this case, yielding is more pronounced than fracture. The influence of $(\Delta W/\Delta V)_c$ on yielding and fracture is clearly demonstrated.

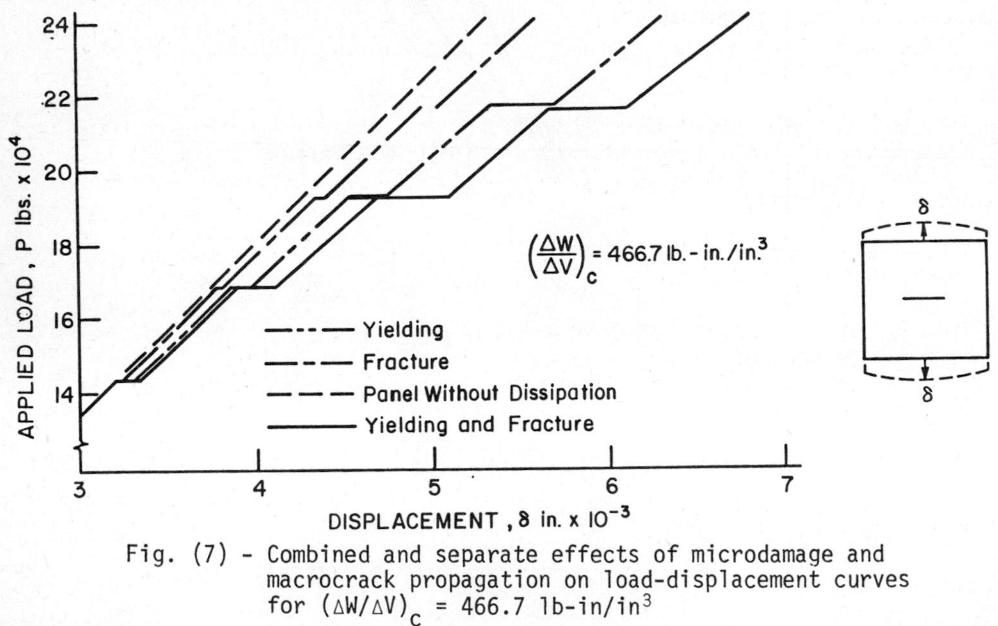

Fig. (7) - Combined and separate effects of microdamage and macrocrack propagation on load-displacement curves for $(\Delta W/\Delta V)_c$ = 466.7 lb-in/in^3

CONCLUDING REMARKS

The global nonlinear response of a ductile center crack panel has been obtained by applying the strain energy density criterion to both yielding and fracture. This approach is fundamentally different from current approaches to the ductile fracture problem which apply separate criteria for yielding and fracture.

It is assumed that the true stress-true strain curve of the material is representative of the damage processes in the larger size structural component. The load rate, temperature, and size of the tensile specimen must be consistent with the material damage in the component for the constitutive relations of the material to be valid.

This application of the strain energy density theory is computationally attractive. It employs a pair of linearly elastic finite element calculations for each set of yield and crack growth increments. Furthermore, the net effect of yielding and crack growth can be determined by suppressing one damage mechanism for a given material and geometry.

REFERENCES

[1] Sih, G. C., "Fracture Toughness Concept", ASTM Special Technical Publication, STP 605, pp. 3-15, 1976.

[2] Linear Fracture Mechanics, edited by G. C. Sih, R. P. Wei and F. Erdogan, Envo Publishing Co., Inc., Bethlehem, Pa., 1974.

[3] Sih, G. C. and Kiefer, B. V., "Nonlinear Response of Solids Due to Crack Growth and Plastic Deformation", Nonlinear and Dynamic Fracture Mechanics, edited by N. Perrone and S. N. Atluri, ASME Applied Mechanics Division, Vol. 35, pp. 136-156, 1979.

[4] Sih, G. C., "Mechanics of Ductile Fracture", Proceedings of International Conference on Fracture Mechanics Technology, edited by G. C. Sih and C. L. Chow, Sijthoff and Noordhoff International Publishers, The Netherlands, pp. 767-784, 1977.

[5] Sih, G. C., "Phenomena of Instability: Fracture Mechanics and Flow Separation", Naval Research Reviews, Vol. 32, No. 3, pp. 30-42, 1980.

[6] Sih, G. C., "A Special Theory of Crack Propagation", Mechanics of Fracture Vol. I: Methods of Analysis and Solutions of Crack Problems, edited by G. C. Sih, Noordhoff International Publishing, The Netherlands, pp. 21-45, 1973.

[7] Sih, G. C., "Experimental Fracture Mechanics: Strain Energy Density Criterion", Mechanics of Fracture Vol. VII: Experimental Evaluation of Stress Concentration and Intensity Factors, edited by G. C. Sih, Martinus Nijhoff Publishers, The Netherlands, pp. XVIII-LVI, 1981.

[8] Hilton, P. D., Gifford, L. N. and Lomacky, O., "Finite Element Fracture Mechanics of Two Dimensional and Axisymmetric Elastic and Elastic-Plastic Cracked Structures", Naval Ship Research and Development Center Report No. 4493, 1975.

ACKNOWLEDGEMENT

The authors gratefully acknowledge the financial support of AFOSR under Contract No. F49620-81-K005 with the Institute of Fracture and Solid Mechanics at Lehigh University.

NONLOCAL MECHANICS OF CRACK CURVING*

A. Cemal Eringen and A. Suresh

Princeton University
Princeton, New Jersey 08544

ABSTRACT

The nonlocal elasticity solution is obtained for the Griffith crack problem under combined loadings, Modes I and II. It is shown that the fracture begins at a critical point in the neighborhood of the crack tip. The location of the critical point is determined and the critical angle at which the crack begins to deviate from its straight line path is determined. These results are in good agreement with atomic lattice theory, and classical predictions.

INTRODUCTION

Presently, there exists several criteria for the static and dynamic curving of a line crack under combined loadings. These are based on either a maximum circumferential stress (Erdogan and Sih [1], Cotterell and Rice [2]), or the Griffith energy release rate (Hussain, et al [3]). Experimental assessment of these criteria was carried out by Kobayashi, et al, in a series of papers of which we mention [4]. Because of the usual crack tip singularity, these criteria are to be applied at a critical distance r_c from the crack tip. It is speculated that r_c is a material property.

The main purpose of the present paper is: (i) to determine r_c, theoretically; (ii) to obtain the direction of crack curving under combined loadings, Modes I and II; (iii) to give a crack curving criterion based on the maximum circumferential stress field.

We note that such a program cannot be carried out by means of classical elasticity theory since according to this theory, the maximum stress is infinite and it is located at the crack tip.

In several previous papers ([5-7]), it was shown that the nonlocal elasticity solutions of crack problems do not contain a stress singularity. Moreover, the maximum stress occurs not at the tip, but at an exterior point to the crack surfaces - in the close vicinity of the crack tip.

---

*Supported by the Office of Naval Research.

For brittle solids, a fracture criterion based on the maximum tensile stress, was established ([5], [6], [8]). Accordingly, when the maximum tensile stress exceeds the cohesive stress that holds bonds together, fracture will occur. Calculations based on this hypothesis proved to be in accordance with the Griffith fracture criterion with the additional dividend that the Griffith constant is fully determined. Cohesive stress calculations showed excellent agreement with the results known to metallurgists [5-10].

Motivated by these findings, we proceed to examine here the problem of crack instability and curving for the plane problem under combined loads.

RESUME OF BASIC EQUATIONS

Basic equations of nonlocal, linear, homogeneous, isotropic elastic solids consist of [5], [11,12]:

$$t_{k\ell,k} + \rho(f_\ell - \ddot{u}_\ell) = 0 \tag{1}$$

$$t_{k\ell}(\underset{\sim}{x},t) = \int_V \alpha(|\underset{\sim}{x}'-\underset{\sim}{x}|,\varepsilon)\sigma_{k\ell}(\underset{\sim}{x}',t)dv(\underset{\sim}{x}') \tag{2}$$

$$\sigma_{k\ell} = \lambda e_{rr}\delta_{k\ell} + 2\mu e_{k\ell} \tag{3}$$

$$e_{k\ell} = \frac{1}{2}(u_{k,\ell}+u_{\ell,k}) \tag{4}$$

where $t_{k\ell}$, ρ, f_ℓ and u_ℓ are respectively, the stress tensor, mass density, body force and the displacement field. We employ a superposed dot to indicate partial derivative with respect to time and a comma to indicate partial derivative with respect to rectangular coordinates x_k, i.e.,

$$\dot{u}_k = \frac{\partial u_k(\underset{\sim}{x},t)}{\partial t}, \quad u_{k,\ell} = \frac{\partial u_k}{\partial x_\ell}$$

As usual, repeated indices indicate summation.

Except for the constitutive equations (2), these equations are identical to those of the classical (local) elasticity theory. Equation (2) replaces the classical Hooke's law. It expresses the physical fact that the stress at a point $\underset{\sim}{x}$ depends on strains at <u>all</u> other points $\underset{\sim}{x}'$ in the body. Naturally, the influence of strain at $\underset{\sim}{x}$ is expected to be greatest in the stress at $\underset{\sim}{x}$ and the influence of strains at any other point $\underset{\sim}{x}'$, at a distance $|\underset{\sim}{x}'-\underset{\sim}{x}|$ from $\underset{\sim}{x}$, must diminish with distance. Hence, the nonlocal modulus $\alpha(|\underset{\sim}{x}'-\underset{\sim}{x}|,\varepsilon)$ must die out with $|\underset{\sim}{x}'-\underset{\sim}{x}|$ having a maximum at $\underset{\sim}{x}' = \underset{\sim}{x}$.

The nonlocal attenuation modulus α has the dimension of length^{-3} so that it depends on an internal characteristic length a. This is indicated by $\varepsilon = \beta a/\ell$ where ℓ is an external characteristic length (e.g., wave length, crack length), a is an internal characteristic length (e.g., lattice parameter, granular distance) and β is a non-dimensional constant appropriate to each material.

Nonlocal elasticity reduces to the classical theory in the limit

$$\lim_{\varepsilon \to 0} \alpha(|\underset{\sim}{x}'-\underset{\sim}{x}|,\varepsilon) = \delta(|\underset{\sim}{x}'-\underset{\sim}{x}|) \tag{5}$$

where δ is the Dirac delta measure. Based on these and other considerations, Eringen [11,12], proposed that α must be a Dirac delta sequence, and obtained several kernels which produce excellent agreement with the dispersion curves of plane waves in lattice dynamics, in the entire Brillouin zone.

Here we give one of these kernels, suitable for the treatment of two-dimensional problems, Ari and Eringen [7]:

$$\alpha(|\underset{\sim}{x}|,\varepsilon) = (2\pi\ell^2\varepsilon^2)^{-1}\; K_0(\sqrt{\underset{\sim}{x}\cdot\underset{\sim}{x}}/\ell\varepsilon), \quad \varepsilon = \beta a/\ell \tag{6}$$

where K_0 is the modified Bessel's function. We note that equation (6) satisfies the differential equation

$$(1 - \varepsilon^2\ell^2\nabla^2)\alpha = \delta(|\underset{\sim}{x}'-\underset{\sim}{x}|) \tag{7}$$

a property which is useful in the treatment of boundary-value problems. If we apply the operator (7) to equation (2), we obtain

$$(1 - \varepsilon^2\ell^2\nabla^2)t_{k\ell} = \sigma_{k\ell} \tag{8}$$

From this, by taking the divergence of both sides, it follows that

$$\sigma_{k\ell,k} + (1 - \varepsilon^2\ell^2\nabla^2)(\rho f_\ell - \rho\ddot{u}_\ell) = 0 \tag{9}$$

where we have used equation (1). If we further substitute from equations (3) and (4), we will have

$$(\lambda+\mu)u_{k,\ell k} + \mu\nabla^2 u_\ell - (1 - \varepsilon^2\ell^2\nabla^2)(\rho\ddot{u}_\ell - \rho f_\ell) = 0 \tag{10}$$

valid in rectangular coordinates. In this way, the integro-differential equations (1) to (4) are reduced to singularly perturbed partial differential equations.

Particularly simple results are obtained for the *static case* and vanishing body forces. In this case, we have

$$\sigma_{k\ell,k} = 0 \tag{11}$$

which is identical to the equilibrium equations of classical elasticity theory.

GRIFFITH CRACK, MODE I AND II

A line crack $|x_1| < c$, $x_2 = 0$ located in an infinite plane subject to a uniform tensile loading, perpendicular to the crack line at infinity, is known as the Griffith crack problem, Mode I. We consider the superposition to Mode I a constant shear loading which is known as the Mode II. The solution of both problems in nonlocal elasticity were given previously by Eringen [8] and his coworkers [6]. Here, we employ an alternative method of solution using a different kernel, namely equation (6).

For the static case with vanishing body forces, equation (10) reduces to the classical Navier's equation

$$(\lambda+\mu)u_{k,\ell k} + \mu\nabla^2 u_\ell = 0 \tag{12}$$

whose solution is well-known, [13]. The classical stress field $\sigma_{k\ell}$ in the neighborhood of the crack tip is of the form

$$\begin{bmatrix}\sigma_{11}\\ \sigma_{22}\\ \sigma_{12}\end{bmatrix} = \frac{K_I}{\sqrt{2\pi r}}\begin{bmatrix}\frac{3}{4}\cos\frac{\theta}{2} + \frac{1}{4}\cos\frac{5\theta}{2}\\ \frac{5}{4}\cos\frac{\theta}{2} - \frac{1}{4}\cos\frac{5\theta}{2}\\ -\frac{1}{4}\sin\frac{\theta}{2} + \frac{1}{4}\sin\frac{5\theta}{2}\end{bmatrix} + \frac{K_{II}}{\sqrt{2\pi r}}\begin{bmatrix}-\frac{7}{4}\sin\frac{\theta}{2} - \frac{1}{4}\sin\frac{5\theta}{2}\\ -\frac{1}{4}\sin\frac{\theta}{2} + \frac{1}{4}\sin\frac{5\theta}{2}\\ \frac{3}{4}\cos\frac{\theta}{2} + \frac{1}{4}\cos\frac{5\theta}{2}\end{bmatrix} \tag{13}$$

where K_I and K_{II} are classical stress intensity factors and (r,θ) are plane polar coordinates with the origin at the right crack tip.

In nonlocal theory, $\sigma_{k\ell}$ given by equation (13) is not the stress field. The stress field $t_{k\ell}$ is obtained by solving equation (8), subject to regularity conditions, i.e., $t_{k\ell}$ must be bounded at the crack tip and at infinity. This is borne out also from the previous solution given in [5] and [7].

We expect that, at large distances from the crack tip, the classical solution will approximate the stress field well. Moreover, as $\varepsilon\to 0$, equation (8) gives $t_{k\ell} \to \sigma_{k\ell}$. Therefore, there exists a boundary layer in the neighborhood of the crack. This then suggests that we may obtain an inner solution of equation (8) and match it to the outer solution $\sigma_{k\ell}$. In fact, this is why the approximate expressions, equation (13), which are valid in the vicinity of the crack tip, are adequate for the determination of $t_{k\ell}$ in the vicinity of the crack tip.

Introducing the complex stress field for any second-order symmetric tensor, $\tau_{k\ell}$ by

$$\Theta_{\underset{\sim}{\tau}} = \tau_{11} + \tau_{22}, \quad \Phi_{\underset{\sim}{\tau}} = \tau_{22} - \tau_{11} + 2i\tau_{12} \tag{14}$$

The differential equations (8) may be replaced by equivalent equations

$$(1 - \varepsilon^2\ell^2\nabla^2)\Theta_{\underset{\sim}{t}} = \Theta_{\underset{\sim}{\sigma}}$$

$$(1 - \varepsilon^2\ell^2\nabla^2)\Phi_{\underset{\sim}{t}} = \Phi_{\underset{\sim}{\sigma}} \tag{15}$$

where, by using equations (14) and (13),

$$\Theta_{\underset{\sim}{\sigma}} = \frac{1}{\sqrt{2\pi r}}[(K_I + iK_{II})e^{i\theta/2} + (K_I - iK_{II})e^{-i\theta/2}]$$

$$\Phi_{\underset{\sim}{\sigma}} = \frac{1}{2\sqrt{2\pi r}}[(K_I + 3iK_{II})e^{-i\theta/2} + (-K_I + iK_{II})e^{-i5\theta/2}] \tag{16}$$

Consequently, the integration of equation (15) requires finding the solution of a differential equation of the form

$$\frac{\partial^2 g_n}{\partial\rho^2} + \frac{1}{\rho}\frac{\partial g_n}{\partial\rho} + \frac{1}{\rho^2}\frac{\partial^2 g_n}{\partial\theta^2} - g_n = -\rho^{-1/2}e^{in\theta/2}, \quad n = \pm 1, \pm 5 \tag{17}$$

where

$$g_n(\rho,\theta) = f_n(\rho)e^{in\theta/2} \qquad \rho = r/\varepsilon\ell \tag{18}$$

The general solution of equation (17) is

$$f_n(\rho) = AI_{n/2}(\rho) + BK_{n/2}(\rho) + \int_0^\rho [I_{n/2}(z)K_{n/2}(\rho) - I_{n/2}(\rho)K_{n/2}(z)]z^{1/2}dz \tag{19}$$

where I_ν and K_ν are modified Bessel functions. Constants of integration A and B are determined by using the regularity conditions at $r=0$ and $r=\infty$, namely f_n must be bounded at $\rho=0$ and as $p\to\infty$.

$$f_n = \int_0^\rho I_{n/2}(z)K_{n/2}(\rho)z^{1/2}dz + \int_\rho^\infty I_{n/2}(\rho)K_{n/2}(z)z^{1/2}dz, \quad n = \pm 1, \pm 5 \tag{20}$$

Employing well-known expressions of $I_{n/2}$ and $K_{n/2}$ [14], we find that

$$f_{\pm 1} = \rho^{-1/2}(1-e^{-\rho})$$

$$f_{\pm 5} = \rho^{-1/2}e^{-\rho}(1 + \frac{3}{\rho} + \frac{3}{\rho^2}) \int_0^\rho [(1 + \frac{3}{z^2})\sinh z - \frac{3}{z}\cosh z]dz \tag{21}$$

$$+ \rho^{-1/2} [(1 + \frac{3}{\rho^2})\sinh\rho - \frac{3}{\rho}\cosh\rho] \int_\rho^\infty e^{-z}(1 + \frac{3}{z} + \frac{3}{z^2})dz$$

Consequently,

$$\Theta_{\underset{\sim}{t}} = (2\pi\varepsilon\ell)^{-1/2}[(K_I + iK_{II})e^{i\theta/2} + (K_I - iK_{II})e^{-i\theta/2}]f_1(\rho)$$

$$\Phi_{\underset{\sim}{t}} = \frac{1}{2} (2\pi\varepsilon\ell)^{-1/2}[(K_I + 3iK_{II})f_1(\rho)e^{-i\theta/2} + (-K_I + iK_{II})f_5(\rho)e^{-5i\theta/2}] \tag{22}$$

But, in polar coordinates, we have

$$t_{rr} + t_{\theta\theta} = \Theta_{\underset{\sim}{t}}$$

$$t_{\theta\theta} - t_{rr} + 2it_{r\theta} = \Phi_{\underset{\sim}{t}}e^{2i\theta} \tag{23}$$

from which we determine the stress field.

$$t_{rr} = (2\pi\varepsilon\ell)^{-1/2} \{[K_I(\cos\frac{\theta}{2} - \frac{1}{4}\cos\frac{3\theta}{2}) + K_{II}(-\sin\frac{\theta}{2} + \frac{3}{4}\sin\frac{3\theta}{2})]f_1(\rho)$$
$$+ \frac{1}{4}(K_I\cos\frac{\theta}{2} - K_{II}\sin\frac{\theta}{2})f_5(\rho)\}$$

$$t_{\theta\theta} = (2\pi\varepsilon\ell)^{-1/2} \{[K_I(\cos\frac{\theta}{2} + \frac{1}{4}\cos\frac{3\theta}{2}) - K_{II}(\sin\frac{\theta}{2} + \frac{3}{4}\sin\frac{3\theta}{2})]f_1(\rho) \tag{24}$$
$$+ \frac{1}{4}(-K_I\cos\frac{\theta}{2} + K_{II}\sin\frac{\theta}{2})f_5(\rho)\}$$

$$t_{r\theta} = \frac{1}{4}(2\pi\varepsilon\ell)^{-1/2} [(K_I\sin\frac{3\theta}{2} + 3K_{II}\cos\frac{3\theta}{2})f_1(\rho) + (K_I\sin\frac{\theta}{2} + K_{II}\cos\frac{\theta}{2})f_5(\rho)]$$

These results are valid in the vicinity of the crack tip.

FRACTURE AND CRACK CURVING

Based on the physics of matter, the fracture must occur when the maximum tensile stress exceeds the cohesive stress which holds bonds together, Eringen [5,8] Consequently, fracture will begin at a point (r_c, θ_c) which are the roots of

$$\frac{\partial t_{\theta\theta}}{\partial r} = 0, \frac{\partial t_{\theta\theta}}{\partial \theta} = 0 \tag{25}$$

provided that $t_{\theta\theta}(r_c, \theta_c)$ is the maximum tensile stress.

Note that unlike classical elasticity, $t_{\theta\theta max}$ is not at the crack tip. Thus, fracture begins ahead of the crack tip at some location, determined by equation (25).

First, consider the case of Mode I only. In this case, $K_{II} = 0$ and equation (25) gives

$$\theta_c = 0, 5 \frac{df_1}{d\rho} = \frac{df_5}{d\rho} \tag{26}$$

so that the fracture is along the crack line at a point ρ_c satisfying (26). Computations give

$$\rho_c = r_c/\varepsilon\ell = 1.095076 \tag{27}$$

This result is in excellent agreement with that calculated by means of atomic lattice theory by Elliot [15], see Table 1.

TABLE 1 - HOOP STRESS ALONG THE CRACK LINE

	t_{max}/t_0	x_c/ℓ
Elliot [15]	27.62	$1 + 0.2(a/\ell)$
Nonlocal (Present) $\varepsilon\ell = 0.22a$*	25.41	$1 + 0.2409(a/\ell)$
$\varepsilon\ell = 0.31a$	21.40	$1 + 0.3394(a/\ell)$

*Values of $\varepsilon\ell/a$ are from [7].

Next, we consider the combined Modes I and II. From equations (24) and (25), it is clear that

$$\rho_c = r_c/\varepsilon\ell = f(K_{II}/K_I)$$
$$\tag{28}$$
$$\theta_c = g(K_{II}/K_I)$$

Following common practice, these functions are plotted against crack angle γ, defined by

$$\cot\gamma = K_{II}/K_I \tag{29}$$

In Figure 1, critical distance r_c is given as a function of the crack angle γ. We notice that r_c decreases with the crack angle. The closest distance to the crack

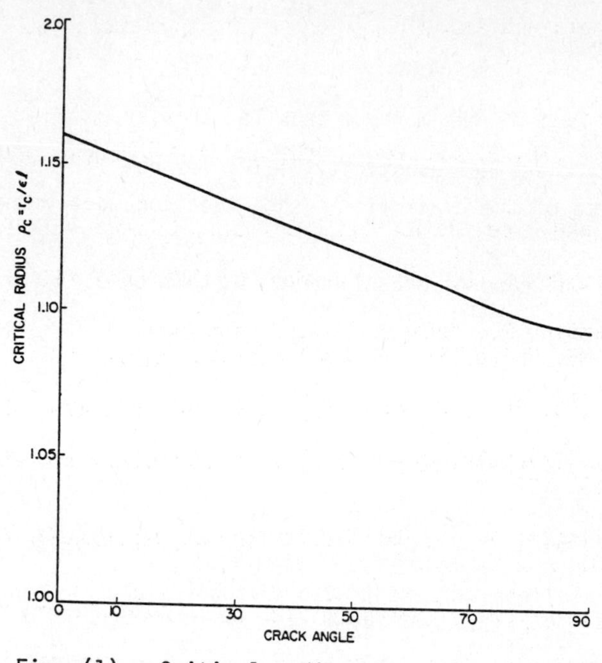

Fig. (1) - Critical radius versus crack angle

tip is obtained when $K_I = 0$ and farthest when $K_{II} = 0$. The change between these two cases is almost a straight line. In Figure 2, values of θ_c are compared with

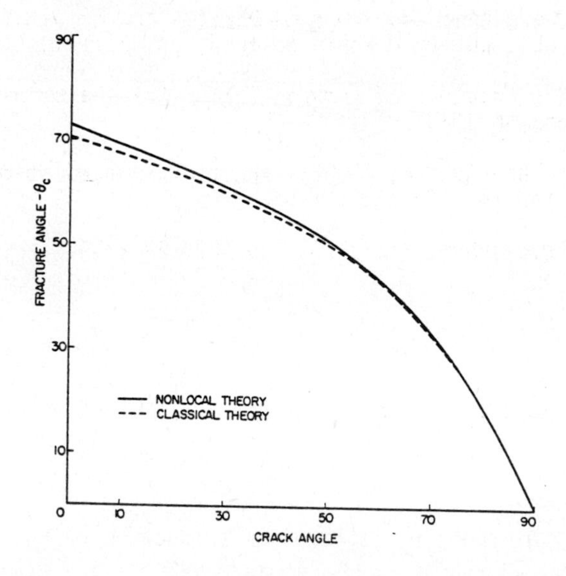

Fig. (2) - Fracture angle versus crack angle

classical results obtained from [13, p. 99]. The agreement is, in general, excellent.

REFERENCES

[1] Erdogan, F. and Sih, G. C., Trans. ASME, J. Basic Engng., 85(D), pp. 519-527, 1963.

[2] Cotterell, B. and Rice, J. R., Int. J. Frac., II, pp. 155-164, 1981.

[3] Hussain, M. A., Pu, S. L. and Underwood, J., ASTM STP 560, pp. 2-28, 1974.

[4] Sun, Y.-J., Ramulu, M., Kobayashi, A. S. and Kang, B.-J., ONR Report #42, Department of ME, University of Washington, November 1981.

[5] Eringen, A. C., J. Phys. D. Appl. Phys., 10, pp. 671-678, 1977.

[6] Eringen, A. C., Speziale, C. and Kim, B. S., J. Mech. and Phys. of Solids, 25, pp. 339-355, 1977.

[7] Ari, N. and Eringen, A. C., ONR Report #56, CE Department, Princeton University, July 1981.

[8] Eringen, A. C., Int. J. of Fracture, 14, pp. 367-379, 1978.

[9] Eringen, A. C., Engng. Fracture Mechanics, 12, pp. 211-219, 1979.

[10] Eringen, A. C., Crystal Lattice Defects, 8, pp. 73-80, 1979.

[11] Eringen, A. C., Int. J. Engng. Sci., 10, pp. 425-435, 1972.

[12] Eringen, A. C., Nonlinear Equations in Physics and Mathematics, A. O. Barut, ed., D. Reidel Publ. Co. Dordrecht, Holland, pp. 271-318, 1978.

[13] Jayatilaka, A. S., Fracture of Engineering Brittle Materials, Appl. Sci. Publishers, London, p. 22, 1979.

[14] Gradshteyn, I. S. and Ryzhik, I. M., Tables of Integrals, Series and Products, Academic Press, New York, p. 967, 1965.

[15] Elliot, H. A., Proc. Phys. Soc., A59, pp. 208-233, 1947.

MOTION OF THE CRACK UNDER CONSTANT LOADING AND AT HIGH CONSTANT TEMPERATURE

A. Neimitz

Technical University of Kielce
Kielce, Poland

ABSTRACT

In the paper, the model of the crack moving under creep condition and its kinematics are considered. The model of the crack is a modified model of the Leonov-Panasuk-Dugdale type which has fracture zone (FZ) changing in time. The changes inside the fracture zone are the result of existence of the crack, external loads and high temperature. In order to find the parameters describing the kinematics of the crack motion, the mechanisms of the void nucleation and growth under the creep condition are discussed.

INTRODUCTION

In the paper, another point of view for the process of the crack motion is presented. The main purpose of the investigation is to show the connection between phenomena arising inside metal during the creep process of the metal and a crack kinematics described by parameters taken from fracture mechanics.

A. Kinematics of the Crack

1. By analogy to the Dugdale model of the crack where the properties of the body, i.e., plasticity, were programmed in the model of the crack, or more precisely, in the zone ahead of the tip of the crack, in the paper a similar procedure is applied. The narrow zone ahead of the tip of the crack, henceforth referred to as the fracture zone (FZ), simulates visco-plasticity of the body. Viscosity of the body is introduced by assumption that the FZ changes its properties in time. These changes are the main reason of the crack motion. The changes of the FZ are represented by the changes of the geometrical sizes of the FZ but magnitude of the cohesive forces inside the FZ remains constant all the time (Y = const.).

The proposed model can be applied both to long range and short range plasticity but in the paper for the sake of simplicity of the derived formulas, only the case when $R/a^* \leq 0{,}1$ and $a \ll W$ is presented. The change of the rheological properties is introduced in an indirect manner. The laws ruling nucleation and growth of the voids in the metal indirectly decide on the velocity of the tip of the crack. The processes of nucleation and growth of the voids are considerably intensified just ahead of the tip of the crack [1-3] and can be identified with mechanisms discussed in [2,4-11] with respect to the level of temperature and external loads. In the proposed model, the arising voids that locally weaken the structure, are replaced by one computational void (CV).

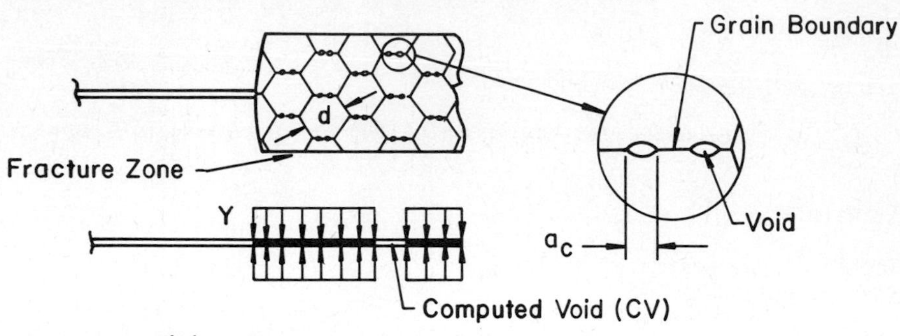

Fig. (1) - The manner of the modeling of the crack

It should be stated that only the second period of the creep (motion of the crack is considered.

2. At the moment t_o^*, the specimen is loaded by the force P (or stress σ) and: $P<P_{max}$; $K<K_{ic}^*<K_{ic}$;

$$R_o<R_o^*<R_{cr} \tag{1}$$

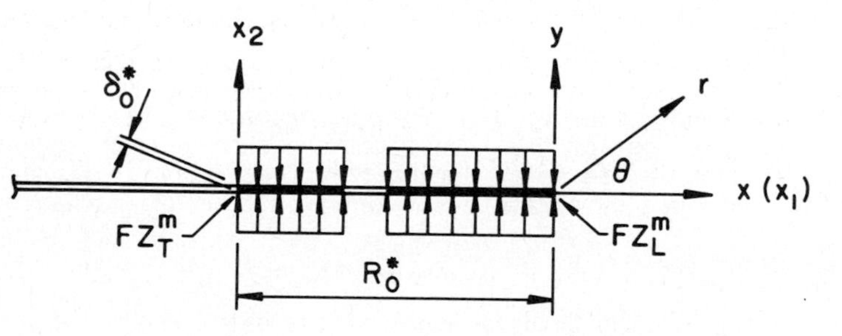

Fig. (2) - The crack together with FZ^m at the moment t_o^*

The cohesive forces Y inside FZ^m are postulated to be known and $\sigma_Y \le Y \le \sigma_m$. The value of R_o^* can be identified with $R_o = \pi K_i^2/8Y^2$ only in the first approximation, when $P/P_{max} << 1$. In general, R_o^* is material constant which can be found analytically or experimentally [12] for material which was creeping in the period of the time:

$$\Delta t = \frac{\Delta t_{IIperiod} R_o}{R_{cr}}$$

In the time $t_o^* < t_t^* < t_{cr}^*$ under influence of the external loading and temperature, the voids inside the FZ nucleate and grow (in the analyzed model inside FZ^m As a result of these phenomena, the geometrical parameters of the FZ^m increase

(R_t^*, δ_t^*). All the time during increase of the CV, the stress intensity factor is equal to zero according to Barenblatt-Dugdale assumption

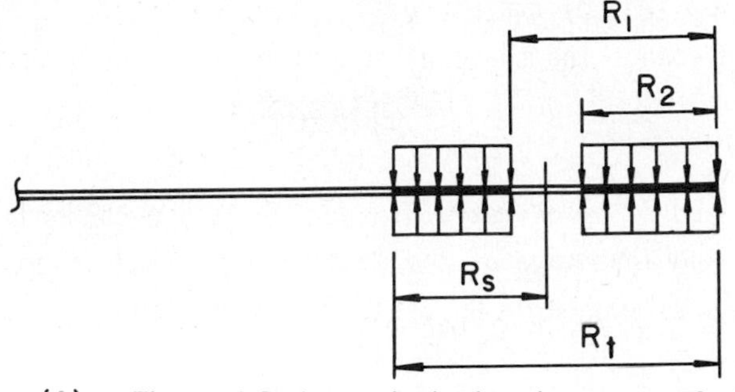

Fig. (3) - The modeled crack during increase of the CV

At the moment $t_t^* = t_{cr}^*$, δ and R reach critical values, as for critical load, $P = P_{max}$ and $K_i = K_{ic}$, but in the model $K_i < K_{ic}$ because $P < P_{max}$, and sudden jump of the trailing edge FZ_T^m occurs until geometrical sizes of the FZ characteristic of material and loading are regained (R_o^* and δ_o^*). Then the process of slow increase of the length of the FZ starts again until the next sudden jump. Thus, the length of the jump of the FZ_T^m is $\Delta R = R_{cr} - R_o^*$ and the next jump takes place after time $\Delta t = t_{cr}^*$. The above mechanism was proposed based on the fact that at high temperature both mechanisms of nucleation and growth of the voids are strongly dependent on the stresses that are considerably higher in FZ than outside. So, phenomena which are outside of the FZ are assumed to be neglected. Assuming that the velocity of the FZ_T^m is very high after crossing over the instability point, it can be written:

$$\beta_T = \frac{R_{cr} - R_o^*}{t_{cr}^*} \tag{2}$$

where R_{cr} and R_o^* are known and only t_{cr}^* is to be found.

3. Superposing the fields of the stresses, the Westergaard function for the semi-infinite crack in the infinite body can be written, see Figure 3.

$$Z_i = K_i(2\pi z)^{1/2} - \frac{1}{\pi} Y \int_{-R_t}^{0} \frac{1}{z-x} \sqrt{-(x/z)}[H(x+R_2) + H(-r-R_1)]dx \tag{3}$$

Developing formula (3) after some calculation, the values of components of the field of the stress can be easily found. But here, it is more important to find this portion of the formula for stresses which is multiplied by $r^{-1/2}$ and to equate it to zero in order to cancel singularity of the field of the stress. It can be written:

$$K_i = 2\sqrt{\frac{2}{\pi}}\, Y(\sqrt{R_t^*} + \sqrt{R_2^*} - \sqrt{R_1^*}) \tag{4}$$

In the formula (4), the value of K_i is known for definite, actual geometry of the specimen and the crack length. For many practical cases, the formulas for K_i are tabularized and can be shown [12] that for $R/a \leq 0{,}1$ and $a+R<<W$, the geometry of specimen does not influence the shape of equation (4). For most of the cases, it can be written:

$$K_i = PF_1(F_2 l)^m \tag{5}$$

where F_1, F_2 are dependent on geometry of specimen and m can be an integer or a fraction. Thus, taking advantage of the above formula and assumption that the motion of the FZ_T^m starts when $R_t^* = R_{cr}$, the following equation can be written:

$$K_{icr} = \frac{PK_{ic}}{P_{max}} \tag{6}$$

Also, it can be written

$$R_1^* = R_t^* - R_s^* + \beta_- t^* \tag{7a}$$

$$R_2^* = R_t^* - R_s^* - \beta_+ t^* \tag{7b}$$

where R_s^* is location of the center of the CV. Assuming that $|\beta_-| = |\beta_+|$, equatio (4) can be transformed for critical moment to the form:

$$K_{icr} = 2\sqrt{\frac{2}{\pi}}\,Y(\sqrt{R_{cr}} + \sqrt{R_{cr}-R_s^*+\beta t_{cr}^*} - \sqrt{R_{cr}-R_s^*-\beta t_{cr}^*}) \tag{8}$$

Subtracting from (8), the formula:

$$K_{ic} = 2\sqrt{\frac{2}{\pi}}\,Y\sqrt{R_{cr}} \tag{9}$$

after some calculation, using (6), (7a) and (7b), the following equation can be found:

$$\bar{\beta}t_{cr}^* = \{(R_{cr}-R_s^*)[\frac{K_{ic}(1-P/P_{max})}{2\sqrt{2/\pi}Y}]^2 - \frac{1}{4}[\frac{K_{ic}(1-P/P_{max})}{2\sqrt{2/\pi}Y}]^4\}^{1/2} \tag{10}$$

or

$$\bar{\beta}t_{cr}^* = [(R_{cr}-R_s^*)R_{cr}(1-P/P_{max})^2 - \frac{1}{4}R_{cr}^2(1-P/P_{max})^4]^{1/2} \tag{10}$$

In the above formulas, R_s^* is the magnitude which is to be postulated. From the physical point of view, it seems most reasonable to postulate that R_s^* is located in the center of the "gravity" of all the voids in the FZ. Thus, when diffusiona mechanism of the voids growth dominates, it can be found:

$$R_s^* = R_{cr} \frac{1}{3} \left(\frac{2V_{oi}^* - V_{ci}^*}{V_{oi}^* - V_{ci}^*}\right) \tag{11}$$

where:

$$V_{ci}^* = \int_0^{t_k} \beta_V d\tau$$ and for V_{oi}^* (see equation (21)).

The functional shape for β_V is discussed in section D. For a more general case, R_s^* can be found as:

$$R_s^* = \frac{R_{cr} - R_o^*}{t_{cr}^*} \frac{\int_0^{t_k} t\left(\int_0^t \beta_V d\tau\right) dt}{\int_0^{t_k} \int_0^t \beta_V d\tau dt} \tag{12}$$

Equation (10a) or (10b) gives some interesting information: (a) There is some similarity of the presented equations to the Monkman-Grant's law for the creep (εt_f = const) and the constant on the right-hand side of the equation (10a) or (10b) is described by the use of the known parameters both of the macroscopic and microscopic type; (b) Equation (10a) or (10b) can be used to estimate the applicability of the proposed model. It occurs that below defined external loading, the radius of the critical void becomes larger than the distance from the center of the CV to the FZ_T^m edge. It may mean that under low loading, the crack motion proceeds in another way. First, the voids increase to the dimensions when they link one to another and then they are linked with the main crack. For example: for the $R_s^* = 0{,}33R_{cr}$, the limit relative loading is $P/P_{max} = 0{,}56$. It can be shown that equations (10a), (10b) are independent of geometry of the specimens and only depend on the external loading when $R/a \leq 0{,}1$ and $a+R<<W$. Details of the calculation for the Griffith's crack can be found in [12]; (c) Equation (10a) or (10b) is the starting equation for calculation of t_{cr}^*. Multiplying equation (10) by ww_m, the volume of the CV can be found. w_m is the equivalent height of the CV and can be found from the formula:

$$w_m = \frac{w_z}{d} a_c = \frac{R_{cr}}{4d} a_c \tag{13}$$

because $w_z \simeq R/4$ [13,14]. Having the volume of the CV, on the basis of the theories of the nucleation and growth of the voids in metal under the creep condition, the sum of the volumes of all the voids inside the FZ in the moment t_{cr}^* can be calculated.

$$2ww_m\{[R_{cr}-R_s^*(t_{cr}^*)]R_{cr}(1-P/P_{max})^2 - 0{,}25R_{cr}^2(1-P/P_{max})^4\}^{1/2}$$

$$= \sum_i V_i(\tau)|_{\tau=t_{cr}} \tag{14}$$

$\sum_i V_i(\tau)|_{\tau=t_{cr}} = V_c^*(t_{cr}^*)$ is being sought in the remaining part of the paper.

B. Simplified Method of Calculation of t_{cr}^*

In many papers concerning creep of the metal [8,15,16], the following relations were observed:

$$-\Delta\rho/\rho = B_i \varepsilon t \sigma^m \tag{15}$$

where $\Delta\rho/\rho$ is the relative change of density of the metal during creep, B_i is empirical constant, ε is the strain in the moment t, t is the time, m is material constant known for many metals (for copper, m=3). Because in equation (15), $\Delta\rho/\rho$ is strongly dependent on σ, there was assumed that only in the FZ, the change of density is significant in the period of the time $t_{cr}^*-t_o^*$. So, the law of mass conservation can be written:

$$(\mathring{V}_1-V_2)\rho_1 + V_2\rho_2 = (\mathring{V}_1 + \Delta\mathring{V}_1 - V_2 - \Delta V_2)\rho_1 + V_2(\rho_2 - \Delta\rho_2)$$

$$+ \Delta V_2(\rho_1 - \Delta\rho_1) \tag{16}$$

and after some calculation (details in [12]), it can be found:

$$t_{cr}^* = \left\{ \frac{-[B_iV_2Y^{n+m} + \Delta V_2(\frac{Y+\sigma_R^*}{2})^{n+m}] + \{[B_iV_2Y^{n+m} + \Delta V_2 (\frac{Y+\sigma_R^*}{2})^{n+m}]^2 +}{2B_i^2V_2(\frac{Y+\sigma_R^*}{2})^{n+m} \quad Y^{n+m}} \right.$$

$$\left. + \frac{4\Delta\mathring{V}_1B_iV_2(\frac{Y+\sigma_R^*}{2})^{n+m} \quad Y^{n+m}\}^{1/2}}{} \right\}^{1/2} \tag{17}$$

where: $V_2 = w\frac{1}{4}(R_o^*)^2$; $\Delta V_2 = w\frac{1}{4}(R_{cr}-R_o^*)R_{cr}$; $\Delta\mathring{V}_1$ - the left-hand size part of equation (14); σ_R^* is the stress in the moment t_o^* ahead of the FZ_L^m in the distance $R_{cr}-R_o^*$, which is to be found from the well-known formulas. The solution of the task is possible with the accuracy to some constant describing location of the R_s^*. It can be assumed that $R_s^* = (0{,}25-0{,}4)R_{cr}$.

C. Calculation of the Summary Volume of the Voids on the Basis of the Nucleation and Growth Laws

In the moment t, the summary volume of the voids in the FZ can be found from equation

$$V_c^* = V_o^* + (\text{surface under adf curve - Figure 4}) \tag{18}$$

or

$$V_c^* = V_o^* + N\{\int_0^{t_1} \beta_v d\tau + \int_0^{t_2} \beta_v d\tau + \ldots + \int_0^{t_k} \beta_v d\tau\} \tag{19}$$

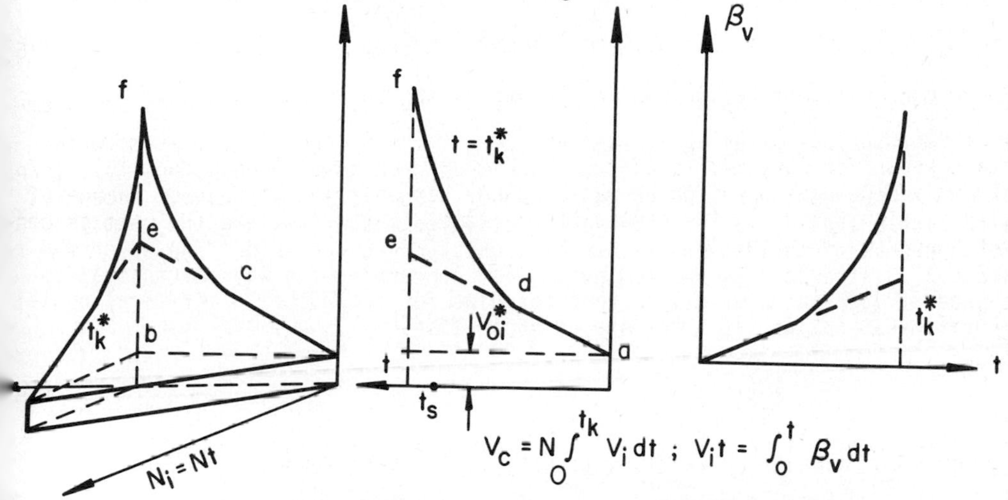

$$V_c = N\int_0^{t_k} V_i\,dt \;;\; V_i t = \int_0^t \beta_v\,dt$$

Fig. (4) - Summary volume of the voids in the FZ

where:

$$V_o^* = N_i V_{oi}^* = V_{oi}\,\frac{wR_{cr}^2}{b^2}\,\frac{1}{4d} \tag{20}$$

and

$$V_{oi}^* = F_v r_c^3 \tag{21}$$

$$\beta_v = \beta_{vd} + H(t_d)\beta_{vp} \tag{22}$$

β_{vp}, β_{vd}, t_d are calculated in the next section.

$$t_k^* = t_{cr}^*\,\frac{R_{cr}}{R_{cr}-R_o^*} \tag{23}$$

Equation (19) can be written in the form:

$$V_c^* = V_o^* + N\{\sum_{i=0}^{i=k} J(t_i) - kJ(t_o)\} \quad (24$$

Multiplying and dividing the portion in the brackets in (24) by Δt and approachin in limit with Δt to zero, the final form of equation (18) is the following:

$$V_c^* = V_o^* + \dot{N}\{\int_0^{t_k} J(\tau)d\tau - t_k J(t_o)\} \quad (25$$

where:

$$J(t) = \int \beta_v d\tau.$$

In equation (32), the β_v is only one unknown.

D. Mechanisms of Nucleation and Growth of the Voids

1. Nucleation of the Voids

Nucleation of the voids is rather a controversial topic. Recently, the opinion prevails that the thermally activated mechanisms are the most import in the process of nucleation of the voids under creep condition. Based on the re sults of [7], the conclusions can be derived concerning the ways of nucleation. the paper [7], authors gave important formulas for the number of critical nuclei per unit area - ρ_c and for the rate of nucleation $\dot{\rho}$:

$$\rho_c = \rho_{max} \exp(-\frac{\Delta G_c}{kT}) = \rho_{max} \exp(-\frac{r_c^3 F_v \sigma}{2kT}) \quad (2$$

$$\dot{\rho} = \frac{4\pi\gamma\delta D_b}{\sigma\Omega\ 4/3} (\rho_{max} - \rho) \exp(\frac{\sigma\Omega - 4\gamma^3 F_v/\sigma^2}{kT}) \quad (2$$

A very important magnitude in the above equations is the shape function F_v which a very sensitive function for the changes of the angle α, Figure 5. The functio F_v was defined for the voids of a type as in Figure 5 in [7].

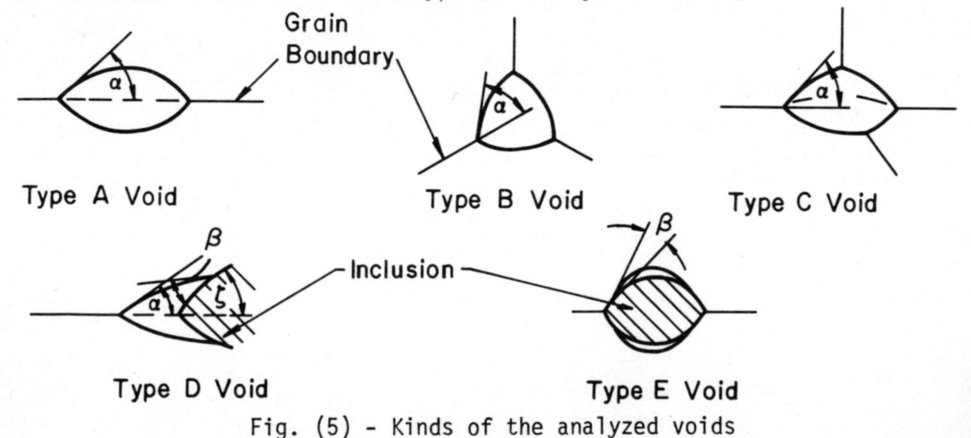

Fig. (5) - Kinds of the analyzed voids

For Type A, B, C voids, the angle α is about 75° and, as is shown in Table 1, it is almost unlikely to find Type A, B, C voids in metal. The calculations were made for copper for $T = 0{,}5T_m$ and $t = 10^6$ and 10^5s. For small angles α, the voids

TABLE 1

	$\sigma = 10$ MN/m²		$\sigma = 60$ NM/m²	
	$\alpha = 2°$	$\alpha = 75°$	$\alpha = 2°$	$\alpha = 75°$
ρ_{max}	$8\ 10^{12}$	$8\ 10^{12}$	$3\ 10^{14}$	$3\ 10^{14}$
ρ_c	$8\ 10^{8,7}$	$8\ 10^{-3.9\ 10^6}$	$3\ 10^{14}$	$3\ 10^{-10^5}$
$\dot{\rho}$	$2{,}4\ 10^{21,4}$	$2{,}4\ 10^{-3,9\ 10^6}$	$5{,}3\ 10^{26,6}$	$5{,}3\ 10^{-10^5}$
N	$2{,}4\ 10^{27,4}$	$2{,}4\ 10^{-3,9\ 10^6}$	$5{,}3\ 10^{31,6}$	$5{,}3\ 10^{-10^5}$

density would have increased considerably. For Type A voids, $\alpha \leq 13$, for Type B voids, $\alpha \leq 30$, for Type C voids, $\alpha \leq 17$, are the values of the angle which could have caused a large number of voids. Another situation is for Type D and E voids. In these two types of voids, the values of the α, β, ζ angles are not known precisely and for many cases, F_v is treated as an adjustable parameter in theory [2] and is estimated as a value of $F_v = 10^{-3}$ to $F_v = 10^{-5}$. Thus from [26] and [27], the following equations can be easily found:

$$\text{For } F_v = 10^{-3}: \quad \rho = \rho_{max} - (\rho_{max} - \rho_o)\exp(-10^{-25}t) \tag{28}$$

It means that $\rho = \rho_o$, where ρ_o is a number of the voids that were created just after loading of the specimen; ρ_{max} is in this case a number of the inclusion per unit area.

$$\text{For } F_v = 10^{-4}: \quad \rho = \rho_{max} - (\rho_{max} - \rho_o)\exp(-10^{13}t) \tag{29}$$

It means that $\rho = \rho_{max}$.

Based on the above consideration, it can be concluded (for details, see [7,19]), that in the creep condition, the voids in metal nucleate in a very short time after loading the body. They nucleate at inclusions and for large stresses, the number of voids is almost equal to the number of inclusions in the plane perpendicular to external loading. In the proposed model of the crack, it is assumed that voids nucleate inside the FZ (because stresses are must higher than outside of the FZ). It is also assumed that velocity of the FZ_L^m edge is constant.

2. Mechanisms of the Void Growth

Because of the limited length of the paper, only final results of the calculation of the β_v are demonstrated and for details, see [19] and the quoted papers.

The voids can increase their volume by diffusional mechanism and by plastic deformation of the metal surrounding the void. There are two parameters which complete each other in order to estimate which of the mechanisms is dominating. First: P_{EA} was given by Edward and Ashby [10]; second: P_{NR} was proposed by Needleman and Rice [11].

$$P_{EA} = \frac{1}{10}\left(\frac{4L^3}{b}\right)^{2/n} \tag{30}$$

$$P_{NR} = r/L \tag{31}$$

where:

$$L = L_o\left(\frac{10^{-3}G}{\sigma}\right)^{\frac{n-1}{3}} \exp\left(\frac{T_m}{T}\right) \tag{32}$$

and authors [10,11] proposed the following range for existence of the acting mechanisms:

$$\begin{array}{lll} 0{,}2 \le P_{NR} \le 20, \quad 10^{-3} \le P_{EA} \le 1 & & \text{- both mechanisms exist} \\ P_{NR} < 0{,}2 & P_{EA} > 1 & \text{- only diffusional mechanism} \\ P_{NR} > 20 & P_{EA} < 10^{-3} & \text{- domination of the plastic deformation} \end{array} \tag{33}$$

Calculations were made for the copper and results are presented in Table 2.

TABLE 2

		$T = 0{,}5T_m$		$T = 0{,}6T_m$		$T = 0{,}8T_m$		Radius* of critical void
		dyf. + plast.		dyf. + plast.		dyf. + plast.		
$\sigma = 10$ MPa	diffusion	$4{,}6<b<1177$ $3{,}7<r<373$	plastic def. diffusion	$1{,}7<b<436$ $1{,}4<r<138$	plastic def. diffusion	$0{,}4<b<109$ $0{,}4<r<35$	plastic def.	$r_c = 0{,}2$
$\sigma = 40$ MPa		$0{,}8<b<203$ $0{,}6<r<64$		$0{,}3<b<75$ $0{,}2<r<22$		$0{,}07<b<19$ $0{,}06<r<6$		$r_c = 0{,}05$
$\sigma = 60$ MPa		$0{,}5<b<122$ $0{,}4<r<39$		$0{,}2<b<45$ $0{,}1<r<14$		$0{,}05<b<13$ $0{,}035<r<4$		$r_c = 0{,}035$

*According to section D1, the critical radius of the void is rather greater and depends on the dimensions of the inclusion.

From Table 2, a conclusion follows that at high temperature, both mechanisms exist from the very beginning of the growth process and for the lower temperature initially voids grow by diffusion of the matter and then by both mechanisms.

A bulk of papers was devoted to the growth of voids by the matter diffusion [8,9,17,22-25]. In most of them, the following formula for the velocity of the void growth was used in the form:

$$\beta_{vd} = \frac{2\pi\Omega D_b\delta}{kT}\,\frac{[\sigma-\sigma_o(1-r^2/b^2)]}{\ln(b/r)-(3-r^2/b^2)(1-r^2/b^2)/4} \tag{34}$$

This formula can be considerably simplified if some conditions are satisfied. In many cases, it can be assumed that $\sigma_o << Y$, $r^2/b^2 << 1$ and $\ln(b/r) - (3-r^2/b^2) \times (1-r^2/b^2)/4 \approx 1,75$ for $10 \leq r/b \leq 30$. So, equation (34) can be written in the simplified form:

$$\beta_{vd} = \frac{2\pi\delta D_b\Omega}{kT}\,\frac{\sigma}{1,75} \tag{35}$$

Velocity of the growth of the voids by the plastic deformation of the surrounding material was found using two different methods. Taking advantage of the simplified void model proposed by Edward and Ashby [10] and the method of the upper limit for the strain rate [20], the differential equation can be found in the form:

$$\frac{d}{2r_c b}\,(\dot{\varepsilon}-\dot{\varepsilon}_{ss}) = \frac{1}{b^2-r^2}\,\frac{dr}{dt} \tag{36}$$

where $\dot{\varepsilon}$ is given by the Norton law and $\dot{\varepsilon}_{ss}$ by dislocation creep power law in the form

$$\dot{\varepsilon}_{ss} = A\,\frac{D_{ov}G\eta}{kT}\,\left(\frac{\sigma}{G}\right)^n \exp\left(-\frac{Q_v}{RT}\right) \tag{37}$$

after solving equation (36) at the initial condition $r=r_c$ at $t=0$; taking advantage of the mass conservation law, it can be found:

$$\beta_{vp} = \frac{\pi d b^2\Delta\dot{\varepsilon}2(\dot{\varepsilon}_{ss}t+1)\,\left|\frac{b+r_c}{b-r_c}\right|\exp\left(\frac{\Delta\dot{\varepsilon}d}{r_c}\,t\right) - 1}{\left|\frac{b+r_c}{b-r_c}\right|\exp\left(\frac{\Delta\dot{\varepsilon}d}{r_c}\,t\right) + 1} \tag{38}$$

Another solution of the problem can be given solving the task of the extension of the spheroidal void in the non-linear viscoelastic body [11,19,21,23]. The result is:

$$\beta_{vp} = 1,42\pi\dot{\varepsilon}r_c^3\exp\left(\frac{1,06}{h(\alpha)}\,\dot{\varepsilon}t\right) \tag{39}$$

Now, we can put together all formulas to write down in the final form equations (18) and (14)

$$V_c^*(t_k^*) = N_c\{\frac{1}{2} D_1 t_k^* + \frac{\overline{R}_1}{\overline{R}_2^2 t_k^*} [\exp(\overline{R}_2 t_k^*) - 1]\} \tag{40}$$

where:

$$N_c = \frac{R_{cr}^2 w}{4db^2} \tag{41}$$

$$\overline{R}_1 = 1,42\pi\dot{\varepsilon}r_c^3 \tag{42}$$

$$\overline{R}_2 = \frac{1,06}{h(\alpha)} \tag{43}$$

$$D_1 = \frac{\pi\Omega D_b \delta}{1,75kT} \tag{44}$$

Using equations (12), (39), (42)-(44), the formula for R_s^* can be found:

$$R_s^* = \frac{R_{cr}-R_o^*}{t_{cr}^*} \frac{1/6\ D_1 t_k^{*3} + \overline{R}_1/\overline{R}_2^3[\exp(\overline{R}_2 t_k)-1]}{1/2\ D_1 t_k^2 + \overline{R}_1/\overline{R}_2^2[\exp(\overline{R}_2 t_k)-1]} \tag{45}$$

Analogous procedure can be used applying the formula (38). Now, in order to find some values for t_{cr}^*, the numerical procedure should be applied. Examples were mad in [12].

REFERENCES

[1] Schapery, R. A., Int. J. Fract. Mech., 11, p. 141, 1975.

[2] Curran, D. R., Seaman, L. and Shockey, D. A., Poulter Laboratory Technical Report 003-80, 1980.

[3] Neate, G. J., Eng. Fract. Mech., 9, p. 297, 1977.

[4] Raj, R. and Ashby, M. F., Trans. MET. AIME, 3, p. 1932, 1972.

[5] Goods, S. H. and Brown, L. M., Acta Metallurgica, 27, p. 1, 1979.

[6] Ballufi, R. W. and Seigle, L. L., Acta Metallurgica, 3, p. 170, 1955.

[7] Raj, M. and Ashby, M. F., Acta Metallurgica, 23, p. 653, 1975.

[8] Needham, N. G., Wheatley, J. E. and Greenwood, G. W., Acta Metallurgica, 23, p. 23, 1975.

[9] Greenwood, G. W., Phil. Mag. A, 43, p. 281, 1981.

[10] Edward, G. H. and Ashby, M. F., Acta Metallurgica, 27, p. 1505, 1979.

[11] Needleman, A. and Rice, J. R., Acta Metallurgica, 28, p. 1315, 1980.

[12] Neimitz, A., "Application of Dugdale model to the crack in the Green conditions", paper in preparation.

[13] Hahn, G. T. and Rosenfield, R. A., Acta Metallurgica, 13, p. 293, 1965.

[14] Vosikovsky, O., Int. J. Fract. Mech. 4, p. 321, 1968.

[15] Cane, B. J. and Greenwood, G. W., Metal Science, 9, p. 55, 1975.

[16] Greenwood, G. W., Phil. Mag., 19, p. 423, 1969.

[17] Raj, R., Acta Metallurgica, 26, p. 341, 1978.

[18] Krzeminski, J., IFTR Reports, 58, 1972.

[19] Neimitz, A., "Nucleation and growth of the voids in the creep condition", paper in preparation.

[20] Ashby, M. F., Edward, G. H., Davenport, J. and Verral, R. A., Acta Metallurgica, 26, p. 1379, 1979.

[21] Budiansky, B., Hutchinson, J. W. and Slutsky, S., Mechanics of Solids, M. J. Sewell, Pergamon Press, Oxford.

[22] Hull, D. and Rimmer, D. E., Phil. Mag., 4, p. 673, 1959.

[23] Rice, J. R., Acta Metallurgica, 29, p. 675, 1981.

[24] Chuang, T.-J., Kagawa, K. J., Rice, J. R. and Sills, L., Acta Metallurgica, 29, p. 265, 1979.

[25] Chen, J. W. and Argon, A. S., Acta Metallurgica, 29, p. 1759, 1981.

NOMENCLATURE

FZ - fracture zone

FZ^m - fracture zone in Dugdale model

FZ^m_L, FZ^m_T - leading and trailing edges of the FZ^m, respectively

$(\ )_o$ - index "o" describes the value of the analyzed parameter at the initial moment

$(\ )_{cr}$ - index "cr" describes the value of the analyzed parameter at the moment of the stability loss by FZ^m_T

$(\ )_t$ - index "t" describes the value of the analyzed parameter at the actual moment

$( )^*$ - index "*" means that analyzed parameter is related to the second period of the creep

a - length of the crack without the FZ

l - length of the crack with the FZ

t - time

w - thickness of the specimen

w_z - height of the FZ

w_m - computational height of the FZ

a_c - $a_c = 2r_c$, critical value of the void diameter when the surface of the void becomes free of stresses

b - half of the distance between two voids

d - grain size

k - Boltzman constant

K_i - stress intensity factor

K_{ic} - critical value of K_i - material constant

n - material constant

R - length of the FZ

Y - the forces closing the crack inside the FZ

W - width of the specimen

N^* - number of the voids in the unit volume of the FZ

V_o^* - summary volume of all the voids of the a_c size in the FZ

V_i^* - the volume of the individual void

V_c^* - summary volume of all voids inside the FZ

$\dot{N}$ - velocity of the void nucleation

ΔG_c - the maximal value of the free energy existing in the body during nucleation of the void of the a_c size

D_b - grain boundary self-diffusion coefficient

D_v - lattice self-diffusion coefficient

F_v - function of energy angles which provides the void volume

T - temperature K

T_m - melting point

A - nondimensional material constant

D_{bo}, D_{vo} - material constant in the process of diffusion

Q_v, Q_b - activation energy of the lattice and grain boundary diffusion, respectively

$\dot{\varepsilon}$ - strain rate

σ - stress acting inside the body

σ_y - yielding point

σ_m - tensile strength

δ^* - opening stretch of the crack

β - mean velocity of the void growth measured with respect to the center of the CV

β_-, β_+ - velocities of the motion of the void edges measured with respect to the center of the CV

β_v - velocity of the volume growth of the individual void

β_{vd} - velocity of the volume growth of the individual void by diffusional mechanism

β_{vp} - velocity of the volume growth of the individual void by plastic deformation of the surrounding material

ρ_c - number of critical nuclei per unit area

ρ_{max} - maximum number of possible nucleation sites per unit area

$\dot{\rho}$ - nucleation rate of voids per unit area

γ - surface free energy per unit area of the matrix material

δ - boundary thickness

Ω - atomic volume

η - distance between atoms

CREEP CAVITATION AND FRACTURE DUE TO A STRESS CONCENTRATION IN 2¼ Cr-1 Mo

T.-S. Liu, T. J. Delph

Lehigh University
Bethlehem, Pennsylvania 18015

and

R. J. Fields

National Bureau of Standards
Washington, D.C. 20234

ABSTRACT

Long-time creep tests of 2¼ Cr-1 Mo specimens containing a circular, central hole are analyzed with regard to their creep cavitation behavior. Experimental results and analytical simulation show broad overall agreement, although there are some unexplained disparities on a smaller scale.

INTRODUCTION

In an effort to obtain improved efficiency, recent years have seen a trend towards higher and higher operating temperatures for energy conversion devices such as gas turbines, coal gasification systems, and the like. Since such devices are typically expected to give many years of useful service, reliable lifetime predictions are a necessity. Unfortunately, for the designer, elevated temperature failure modes are usually quite different from those experienced at lower temperatures and are, at the present, not at all well understood.

One of the principal phenomena responsible for the elevated temperature failure of metals is that of intergranular creep cavitation. This is brought about by the nucleation of tiny cavities, or voids, at hard inclusions on grain boundaries or near grain boundary triple points. The cavities grow with time until eventually they begin to coalesce, forming an array of microcracks which lead to failure. Although other failure modes may exist [1], depending upon the temperature and applied loading, intergranular creep cavitation is usually the primary failure mode in structures operating at high temperatures under relatively static loads.

Due to its practical significance, creep cavitation has been the focus of extensive study. Most of this work has been aimed at obtaining a better understanding of the micromechanical processes of cavity growth in a stress field which, on the macroscale, is both homogeneous, or uniform, and uniaxial. A complete and reasonably up-to-date summary of such micromechanical models has been given by Svensson and Dunlap [2]. In the present work, we take a somewhat different approach. Instead of relying upon micromechanical modeling techniques, we utilize an empirically-derived description of cavity nucleation and growth to investigate both analytically and experimentally the distribution of creep cavitation in the neighborhood of a stress concentration. This is a problem of considerable practical importance since structural failures often originate at locations such as fastener holes, bends, and fillets where stress concentrations are known to be present.

The particular problem we wish to examine is that of creep cavitation about a central, circular hole in a plane stress tensile strip. This configuration is sufficiently simple so as to be quite tractable from both an experimental and analytical viewpoint, but contains many of the features present in more complex problems. In particular, the presence of the circular hole results in a nonuniform, multiaxial, time-dependent stress field in the neighborhood of the hole.

The material used in the investigation was the ferritic stainless steel 2¼ Cr-1 Mo. This metal is widely used in elevated temperature applications. In the ferritic condition, 2¼ Cr-1 Mo is highly resistant to creep cavitation, intergranular cavities being observed only after several thousand hours under load. For this reason, not a great deal is known about its cavitation behavior under creep conditions. Furthermore its elevated temperature mechanical behavior is relatively complex, making it somewhat difficult to characterize analytically.

EXPERIMENTS

Two experimental specimens were fabricated from a commercial heat of 2¼ Cr-1 Mo steel which conformed to ASTM A542 Class 2 requirements. Prior to machining, the materials were subjected to the following heat treatment: 10 minutes at 1000°C cooling to 710°C at 35°C/hr, 2 hours at 710°C, furnace cooling to room temperature at 50°C/hr. This heat treatment is identical to the one employed by Matera et al [3], and was utilized in order to be able to make subsequent use of the cavitation data obtained by these investigators.

After heat treatment, the average grain diameter was found to be 40.4 μm, in quite good agreement with the value reported in [3]. The resulting microstructure was a ferrite-pearlite structure similar to that reported in [3]. Microscopic examination revealed the presence of large numbers of carbide inclusions on the grain boundaries, probably Cr-Mo dual carbides. Such a structure can be expected to be very stable at temperatures of 600°C and below, and hence is often sought in industrial applications.

The specimens were machined to the final configuration shown in Figure 1, which is basically that of a plane stress tensile strip containing a central circular hole whose diameter is one-quarter of the specimen width.

The specimens were tested at 565°C in standard creep machines under piecewise constant load conditions. Specimen no. 1 was loaded at a nominal stress of 40 MPa for 4520 hours, the load level then being increased to 114 MPa for an additional 196 hours until failure occurred. Failure was brought about by the initiation and propagation of creep cracks from either side of the hole. Figure 2 shows a magnified view of one half of the fractured specimen. The darker band around the interior of the hole is an oxide layer.

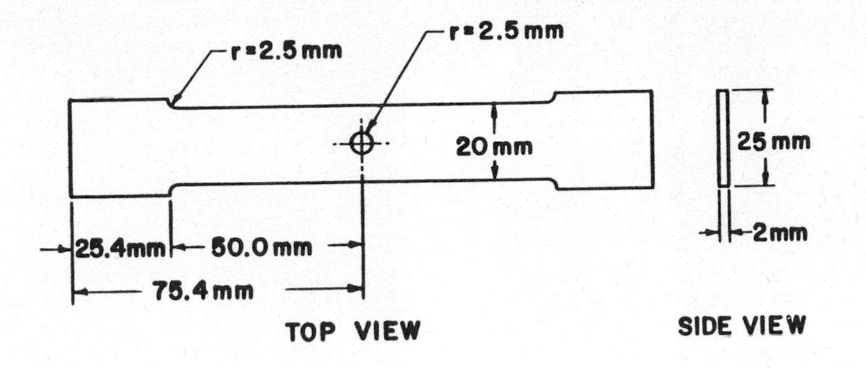

Fig. (1) - Specimen configuration

Fig. (2) - Specimen no. 1 after failure

It can be seen that cavitation is fairly extensive near the point of crack initiation and to one side of the fracture surface. However, little or no cavitation was observed elsewhere, in particular, in the area of the specimen remote from the hole where the stress field would be expected to be approximately uniaxial.

Figure 3 shows an etched section at high magnification in the neighborhood of the central hole. The cavitation in this region is intergranular in nature. Figure 4 shows an etched section taken in an area near the fractured surface, but away from the central hole. Here the cavitation appears to be primarily of a transgranular nature. It is surmised that this type of cavitation was produced by the fairly rapid increase in stress as the minimum cross-sectional area decreased due to the propagation of creep cracks from both sides of the hole. Thus,

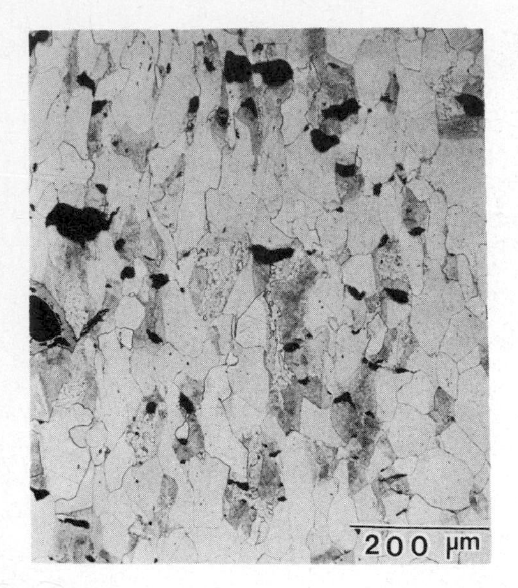

Fig. (3) - Intergranular cavitation near hole

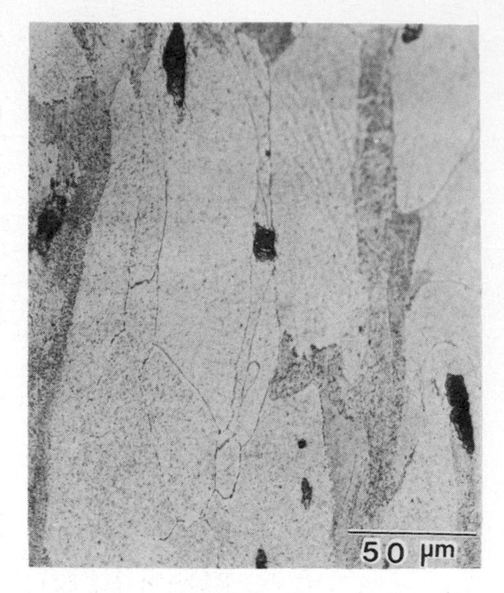

Fig. (4) - Transgranular cavitation away from hole

initially the failure was of a relatively slow intergranular nature, accompanied by creep cracking, in the immediate neighborhood of the hole. Once the cracks had reduced the cross-sectional area sufficiently, it appears that fracture then proceeded in a faster transgranular ductile mode.

Specimen no. 2 was loaded at a nominal stress of 42 MPa for 3000 hours and then removed from the oven for metallographic examination. A careful optical examination revealed no apparent cavitation, and the specimen was returned to the testing machine with an increase in load to a nominal stress of 87.4 MPa. After an additional 1324 hours, the test was terminated prior to failure. The specimen was again carefully polished using 1 micron diamond compound on a Politex ASD-125 lap followed by etching in a 2% nital solution. It was found that this process of polishing and etching had to be rather painstakingly repeated a number of times before the full extent of cavitation became visible.

As with specimen no. 1, a fairly abundant number of cavities were observed nea the edge of the hole covering an area about 30° to either side of the axis of mini mum cross-sectional area. For brevity, this axis will henceforth be called the horizontal axis, the tensile axis being assigned the vertical direction. Figures 5 and 6 show the sides of the hole immediately adjacent to the horizontal axis. As before, the cavitation is almost exclusively of an intergranular nature. The larger cavities display an elongated shape with the long axis roughly perpendicula to the tensile axis. Figure 7 shows an enlarged view of a portion of the specimen depicted in Figure 6.

Although the bulk of the cavitation was in the neighborhood of the central hole, some isolated cavitation was observed out to a distance midway between the

Fig. (5) - Side of hole

Fig. (6) - Side of hole

hole and the edge of the specimen. Figure 8 shows several cavities located at about one-third the way between the hole and the edge. The cavities observed here appear to be in the process of coalescing to form larger cavities or microcracks.

ANALYSIS

In order to assess the effects of the time-dependent stress and strain fields upon the creep cavitation behavior, an inelastic finite element analysis of the test of specimen no. 2 was conducted. At the heart of any finite element analysis is a set of constitutive equations which is presumed to describe the mechanical behavior of the material under consideration. The analysis carried out in the present case is noteworthy in that it employs the recently developed constitutive model of Robinson [4,5]. This model was in fact specifically developed to describe the rather complex mechanical behavior of 2¼ Cr-1 Mo, although it has applicability to other metals as well. Robinson's model is one of a class of "unified" constitutive theories, so-called because of the efforts of such theories to describe both short-term "plastic" deformation as well as long-term "creep" deformation. This feature makes it particularly attractive for our application because it is known that substantial inelastic deformation may occur in the neighborhood of a stress concentration over a very short period of time after load application.

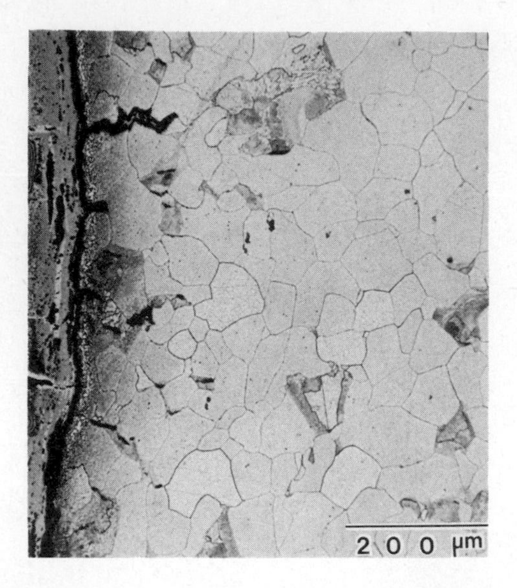

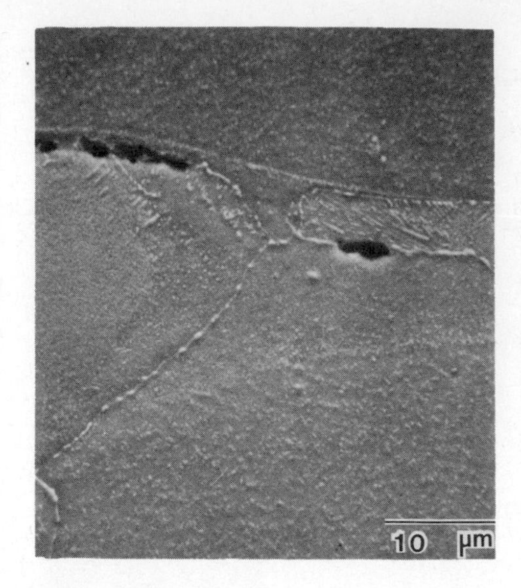

Fig. (7) - Intergranular cavitation near hole

Fig. (8) - Intergranular cavitation further away from hole

In the absence of load reduction, reversals, or other complicating factors, Robinson's constitutive equations may be written as

$$2\mu\dot{\varepsilon}^{I}_{ij} = F^{(n-1)/2}\,\Sigma_{ij} \tag{1}$$

$$\dot{\alpha}_{ij} = 2\mu H\dot{\varepsilon}^{I}_{ij}/G^{\beta/2} - [RG^{(m-\beta-1)/2}]\alpha_{ij} \tag{2}$$

where

$$F = \Sigma_{ij}\Sigma_{ij}/2K^2 - 1$$

$$G = \alpha_{ij}\alpha_{ij}/2K^2$$

$$\Sigma_{ij} = \sigma'_{ij} - \alpha_{ij}$$

The total strain rate is assumed to be the sum of elastic and inelastic parts, $\dot{\varepsilon}^{I}_{ij}$ being the inelastic contribution. The deviatoric tensor quantity α_{ij} has the character of a state variable and σ'_{ij} is the deviatoric part of the stress tensor. Here, μ, n, H, β, R, K, and m are material constants under isothermal conditions. To avoid a singularity in equation (2) when $G \to 0$, G is replaced by a constant G_o in equation (2) whenever $G < G_o$. Values of these constants appropriate to a particular heat and heat treatment of 2¼ Cr-1 Mo were taken from [5]. Typically, the constant G_o is quite small, so that small values of G such as encountered on initial loading of a virgin material lead to quite high values of $\dot{\alpha}_{ij}$. This fact leads to considerable difficulties in numerical solutions of Robinson's equations, as extremely small time steps are initially required in order to integrate the equations forward in time. We shall return to this point subsequently.

The finite element formulation of Robinson's constitutive model follows from an application of the Galerkin technique [6] to the stress equations of equilibrium for quasi-static conditions in the absence of body forces,

$$\sigma_{ij,j} = 0 \tag{3}$$

Standard manipulations yield the finite element equations in the form

$$[K]\{\delta\} = \int_V [B]^T[C]\{\varepsilon\}^I dV + \int_S [N]^T\{T\}dS \tag{4}$$

where, in the usual notation of the finite element method, $[K]$ is the elastic stiffness matrix, $\{\delta\}$ the vector of nodal point displacements, $[B]$ the strain-displacement matrix, $[C]$ the matrix of elastic constants, $\{\varepsilon\}^I$ the inelastic strain vector, $[N]$ the matrix of interpolating functions, and $\{T\}$ the traction vector. Furthermore the stresses are related to the elastic strains by Hooke's law.

Assuming the total strains to be the sum of an elastic and an inelastic part, we may write Hooke's law in finite element notation as

$$\{\sigma\} = [C]([B]\{\delta\} - \{\varepsilon\}^I) \tag{5}$$

Equations (1) and (2), in conjunction with equations (4) and (5), may now be integrated forward in time from a given set of initial conditions, which, for a virgin material, we take to be $\varepsilon^I_{ij} = \alpha_{ij} = 0$. Briefly, this is done as follows. Equation (4) determines the nodal point displacements for a given inelastic strain field and specified surface tractions. Equations (1) and (2) then yield the inelastic strain and state variable rates, given the state variable field. This information can then be used to implement any standard numerical integration scheme.

In practice, considerable care must be used in selecting an integration scheme. The primary difficulty, mentioned earlier, is the necessity of using extremely small initial time step sizes, often on the order of 10^{-5} hours, because of the high initial values of $\dot{\alpha}_{ij}$. However, to be able to handle problems involving long-term creep behavior, the integration must be continued out to thousands of hours. Hence, some method of efficiently controlling the time step size must be present in any integration scheme selected, as well as the usual requisites of accuracy and stability. In the present code, we have utilized the well-known Adams-Bashforth-Moulton method, a fourth-order accurate predictor-corrector algorithm with excellent stability properties (though not unconditionally stable). The attractive feature of this method is that the difference between the predicted and corrected solution at any time step may be used to estimate the error incurred at that step, and the time step size adjusted accordingly. The method has the drawback that it is somewhat expensive of core storage, but this is not expected to prove too much of a difficulty for the moderate-sized problems we envision. Experience to date with this method has been quite favorable.

The analysis was carried out for the loading history experienced by specimen no. 2, assuming the load to be applied instantaneously at t=0. The finite element discretization for the problem is shown in Figure 9, having 10 QUAD-12 ele-

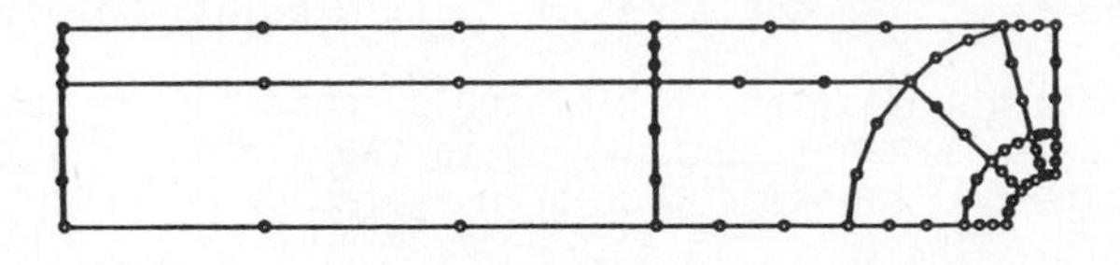

Fig. (9) - Finite element discretization

ments and 124 degrees of freedom. Since the specimen geometry and loading are both symmetric about horizontal and vertical axes through the center of the hole, it is necessary only to model a quarter of the specimen. The time integration procedure discussed previously proved to be quite efficient, requiring approximately 900 system seconds on a CDC CYBER 720 computer, even though the initial time step size was 10^{-5} hours.

The purpose of the analysis was to allow analytical predictions of creep damage and to compare these to the observed pattern of cavitation in specimen no. 2. To this end, we made use of some empirical cavitation data generated by Piatti and co-workers [3], [7] for 2¼ Cr-1 Mo. This data was obtained from density change measurements made on 2¼ Cr-1 Mo specimens crept under constant uniaxial load for various periods of time and at various temperatures. The ratio of the change in density at the end of a particular period to the initial density, $\Delta\rho/\rho_o$, was taken to be directly related to the cavity void volume fraction and hence was considered a measure of the creep damage D incurred by the specimen. The resulting data was correlated statistically with the elapsed time, the applied stress, and the accumulated inelastic strain [3]. For uniaxial loading histories in which the stress is not constant, the result may be expressed in the form [7]

$$D = -\Delta\rho/\rho_o = H(\varepsilon^I)^\alpha [\int_o^t \sigma^{\gamma/\delta} d\tau]^\delta \tag{6}$$

Under isothermal conditions, the quantities H, α, γ, and δ are material constants. These were obtained for the temperature of interest in the present case, 565°C, by interpolation from tabular data presented in [3]. For stress in units of megapascals and time in hours, these were $H = 2.59 \times 10^{-7}$, $\alpha = 0.47$, $\gamma = 3.76$, and $\delta = 0.33$.

The presence of the central hole in the specimen leads to a multiaxial state of stress in the neighborhood of the hole, so equation (6) must be appropriately generalized to multiaxial conditions. Not a great deal is known about the effects of multiaxial stress states upon creep cavitation. As noted in [2], most theoretical models of this phenomenon consider cavity growth to be influenced solely by the maximum positive principal normal stress. However, there is evidence to indicate [8-10] that the von Mises effective stress and the hydrostatic stress may also play a role. In carrying out the present analysis, two possible generalizations of equation (6) to multiaxial conditions were considered: one based solely on maximum positive principal stress (and corresponding inelastic strain) and the other solely on von Mises effective stress and inelastic strain. The results of the calculations using the maximum principal stress criterion are shown in Figure 10 in the form of contours along which the creep damage is constant over

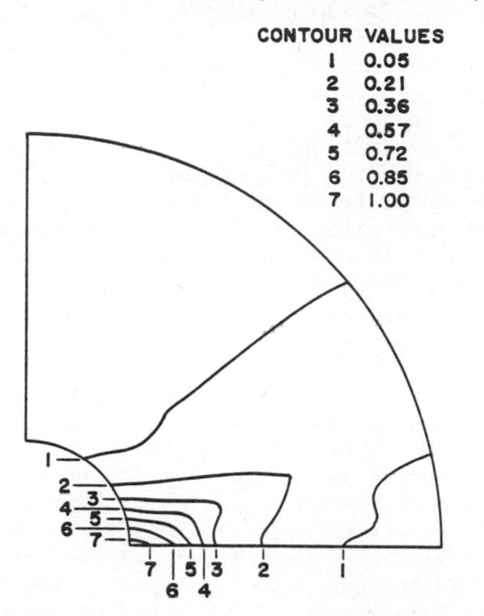

Fig. (10) - Constant damage contours using maximum principal stress criterion

a portion of the specimen area in the neighborhood of the central hole. The values of creep damage have been normalized by the largest value for easy comparison. Surprisingly enough, contours of constant creep damage obtained using an effective stress criterion showed little difference in spatial distribution from those shown in Figure 10, and differed by about only 15% in numerical value. This is a result which might not be readily expected. However, it is interesting to note in this context that Blass [11], in analyzing failure data for type 304 stainless steel specimens under multiaxial conditions, found little difference between a criterion based upon maximum principal stress and one based upon von Mises effective stress.

It can be seen from Figure 10 that the region of greatest damage is predicted to lie in the immediate neighborhood of the intersection of the central circular hole and the horizontal axis. Not coincidentally, this is the area of highest stress concentration. The predicted damage level drops off rather sharply with angular distance along the circumference of the hole, and is about a third of the maximum value at 30° from the horizontal axis. It is interesting to note that the contours seem to develop a banded structure with increasing distance from the central hole, the center of the band having about a 30° orientation to the horizontal.

DISCUSSION

As we noted in the Introduction, the purpose of the present study was to make qualitative comparisons of observed and calculated patterns of creep cavitation in the neighborhood of a stress concentration in 2¼ Cr-1 Mo. For reasons to be discussed subsequently, quantitative comparisons were not felt to be meaningful. However, the qualitative trends in both the calculated and observed void distributions are fairly clear.

In an overall sense, the calculations of creep damage, or cavitation, for the test of specimen no. 2 are in rough agreement with experimental observations, in that the most heavily cavitated region was in the neighborhood of the central hole about the point of greatest stress concentration. However, on a smaller scale, there was a marked dissimilarity between experiment and analytical simulation. Experimentally, the cavitation was fairly evenly distributed about the circumference of the hole to about 30° to either side of the horizontal axis. Furthermore, although the indications are not completely unambiguous, the cavitation seemed to be somewhat heavier in a region 20°-30° to either side of the horizontal axis. This is in contrast to the numerical simulation, which predicted the greatest damage at the intersection of the hole and the horizontal axis and a rapid drop in damage level with angular distance around the hole. Away from the immediate neighborhood of the hole, observed levels of cavitation were too low to permit any meaningful comparisons.

The reasons for the discrepancy between calculated and observed cavity distributions near the hole are not known at present. One possible explanation is the presence of inaccuracies in the analytical modeling, which incorporated several simplifying assumptions. One of these was the set of material constants used in the constitutive relations. These were taken from [5] and were selected so as to typify the mechanical behavior of a particular heat and heat treatment of 2¼ Cr-1 Mo at 565°C. It is known, however, that differing heats and heat treatments can substantially affect elevated temperature deformation behavior, so that the material constants used in the present study may not be the most representative for the present heat and heat treatment. However it does not seem likely that the use of a somewhat different set of material constants in the analysis would qualitatively affect to any great extent the calculated damage distribution.

Another simplifying assumption made in the analysis was the use of the linearized strain-displacement relations, which effectively ignores any changes in specimen geometry. In point of fact, the maximum strain component calculated in the analysis was approximately 6%, which is somewhat outside the range in which the linearized strain-displacement relationship is generally felt to offer a good approximation. Thus the use of this relationship resulted in some error in the analysis, but again this particular assumption should not greatly upset the analytically predicted damage distribution. It is worth noting in this context that the deformations experienced by the experimental specimen were sufficiently small so that only a slight elongation of the hole in the direction of the tensile axis could be perceived.

The final assumption employed was that the mechanical behavior of the specimen, as modeled by the constitutive relations, was unaffected by the presence of creep cavitation or damage. The amount of error incurred by the use of this assumption is difficult to assess without a better knowledge of the effects of creep damage upon mechanical behavior. However in view of the fact that the cavitation in the specimen is nowhere particularly extensive, we surmise that in the present case these effects are not very large. In any case, it appears that modification of the constitutive relations to account for the effects of damage would lead to damage predictions which are qualitatively similar to those shown in Figure 10, that is, with the damage rather narrowly concentrated at the point of greatest stress concentration.

Despite any inaccuracies resulting from analytical simplifications, we feel that the lack of agreement between analysis and experiment is more likely due to the multiaxial forms assumed for the damage law, equation (6). Here we alternatively assumed that in a multiaxial situation the damage depends upon either maximum principal or von Mises effective quantities. The former criterion is favored by most existing micromechanical models, while the later is recommended for making creep damage calculations for design use [12]. However, it has been reported [10], [13] that the hydrostatic stress component may have a significant effect upon cavity growth. Hayhurst [8] has in fact suggested that the most general multiaxial creep damage criterion consists of an appropriate linear combination of maximum principal, von Mises effective, and hydrostatic quantities. Thus, it is quite possible that the inclusion of a hydrostatic component into the multiaxial generalization of the damage law would significantly alter the analytically predicted damage distribution.

Another and considerably more speculative explanation for the observed discrepancy between analysis and experiment arises from the two stage nature of the cavitation phenomenon; that is, initial nucleation followed by subsequent growth. It is possible that the multiaxial criterion for cavity nucleation differs significantly from that for cavity growth. The bulk of previous cavitation experiments, including those of Matera et al [3] which were utilized in this study, were carried out in a homogeneous stress field where the effects of such a difference could not be noticed. However in a nonuniform stress field such as considered in the present study, cavities might nucleate preferentially in one area, but grow at a more rapid rate in another. Such an effect could easily result in substantial modifications to the picture shown in Figure 10.

Though the results obtained in the present study are far from conclusive, they do serve to demonstrate the complex nature of the problems involved in understanding cavitation behavior. This field is today a quite active one due to its practical significance in making failure predictions. However it appears that there still exist abundant opportunities for further work in this area.

ACKNOWLEDGEMENTS

The authors gratefully acknowledge the initial support of Lehigh University and the Fracture and Deformation Section of the National Bureau of Standards, and the later support of the U.S. Department of Energy under Contract DE-AC01-81ER10816.

REFERENCES

[1] Ashby, M. F., Fields, R. J. and Weerasooriya, T., "Fracture mechanisms in pure iron, two austenitic steels and one ferritic steel", Met. Trans. A, 11, p. 333, 1980.

[2] Svensson, L.-E. and Dunlap, G. L., "Growth of intergranular creep cavities", Int. Metals Reviews, 26, p. 109, 1981.

[3] Matera, R., Piatti, G., Belloni, G. and Mateazzi, S., "Dannegiemento a scorrimento viscaso dell'acciaio 2.25 Cr-1 Mo", 4th Congress Nazionale de Meccanica Teorica ed Applicata, p. 187, Firenze, 1978.

[4] Robinson, D. N., "A unified creep-plasticity model for structural metals at high temperatures", ORNL/TM-5969, Oak Ridge National Laboratory, 1978.

[5] Robinson, D. N., "Development of refined constitutive laws" in High-Temperature Structural Design Program Semiannual Progress Report for Period Ending December 31, 1978, ORNL-5540, Oak Ridge National Laboratory, 1979.

[6] Zienkiewicz, O. C., The Finite Element Method, McGraw-Hill, 1977.

[7] Piatti, G., Bernasconi, G. and Cozzarelli, F. A., "Damage equations for creep rupture in steel", Proc. Fifth Int. Congress on Structural Mechanics in Reactor Technology (SMIRT), Paper No. L 11/4, Berlin, 1980.

[8] Hayhurst, D. R., "Creep rupture under multiaxial states of stress", J. Mech. Phys. Solids, Vol. 20, p. 381, 1972.

[9] Cane, B. J., "Interrelationship between creep deformation and creep rupture in 2¼ Cr-1 Mo steel", Metal Science, Vol. 13, p. 287, 1979.

[10] Cane, B. J., "Mechanistic control regimes for intergranular cavity growth in 2.25 Cr-1 Mo steel under various stresses and stress states", Metal Science Vol. 15, p. 302, 1981.

[11] Blass, J. J., "An assessment of multiaxial creep-rupture criteria for type 304 stainless steel", in High-Temperature Structural Design Program Semiannual Progress Report for Period Ending June 30, 1978, ORNL-5433, Oak Ridge National Laboratory, 1979.

[12] Code Case N-47-12, Cases of the A.S.M.E. Boiler and Pressure Vessel Code, American Society of Mechanical Engineers, New York.

[13] Hull, D. and Rimmer, D. E., "The growth of grain-boundary voids under stress", Phil. Mag., Vol. 4, p. 673, 1959.

SECTION V

STRESS AND FAILURE ANALYSIS

STRESS INTENSITY FACTORS FOR RADIAL CRACKS IN BIMATERIAL MEDIA

O. Aksogan*

Gannon University
Erie, Pennsylvania 16541

ABSTRACT

The plane elasticity problem of two bonded fractured semi-infinite planes of different materials is considered. The problem is treated as a mixed boundary value problem of linear elasticity. The lines, on which the arrays of cracks are situated, being chosen to be concurrent with the interface, the Mellin transform is used in conjunction with the Green's function technique to obtain the integral equations of the problem. These equations, which are singular integral equations of the first kind with Cauchy type singularities, are solved numerically by an effective procedure depending on the properties of the Chebyshev polynomials.

The stress intensity factors at all crack tips are determined using the results of the above-mentioned numerical solution. Despite the fact that the method given here is applicable to any number of arrays of radial cracks, there being results only for cases with two cracks in the literature, comparisons could be made only for such cases and the corresponding values matched perfectly. Some new cases not found in the literature were treated and the results found were presented in graphical form. It deserves mentioning that the procedure used in this study can be extended, in a straightforward manner, to the problems of the debonding of two bonded semi-infinite planes with cracks.

INTRODUCTION

In this study, the plane elasticity problem of two perfectly bonded semi-infinite planes of different materials with internal cracks is considered. The problem is treated as a mixed boundary value problem of linear elasticity. The lines, on which the collinear arrays of cracks are situated, having been chosen to be concurrent with the interface, the most suitable choice of analytical approach would be the application of the Mellin transform. Making use of the Green's function in conjunction with the Mellin transform, the analytical formulation of the problem ends up with a system of singular integral equations of the first kind with simple Cauchy type kernels.

*On leave from Middle East Technical University, Ankara, Turkey.

The unknowns of the above-mentioned system of singular integral equations are the density functions of the dislocations of the two planar types on all crack lines in the semi-infinite planes forming the medium. The dislocation density functions being determined numerically by means of an effective procedure employing the Gauss-Chebyshev quadrature formulas [1,2], the stress intensity factors of the two planar types are evaluated by straightforward expressions in terms of the dislocation density functions [3]. Actually, despite the fact that the stress intensity factors are the main concern of this study, it should be noted that it would suffice to apply widely known relations [4-6] to determine the stresses, the crack opening displacements, and the probable directions of crack propagation for the complete medium once the dislocation density functions are determined.

The formulation of the problem has been carried for two semi-infinite planes, the complete loading composing of the self-equilibrating tractions on the surfaces of the cracks. However, if the superposition and Saint Venant principles are employed, the foregoing restrictions do not harm the general applicability of the results to comparatively small regions with cracks around bimaterial interfaces of composite planes with almost any possible loading at points remote from the region of concern.

It deserves mentioning, concerning the previous literature about the subject, that there are three main closely related studies on the problem of two bonded semi-infinite planes with cracks [4-6]. The present study is a generalization of those three studies to arbitrary number of radial arrays of cracks, none of them terminating at the interface. Actually, the analytical formulation of the present work applies to cracks terminating at the interface as well. However, the numerical solution of the singular integral equations necessitates the determination of the powers of singularity at the crack tips lying on the interface which show up in the so-called generalized Cauchy kernels in some of the singular integral equations.

Needless to say, the results of the present study hold as special cases for the problems of a semi-infinite plane, both when its edge is free and when it is bonded to a rigid plane. In the former case, which has been previously published by the author [7], some problems of edge cracks with nearby internal cracks can be solved by a straightforward application of the so-called reflection procedure used by Gupta and Erdogan [8].

ANALYSIS

For solving boundary value problems of this type, in which all boundaries are on radial lines, the suitable type of transform is the Mellin transform which is defined as follows:

$$\tilde{f}(s) = \int_0^\infty f(r) r^{s-1} dr, \quad \tilde{f}(r) = \frac{1}{2\pi i} \int_{c-i\infty}^{c+i\infty} f(s) r^{-s} ds \qquad (1)$$

where c is determined the regularity conditions of the problem in hand and i is the unit of imaginary numbers.

In polar coordinates, in the absence of body forces, the stresses and the displacements in the plane problems of elastostatics can be given by [9]

$$\sigma_r = \frac{1}{r^2}\frac{\partial^2\Phi}{\partial\theta^2} + \frac{1}{r}\frac{\partial\Phi}{\partial r},\quad \sigma_\theta = \frac{\partial^2\Phi}{\partial r^2},\quad \tau_{r\theta} = -\frac{\partial}{\partial r}\left(\frac{1}{r}\frac{\partial\Phi}{\partial\theta}\right)$$

$$2\mu u_r = -\frac{\partial\Phi}{\partial r} + (1-\eta)r\frac{\partial\Psi}{\partial\theta},\quad 2\mu u_\theta = -\frac{1}{r}\frac{\partial\Phi}{\partial\theta} + (1-\eta)r^2\frac{\partial\Psi}{\partial r} \tag{2}$$

in which μ is the shear modulus, $\eta = \nu/(1+\nu)$ for plane stress and $\eta=\nu$ for plane strain where ν is the Poisson's ratio. The Airy stress function $\Phi(r,\theta)$ and the displacement function $\Psi(r,\theta)$ satisfy the following equations:

$$\nabla^2(\nabla^2\Phi) = 0,\ \nabla^2\Psi = 0,\ \frac{\partial}{\partial r}\left(r\frac{\partial\Psi}{\partial\theta}\right) = \nabla^2\Phi \tag{3}$$

in which ∇^2 stands for the harmonic operator.

In plane elasticity problems in which the medium is composed of wedge-shaped domains, the solution of (3) for each domain yields [10]

$$\tilde{\phi}(s,\theta) = Ae^{is\theta} + \overline{A}e^{-is\theta} + Be^{i(s+2)\theta} + \overline{B}e^{-i(s+2)\theta},$$

$$(r^2\tilde{t}) = 2i(s+1)[Ase^{is\theta} + B(s+1)e^{i(s+2)\theta} - \overline{B}e^{-i(s+2)\theta}] \tag{4}$$

$$(r^2\tilde{v}) = -\frac{s+1}{\mu}[Ase^{is\theta} + B(s+1)e^{i(s+2)\theta} + \kappa\overline{B}e^{-i(s+2)\theta}]$$

in which

$$t = \tau_{r\theta} + i\sigma_\theta,\quad v = \frac{\partial u_r}{\partial r} + i\frac{\partial u_\theta}{\partial r}$$

$$\kappa = 3-4\eta \tag{5}$$

and A and B, with their complex conjugates $\overline{A}$ and $\overline{B}$, are independent of θ.

In the present study, the infinite plane is divided into h infinite wedges by the interface and the lines of cracks (see Figure 1). Let the union of all straight line segments representing the cracks along one radial line be called L and the remainder L', the former being finite and the latter infinite. Then the boundary conditions of this problem can be expressed as follows:*

---

*The derivatives of the equations concerning displacements have been used purely for the sake of convenience in computations. This change will be accounted for by the side conditions of the integral equations later.

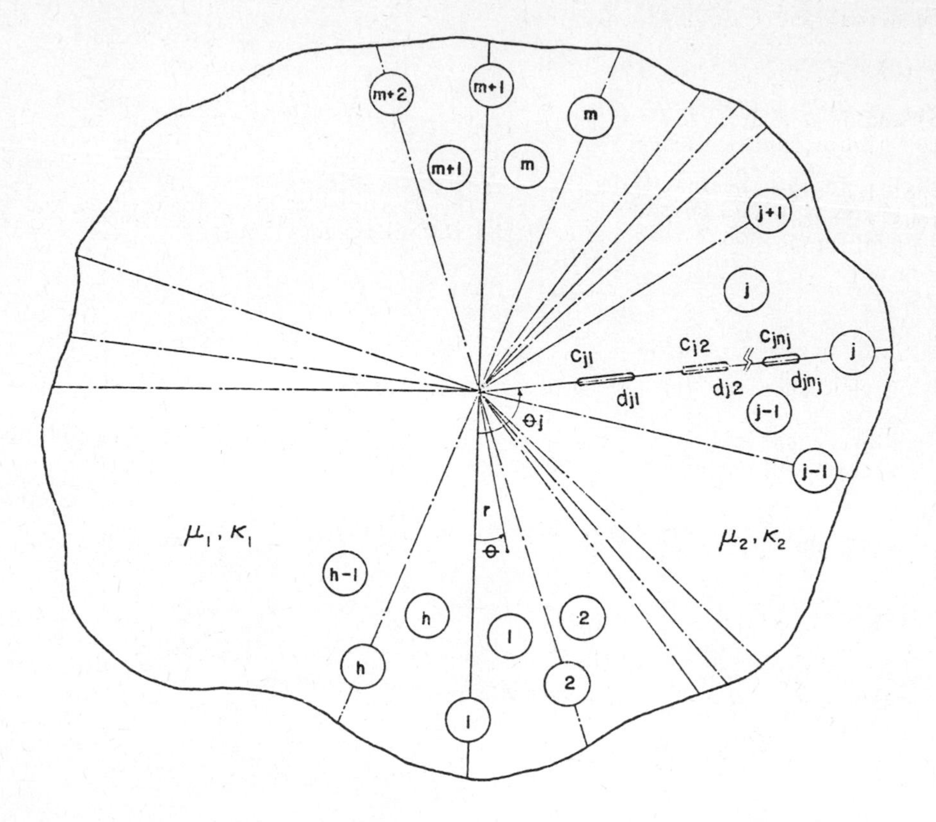

Fig. (1) - General setting of the problem

$$
\begin{aligned}
&t_1(r,0) = t_h(r,2\pi) \text{ in } 0<r<\infty \\
&v_1(r,0) = v_h(r,2\pi) \text{ in } 0<r<\infty \\
&t_{m+1}(r,\pi) = t_m(r,\pi) \text{ in } 0<r<\infty \\
&v_{m+1}(r,\pi) = v_m(r,\pi) \text{ in } 0<r<\infty \\
&t_{j-1}(r,\theta_j) = t_j(r,\theta_j) \text{ on } L_j+L'_j \\
&v_{j-1}(r,\theta_j) = v_j(r,\theta_j) \text{ on } L'_j \\
&t_j(r,\theta_j) = w_j(r) \text{ on } L_j
\end{aligned}
\tag{6}
$$

where $j = 2,3,\ldots,h$ $(j \neq m+1)$ and the loading functions $w_j(r)$ are defined in terms of the normal and the shear tractions as follows:

$$w_j(r) = q_j(r) + ip_j(r),\ j = 2,3,\ldots,h\ (j \neq m+1) \tag{7}$$

In (6) and (7), and in the sequel, the subscripts show the region or the boundary to which a certain quantity pertains.

Applying the Green's function technique to this problem as has been done previously by the author [10,11], the integral equations of the problem are obtained as follows:

$$\frac{1}{\pi}\sum_{\ell=2}^{m}\int_{L_\ell}\frac{\mu_2}{K_3K_5r_\ell}\left(\frac{r_\ell}{r}\right)\{[Ne^{-2i\alpha}T_1(\alpha) - N(e^{-2i\alpha} - e^{-2i\theta_j})T_2(\alpha) + (Ne^{2i\alpha}$$

$$- Ne^{2i\theta_\ell} - K_7e^{2i\theta_\ell})T_1(-\alpha) - N(2e^{2i\alpha} - e^{2i\theta_j} - 2e^{2i\theta_\ell} + 1) \times$$

$$\times T_2(-\alpha) + N(e^{2i\alpha} - e^{2i\theta_j} - e^{2i\theta_\ell} + 1)T_3(-\alpha) + K_5e^{-2i\beta}T_1(\beta)$$

$$+ K_5e^{2i\beta}T_1(-\beta) - (K_5e^{2i\beta} - 1)T_2(-\beta)]g_\ell(r_\ell) + [-N(e^{-2i\alpha}$$

$$- 2e^{-2i\theta_j})T_1(\alpha) + N(e^{-2i\alpha} - e^{-2i\theta_j})T_2(\alpha) + (Ne^{2i\alpha} - 2Ne^{2i\theta_j}$$

$$- Ne^{2i\theta_\ell} - K_7e^{2i\theta_\ell} + 2N)T_1(-\alpha) - N(2e^{2i\alpha} - 3e^{2i\theta_j} - 2e^{2i\theta_\ell}$$

$$+ 3)T_2(-\alpha) + N(e^{2i\alpha} - e^{2i\theta_j} - e^{2i\theta_\ell} + 1)T_3(-\alpha) + K_5e^{-2i\beta}T_1(\beta)$$

$$- K_5(e^{2i\beta} - 2)T_1(-\beta) + K_5(e^{2i\beta} - 1)T_2(-\beta)]if_\ell(r_\ell)\}dr_\ell$$

$$+ \frac{1}{\pi}\sum_{\ell=m+2}^{h}\int_{L_\ell}\frac{\mu_2}{K_3K_5r_\ell}\left(\frac{r_\ell}{r}\right)^2\{[K_4e^{-2i\beta}T_1(\beta) + (K_4e^{2i\beta} + K_9e^{-2i\theta_\ell})T_1(-\beta)$$

$$- (K_4e^{2i\beta} + K_9e^{-2i\theta_\ell} - K_8)T_2(-\beta)]g_\ell(r_\ell)$$

$$+ [K_4e^{-2i\beta}T_1(\beta) - (K_4e^{2i\beta} + K_9e^{-2i\theta_\ell} - 2K_8)T_1(-\beta)$$

$$+ (K_4 e^{2i\beta} + K_9 e^{-2i\theta_\ell} - K_8)T_2(-\beta)]if_\ell(r_\ell)\}dr_\ell$$

$$= q_j(r) + ip_j(r),\ j = 2,3,\ldots,h\ (j\neq m+1) \tag{8}$$

in which f and g express the density functions of the opening and the in-plane shearing mode crack opening displacements, respectively, and

$$\begin{aligned}
&M = \frac{\mu_1}{\mu_2}, \quad N = M-1, \quad K_1 = \kappa_1+M \\
&K_2 = \kappa_1 - M\kappa_2, \quad K_3 = \kappa_2+1, \quad K_4 = MK_3 \\
&K_5 = M\kappa_2+1, \quad K_6 = M^{-1}\kappa_1+1, \quad K_7 = K_2K_5/K_1 \\
&K_8 = K_3K_5/K_6, \quad K_9 = K_8-K_4, \\
&\alpha = \theta_j+\theta_\ell, \quad \beta = \theta_j-\theta_\ell, \\
&T_{n+1}(\gamma) = \sum_{k=2}^{\infty} k^n e^{ik\gamma} \left(\frac{r}{r_\ell}\right)^k,\ n = 0,1,2
\end{aligned} \tag{9}$$

It should be noted that instead of the displacement conditions, the conditions on the displacement derivatives have been used for convenience. This may cause discontinuities in the displacements which can be prevented by stipulating the appropriate integral conditions on the density functions, i.e.,

$$\int_{L_{j\ell}} \{g_j(r_j) + if_j(r_j)\}dr_j = 0,\ j = 2,3,\ldots,h\ (j\neq m+1) \quad \ell = 1,2,\ldots,n_j \tag{10}$$

where n, with the subscript j, is the number of cracks on line j and L, with the subscripts j and ℓ, expresses one of the cracks on that line. The so-called side conditions (10) render the solution of (8) unique.

After replacing the infinite series T_i, (i=1,2,3), in (8) by their equivalents having finite number of terms (see [10]), it is observed that the system of integral equations in hand is a system of singular integral equations of the first kind with Cauchy type kernels. When each and every crack in the medium is embedded in a homogeneous section, the kernels are all simple Cauchy type kernel An elegant numerical method of solution for such systems of singular integral equations was developed by Erdogan [1] and was adapted by the author [2] to mult part mixed boundary value problems.

NUMERICAL RESULTS AND DISCUSSION

The numerical method of solution [1-3] used for solving the system of singular integral equations (8) with the side conditions (10) is a collocation technique in which the Gauss-Chebyshev quadrature formulas are employed. The points of collocation and the steps of integration used in the numerical solution are marked by the zeroes of the corresponding type of Chebyshev polynomials.

The results of the present work, for the special cases of one crack with an arbitrary orientation [5] and one or two cracks normal to the interface [4,6], are shown both analytically and numerically to coincide with those in the literature. Moreover, some numerical results for two arbitrarily oriented cracks are presented graphically for uniform stresses at infinity parallel to the interface and perpendicular to it, each being applied separately. The numerical results presented in Figures 2-5 are the normalized stress intensity factors given in the form

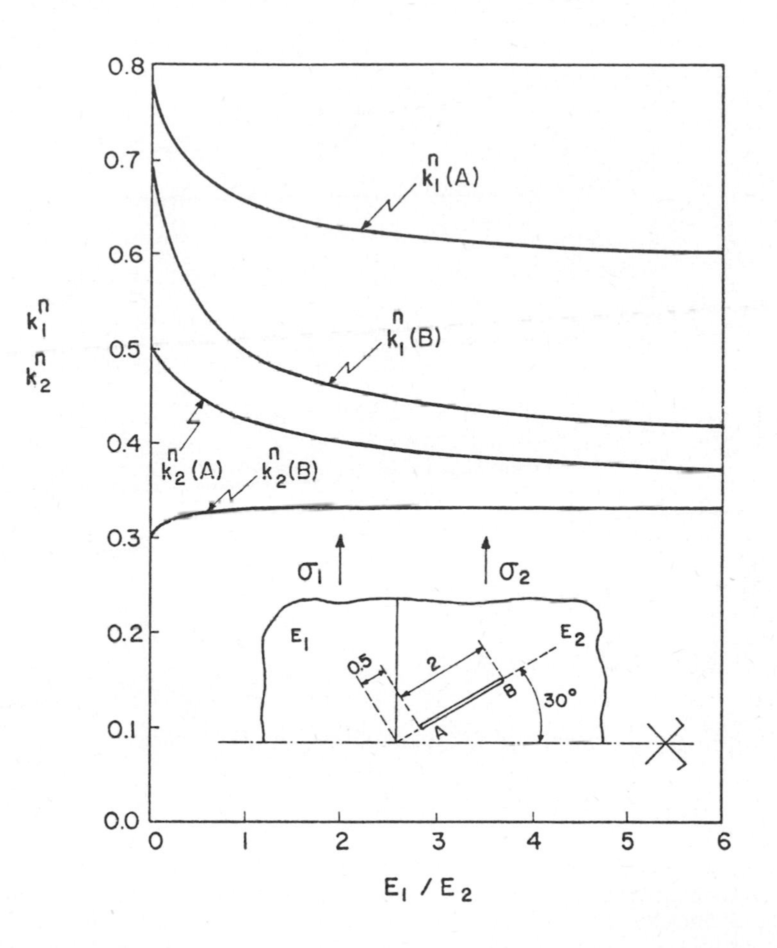

Fig. (2) - Variations of the normalized stress intensity factors with the ratio of the elastic moduli under tension at infinity parallel to the interface ($\nu_1 = \nu_2 = 1/3$)

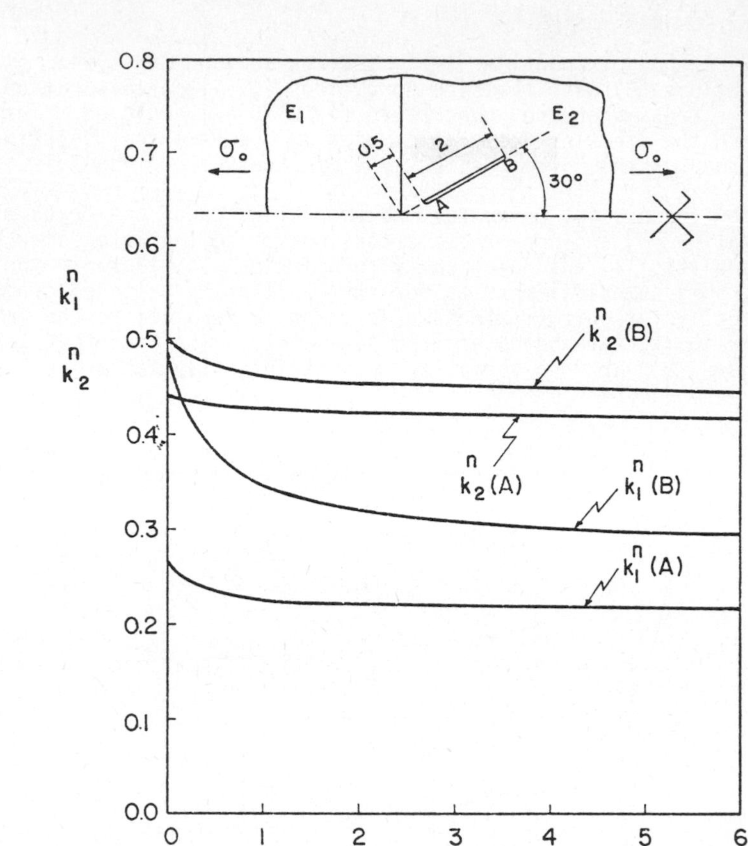

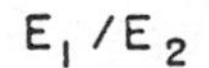

Fig. (3) - Variations of the normalized stress intensity factors with the ratio of the elastic moduli under tension at infinity normal to the interface ($\nu_1 = \nu_2 = 1/3$)

$$k_1^n = k_1/(\sigma_i a_o^{1/2}),\ k_2^n = k_2/(\sigma_i a_o^{1/2}) \tag{11}$$

in terms of the stress intensity factors which are in turn defined as

$$k_1 = \lim_{r \to 0} \sigma_\theta (2r)^{1/2},\ k_2 = \lim_{r \to 0} \tau_{r\theta} (2r)^{1/2} \tag{12}$$

In the last two equations above, r is the distance from the crack tip of concern a_o is the half length of the crack which is being considered and σ_i $(i = 0,1,2)$ are the tensile stresses in the region in which that particular crack is embedde

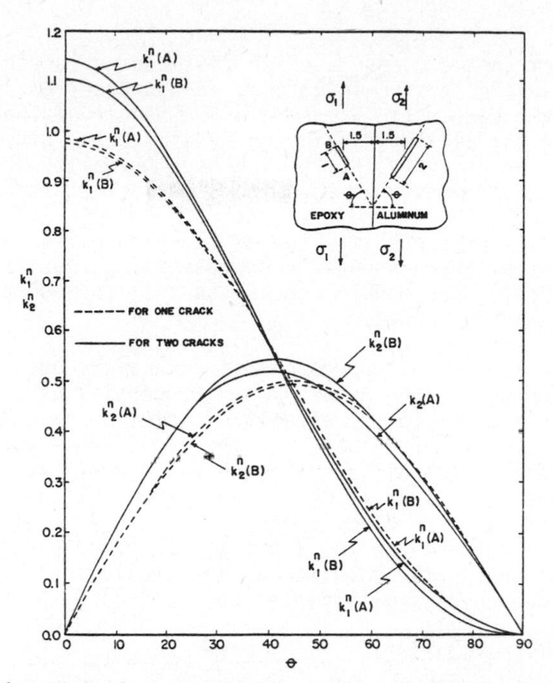

Fig. (4) - Variations of the normalized stress intensity factors with the inclinations of one and two cracks under tension parallel to the interface (epoxy: $E = 4.45 \times 10^5$ psi, $\nu = 0.35$; aluminum: $E = 10^7$ psi, $\nu = 0.3$)

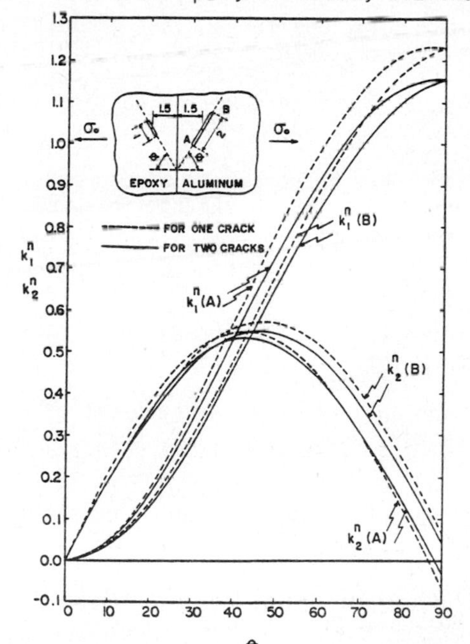

Fig. (5) - Variations of the normalized stress intensity factors with the inclinations of one and two cracks under tension normal to the interface (epoxy: $E = 4.45 \times 10^5$ psi, $\nu = 0.35$; aluminum: $E = 10^7$ psi, $\nu = 0.3$)

It should be noted that in the cases in which the loading considered is composed of tensile stresses parallel to the interface, the stresses on the two sides of the interface are in the same ratio as the elastic moduli of the corresponding materials. The reason for all lengths in the inserts being given dimensionless is because the stress intensity factors having been normalized as they are, a magnification of the geometry of the medium does not affect their values at all.

The numerical results presented in Figures 2-5 are self-explanatory with thei inserts. In Figures 4 and 5 are found the corresponding results for a single crack, as well as those for the double cracks, so that an assessment of the inter action of two such cracks can be made.

Actually, the numerical results presented in the aforementioned figures do not contribute anything new to the widely known properties of the interactions of cracks [10], which are similar to the effects of holes on cracks, and the similar trends concerning the effects of a change in the rigidity of a nearby region [5] (as of a bonded half plane with different elastic properties). However, the results of this study are worth a great deal due to the applicability of its comput er program to a wide class of problems of the same kind in a straightforward manner. More than that, it offers the tools and presents the procedure that will constitute the foundation for the solution of the problems of two imperfectly bonded (or partially debonded) semi-infinite half planes with cracks.

REFERENCES

[1] Erdogan, J., "Approximate solutions of systems of singular integral equations", SIAM J. Appl. Math., 17, p. 1041, 1969.

[2] Aksogan, O., "Numerical solutions for a class of multi-part mixed boundary value problems", Int. J. Num. Meth. Engng., 12, p. 1025, 1978.

[3] Krenk, S., "On the use of the interpolation polynomial for solutions of sin lar integral equations", Quart. Appl. Math., 32, p. 479, 1975.

[4] Cook, T. S. and Erdogan, F., "Stresses in bonded materials with a crack per pendicular to the interface", Int. J. Engng. Sci., 10, p. 667, 1972.

[5] Erdogan, F. and Aksogan, O., "Bonded half planes containing an arbitrarily oriented crack", Int. J. Solids Struct., 10, p. 569, 1974.

[6] Erdogan, F. and Biricikoglu, V., "Two bonded half planes with a crack going through the interface", Int. J. Engng. Sci., 11, p. 745, 1973.

[7] Aksogan, O., "Plane and anti-plane shear problems of a semi-infinite half plane with radial cracks", Proc. First Int. Conf. on Num. Meth. in Frac. Mech., Swansea, p. 95, 1978.

[8] Gupta, G. D. and Erdogan, F., "The problem of edge cracks in an infinite strip", J. Appl. Mech., Trans. ASME, 96, p. 1001, 1974.

[9] Coker, E. G. and Filon, L. G. N., A Treatise on Photoelasticity, Cambridge University Press, 1931.

[10] Aksogan, O., "The interaction of collinear arrays of Griffith cracks on two radial lines", J. Engng. for Industry, Trans. ASME, 98, p. 1086, 1976.

[11] Aksogan, O., "The stress intensity factors for X-formed arrays of cracks", Proc. 4th Int. Conf. on Frac., 3, Waterloo, p. 177, 1977.

SUDDEN TWISTING OF PARTIALLY BONDED CYLINDRICAL RODS

M. K. Kassir

City College of New York
New York, New York 10031

and

K. K. Bandyopadhyay

Gibbs and Hills, Inc.
New York, New York 10001

ABSTRACT

This study is concerned with the axially-symmetric transient problem of a torque applied suddenly to the end of a composite consisting of two concentric partially bonded cylinders. The cylinders are assumed to be perfectly bonded along their contact surface except for a cylindrical crack emanating from the horizontal boundary. The dynamic stress field at the periphery of the crack is found to possess the same essential features as other singular problems in LEFM. Of particular interest is the magnitude of the overshoot in the dynamic stress-intensity factor compared to the corresponding static value and the time interval in which it occurs. This has been displayed graphically for several composites made of different elastic materials.

INTRODUCTION

Structural components are often subjected to impulsive loads which generate transient stress waves. At a certain time, the propagation of these waves can cause high stress elevation in local regions surrounding mechanical defects or cracks. In addition, the magnitude of the dynamic stress-intensity factor is considerably larger than the corresponding statical one, and may lead to crack extension and eventual failure [1]. Thus, it is desirable to obtain a knowledge of the dynamic amplification of the stress-intensity factor and the time interval in which it occurs.

In this paper, the axially-symmetric transient problem of a torque applied suddenly to the plane surface of two concentric partially bonded semi-infinite cylinders is considered. The two rods are assumed to be perfectly bonded along their contact surface except for a cylindrical crack emanating from the horizontal boundary. A time-dependent torque is applied suddenly to the end of the inner rod and the ensuing near field is determined by solving the wave equation. To this end, a Laplace transform is introduced to eliminate the time variable, and in the transformed plane, the displacement and stress components are expressed in terms of infinite integrals containing arbitrary functions and Bessel functions which satisfy the field equations of motion and give the proper behavior throughout the medium. Application of the boundary conditions reduce the determination of the arbitrary functions to the solution of a Fredholm equation of the second kind in the transformed plane. The kernel of this equation has a removable singularity, and in connection with its numerical treatment, it is found convenient to isolate the singular portion of the integrand in closed form and then carry out the numerical solution. The displacement and stress components near the crack border, and in particular, the stress-intensity factor, are obtained by inverting the solution of the Fredholm equation by means of numerical procedure [2].

The dynamic stress field at the periphery of the crack border is found to possess the same basic characteristics as other singular problems in linear elastic fracture mechanics. Moreover, the stress-intensity factor varies with time, reaching a peak value higher than the static one and then oscillating about the static value with decreasing magnitude. For the example of a composite with $a/h = 1$ and modular ratios 1, 3 and 10, the overshoot in the stress-intensity factor is about 40% and it takes place in a time interval of $\bar{t} = 2.5$ (see Figures 3 and 4).

BASIC EQUATIONS AND FORMULATION

Figure 1 shows a composite consisting of two elastic semi-infinite rods of shearing moduli G_1 and G_2 and radii a and b (a>>b), respectively. The rods are

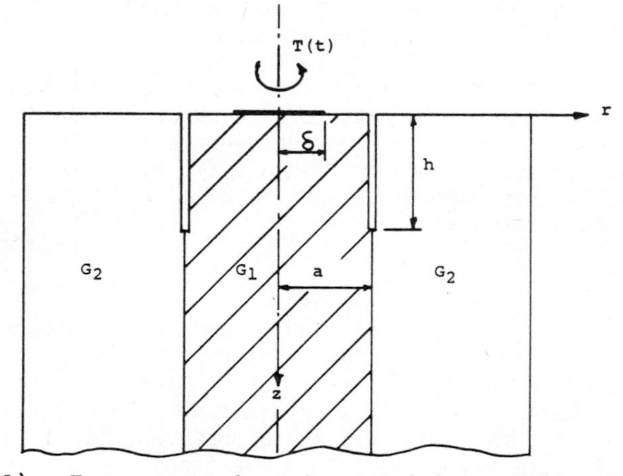

Fig. (1) - Two concentric rods containing a cylindrical crack

completely bonded along their length except for a cylindrical crack of depth h emanating from the horizontal boundary. A cylindrical coordinate system (r,θ,z) is used with the z-axis coinciding with the longitudinal axes of the cylinders. In this coordinate system, the crack occupies the region $r=a$, $0<\theta<2\pi$, $0<z<h$. Suppose that at time $t=0$, an impulsive torque, $T(t)$, is applied to the end of the inner cylinder and is generated by time-dependent shearing stress $\tau_{\theta z}(r,z,t)$ whose spatial distribution is arbitrary

$$\tau_{z\theta}(r,0,t) = \begin{cases} \tau_o(r)H(t), & 0 \le r \le \delta \quad (1a) \\ 0 & , r > \delta \quad (1b) \end{cases}$$

Here, $H(t)$ is the Heaviside unit step function and $\tau_o(r)$ is a known function. Because of the axisymmetric nature of the problem, the radial and axial displacement components vanish throughout the medium and the component $u_\theta(r,z,t)$ is governed by the equations of motion

$$\frac{\partial^2 u_\theta}{\partial r^2} + \frac{1}{r}\frac{\partial u_\theta}{\partial r} - \frac{u_\theta}{r^2} + \frac{\partial^2 u_\theta}{\partial z^2} = \frac{1}{c_j^2}\frac{\partial^2 u_\theta}{\partial t^2}, \; j = 1,2 \quad (2)$$

In equations (2), $c_j = (G_j/\rho_j)^{1/2}$ denotes the shear wave velocities in the composite with ρ_j being the mass density of material j (j = 1,2).

The appropriate boundary conditions can be described as follows: Since the crack surfaces are free from tractions at all times, it follows that

$$\tau_{r\theta 1}(a,z,t) = \tau_{r\theta 2}(a,z,t) = 0, \; 0 \le z \le h \quad (3)$$

while along the interface between the two cylinders, the displacement and radial shears must be continuous, i.e.,

$$\left.\begin{aligned} u_{\theta 1}(a,z,t) &= u_{\theta 2}(a,z,t) \quad (4a) \\ \tau_{r\theta 1}(a,z,t) &= \tau_{r\theta 2}(a,z,t) \quad (4b) \end{aligned}\right\} z > h$$

In addition, at remote distances from the loading region, the displacement and stresses must vanish. The initial conditions in time for u_θ and $\partial u_\theta/\partial t$ are assumed to be zero.

In order to eliminate the time variable in the equations of motion, use is made of the Laplace transform of u_θ defined as

$$u_\theta^*(r,z;p) = \int_0^\infty u_\theta(r,z,t)e^{-pt}dt \quad (5a)$$

$$u_\theta(r,z,t) = \frac{1}{2\pi i}\int_{Br} u_\theta^*(r,z;p)e^{pt}dp \quad (5b)$$

where the path of integration in equation (5b) is the usual Bromwich path which is a line on the right-hand side and parallel to the imaginary axis in the p-plane. With the aid of equations (5) and the initial conditions, equations (2) can be written in the transformed plane as

$$\frac{\partial^2 u_\theta^*}{\partial r^2} + \frac{1}{r}\frac{\partial u_\theta^*}{\partial r} + \frac{\partial^2 u_\theta^*}{\partial z^2} - \left(\frac{1}{r^2} + \frac{p^2}{c_j^2}\right)u_\theta^* = 0 \quad (6)$$

Making use of the finiteness requirement of u_θ^* as $r\to 0$ and $r\to\infty$ and introducing the abbreviation

$$\alpha_j(s,p) = (s^2 + \frac{p^2}{c_j^2})^{1/2} \tag{7}$$

the proper expressions for the displacement and stresses in the inner rod are:

$$u_{\theta 1}^* = \frac{1}{p}\int_0^\infty [A(s,p)J_1(rs)e^{-\alpha_1 z} + B(s,p)I_1(r\alpha_1)\cos(sz)]ds \tag{8a}$$

$$\tau_{r\theta 1}^* = \frac{G_1}{p}\int_0^\infty [-sA(s,p)J_2(rs)e^{-\alpha_1 z} + B(s,p)I_2(r\alpha_1)\cos(sz)]ds \tag{8b}$$

$$\tau_{\theta z1}^* = -\frac{G_1}{p}\int_0^\infty [A(s,p)J_1(rs)e^{-\alpha_1 z} + sB(s,p)I_1(r\alpha_1)\sin(sz)]ds \tag{8c}$$

while for the outer medium

$$u_{\theta 2}^* = \frac{1}{p}\int_0^\infty C(s,p)K_1(r\alpha_2)\cos(sz)ds \tag{9a}$$

$$\tau_{r\theta 2}^* = -\frac{G_2}{p}\int_0^\infty \alpha_2 C(s,p)K_2(r\alpha_2)\cos(sz)ds \tag{9b}$$

$$\tau_{\theta z2}^* = -\frac{G_2}{p}\int_0^\infty sC(s,p)K_1(r\alpha_2)\sin(sz)ds \tag{9c}$$

In equations (8) and (9), J_n denotes the Bessel function of the first kind of order n, I_n and K_n are, respectively, the modified Bessel functions of the first and second kinds and A(s,p), B(s,p) and C(s,p) are arbitrary functions to be determined from the boundary conditions. In the p-plane, conditions (1) read

$$\tau_{\theta z}^* = \begin{cases} (1/p)\tau_o(r), \; z=0, \; 0\leq r\leq\delta & (10a) \\ 0 \qquad , \; z=0, \; r>\delta & (10b) \end{cases}$$

while the continuity conditions in equations (3) and (4) become

$$\tau_{r\theta 1}^* = \tau_{r\theta 2}^* = 0, \; r=a, \; 0\leq z\leq h \tag{11a}$$

$$u_{\theta 1}^* = u_{\theta 2}^* \qquad , \; r=a, \; z>h \tag{11b}$$

$$\tau_{r\theta 2}^* = \tau_{r\theta 2}^* \qquad , \; r=a, \; z>h \tag{11c}$$

METHOD OF SOLUTION

In the transformed plane, equation (8c) is used in conjunction with conditions (10) and the Hankel inversion theorem to obtain the following relation for the function A(s,p)

$$\alpha_1(s,p)A(s,p) = -\frac{1}{pG_1}\int_0^{\delta} r\tau_o(r)J_1(rs)dr \tag{12}$$

On the other hand, application of conditions (11) to equations (8) and (9) renders two sets of simultaneous dual integral equations to determine B(s,p) and C(s,p)

$$\left.\begin{aligned}\int_0^\infty \alpha_1 BI_2(a\alpha_1)\cos(sz)ds &= \int_0^\infty sA(s,p)J_2(as)e^{-\alpha_1 z}ds \quad (13a)\\ \int_0^\infty \alpha_2 CK_2(a\alpha_2)\cos(sz)ds &= 0 \quad (13b)\end{aligned}\right\} 0\le z\le h$$

and

$$\int_0^\infty [BI_1(a\alpha_1) - CK_1(a\alpha_2)]\cos(sz)ds = -\int_0^\infty A(s,p)J_1(as)e^{-\alpha_1 z}ds \tag{14a}$$

$$\int_0^\infty [G_1\alpha_1 BI_2(a\alpha_2) + G_2\alpha_2 CK_2(a\alpha_2)]\cos(sz)ds = G_1\int_0^\infty sA(s,p)J_2(as)e^{-\alpha_1 z}ds,\ z>h \tag{14b}$$

For the purpose of reducing equations (13) and (14) into a single set of dual integral equations, it is observed from equations (13a), (13b) and (14b) that the following relation holds for all values of z

$$\int_0^\infty [\alpha_1 BI_2(a\alpha_2) + \frac{\alpha_2}{G}CK_2(a\alpha_2)]\cos(sz)ds = f(z,p),\ G_1 \neq 0 \tag{15}$$

in which $G = G_1/G_2$ and f(z,p) stands for

$$f(z,p) = \int_0^\infty tA(t,p)J_2(at)e^{-\alpha_1 z}dt \tag{16}$$

In view of these observations, equation (15) can be inverted by Fourier theorem to yield

$$C(s,p) = \frac{G}{\alpha_2 K_2(a\alpha_2)}[\frac{2}{\pi}f_c(s,p) - \alpha_1 I_2(a\alpha_1)B(s,p)] \tag{17}$$

Here, $f_c(s,p)$ designates the Fourier cosine transform of f(z,p)

$$f_c(s,p) = \int_0^\infty f(z,p)\cos(sz)dz \tag{18}$$

Inserting relation (16) into (18), making a permissible change in the order of integration and noting the identity [3]

$$\int_0^\infty e^{-\alpha_1 z} \cos(sz)dz = \frac{\alpha_1}{s^2+\alpha_1^2} \tag{19}$$

it follows that

$$f_c(s,p) = \int_0^\infty \frac{tJ_2(at)}{s^2+\alpha_1^2} \alpha_1 A(t,p)dt \tag{20}$$

Moreover, the function A(t,p) in equation (20) can be eliminated by first performing an integration by parts on the expression in relation (12) and utilizing the integral [3]

$$\int_0^\infty \frac{tJ_2(at)J_2(bt)}{t^2+\theta^2} dt = \begin{cases} I_2(b\theta)K_2(a\theta), & 0<b<a \\ I_2(a\theta)K_2(b\theta), & b>a \end{cases} \tag{21}$$

to arrive at the following formula for $f_c(s,p)$

$$f_c(s,p) = -\frac{\alpha_1}{pG_1} K_2(a\alpha_1)I^*(a,\delta,p) \tag{22}$$

where I^* stands for the integral

$$I^*(a,\delta,p) = \int_0^\delta r\tau_0(r)I_1(r\alpha_1)dr \tag{23}$$

which can be evaluated once the function $\tau_0(r)$ is specified.

The remaining boundary conditions provide a set of dual integral equations to determine the function B(s,p). Toward this end, equations (13a), (14a) and (17) may be combined to give

$$\int_0^\infty \alpha_1 BI_2(a\alpha_1)\cos(sz)ds = f(z,p), \quad 0\underline{<}z\underline{<}h \tag{24a}$$

$$\int_0^\infty BD(s,p)\cos(sz)ds = g(z,p), \quad z>h \tag{24b}$$

where f(z,p) is defined in equation (16) and the following contractions have been introduced

$$D(s,p) = I_1(a\alpha_1) + G\,\frac{\alpha_1 K_1(a\alpha_2)}{\alpha_2 K_2(a\alpha_1)}\,I_2(a\alpha_1) \tag{25a}$$

$$g(z,p) = -\int_0^\infty A(s,p)J_1(as)e^{-\alpha_1 z}ds + \frac{2}{\pi}G\int_0^\infty \alpha_2^{-1} f_c(s,p)\,\frac{K_1(a\alpha_2)}{K_2(a\alpha_2)}\cos(sz)ds \tag{25b}$$

The first step in the solution of the pair of equations (24) is to set

$$B(s,p) = B_1(s,p) + B_2(s,p) \tag{26}$$

such that

$$\int_0^\infty B_2(s,p)D(s,p)\cos(sz)ds = g(z,p),\quad 0\le z<\infty \tag{27}$$

Hence, it immediately follows that

$$B_2(s,p) = \frac{2}{\pi D(s,p)}\int_0^\infty g(z,p)\cos(sz)dz \tag{28}$$

Substituting for $g(z,p)$ from equation (25b), utilizing the results in equations (12), (19) and (23) and the identity [3]

$$\int_0^\infty \frac{J_1(a\theta)J_2(r\theta)}{\theta^2+s^2}\,d\theta = \frac{1}{s}\,I_2(rs)K_1(as),\quad r<a<\infty \tag{29}$$

the function $B_2(s,p)$ can be evaluated in terms of the specified shear stress as

$$B_2(s,p) = \frac{2}{\pi}\,\frac{K_1(a\alpha_2)}{G_1 D}\left[1-G\,\frac{\alpha_1 K_2(a\alpha_1)}{\alpha_2 K_2(a\alpha_2)}\right]I^*(\alpha_1,\delta) \tag{30}$$

Having established the value of $B_2(s,p)$, equations (24) can be rearranged with the help of equations (26), (27) and (30) to form the pair of equations

$$\int_0^\infty sB_1(s,p)D(s,p)H(s,p)\cos(sz)ds = (1+G)W(z,p),\quad 0\le z\le h \tag{31a}$$

$$\int_0^\infty B_1(s,p)D(s,p)\cos(sz)ds = 0,\quad z>h \tag{31b}$$

where the following abbreviations have been adopted

$$H(s,p) = (1+G)\,\frac{\alpha_1 I_2(a\alpha_1)}{sD(s,p)} \tag{32}$$

and

$$W(z,p) = f(z,p) - \int_0^\infty \alpha_1 B_2(s,p) I_2(a\alpha_1)\cos(sz)ds \tag{33}$$

Equations (31) constitute a standard set of dual integral equations with arbitrary weight kernel to determine the remaining unknown function B_1. This is accomplished by writing

$$B_1(s,p)D(s,p) = \int_0^\infty t^{1/2}\phi^*(t,p)J_o(st)dt \tag{34}$$

and the auxiliary function, ϕ^*, is governed by the Fredholm integral equation [4]

$$\phi^*(t,p) + \int_0^h \phi^*(\theta,p)K(\theta,t,p)d\theta = \frac{2}{\pi}(1+G)t^{1/2}\int_0^t \frac{W(z,p)dz}{(t^2-z^2)^{1/2}} \tag{35}$$

In equation (35), the symmetric kernel is given by

$$K(\theta,t,p) = (\theta t)^{1/2}\int_0^\infty s[H(s,p)-1]J_o(st)J_o(s\theta)ds \tag{36}$$

Sufficient is known about the function W(z,p) so that the load term in equation (35) can be expressed in terms of the applied stress distribution. For this purpose, making use of equation (33) and the result [3]

$$\int_0^t \frac{\cos(sz)dz}{(t^2-z^2)^{1/2}} = \frac{\pi}{2}J_o(st) \tag{37}$$

it follows that

$$\int_0^t \frac{W(z,p)dz}{(t^2-z^2)^{1/2}} = \int_0^t \frac{f(z,p)dz}{(t^2-z^2)^{1/2}} - \int_0^\infty \alpha_1 B_2 I_2(a\alpha_1)J_o(st)ds \tag{38}$$

It remains to evaluate the first integral in equation (38). By virtue of the fact that

$$f(z,p) = \frac{2}{\pi}\int_0^\infty f_c(s,p)\cos(sz)ds \tag{39}$$

and making use of the results in equations (22) and (37), it may be shown that

$$\int_0^t \frac{f(z,p)dz}{(t^2-z^2)^{1/2}} = -\frac{1}{pG_1}\int_0^\infty \alpha_1 J_o(st)K_2(as)I^*(\alpha_1,\delta)ds \tag{40}$$

Combining the results of equations (30), (38) and (40), one may finally express the load term in the integral equation as

$$\int_0^t \frac{W(z,p)dz}{(t^2-z^2)^{1/2}} = -\int_0^\infty [\frac{\alpha_1 K_2(a\alpha_1)}{pG_1} + \alpha_1 \frac{K_1(a\alpha_1)I_2(a\alpha_1)}{D(s,p)} (\frac{1}{G_1}$$
$$- \frac{\alpha_1 K_2(a\alpha_1)}{G_2\alpha_2 K_2(a\alpha_2)})]I^*(\alpha_1,\delta)ds \quad (41)$$

which is dependent on the applied loading at the end of the cylinders.

STRESS-INTENSITY FACTOR

The theory of Linear Elastic Fracture Mechanics is concerned with the determination of the stress field in the neighborhood of a sharp crack where extension of the crack is imminent. Let r_1 and θ_1 be a set of local coordinates measured from the periphery of the crack border as shown in Figure 2. In the transformed

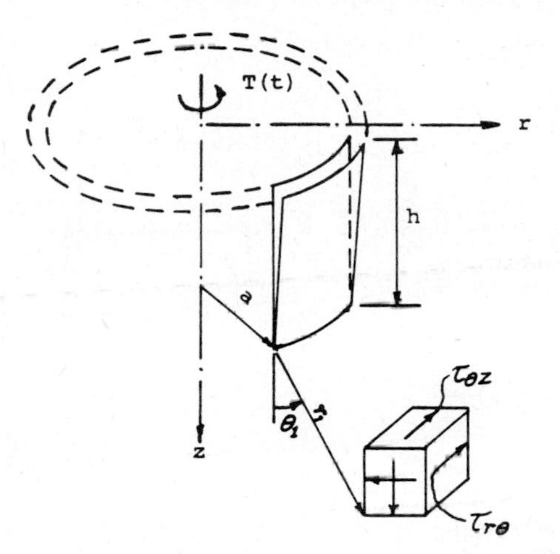

Fig. (2) - Stress field in the vicinity of crack tip

plane, the singular part of the stress expressions near the crack border can be extracted by noting that the infinite integrals appearing in the stress expressions are finite throughout their range except at the upper limit. Thus, by expanding these integrals for large argument and retaining the singular terms, it can be shown that the local stresses are

$$\tau^*_{r\theta} = -\frac{k^*_3}{(2r_1)^{1/2}} \sin(\theta_1/2) + O(r_1^0) \quad (42a)$$

$$\tau^*_{z\theta} = \frac{k^*_3}{(2r_1)^{1/2}} \cos(\theta_1/2) + O(r_1^0) \quad (42b)$$

where

$$k_3^* = -\frac{G_1}{1+G}\frac{\phi^*(h,p)}{p} \tag{43}$$

and ϕ^* is the solution of equation (35) at t=h. Applying the Laplace inversion theorem to the relation (43), the dynamic stress-intensity factor becomes

$$k_3(t) = -\frac{G_1}{1+G}\phi(h,t) \tag{44}$$

with

$$\phi(h,t) = \frac{1}{2\pi i}\int_{Br}\phi^*(h,p)\frac{e^{pt}}{p}\,dp \tag{45}$$

NUMERICAL RESULTS AND CONCLUSIONS

Consider the example of a constant shearing stress, $\tau_o(r) = \tau_o$, applied to the inner cylinder. Then

$$T(t) = \frac{2}{3}\pi\tau_o\delta^3 H(t) \tag{46}$$

By letting $\delta \to 0$, the deformation can be considered to be caused by a concentrated torque, $T_o H(t)$, applied at the origin of the coordinates. In this case, the integral, I^*, in equation (23) can be evaluated in closed form. Toward this end, one may write

$$I^* = \frac{3T}{2\pi\delta^3}\frac{d}{d\alpha_1}\left[\int_0^\delta I_o(r\alpha_1)dr\right] \tag{47}$$

Then, upon expanding the modified Bessel function, I_o, in infinite series, performing the integration term by term and carrying out the limiting procedure mentioned earlier, one arrives at

$$I^* = \frac{T_o\alpha_1}{4\pi} \tag{48}$$

For purposes of obtaining numerical solution of equation (35), it is found convenient to introduce the following non-dimensional parameters:

$$\left.\begin{aligned} &t = h\xi,\ \theta = h\eta,\ sh = \lambda,\ \bar{a} = (a/h),\ \gamma_j = \alpha_j h,\ j = 1,2 \\ &\bar{p} = (hp/c_1) \\ &\phi^*(t,p) = \frac{(1+G)T_o}{2\pi^2 G_1 h^{5/2}}\Phi^*(\xi,p) \end{aligned}\right\} \tag{49}$$

In view of this, the Fredholm equation in relation (35) takes the form

$$\Phi^*(\xi,\bar{p}) + \int_0^1 \Phi^*(\xi,\eta,\bar{p})L(\xi,\eta,\bar{p})d\eta = \xi^{1/2}M(\xi,\bar{p}) \tag{50}$$

and the symmetric kernel

$$L(\xi,\eta,\bar{p}) = (\xi\eta)^{1/2} \int_0^\infty \lambda[(1+G)\frac{\gamma_1 I_2(\bar{a}\gamma_1)}{\lambda D(\lambda,\bar{p})} - 1]J_o(\lambda\xi)J_o(\lambda\eta)d \tag{51}$$

In equation (50), the load term is

$$M(\xi,\bar{p}) = -\int_0^\infty \gamma_1^2\{K_2(\bar{a}\gamma_1) + I_2(\bar{a}\gamma_1)\frac{K_1(\bar{a}\gamma_1)}{D(\lambda,\bar{p})}[1-G\frac{\gamma_1 K_2(\bar{a}\gamma_1)}{\gamma_2 K_2(\bar{a}\gamma_2)}]\}J_o(\lambda\xi)d\lambda \tag{52}$$

Similarly, the stress-intensity factor is obtained from the relation

$$k_3(\bar{t}) = -\frac{T_o h^{1/2}}{2\pi^2 h^3}\frac{\Phi(1,\bar{t})}{\bar{p}} \tag{53}$$

where

$$\Phi(1,\bar{t}) = \frac{1}{2\pi i}\int_{Br}\frac{\Phi^*(1,\bar{p})}{\bar{p}}e^{\bar{p}\bar{t}}d\bar{p}, \quad \bar{t} = (c_1 t/h) \tag{54}$$

In carrying out the numerical solution of the Fredholm equation by the method described by Kantorovich and Krylov [5], it was found that the kernel has a removable singularity at the upper limit of integration. This behavior can be improved by isolating the slowly convergent part through a "closed form" solution. It can be shown that for large values of the argument

$$(1+G)\frac{\gamma_1 I_2(\bar{a}\gamma_1)}{\lambda D(\lambda)} - 1 = \frac{G-1}{G+1}\frac{3}{2\bar{a}\lambda} + O(\lambda^{-3}) \tag{55}$$

which indicates that for G=1 (homogeneous body), the kernel in equation (51) is well-behaved and when $G\neq 1$, the singularity can be removed by utilizing the result

$$\int_0^\infty J_o(\lambda\xi)J_o(\lambda\eta)d\lambda = \frac{2}{\pi}\begin{cases}\frac{1}{\xi}\kappa(\frac{\eta}{\xi}) \\ \\ \frac{1}{\eta}\kappa(\frac{\xi}{\eta})\end{cases} \tag{56}$$

where κ stands for the complete elliptic integral of the first kind. In view of these observations, the kernel in equation (51) is written in the equivalent form

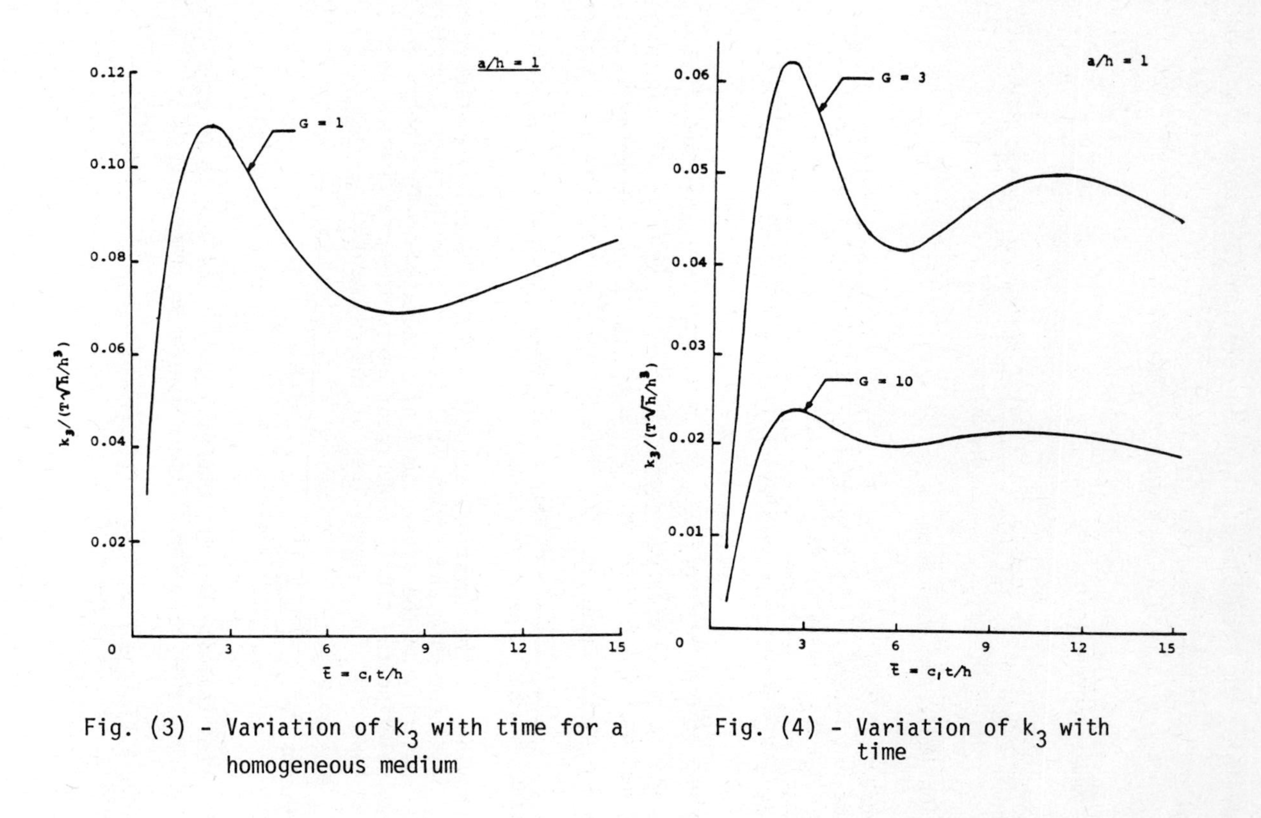

Fig. (3) - Variation of k_3 with time for a homogeneous medium

Fig. (4) - Variation of k_3 with time

$$L(\xi,\eta) = (\xi\eta)^{1/2}\{\frac{3}{2\pi}\,\frac{G-1}{G+1}\int_0^\infty J_o(\lambda\xi)J_o(\lambda\eta)d\lambda + \int_0^{\lambda_o} \lambda[(1+G)\,\frac{\gamma_1 I_2(\bar{a}\gamma_1)}{\lambda D(\lambda)}$$

$$- 1 - \frac{G-1}{G+1}\,\frac{3}{2\bar{a}\lambda}]J_o(\lambda\xi)J_o(\lambda\eta)d\lambda\} \qquad (57)$$

and the parameter, λ_o, is chosen to be large enough to satisfy the relation

$$(1+G)\,\frac{\gamma_1 I_2(\bar{a}\gamma_1)}{\lambda D(\lambda)} - 1 = \frac{G-1}{G+1}\,\frac{3}{2\bar{a}\lambda} \qquad (58)$$

The numerical solution of the integral equation was carried out by the modified method described in [5] to obtain the function $\Phi^*(1,\bar{p})$. In order to compute the dynamic stress-intensity factor, the Laplace inversion indicated in equation (54) was numerically performed by the method described in [2] for a/h = 1 and modular ratios G_1/G_2 = 1, 3 and 10. The results are shown in Figures 3 and 4.

The transient effect is observed as k_3 first reaches a peak and then oscillates about the static value with decreasing amplitude. The dynamic overshoot is observed to be about 40% higher than the static value and to occur at $\bar{t}$ = 2.5(G_1 =G_2). As G = G_1/G_2 increases, the dynamic overshoot decreases in magnitude. For G=10, it is about 20%. In all cases examined as $\bar{t}$ increases, the value of k_3 approaches that of the static case.

REFERENCES

[1] Chen, E. P. and Sih, G. C., Elastodynamic Crack Problems, Noordhoff International Publishing, Leyden, The Netherlands, 1977.

[2] Miller, M. K. and Guy, K. T., "Numerical inversion of the Laplace transform by use of Jacobi polynomials", SIAM Journal of Numerical Analysis, Vol. 3, pp. 624-635, 1966.

[3] Erdelyi, A., ed., Tables of Integral Transforms, Vols. I and II, McGraw-Hill, New York, 1954.

[4] Sneddon, I. N., Mixed Boundary Value Problems in Potential Theory, North-Holland Publishing Co., Amsterdam, The Netherlands, 1966.

[5] Krylov, V. I. and Kantorovich, L. V., Approximate Methods of Higher Analysis, Interscience, New York, 1958.

AN AXISYMMETRIC-ELASTODYNAMIC ANALYSIS OF A CRACK IN ORTHOTROPIC MEDIA USING A PATH-INDEPENDENT INTEGRAL

D. A. Scarth

W. L. Wardrop & Associates, Ltd.
Winnipeg, Manitoba, Canada

T. R. Hsu and G. S. Pizey

University of Manitoba
Winnipeg, Manitoba, Canada

ABSTRACT

The effects of material orthotropy were considered in an elastodynamic analysis of a stationary crack under axisymmetric conditions. Although closed-form solutions to transient problems involving a stationary crack are not available, static solutions indicate that the dynamic energy release rate would be a more meaningful crack-tip parameter than the dynamic stress intensity factor. Computation of the dynamic energy release rate was performed by means of a path-independent energy-flux integral in conjunction with the finite-element technique. A case involving an outer-edge cracked hollow thick-walled circular cylinder possessing orthotropic properties typical of real materials is presented. The variation of the maximum value of the dynamic energy release rate with orthotropy was relatively small with the maximum being largest for the material with the lowest averaged elastic stiffness.

INTRODUCTION

The presence of orthotropy in some real engineering materials has led to the need to determine the effects of such behavior in dynamic crack analyses. As well, some geometries and loadings may be best treated as being axisymmetric in nature. Applications include girth welds in pipes, irradiated zirconium-alloy components and some geological structures*, where crack growth may be desired [1]. The calculation of some parameter characterizing the intensity of the time-dependent near-tip stress field is necessary in order to predict if crack growth will occur. Since material ductility is lowest at high loading rates, the assumption of linear elastic material behavior may be justified in the investigation of problems where the loading is dynamic.

---

*Some geological structures may be approximated as being transversely isotropic.

It will be shown that the dynamic energy release rate is a parameter suitable as a fracture criterion. Path-independent integrals describing the dynamic energy release rate, or energy flux to the crack tip, by nature need not be restricted to isotropic materials. Such integrals have been formulated for the elastodynamic analysis of cracks under planar conditions [2-5]. Results from isotropic case studies show this approach to be promising [6,7]. These integrals are an extension of the J-integral by Rice [8] and it appears that the adverse effect of the theoretical crack-tip singularity is only minor [7]. A path-independent integral had been formulated for axisymmetric crack problems involving static loads [9,10].

Described here is a path-independent energy-flux integral for the elastodynamic analysis of stationary cracks under axisymmetric Mode I conditions. This integral is evaluated from the field solution from a conventional elastodynamic finite-element computer code. An application to a problem involving an outer-edge cracked hollow thick-walled circular cylinder subjected to sudden end loadings is presented. Cases were run for isotropic and orthotropic materials and the ensuing results were compared.

CRACK TIP PARAMETER

Solutions for the elastodynamic response of cracked isotropic solids show the transient stress field to be directly proportional to the dynamic stress intensity factor, K(t) [11,12]. For cracks under plane strain Mode I conditions, the dynamic energy release rate, G(t), is given by [5]

$$G(t) = K_I(t)^2 \frac{(1-\nu^2)}{E} \tag{1}$$

where E is Young's modulus and ν is Poisson's ratio. The two parameters are equivalent in describing the intensity of the near-tip stress field for a given mode.

Closed-form solutions to transient problems involving stationary cracks in anisotropic materials are not known to the authors. However, solutions to static problems are found in, for example, [13-16]. Consider a cracked orthotropic infinite plate in an (x,y,z) cartesian coordinate system in which the y-axis is perpendicular to the crack plane. For Mode I loadings, the static stress field is typified by the relation between the component σ_{yy} and the static stress intensity factor, K_I [13]:

$$\sigma_{yy} = \frac{K_I}{\sqrt{2\pi R}} \text{Re} \left[\frac{1}{s_1-s_2} \left\{\frac{s_1}{\sqrt{\cos\phi+s_2\sin\phi}} - \frac{s_2}{\sqrt{\cos\phi+s_1\sin\phi}}\right\}\right] \tag{2}$$

The variables R,ϕ are local polar coordinates at the crack tip and s_1 and s_2 are equal to two complex or purely imaginary roots of a characteristic polynomial whose coefficients are the elastic constants. It is apparent that the magnitude of the stress field is dependent upon the material constants through s_1 and s_2, as well as K_I. Thus, as has been criticized in [17], the stress intensity factor by definition does not describe completely the intensity of the static stress field in anisotropic materials. In the case of self-equilibrating or zero crack-surface tractions, the stress intensity factor for infinite geometries generally is equal to the material-property-independent isotropic value [13].

As an alternative, consider the relation between K_I and the strain energy release rate, G, for the previously mentioned cracked plate under plane strain conditions [13];

$$G = K_I^2 \sqrt{\frac{b_{11}b_{22}}{2}} \left[\sqrt{\frac{b_{22}}{b_{11}}} + \frac{2b_{12}+b_{66}}{2b_{11}}\right]^{1/2} \tag{3}$$

where b_{ij} are compliance functions of the elastic constants. In this expression, the effect of directional material properties on the field solution, such as that given by equation (2), is incorporated into the calculation of G. An assumption implicit in equation (3) is that the crack will propagate in a coplanar fashion. This is the case for an orthotropic solid but not so for solids exhibiting general anisotropy [18].

It is noted that since plane strain conditions prevail near the edge of a crack under axisymmetric conditions [11], equations (1) and (3) apply to axisymmetric problems as well.

In extending these concepts to elastodynamic-orthotropic crack analyses, it is apparent that the dynamic energy release rate would be more descriptive of the intensity of the near-tip stress field. The use of this parameter will encompass both the widely accepted energy-release-rate fracture criterion used for isotropic materials and the effects of material orientation.

PATH-INDEPENDENT INTEGRAL FOR AXISYMMETRIC ELASTODYNAMIC CRACK ANALYSES

The formulation presented here of this path-independent integral shall be brief and shall emphasize its physical relevance in this study. Consider a homogeneous axisymmetric solid containing a stationary axisymmetric inner-edge crack of tip radius a under Mode I loading. A fixed (r,θ,z) cylindrical coordinate system is located at the center of the solid and is aligned such that the z-axis coincides with the axis of symmetry. In Figure 1 is presented a view in the r-z plane of the crack details. It is assumed that a fracture-process region surrounds the crack tip in which the mechanisms of void growth and coalescence occur [19]. The counterclockwise contour Γ_p surrounds the area of the fracture-process region (in the r-z plane) and the contours Γ_s join the arbitrary contour Γ to Γ_p. The area A is enclosed by a closed loop composed of the contours Γ_p, Γ and Γ_s. The direction of the outward unit surface-normal vector, n_i, relative to the r-axis is specified by the angle ψ. The subscript i = 1,2,3 corresponds to the r,θ,z components respectively.

The axisymmetric solid shall be represented by a segment of the solid subtending an angle α in the θ-coordinate. The boundary of the fracture-process region in this segment, denoted as S_p, is created by rotating the contour Γ_p through the angle α.

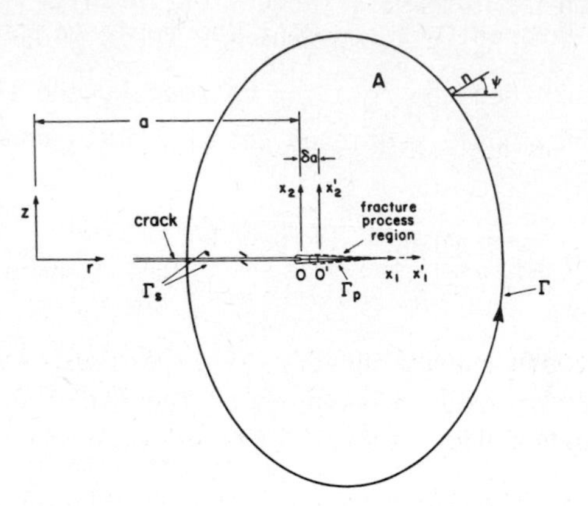

Fig. (1) - Crack details showing path-independent-integral model

The model used for outer-edge cracks is essentially the same as that for inner-edge cracks and the following discourse applies to both situations.

The energy flux into the crack tip is modelled as the energy flux into the fracture-process region. This energy flux is assumed to be the rate of work done by the stress vector, T_i, on the boundary of the fracture-process region, with respect to the extension of a crack face. For an inner-edge crack, one assumes a first-order virtual crack extension, $+\delta a$, while for an outer-edge crack, the assumed virtual crack extension is $-\delta a$. By neglecting second-order virtual quantities, one may write an expression for the time-dependent energy flux into the fracture-process region or dynamic energy release rate, denoted as $J_{rg}(t)$, during this virtual extension as [20]

$$J_{rg}(t)\Big|_{\substack{\text{inner}\\ \text{outer}}} = \underset{\delta a,\delta u_i \to 0}{\text{Limit}} \pm \frac{1}{\alpha a}\int_{S_p} T_i \frac{\delta u_i}{\delta a}\, dS_p \quad \text{sum over } i = 1,3 \tag{4}$$

In this equation, t is time, δu_i are first-order virtual displacements and the + and - signs of the right-hand side correspond to inner- and outer-edge cracks, respectively. By assuming the characteristic dimension of the fracture-process region to be small, the mechanisms on its boundary are autonomous with respect to geometry and loading [19]. Under this assumption, a path-independent expression for $J_{rg}(t)$ can be obtained through a structural-energy-rate balance. This balance involves setting $J_{rg}(t)$ to be the difference between the rate of work done by T_i on a surface composed of the boundaries of a finite control volume (formed by rotating the area A through the angle α) minus S_p, and the rate of work done (including that by inertia and body forces) in the control volume. Through the use of Gauss' divergence theorem, one obtains the final result [20];

$$J_{rg}(t)\Big|_{\substack{inner\\outer}} = \pm \int_{\Gamma} ((W - \sigma_{rr} \frac{\partial u_r}{\partial r} - \sigma_{rz} \frac{\partial u_z}{\partial r})\cos\psi - (\sigma_{rz} \frac{\partial u_r}{\partial r}$$

$$+ \sigma_{zz} \frac{\partial u_z}{\partial r})\sin\psi)\frac{r}{a}\, d\Gamma \pm \int_{\Gamma_s} P \frac{\partial u_z}{\partial r} \sin(\psi) \frac{r}{a}\, d\Gamma$$

$$\pm \int_{A} ((\rho(\ddot{u}_i - f_i) \frac{\partial u_i}{\partial r})\frac{r}{a} + \frac{(\sigma_{\theta\theta}\varepsilon_{\theta\theta} - W)}{a})dA \quad \text{sum over } i = 1,3 \qquad (5)$$

where W is the elastic strain energy density, P is pressure applied to the crack surfaces, ρ is the material's mass density, f_i are body forces per unit mass, $\varepsilon_{\theta\theta}$ is hoop strain and an overhead dot denotes differentiation with respect to time.

It can be shown that in the limit as $\Gamma \to \Gamma_p$, $J_{rg}(t)$ approaches the J-integral by Rice [8].

NUMERICAL ILLUSTRATION

A. Description

A homogeneous hollow thick-walled circular cylinder of inner radius r_1, outer radius r_2 and length 2L is illustrated in Figure 2. The cylinder contains

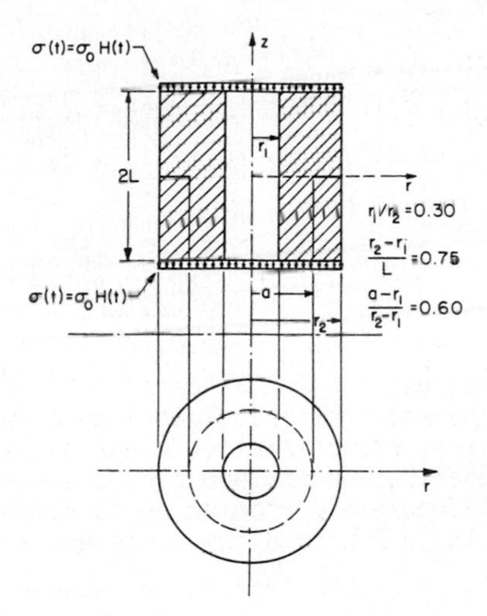

Fig. (2) - A thick-walled circular cylinder with a circumferential outer-edge crack subjected to sudden end loadings

an axisymmetric circumferential outer-edge crack of zero thickness and tip radius a. The shape of the cylinder is given by $r_1/r_2 = 0.30$ and $(r_2-r_1)/L = 0.75$, with $(a-r_1)/(r_2-r_1) = 0.60$. The ends of the cylinder, at $z = \pm L$, were subjected to the loading

$$\sigma(t) = \sigma_o H(t) \tag{6}$$

where σ_o is a constant tensile stress and H(t) is the unit Heaviside step function. The problem is symmetric with respect to the crack plane. Three cases were run by considering an isotropic and two orthotropic materials. The elastic constants used are typical of real orthotropic materials and are expressed in terms of the two isotropic elastic constants in Table 1. The mass density was the same for all three materials.

TABLE 1 - ELASTIC PROPERTIES OF THE TWO ORTHOTROPIC MATERIALS IN TERMS OF THE ISOTROPIC PROPERTIES

Material	E_r	E_θ	E_z	$\nu_{r\theta}=\nu_{rz}=\nu_{\theta z}$	μ_{rz}
Orthotropic 1	.5E	.75E	E	ν (=.35)	$\frac{.5E}{2(1+\nu)}$
Orthotropic 2	2E	1.5E	E	ν	$\frac{E}{2(1+\nu)}$

B. Solution Techniques

The finite-element modelling involved the use of linear-elastic constant-strain toroidal elements [21] having a lumped-mass matrix. The economical lumped-mass matrix was used in place of the variationally-formulated consistent-mass matrix due to the suitability of the lumped-mass matrix in modelling wave propagation. Damping was not included.

The assembled set of simultaneous differential equations of motion were integrated numerically by using the Newmark β direct time-integration technique. The Newmark β parameters were $\beta = 0.25$ and $\gamma = 0.50$. The time-step size was chosen such that the fastest dilatational wave would traverse the smallest element during one time increment.

The finite-element mesh used in the $J_{rg}(t)$-integral computations is presented in Figure 3. The mesh consists of 300 uniform square elements with 336 nodes. The element size is equal to $(r_2-a)/6$ which is relatively coarse in view of the theoretically large stress gradients near the crack tip. The coarse mesh is a consequence of maintaining a constant element size for all contours to minimize path dependence of spacial-discretization errors, and the storage limitations of the computer code. However, a coarse mesh did not adversely affect the results presented in [7]. Computations were performed by using the four contours, Γ, illustrated by the dashed lines in Figure 3, where a contour is passed through the centroids of the elements.

The linear extrapolation technique [22] combined with equation (1) was also applied to the isotropic case to compare with the results from the $J_{rg}(t)$-integral. By inverting the analytical near-tip stress-field solution [12], a series of values of a parameter $K_I^*(R,t)$ were calculated from corresponding values of $\sigma_{zz}(R,\phi,t)$ computed by the finite-element code. The dynamic stress intensity factor was obtained by extrapolating $K_I^*(R,t)$ to the crack tip. The finite element mesh employed in this technique is presented in Figure 4 (a and b). The mesh con-

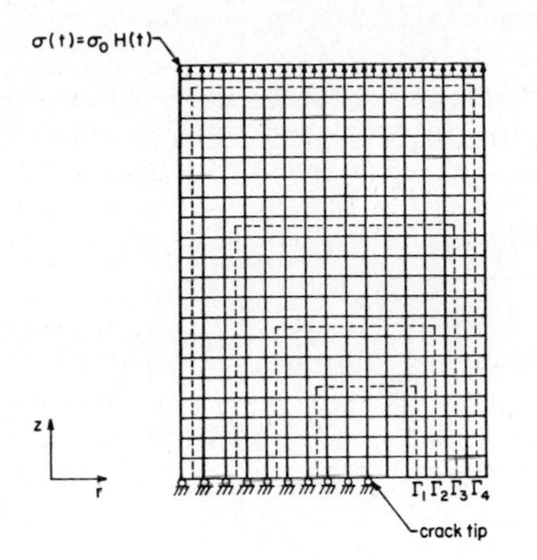

Fig. (3) - Finite-element mesh used in the $J_{rg}(t)$-integral computations, with contours as illustrated

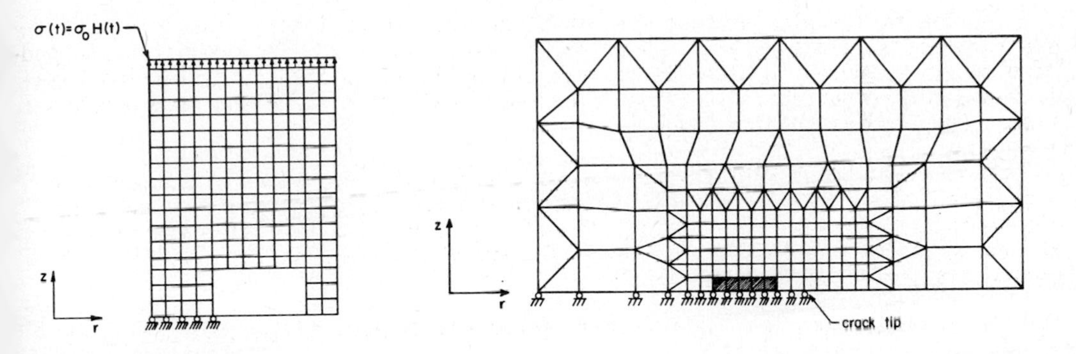

(a) Overall mesh

(b) Near-tip region

Fig. (4) - Finite-element mesh used in the linear extrapolation technique

sists of 377 elements with 375 nodes, where the size of the elements in the crack-tip region is $(r_2-a)/30$ to model effectively the large stress gradients. The five elements shown shaded in Figure 4(b) were used in the extrapolation.

C. Results and Discussion

The dynamic energy release rate was nondimensionalized by using

$$\tilde{G}(t) = \frac{J_{rg}(t)}{G_{static}} \tag{7}$$

where $\tilde{G}(t)$ is the normalized dynamic energy release rate and G_{static} is the strain energy release rate of the isotropic cylinder in static equilibrium. Although a closed-form solution for the static stress intensity factor (and hence G_{static}) for an isotropic outer-edge cracked cylinder is presented in [23], numerical values were presented for a different relative wall thickness. Alternatively, the static versions of the $J_{rg}(t)$-integral and the linear-extrapolation technique wer applied to obtain an averaged result of

$$G_{static} = 6.3\sigma_0^2(r_2-a)\,\frac{(1-\nu^2)}{E} \tag{8}$$

with G_{static} from the two techniques differing by 12%.

Figure 5 shows the variation of $\tilde{G}(t)$ with nondimensionalized time, c_1t/L, for the isotropic cylinder. The constant c_1 is the dilatational wave speed in th

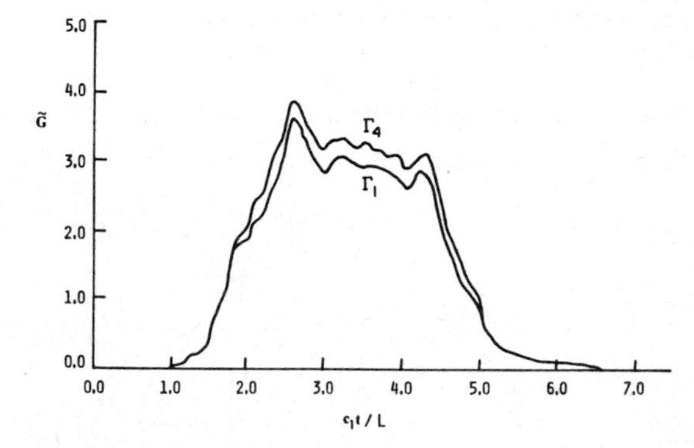

Fig. (5) - Variation of the normalized dynamic energy release rate with nondimensionalized time for the isotropic cylinder

isotropic material. The lower- and upper-bounds of the results from the four co tours are presented, arising from the contours Γ_1 and Γ_4, respectively. The history of $\tilde{G}(t)$ is approximately independent of the path of integration. The maxim in $\tilde{G}(t)$, denoted as $\tilde{G}_{max}$, computed using Γ_1 and Γ_4 varied by 7%.

Figure 6 indicates the lower- and upper-bounds on the results of $\tilde{G}(t)$ versus c_1t/L for the cylinder composed of orthotropic material no. 1. To allow a direct comparison of isotropic and orthotropic results, the isotropic values of G_{static} and c_1 were used in the normalization of all the orthotropic results. A in the previous figure, the lower- and upper-bounds of the results arose from the contours Γ_1 and Γ_4, respectively. The history of $\tilde{G}(t)$ is again approximately inc pendent of the path of integration, with $\tilde{G}_{max}$ obtained from Γ_1 and Γ_4 differing 8%.

The results from the analysis of the cylinder composed of orthotropic material no. 2 displayed path independence similar to that for the previous two materials.

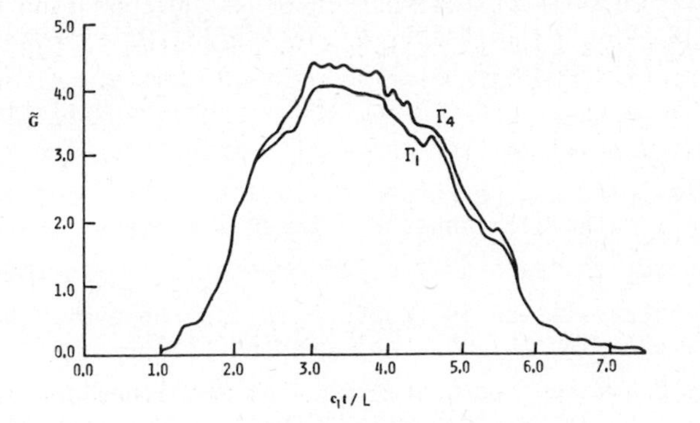

Fig. (6) - Variation of the normalized dynamic energy release rate with nondimensionalized time for the cylinder composed of orthotropic material no. 1

The nondimensionalized time histories of $\tilde{G}(t)$ computed using Γ_2 for the three materials are compared in Figure 7. Also shown is $\tilde{G}(t)$ for the isotropic

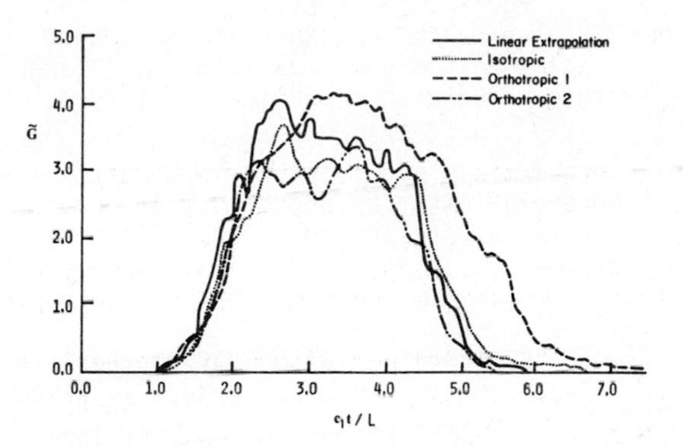

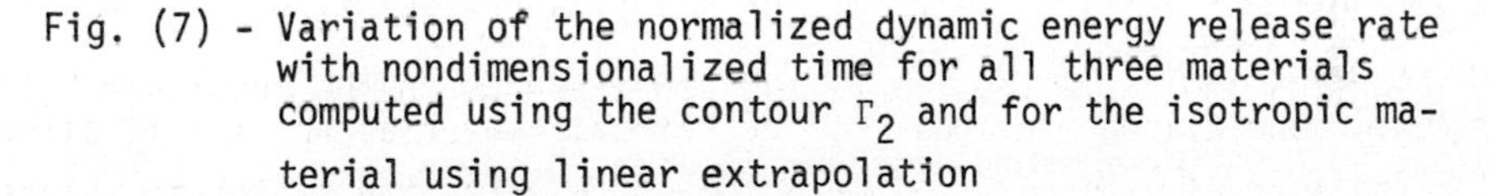

Fig. (7) - Variation of the normalized dynamic energy release rate with nondimensionalized time for all three materials computed using the contour Γ_2 and for the isotropic material using linear extrapolation

case obtained from the linear-extrapolation technique. In comparing the two curves obtained for the isotropic material, the generally lower values of $\tilde{G}(t)$ computed by the $J_{rg}(t)$-integral is attributed to the relatively coarse mesh employed in this technique. The linear-extrapolation technique produced a maximum in $\tilde{G}(t)$ which was 10% larger than that found using $J_{rg}(t)$.

In the isotropic case, the decline in $\tilde{G}(t)$ from the initial rise occurred as a result of the initial tensile wavefront, having reflected off the plane of symmetry and subsequently the free end, returning to the crack as a compression wave. In the orthotropic cases, two dilatational wave speeds were present and interpretation of the histories is less clear. However, as expected, the overall effect of using the less-stiff orthotropic material no. 1 was a slower transient than in the isotropic case, while using the more-stiff orthotropic material no. 2 resulted in a faster transient.

The effect of orthotropy on the maximum in $\tilde{G}(t)$ can be perceived by comparing the three curves obtained from the $J_{rg}(t)$-integral. The $\tilde{G}_{max}$ values for orthotropic materials 1 and 2 were respectively 13% greater and 11% less than G_{max} for the isotropic material. Thus, for these materials, the effect of orthotropy is small and about the same as the difference in $\tilde{G}_{max}$ between the $J_{rg}(t)$-integral and linear-extrapolation isotropic results. However, these results are qualitatively consistent with the physical meaning of the $J_{rg}(t)$-integral as an energy flux into the fracture-process region. As mentioned in the section, "Path Independent Integral for Axisymmetric Elastodynamic Crack Analyses", the formulation of $J_{rg}(t)$ is based on the rate of work done by the stress vector on the boundary of the fracture-process region. If the material possesses some less-stiff elastic properties (e.g., orthotropic material no. 1), the displacements on the boundary will be relatively larger. By assuming the magnitude of the stress vector to have a weak dependence on the elastic constants, the corresponding rate of work done and hence energy flux will be relatively larger. Or, from the standpoint of the computation of $J_{rg}(t)$, it is apparent that for the same less-stiff material the rates of work done on the boundary of and inside the finite control volume mentioned in the section cited above, will also be larger. One would expect the magnitude of the difference between these two quantities, namely $J_{rg}(t)$, to attain a larger value. The converse reasoning applies for a material possessing some more-stiff elastic properties (e.g., orthotropic material no. 2). These trends can be reasoned more explicitly by deducing the result of increasing (or decreasing) the displacements and strains in equation (5).

CONCLUSIONS

The $J_{rg}(t)$-integral was approximately path-independent in this case for both the isotropic and the orthotropic materials. The variation of the maximum of the dynamic energy release rate with material orthotropy was small and the maximum was largest for the material with the lowest averaged elastic stiffness.

REFERENCES

[1] Lamb, G. H., "Underground coal gasification", Noyes Data Corp., 1977.

[2] Nilsson, F., "A path-independent integral for transient crack problems", Int. J. Solids Struct., 9, pp. 1107-1115, 1973.

[3] Gurtin, M. E., "On a path-independent integral for elastodynamics", Int. J. Fract., 12, pp. 643-644, 1976.

[4] Bui, H. D., "Stress and crack-displacement intensity factors in elastodynamics", Advances in Research on the Strength and Fracture of Materials, D. M. R. Taplin, ed., Pergamon Press, 3A, pp. 91-95, 1977.

[5] Kishimoto, K., Aoki, S. and Sakata, M., "On the path-independent integral - J", Engng. Fract. Mech., 13, pp. 841-850, 1980.

[6] Mall, S., "A finite element analysis of transient crack problems with a path-independent integral", in Advances in Fracture Research, 5th International Conference on Fracture, Cannes, France, March 29 - April 3, 1981.

[7] Kishimoto, K., Aoki, S. and Sakata, M., "Dynamic stress intensity factors using Ĵ-integral and finite element method", Engng. Fract. Mech., 13, pp. 387-394, 1980.

[8] Rice, J. R., "Mathematical analysis in the mechanics of fracture", in Fracture - An Advanced Treatise II Mathematical Fundamentals", H. Liebowitz, ed., Academic Press, pp. 191-311, 1968.

[9] Astiz, M. A., Elices, M. and Sánchez Galvez, V., "On energy release rates in axisymmetrical problems", in Advances in Research on the Strength and Fracture of Materials", D. M. R. Taplin, ed., Pergamon Press, 3A, pp. 395-400, 1977.

[10] Bergkvist, H. and Huong, G. L., "J-integral related quantities in axisymmetric cases", Int. J. Fract., 13, pp. 556-558, 1977.

[11] Chen, E. P. and Sih, G. C., "Transient response of cracks to impact loads", Mechanics of Fracture 4 - Elastodynamic Crack Problems, G. C. Sih, ed., Noordhoff Int. Publ., pp. 1-58, 1977.

[12] Chen, E. P. and Sih, G. C., "Scattering waves about stationary and moving cracks", in Mechanics of Fracture 4 - Elastodynamic Crack Problems, G. C. Sih, ed., Noordhoff Int. Publ., pp. 119-212, 1977.

[13] Sih, G. C., Paris, P. C. and Irwin, G. R., "On cracks in rectilinearly anisotropic bodies", Int. J. Fract. Mech., 1, pp. 189-203, 1965.

[14] "Cracks in anisotropic materials", in Mechanics of Fracture 2 - Three-Dimensional Crack Problems, M. K. Kassir and G. C. Sih, eds., Noordhoff Int. Publ., pp. 336-381, 1975.

[15] Bowie, O. L. and Freese, C. E., "Central crack in plane orthotropic rectangular sheet", Int. J. Fract. Mech., 8, pp. 49-58, 1972.

[16] Atsumi, A. and Shindo, Y., "Singular stresses in a transversely isotropic circular cylinder with circumferential edge crack", Int. J. Engng. Sci., 17, pp. 1229-1236, 1979.

[17] Cook, T. S. and Rau, C. A., Jr., "A critical review of anisotropic fracture mechanics", Prospects of Fracture Mechanics, G. C. Sih, H. C. van Elst and D. Broek, eds., Noordhoff Int. Publ., pp. 509-523, 1974.

[18] Sih, G. C. and Liebowitz, H., "Mathematical theories of brittle fracture", in Fracture - An Advanced Treatise, H. Liebowitz, ed., Academic Press, pp. 67-190, 1968.

[19] Broberg, K. B., "Crack-growth criteria and non-linear fracture mechanics", J. Mech. Phys. Solids, 19, pp. 407-418, 1971.

[20] Scarth, D. A., "The elastodynamic analysis of cracks under axisymmetric conditions by a path-independent integral", Msc. Thesis, Univ. of Manitoba, Winnipeg, Canada, 1982.

[21] Hsu, T. R., Bertels, A. W. M., Ayra, B. and Banerjee, S., "Application of the finite element method to the non-linear reactor fuel behaviour", Proc. of First Int. Conf. on Computational Methods in Non-Linear Mechanics, pp. 531-540, 1974.

[22] Chen, Y. M. and Wilkins, M. L., "Numerical analysis of dynamic crack problems", in Mechanics of Fracture 4 - Elastodynamic Crack Problems, Noordhoff Int. Publ., pp. 295-345, 1977.

[23] Erdol, R. and Erdogan, F., "A thick-walled cylinder with an axisymmetric internal or edge crack", Trans. ASME, J. Appl. Mech., 45, pp. 281-286, 1978.

UNSTABLE GROWTH OF BRANCHED CRACKS

E. E. Gdoutos*

Lehigh University
Bethlehem, Pennsylvania 18015

ABSTRACT

The growth behavior of branched cracks embedded in an elastic plate subjected to some simple loading conditions is studied. The strain energy density criterion is applied to determine the conditions of unstable fracture starting from the crack tips. Considered are symmetrically and asymmetrically branched cracks, and crack branching from an elliptical crack. The most vulnerable crack branch to initiate unstable fracture, the corresponding critical load and the initial fracture path are determined. The severity of a branched crack is also discussed and compared to the results of a single crack with the same overall length. Comparisons on the critical loads are made for the various branched crack configurations.

INTRODUCTION

The phenomenon of crack branching has been observed during the rapid unstable fracture of various brittle materials. The crack when traveling at a high velocity under Mode I deformation divides into two branches. This process may continue until a pattern of multiple crack divisions is obtained. Many attempts have been made to analyze the crack branching phenomenon. The first study is due to Yoffé [1] who assumed that the crack moved toward the direction of maximum circumferential stress which occurs at some angle off the crack axis. Thus, he predicted the half branch angle to be 26° for $\nu = 0.25$ which deviates from the experimentally observed value of ≃15° in glass plates. This deviation is attributed to the use of the maximum circumferential stress criterion which assumes failure to occur when only one of the stress components reaches a maximum. However, a fracture process cannot be controlled by only one of the stress components.

A satisfactory explanation of the crack branching phenomenon was made by Sih [2] who used a criterion based on the strain energy density in the vicinity of the crack tip [3,4]. Based on the observation that a macrocrack runs in the direction of excessive dilatation than distortion, he obtained a half branch angle of 15.06° for a Poisson's ratio $\nu = 0.25$ and $c/c_2 = 0.25$, where c is the

* On leave from Democritus University of Thrace, School of Engineering, Xanthi, Greece.

crack velocity and c_2 is the shear wave velocity. This angle agrees well with the experimental observation. The positions of excessive dilatation can be obtained from the stationary values of the strain energy density factor.

A number of studies aiming to determine the stress intensity factors at the tips of branched cracks have appeared in the literature. Andersson [5] first studied the problem of an asymmetric branched crack using the method of complex potentials together with the conformal mapping technique. This method was used afterward by Chatterjee [6] who solved the problem of a branched crack consisting of a main crack and a straight branch crack starting from one of its tips. Vitek [7] calculated the stress intensity factors of branched cracks of various configurations. Theocaris [8] and Theocaris and Ioakimidis [9] solved the problem of axymmetric and symmetric crack branching using the method of complex potentials. Experimental studies of crack branching have been made by Schardin [10], Clark an Irwin [11], Kalthoff [12] and Theocaris [13].

In the present work, the problem of a symmetric or asymmetric branched crack is considered and the conditions of crack propagation under a monotonically increasing tensile load are determined. The study addresses the case when fracture is accompanied by small plastic deformation. The strain energy density criterion [3,4] is used to determine the most vulnerable crack branch to initiate unstable fracture, the corresponding critical load and the initial fracture path. Comparisons on the critical load for the various branched crack configurations and the single crack of the same overall length are made.

THE STRAIN ENERGY DENSITY CRITERION FOR UNSTABLE CRACK GROWTH

The strain energy density criterion is a general theory of fracture and has been successfully applied to the solution of problems of fast and slow crack grow The merit of the theory lies in its generality and its unique character in analyzing different failure modes including both fracture and yielding. It can successfully cope with the problem of mixed mode fracture where other theories present serious limitations. A number of mixed mode crack problems under a variety of geometrical configurations and loading conditions has been solved by the author usi the strain energy density criterion [14]. In applying to problems of rapid crac extension, the fundamental parameter is the strain energy density factor, S, defined as

$$S = \left(\frac{dW}{dV}\right)r \quad (1$$

where (dW/dV) is the strain energy density function and r is the radial distance referencing the site of fracture initiation. It is observed that S is direction sensitive.

The theory assumes that fracture initiation starts when a material element absorbs a critical amount of strain energy. The location of fracture coincides with the element with the minimum strain energy density as compared to other elements on the same radial distance from the crack tip. Breakage of the element a therefore crack initiation starts when S reaches a critical value, S_{cr}, characte istic of the material. Mathematically speaking, in the general case of a mixed mode stress field governed by the values of the opening-mode and sliding-mode, stress intensity factors k_1 and k_2, S is given by [3,4]

$$S = a_{11}k_1^2 + 2a_{12}k_1k_2 + a_{22}k_2^2 \tag{2}$$

in which the coefficients a_{ij} $(i,j = 1,2)$ stand for

$$\begin{aligned} 16\mu a_{11} &= (1+\cos\theta)(\kappa-\cos\theta) \\ 16\mu a_{12} &= [2\cos\theta - (\kappa-1)]\sin\theta \\ 16\mu a_{22} &= (\kappa+1)(1-\cos\theta) + (1+\cos\theta)(3\cos\theta-1) \end{aligned} \tag{3}$$

In these relations, μ is the shear modulus of elasticity, ν the Poisson's ratio and $\kappa = 3-4\nu$ for plane strain and $(3-\nu)/(1+\nu)$ for plane stress. The requirement of minimum S with respect to the polar angle θ leads to the relations:

$$\begin{aligned} &\sin\theta[2\cos\theta - (\kappa-1)]k_1^2 + 2[2\cos 2\theta - (\kappa-1)\cos\theta]k_1k_2 \\ &\qquad + \sin\theta(\kappa-1 - 6\cos\theta)k_2^2 = 0 \end{aligned} \tag{4}$$

$$\begin{aligned} &[2\cos 2\theta - (\kappa-1)\cos\theta]k_1^2 + 2[(\kappa-1)\sin\theta - 4\sin 2\theta]k_1k_2 \\ &\qquad + [(\kappa-1)\cos\theta - 6\cos 2\theta]k_2^2 > 0 \end{aligned} \tag{5}$$

Relations (4) and (5) yield the angle θ_0 of initial crack extension. The value S_{min} is obtained by substituting θ_0 in equation (2). The onset of rapid crack extension is obtained when S_{min} reaches S_{cr}, which is a material constant. Values of S_{cr} for various metal alloys can be found in [15].

In the following, the strain energy density criterion will be used to determine the critical fracture characteristic quantities of a plate with a symmetric or asymmetric branched crack and crack branches emanating from an elliptical crack.

DOUBLY SYMMETRIC BRANCHED CRACK

Consider a doubly symmetric branched crack composed of the main branch A'A of length 2a and the four equal branches, AB, AC, A'B', A'C' of length a_1 symmetrically located to the crack A'A at an angle α_1. The crack is embedded in an infinite elastic isotropic plate which is subjected to a uniform uniaxial stress σ at infinity perpendicular to the main crack A'A. For this geometrical configuration and under plane strain conditions, Vitek [7] found the values of k_1 and k_2 at the tips of the branches.

Using the strain energy density criterion, the critical value σ_{cr} of the applied stress for crack extension from either of the tips B, C, B', C' is determined. Figure 1 presents the variation of the non-dimensional quantity $\sigma_{cr}(a/16\mu S_{cr})^{1/2}$ versus the branch inclination angle α_1 $(0 \leq \alpha_1 \leq 90°)$ for various values of the

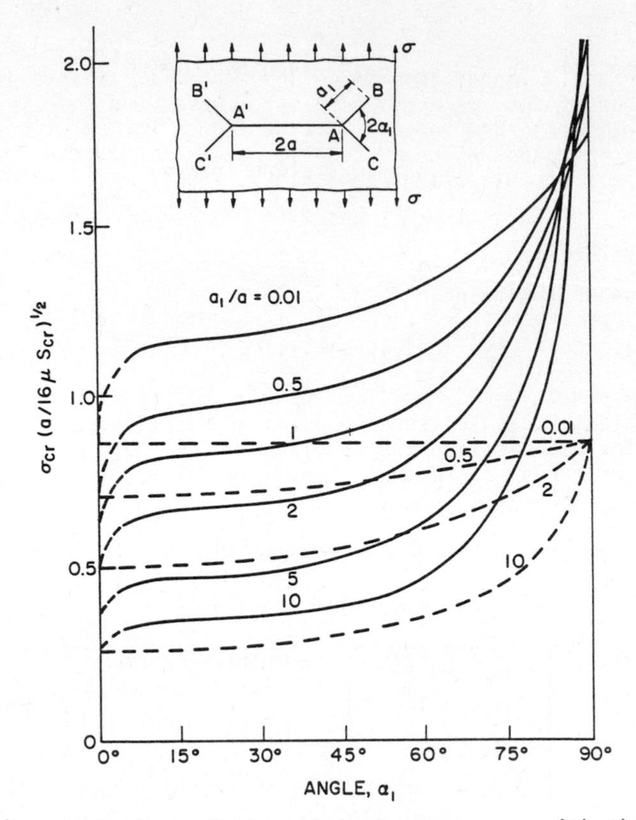

Fig. (1) - Variations of the critical stress σ_{cr} with the angle α_1

ratio a_1/a of the branch crack length to the main half crack length. The Poisso ratio ν is assigned the value 1/3. It is observed that for a given angle α_1, th critical stress σ_{cr} decreases with the branch length a_1 for all angles α_1 not ly in the interval (85°,90°). For the latter case, the stress σ_{cr} increases with t branch length a_1 after a_1 reaches a critical length. Thus, for the H-shaped cra loaded perpendicularly to the main crack, and for branch lengths greater than a critical length, the stress σ_{cr} for extension of the branches increases with the branch length. Furthermore, for a given value of the length of the branches a_1, the critical stress σ_{cr} increases as the angle α_1 increases. Thus, the more sev doubly symmetric branched cracks are those with long branches and smaller angles inclination with respect to the plane of the main crack.

In order to obtain an estimate of the effect that a branched crack has on th strength of a body as compared to a line crack, we define the equivalent crack b the projection of the inclined branches to the plane of the main crack (length c the equivalent crack $2c = 2(a+a_1\cos\alpha_1)$). The values of the critical stress for initiation of the equivalent crack are shown in Figure 1 by dotted lines. It is observed that the critical stress for the equivalent crack is always smaller tha for the branched crack. The difference between the two stresses increases as the angle of inclination of the branches increases. Thus, the doubly symmetric branched crack always retards propagation as compared to the line crack. The critical stress of the branched crack is also compared with the corresponding stress for the main crack of length 2a without branches. The value of the quantity $\sigma_{cr}(a/16\mu S_{cr})^{1/2}$ for the latter is equal to 0.867 and corresponds approximately to the dotted curve of Figure 1 for $a_1/a = 0.01$. We observe that for $a_1 = 0.01a$, the critical stress of the branched crack with branches of any inclination is always greater than for the main crack. This also holds true for $a_1 = 0.5a$ and $\alpha_1 > 3.5°$, for $a_1 = a$ and $\alpha_1 > 37.5°$, for $a_1 = 2a$ and $\alpha_1 > 62°$, for $a_1 = 5a$ and $\alpha_1 > 68°$ for $a_1 = 10a$ and $\alpha_1 > 79°$. It is thus concluded that for branches of either small lengths or large inclination angles, the existence of the branch slows the propagation.

Figure 2 presents the variation of the angle θ_0 that the fracture path starting from the tip B of the branch AB forms with the line AB. Positive angles θ_0 mean

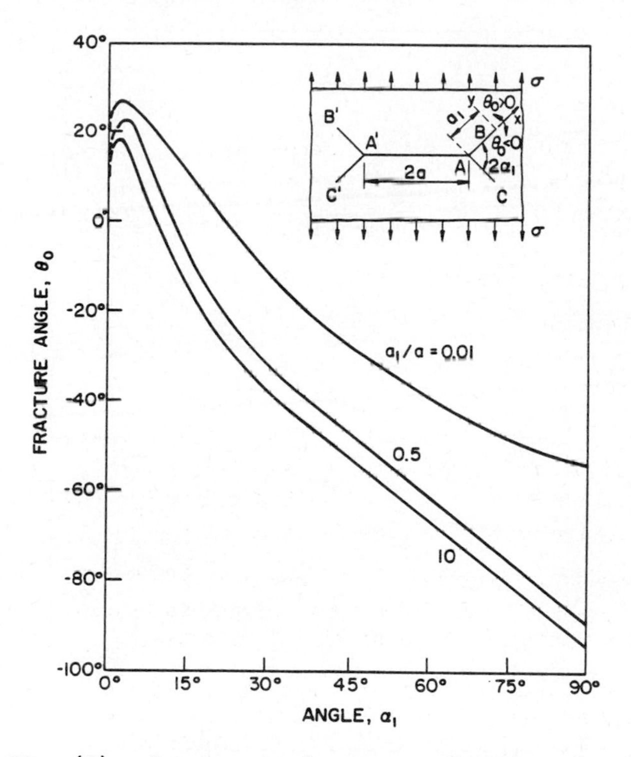

Fig. (2) - Fracture angle θ_0 as a function of α_1

that the branch AB extends in a counterclockwise direction, while negative angles θ_0 stand for a clockwise propagation of AB. It is seen from Figure 2 that θ_0 is positive for $a_1 = 0.01a$ and $\alpha_1 < 2.5°$ for $a_1 = 0.5a$ and $\alpha_1 < 12.5°$ and for $a_1 = 10a$ and $\alpha_1 < 9°$. For these cases, the branches tend to propagate with increasing angle, which means that they have the tendency to repel each other. For the remaining values of α_1, the branches propagate with decreasing angle; that is, they tend to attract each other. For the particular values of the length a_1 and inclination angle α_1 of the branches given above, the crack propagation angle θ_0 is zero and therefore the branches maintain their original direction. This result corroborates the experimental observations of Kalthoff [12], who found that for the case of two symmetric branches emanating from the boundary of a plexiglass plate, the critical angle subtending by the branches for which the branches retain their original direction is approximately 28°. If now the angle formed by the fracture path originating from the tip of the branches and the direction of the applied stress is calculated, we find that the branches extend in a direction which approaches that perpendicular to the applied stress as the branch length increases.

SKEW-SYMMETRIC BRANCHED CRACK

The skew symmetric branched crack consisting of the main crack A'A and the inclined branches AB and A'B' is shown in Figure 3. The crack is embedded in an in

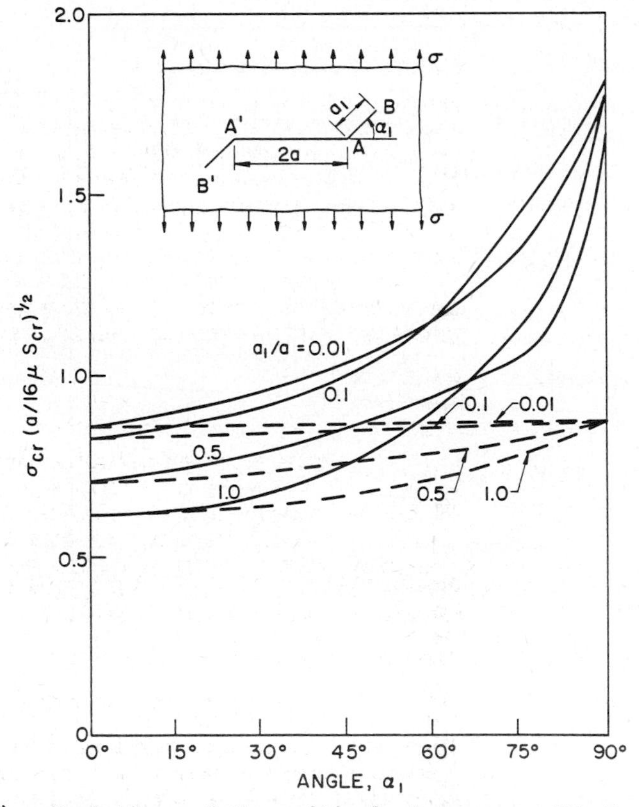

Fig. (3) - Variations of the critical stress σ_{cr} with the angle α_1

finite isotropic elastic plate subjected to an uniform uniaxial stress σ perpendicular to the branch A'A. For the values of the stress intensity factors k_1 and k_2 determined by Vitek [7] and using the strain energy density criterion, the critical fracture stress σ_{cr} and the fracture angle θ_o were determined. Figure 3 presents the variation of the normalized critical stress $\sigma_{cr}(a/16\mu S_{cr})^{1/2}$ versus the angle α_1 of inclination of the branches for $a_1/a = 0.01, 0.1, 0.5$ and 1.0. The Poisson's ratio ν is equal to 1/3. For $\alpha_1 = 0$, the values of the quantity $\sigma_{cr}(a/16\mu S_{cr})^{1/2}$ correspond to those obtained for a line crack of length $2(a+a_1)$ loaded in a direction perpendicular to the crack plane. It is observed from Figure 3 that, as in the previous case of the doubly symmetric branched crack, the critical stress σ_{cr} increases as the inclination angle α_1 of the branches increases. The stress σ_{cr} decreases as the length of the branches increases for angles α_1 smaller than 60°, while for angles $\alpha_1 > 60°$, this rule is not valid. As in the previous case, we define the equivalent crack as the line crack obtained by the projections of the inclined branches to the plane of the main crack. The values of the critical stress for the equivalent crack are given in Figure 3 by dotted lines. It is observed that the fracture stress of the branched crack is always greater than that of the equivalent crack.

Figure 4 presents the variation of the fracture angle $-\theta_o$ formed by the path of the propagating branch AB starting from the tip B with its initial direction AB versus the angle α_1 for $a_1/a = 0.01, 0.1$ and 1.0. It is observed that the angle θ_o is always negative; therefore, the branches propagate with decreasing angle and tend to become perpendicular to the direction of the applied stress.

ASYMMETRICALLY BRANCHED CRACK

The asymmetrically branched crack considered consists of the main branch A'A of length 2a and branches AB, A'B' and AC, A'C' of lengths a_1 and a_2 which are inclined at angles 45° and -45° to the main branch, Figure 5. For this geometrical configuration with a Poisson's ratio ν of the plate equal to 1/2, Vitek [7] gave the values of the k_1 and k_2 stress intensity factors for tips B and C of the branch AB and AC under conditions of plane strain. Using these values of k_1 and k_2 in conjunction with the strain energy density criterion, the critical stresses σ_{cr}^B and σ_{cr}^C for possible crack extension from the tips B and C are determined. It is evident that the crack would extend from the tip with the smaller required critical stress for propagation. Working in this way, it is found that the crack always extends from tip B of the longer branch AB. The variation of the normalized critical stress σ_{cr} versus a_1/a_2 for $a_2/a = 0.01, 0.1, 0.5$ and 10 is displayed in Figure 5. It is observed that the stress σ_{cr} decreases as a_1/a_2 or a_2/a increases. The values of the stress σ_{cr} for a doubly symmetrical branched crack whose branches are equal to the longer branch of the asymmetric crack are displayed in Figure 5 by dotted lines. It is observed that the critical stress for the asymmetric crack is always smaller than for the symmetric crack.

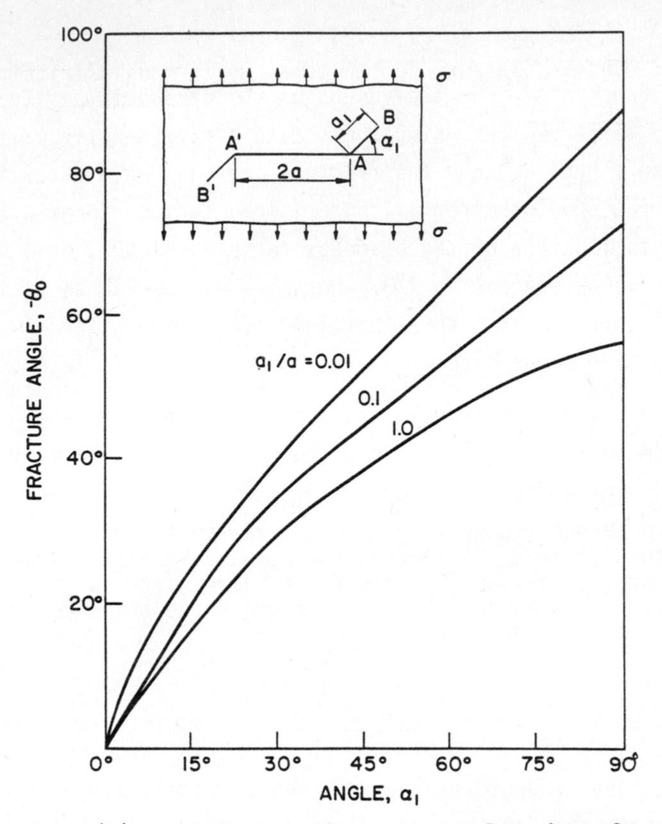

Fig. (4) - Fracture angle $-\theta_0$ as a function of α_1

The values of the angle $-\theta_0$ formed by the fracture path originating from the tip B of the branch AB with the direction of AB are displayed in Figure 6. It is observed that the angle θ_0 is always negative, which means that the crack tends to propagate with decreasing branch angle. Calculating the values of the angle formed by the propagating branch with the direction of the applied stress, we fin that they lie in the interval (73°,98°). Thus, the propagating branch is rather close to the direction perpendicular to the applied stress.

CRACK BRANCHES EMANATING FROM AN ELLIPTICAL CRACK

Two types of crack branches emanating from an elliptical crack will be considered. In the first case, two equal and symmetrically located branches emanate from each end point of the elliptical crack. In the second, one branch emanates from each end point of the elliptical crack, with both these branches having the same length, antisymmetrically located with respect to the elliptical crack. The cracks are embedded in an infinite isotropic elastic plate subjected to a uniform uniaxial stress σ at infinity perpendicular to the major axis of the elliptical crack. These geometrical configurations are analogous to the doubly symmetric branched crack and the skew-symmetric branched crack studied previously. The onl difference noted is that the main branch is elliptical rather than sharp.

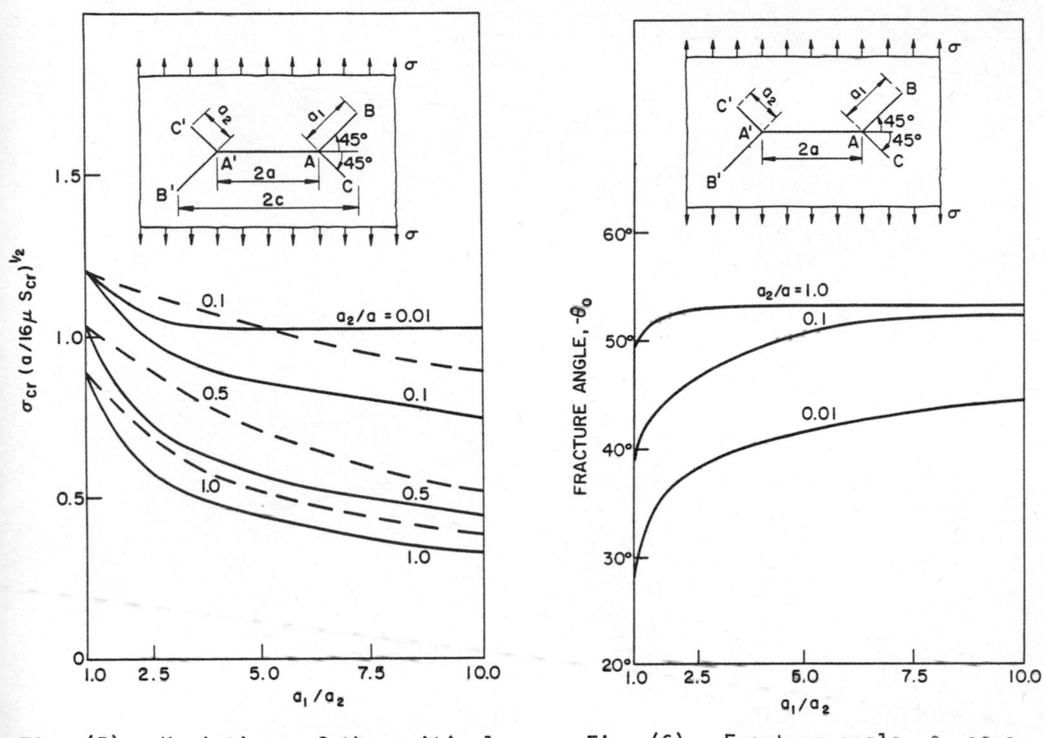

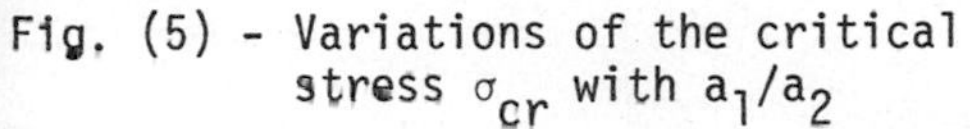
Fig. (5) - Variations of the critical stress σ_{cr} with a_1/a_2

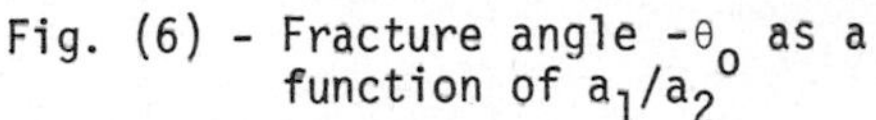
Fig. (6) - Fracture angle $-\theta_0$ as a function of a_1/a_2

Working as in the previous cases using the values of the k_1 and k_2 stress intensity factors given by Vitek [7], the values of the critical stress for unstable growth of the branches are determined. Figure 7 presents the variation of the quantity $\sigma_{cr}(a/16\mu S_{cr})^{1/2}$ versus the ratio a_1/a of the branch length a_1 to the half length of the major axis of the elliptical crack for α_1 = 30°, 45° and 60°. The root radius of the elliptical crack is equal to $\rho/a = b/a = 0.5$. As in the case of the doubly symmetric branched crack, the critical stress decreases as either the length of the branches increases or their angle of inclination decreases. The values of the critical stress for the doubly symmetric branched crack are shown in Figure 7 by dotted lines for the same angles of inclination as the elliptical main crack. It is observed that for α_1 = 30° and 45°, the critical stress for the elliptical crack is smaller than the corresponding stress for the sharp crack, while for α_1 = 60°, this rule is reversed. Thus, it is concluded that for large branch angles, the existence of a radius of curvature at the root of the main branch increases the strength of the cracked plate, while the contrary occurs for small branch angles.

Figure 8 presents the values of the fracture angle $-\theta_0$ for the tip B of the branch AB. As in the case of the doubly symmetric branched crack, the angle θ_0

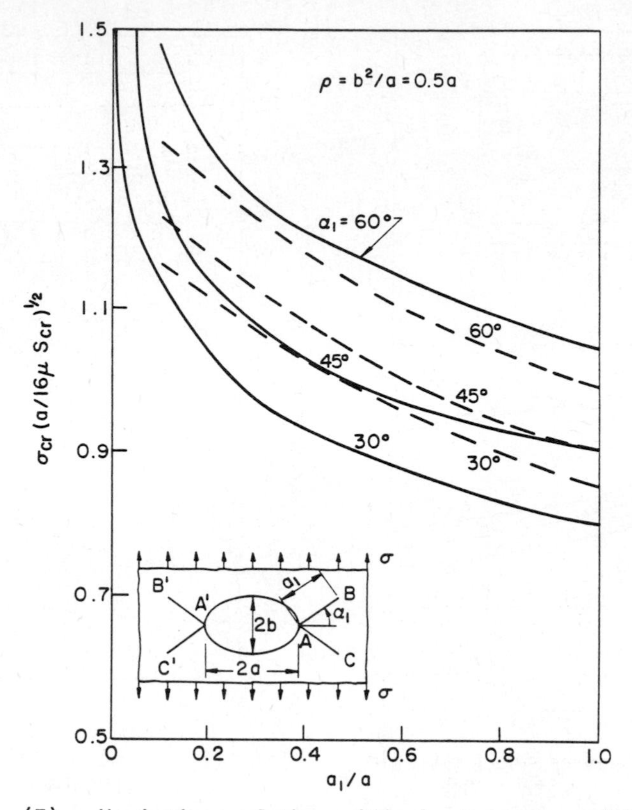

Fig. (7) - Variations of the critical stress σ_{cr} with a_1/a

is negative, that is, branches tend to propagate with decreasing angle, as if they attract each other. The values of the angle $-\theta_0$ for the doubly symmetric branched crack are shown in Figure 8 by dotted lines. It is observed that the angles $-\theta_0$ for the elliptical crack are always larger than for the sharp crack.

For the skew-symmetric elliptical branched crack, Figure 9 presents the variation of the quantity $\sigma_{cr}(a/16\mu S_{cr})^{1/2}$ versus the branch angle α_1 for the values of the root radius of the elliptical crack equal to ρ/a = 0, 0.1, 0.5 and 1.0. The value ρ=0 corresponds to the sharp bent crack studied previously, while the value ρ=a corresponds to the case of a circular hole. The curves of Figure 9 correspond to a_1 = 0.01a. Figure 10 presents the same results for a_1 = 0.5a. It is observed that for small branch lengths (a_1 = 0.01a), the critical stress increases as the root radius ρ increases. Therefore, the most dangerously shaped main crack on the strength of the body is the sharp crack. However, when the branch length a_1 increases, no such rule can be stated. As in the case of the sharp bent crack, the critical stress σ_{cr} increases with the inclination of the branches for all root radii of the elliptical crack. Furthermore, the previously established effect of the decrease of the critical stress with the increase of the branch length is also

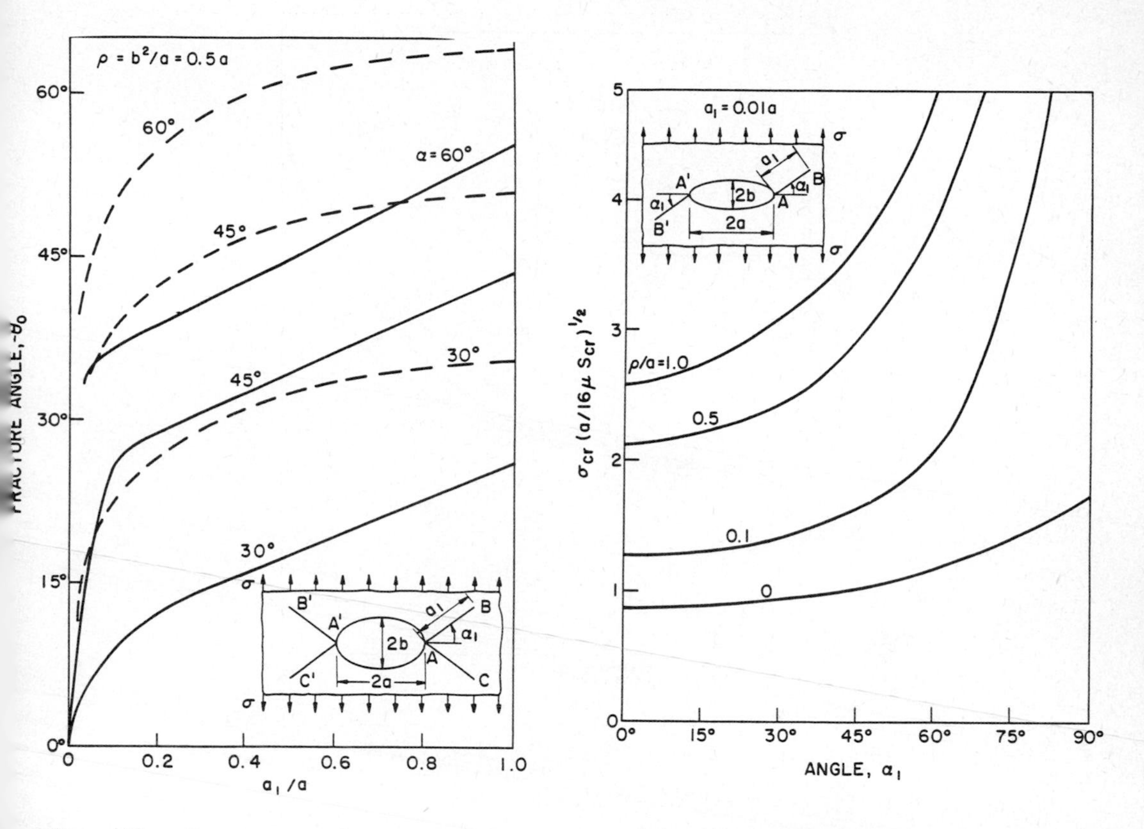

Fig. (8) - Fracture angle $-\theta_0$ as a function of a_1/a

Fig. (9) - Variations of the critical stress σ_{cr} with angle α_1 for $a_1 = 0.01a$

observed. The fracture path originating from the tips of the branches presents the same general characteristics as in the case of the sharp crack. Thus, the fracture angle θ_0 is negative, it increases with the branch inclination and the branches have a tendency to propagate in a direction perpendicular to the applied stress. Thus, diagrams concerning the fracture angles are not reported for space saving.

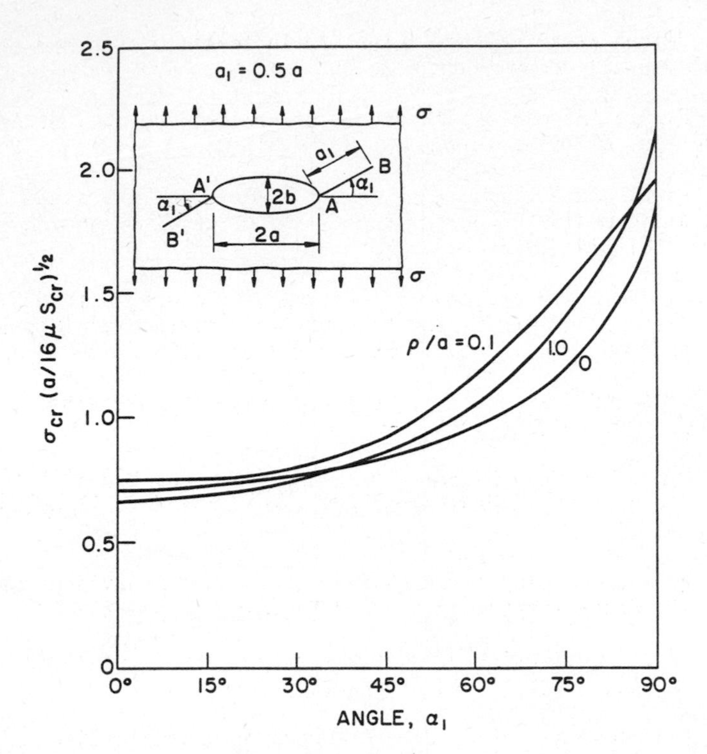

Fig. (10) - Variations of the critical stress σ_{cr} with angle α_1 for $a_1 = 0.5a$

REFERENCES

[1] Yoffé, E. H., "The moving Griffith crack", Philosophical Magazine, 42, pp. 739-750, 1951.

[2] Sih, G. C., "Dynamic crack problems - strain energy density fracture theory", Mechanics of Fracture 4, edited by G. C. Sih, Noordhoff International Publishing, pp. IX-XLVII, 1977.

[3] Sih, G. C., "A special theory of crack propagation: methods of analysis and solutions of crack problems", Mechanics of Fracture 1, edited by G. C. Sih, Noordhoff International Publishing, Leyden, pp. XXI-XLV, 1973.

[4] Sih, G. C., "Strain-energy-density factor applied to mixed mode crack problems", International Journal of Fracture, 10, pp. 305-321, 1974.

[5] Andersson, H., "Stress-intensity factors at the tips of a star-shaped contour in an infinite tensile sheet", Journal of the Mechanics and Physics of Solids, 17, pp. 405-417, 1969.

[6] Chatterjee, S. N., "The stress field in the neighborhood of a branched crack in an infinite elastic sheet", International Journal of Solids and Structures 11, pp. 521-538, 1975.

[7] Vitek, V., "Plane strain stress intensity factors for branched cracks", International Journal of Fracture, 13, pp. 481-501, 1977.

[8] Theocaris, P. S., "Asymmetric branching of cracks", Journal of Applied Mechanics, 44, pp. 611-618, 1977.

[9] Theocaris, P. S. and Ioakimidis, N., "The symmetrically branched crack in an infinite elastic medium", Journal of Applied Mathematics and Physics (ZAMP), Vol. 27, pp. 801-814, 1976.

[10] Schardin, H., "Velocity effects in fracture", Fracture, edited by B. L. Averbach et al, Technology Press of Massachusetts Institute of Technology and John Wiley and Sons, Inc., New York, pp. 297-330, 1959.

[11] Clark, A. B. J. and Irwin, G. R., "Crack-propagation behaviors", Experimental Mechanics, 6, pp. 321-330, 1966.

[12] Kalthoff, J. F., "On the characteristic angle for crack branching in brittle materials", International Journal of Fracture Mechanics, 7, pp. 478-480, 1971.

[13] Theocaris, P. S., "Complex stress-intensity factors at bifurcated cracks", Journal of the Mechanics and Physics of Solids, 20, pp. 265-279, 1972.

[14] Gdoutos, E. E., "Application of strain energy density theory to engineering problems", Martinus Nijhoff Publishers (in press).

[15] Sih, G. C. and Macdonald, B., "Fracture mechanics application to engineering problems - strain energy density fracture criterion", Engineering Fracture Mechanics, 6, pp. 361-386, 1974.

EXPERIMENTAL STRESS INTENSITY DISTRIBUTIONS BY OPTICAL METHODS

C. W. Smith and G. Nicoletto

Virginia Polytechnic Institute and State University
Blacksburg, Virginia 24061

ABSTRACT

For over a decade, the first author and his associates have worked towards the development of an optical experimental modelling technique for predicting both the flaw shape and the stress intensity factor distribution in three dimensional (3D) cracked body problems where neither are known a priori. The application is associated with sub-critical flaw growth, the precursor to most service fractures.

This paper presents an assessment of results obtained by applying the technique which consists of a marriage between frozen stress photoelasticity and Moiré analysis to measure the stress intensity factor distribution across a straight front crack in a body of finite thickness in order to assess constraint effects.

INTRODUCTION

The growth of sub-critical cracks to critical size is the most common precur of service fractures. Since such growth usually starts near stress raisers, the body geometry as well as flaw configuration is often complex, involving curved crack fronts, non self-similar flaw growth and non-planar flaws. These complications have forced analysts to turn to numerical methods in order to achieve some measure of tractability in computing stress intensity factor (SIF) distributions. Such solutions usually require computer code verification. The first author and his associates have been studying ways of developing cost effective experimental modelling techniques for achieving this result for over a decade. First developed for Mode I analysis only [1], the original method, involving a marriage between frozen stress analysis and fracture mechanics, has been extended to mixed mode analysis [2] and recently, has been supplemented with a technique involving Moiré interferometry for extracting SIF estimates from near tip displacement fields for comparison with photoelastic results using linear elastic fracture mechanics (LEFM).

By coupling the photoelastic and Moiré measurements, information relating to the degree of constraint in a three dimensional cracked body problem can be obtained. After reviewing analytical and experimental considerations and citing some deviations of crack shapes from those assumed in numerical models, we present results obtained by applying the method to a simplified 3D problem in order to assess the influence of constraint variation along the flaw border on the SIF distribution.

ANALYTICAL FOUNDATIONS (MODE I)

Sih and Kassir [3] have shown that the near field stresses take the same functional form for a curved crack front as for a straight one.

For the case of Mode I loading, these equations may be written as:

$$\sigma_{ij} = \frac{K_I}{r^{1/2}} f_{ij}(\theta) + \sigma^0_{ij}(r,\theta) \quad (i,j = n,z) \tag{1}$$

for the stresses in a plane mutually orthogonal to the flaw surface and the flaw border referred to a set of local rectangular cartesian coordinates as pictured in Figure 1, where the terms containing K_1, the SIF, are identical to Irwin's

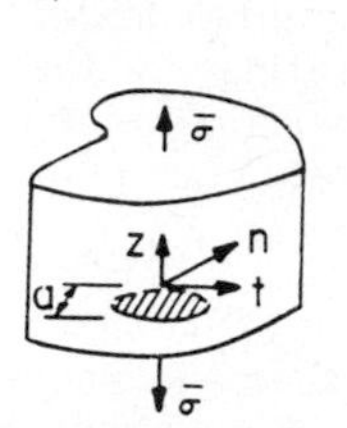

Fig. (1) - Mode I problem geometry and notation

equations for the plane case and σ^0_{ij} are normally taken to be constant for a given point along the flaw border within the measurement zone (normally 0.1 to 1.0 mm measured perpendicularly from the crack plane), but may vary from point to point. Equations (1) do not satisfy boundary conditions on the crack surface since they are only applied in a region near $\theta = \neq \pi/2$. This approach is often used in connection with hybrid finite element analysis. Observing that stress fringes tend to spread approximately normal to the flaw surface, Figure 2, equations (1) are evaluated along $\theta = \pi/2$, Figure 1, and the maximum in-plane shearing stress is computed as:

$$\tau^{nz}_{max} = \frac{1}{2} [(\sigma_{nn} - \sigma_{zz})^2 + 4\sigma^2_{nz}]^{1/2} \tag{2}$$

which, when truncated to the same order as equations (1), leads to the two parameter equation:

$$\tau^{nz}_{max} = \frac{A}{r^{1/2}} + B \text{ where} \begin{cases} A = K/(8\pi)^{1/2} \\ B = f(\sigma_{ij}) \end{cases} \tag{3}$$

which can be rearranged into the normalized form

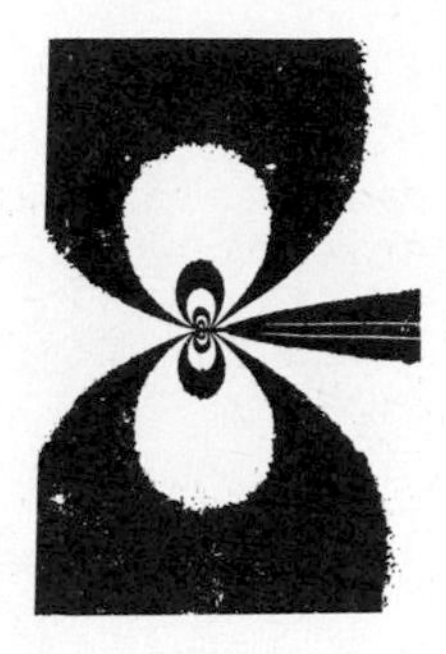

Fig. (2) - Stress fringes for Mode I loading

$$\frac{K_{AP}}{q(\pi a)^{1/2}} = \frac{K_I}{q(\pi a)^{1/2}} + \frac{f(\sigma^0_{ij})(8)^{1/2}}{q} \{\frac{r}{a}\}^{1/2} \tag{4}$$

where

$$K_{AP} = \tau^{nz}_{max}(8\pi r)^{1/2}$$

and, from the stress-optic law, $\tau_{max} = Nf/2t'$ where N is the stress fringe order, f the material fringe value and t' the slice thickness in the t direction, q is the remote loading parameter {such as uniform stress (σ), or pressure (p)}, and a the characteristic flaw depth. Equations (1) with σ^0_{ij} as described above requires that a linear relation exist between the normalized apparent stress intensity factor and the square root of the normalized distance from the crack tip. Thus, one need only locate the linear zone in a set of photoelastic data and extrapolate across a very near field non-linear zone to the crack tip in order to obtain the SIF. An example of this approach using data from a slice removed from a plate with a crack emanating from a hole (a problem discussed in the sequel) is given in Figure 3. Since the method is 3D, several slices are always removed from a given flaw border. The proper location for the linear zone must be common to all such slices and lie approximately in the region located between 0.1 and 1.0 mm from the crack plane.

Once the photoelastic data have been read, a linear grid corresponding to a density of 1200 lines per millimeter is applied to the stress frozen slice surface. This density is achieved by utilizing a virtual master grating developed by D, Post [4]. The slice is then annealed and the resulting deformed grid is superimposed upon the master grid, yielding Moiré displacement fringes. By employing the plane strain near tip displacement field equations of LEFM:

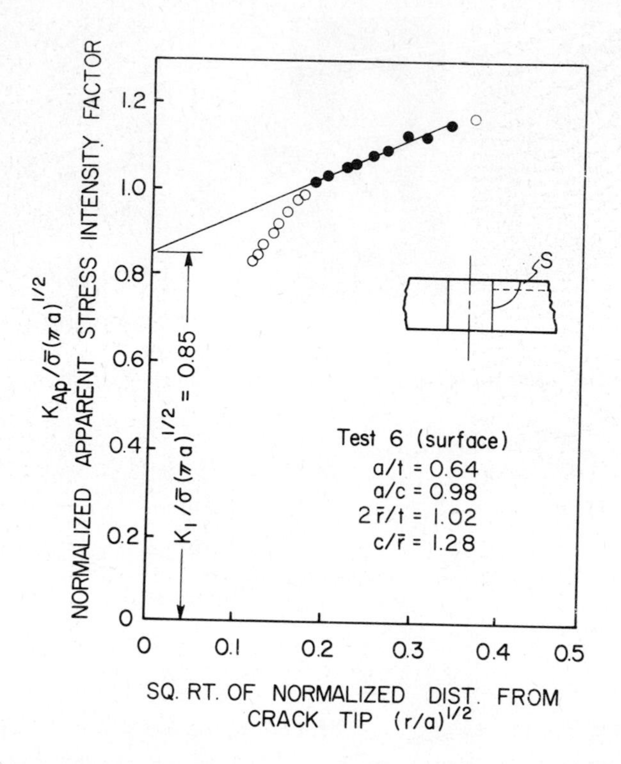

Fig. (3) - Estimation of SIF from photoelastic data from a frozen slice

$$u_z = \frac{2(1+\nu)}{E} K_{AP} \left(\frac{r}{2\pi}\right)^{1/2} \sin\frac{\theta}{2}\left[2 - 2\nu - \cos^2\frac{\theta}{2}\right]$$

$$u_n = \frac{2(1+\nu)}{E} K_{AP} \left(\frac{r}{2\pi}\right)^{1/2} \cos\frac{\theta}{2}\left[1 - 2\nu + \sin^2\frac{\theta}{2}\right] \qquad (5)$$

along the zone of measurement ($\theta = \pi/2$), values of a normalized K_{AP} can be computed and plotted against $(r/a)^{1/2}$ and extrapolated to the origin to obtain a normalized K_1. The form for K_{AP}, using a grating parallel to the crack plane, is:

$$K_{AP} = \frac{Cu_z}{r^{1/2}} \qquad (6)$$

Moiré fringe patterns for u_n and u_z near a crack tip due to Mode I loading are shown in Figure 4. Estimation of the SIF from a set of Moiré data is shown in Figure 5, for an internal slice taken from a Mode I problem.

(a)

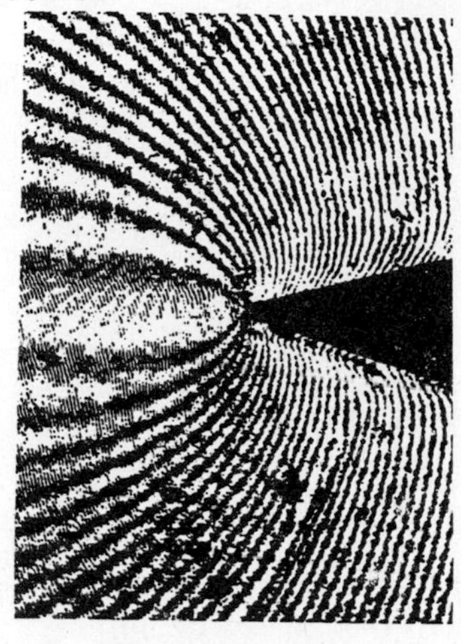

(b)

Fig. (4) - Moiré fringes for (a) u_n and (b) u_z

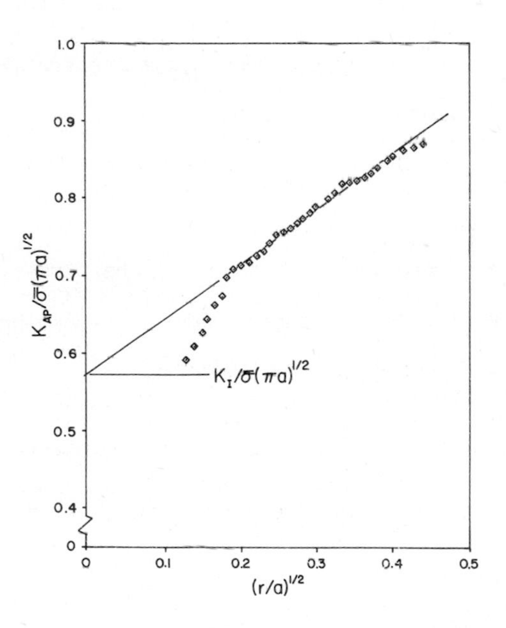

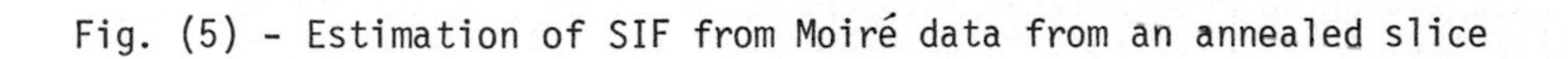
Fig. (5) - Estimation of SIF from Moiré data from an annealed slice

EXPERIMENTAL CONSIDERATIONS

A. Procedures

Before discussing the experiments and their results, it would seem appropriate to briefly outline the experimental procedure when utilizing both the photoelastic and Moiré methods. The procedure is as follows:

1. Cast fringe free parts of the photoelastic model from an appropriate diphase photoelastic material.

2. Insert cracks at desired locations. For simple shapes not to be grown, machined artificial cracks may be used. When it is desired to grow cracks, tiny starter cracks are produced by striking a sharp rounded tip blade which is held normal to and in contact with a surface of the model part. The crack will emanate dynamically from the blade tip and then arrest itself.

3. The parts of the model are glued together with a glue compatible photoelastic model material. Since local self-equilibrating stresses normally accompany the shrinkage of the glue when it dries, it is important to keep the crack tips away from the glue joints.

4. Place the cracked model in an oven in a loading rig and heat to a temperature above critical.

5. Apply live loads. For natural cracks, the loads must be sufficient to grow the cracks to their desired size. Loads are then reduced to stop crack growth.

6. Cool slowly under reduced load to room temperature.

7. Remove slices mutually orthogonal to crack border and crack surface at intervals along crack surface.

8. Analyze slices photoelastically and estimate SIF values for each slice from photoelastic data.

9. Apply a linear grating to one side of the stress frozen model slices.

10. Place slices in oven and anneal them by reversing the stress freezing thermal cycle.

11. Superimpose the resulting deformed grating upon the master grating to obtain Moiré displacement fields near the crack tip. This implies reversibility i the zone dominated by the elastic singularity and is approximately true.

12. Estimate SIF values from local displacement fields through the use of the LEFM plane strain or plane stress field equations.

13. Compare results of 8 and 12.

B. Prior Observations

In studying sub-critical flaw growth in 3D problems over the past decade, the first author and his associates have noted a number of instances in which the growth of real cracks has departed from that assumed in formulating numerical models. For example, suppose we consider the hole-crack problem pictured in Figure 6. This has been recognized as a common problem in the aircraft field for a

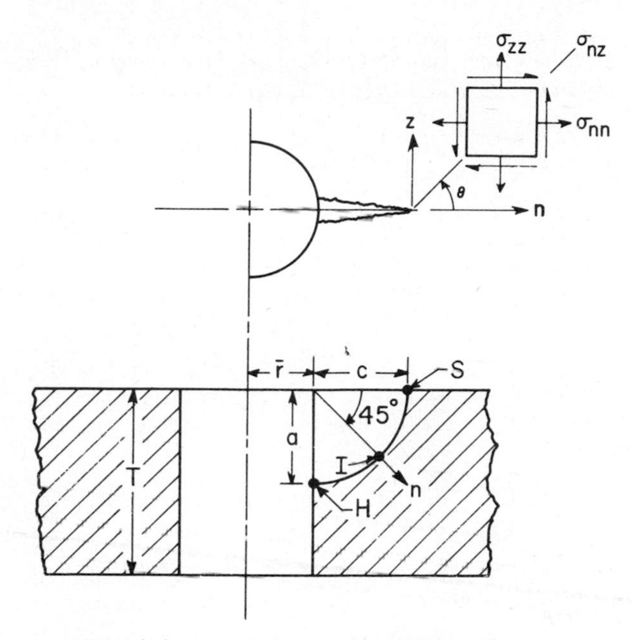

Fig. (6) - Corner crack problem geometry

number of years. Initially, the problem was treated by considering the cracks to be quarter-circular throughout their growth [5]. Subsequently, more precise numerical models were constructed [6] using quarter elliptic shapes and these shapes were found to be reasonably accurate descriptions when real cracks were grown under remote tensile loads [7,8] with fastest growth occurring in the depth direction for remote simple tensile loads. When, however, the plates are flexed prior to tensile loading in order to keep the aspect ratio of the crack near unity, then the flaw border tends to flatten in the central region as shown in the upper part of Figure 7. Crack No. 1, however, is almost a perfect quarter circle. In the lower part of Figure 7, we see a comparison between the SIF distribution predicted for quarter circular cracks [6] compared with experimental values obtained photoelastically for flattened cracks [9]. Differences in normalized SIF values as large as ≈100% are noted. We thus infer the importance of crack shape upon the SIF distribution in 3D problems with complex boundaries. Similar flattening has also been observed in nozzle corner cracks in models of nozzles in thick walled reactor vessels [10-12]. Such effects are believed to be related to constraint influences in the problem. We now consider the results of some simple experiments designed to study such effects.

C. Experiments on Constraint Effects

In order to attempt to clarify constraint effects using techniques described above, a rather simple specimen geometry, pictured in Figure 8, was adopted. A straight front artificial crack was employed in order to eliminate curvature effects and permit us to focus solely upon thickness effects. Upon carrying out the stress

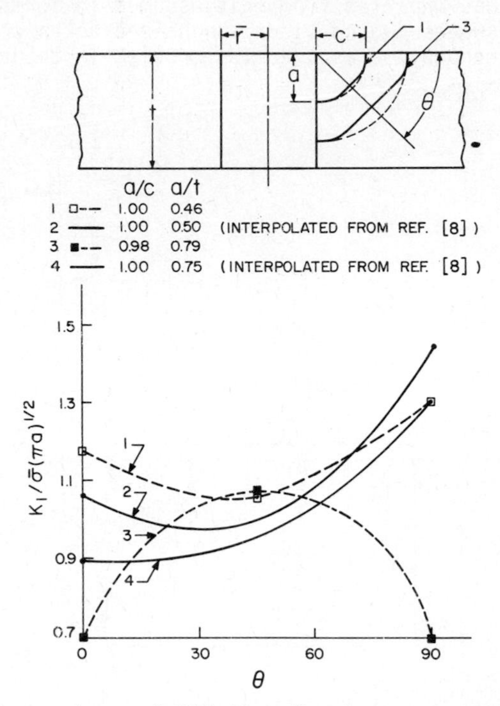

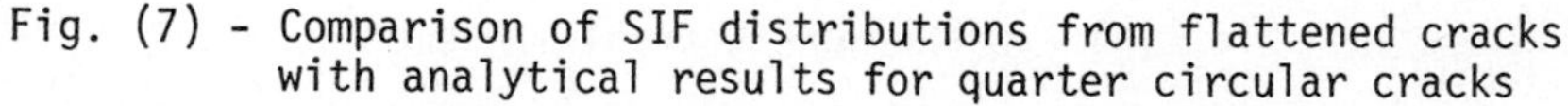

Fig. (7) - Comparison of SIF distributions from flattened cracks with analytical results for quarter circular cracks

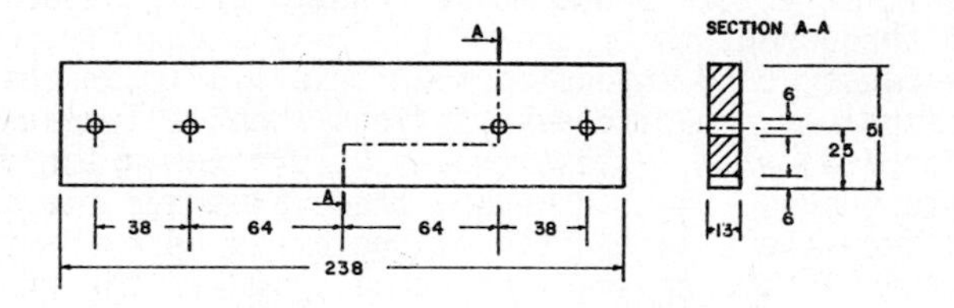

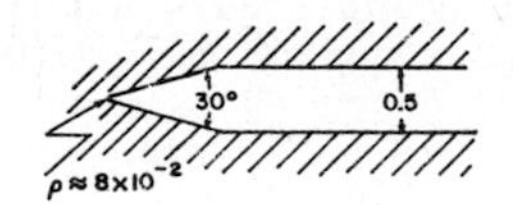

Fig. (8) - Specimen geometry for constraint effect measurements

freezing and annealing procedures described above, SIF estimates were obtained from both photoelastic and Moiré data. Results from a typical internal slice are shown in Figure 9. As seen in the figure, results showed that the difference between the SIF values measured photoelastically and by Moiré using the plane strain displacement equations were well within an experimental scatter of ±5%. However, substantial divergence was observed between the apparent SIF values as one moved away from the crack tip. This is interpreted to mean that the assumed plane strain constraint used in the Moiré Method is only achieved in the limit very near the crack tip. As one moves away from the tip, the real behavior exhibited by the photoelastic data suggests that the plane strain constraint is relaxed substantially within the measurement zone yielding apparent SIF values lower than predicted by the plane strain displacement field equations.

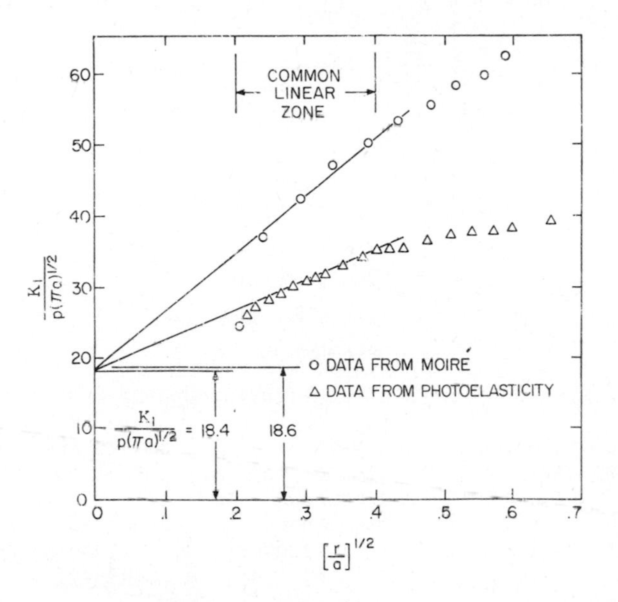

Fig. (9) - Photoelastic versus Moiré results from the same slice

Figure 10 presents a composite of results showing how the SIF varies through the thickness of the beam. The results are normalized with respect to the two dimensional solution (K_T) [19]. These results suggest that, near the center of the beam a state of nearly pure plane strain constraint exists. However, as we approach the boundary where the crack border intersects the plate surfaces, it would appear that the constraint drops off rapidly towards a state of plane stress. The rather large elevation of the interior SIF values above the two dimensional theoretical result and the large difference between the Moiré plane stress and plane strain results at the boundaries is believed due primarily to an increased sensitivity to constraint effects of the incompressible photoelastic material. The results near the outer surface suggest that there is not enough thickness available to produce plane strain constraint in that region.

The experiments described above were interpreted on the basis of the existence of an inverse square root singularity dominated zone as described by equations (1). The linear zones in Figures 3, 5 and 9 provide experimental verification of such an interpretation. However, it should be noted that, mathematically, other complications may exist. For example, if one assumes a plane strain constraint in the central region, a plane stress state at the surface and attempts to match these solutions by asymptotic methods, one finds it necessary to impose a continuously varying order on the stress singularity as one approaches the free boundary [13]. More exact analysis leads to a similar observation [14]. Or, use

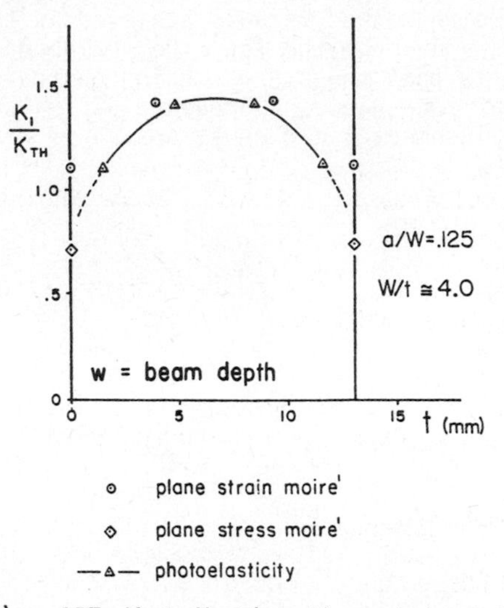

Fig. (10) - SIF distribution through beam thickness

of the eigenfunction method of analysis [15,16] leads to the conclusion that the order of the singularity varies with the boundary condition. Other studies [17, 18] also suggest the presence of a boundary layer effect. In the present work, the authors have not addressed these issues, but rather have tried to show that the classical definition of the SIF is adequate for obtaining reasonably accurate (±5%) results to such problems excepting the free boundary.

SUMMARY

An experimental modelling procedure consisting of a marriage between frozen stress photoelasticity, Moiré analysis and linear elastic fracture mechanics has been briefly reviewed and the results of applying the methods to three dimensional cracked body problems have been discussed. As with other methods, this method has its own limitations. Major ones are:

1. It applies only to LEFM controlled processes.

2. The model material is incompressible, and

3. It does not model fatigue enhanced effects (i.e., overloads, crack closure, stress ratio effects, etc.).

Despite these drawbacks, the method offers the potential for obtaining insight into the behavior of 3D cracked body problems not previously available and a potential for attacking problems where neither flaw shape nor SIF distribution is known a priori.

ACKNOWLEDGEMENTS

The authors wish to acknowledge the contribution to the Moiré work of their colleague, D. Post, and the support of the Solid Mechanics Program of the National Science Foundation under Grant No. MEA-811-3565.

REFERENCES

[1] Smith, C. W., "Use of three dimensional photoelasticity and progress in related areas", Experimental Techniques in Fracture Mechanics - 2, A. S. Kobayashi, ed., SESA Monograph No. 2, pp. 3-58, 1975.

[2] Smith, C. W., Peters, W. H. and Andonian, A. T., "Mixed mode stress intensities for part circular surface flaws", J. of Engineering Fracture Mechanics, Vol. 13, pp. 615-629, 1979.

[3] Sih, G. C. and Kassir, M. K., "Three dimensional stress distribution around an elliptical crack under arbitrary loadings", J. of Applied Mechanics, Vol. 33, pp. 601-611, 1966.

[4] Post, D. and Baracat, W. A., "High-sensitivity Moiré interferometry - a simplified approach", J. of Experimental Mechanics, Vol. 21, No. 3, pp. 100-104, March 1981.

[5] Wilhem, D. P., "Fracture mechanics guidelines for aircraft structural applications", USAFFDL TR-69-111, Wright Patterson Air Force Flight Dynamics Laboratory, February 1970.

[6] Smith, F. W. and Kullgren, T. E., "Part elliptical cracks emanating from open and loaded holes in plates", J. of Engineering Materials and Technology, Vol. 101, No. 1, pp. 12-17, January 1979.

[7] McGowan, J. J. and Smith, C. W., "Stress intensity factors for deep cracks emanating from the corner formed by a hole intersecting the plate surface", ASTM STP 590, pp. 460-476, 1976.

[8] Kullgren, T. E. and Smith, F. W., "Static fracture testing of PMMA plates having flawed fastener holes", J. of Experimental Mechanics, Vol. 20, No. 3, March 1980.

[9] Smith, C. W., Peters, W. H. and Gou, S. F., "Influence of flaw geometries on hole-crack stress intensities", ASTM STP 677, pp. 431-445, 1979.

[10] Broekhoven, M. J. G., "Fatigue and fracture behavior of cracks at nozzle corners: comparisons of theoretical predictions with experimental data", Proceedings of the Third International Conference on Pressure Vessel Technology, Part II, pp. 839-852, 1977.

[11] Smith, C. W. and Peters, W. H., "Experimental observations of 3D geometric effects in cracked bodies", Developments in Theoretical and Applied Mechanics, Vol. 9, pp. 225-234, 1978.

[12] Smith, C. W. and Peters, W. H., "Prediction of flaw shapes and stress intensity distributions in 3D problems by the frozen stress method", Preprints from Sixth International Congress on Experimental Stress Analysis, pp. 861-865, 1978.

[13] McGowan, J. J. and Smith, C. W., "Analysis of a straight front crack", unpublished, 1977.

[14] Folias, E. S., "On the three dimensional theory of cracked plates", J. of Applied Mechanics, pp. 663-674, September 1975.

[15] Benthem, J. P., "The quarter-infinite crack in a half space; alternative and additional solutions", Int. Journal of Solids and Structures, Vol. 16, pp. 119-130, 1980.

[16] Benthem, J. P., "State of stress at the vertex of a quarter infinite crack in a half space", Int. Journal of Solids and Structures, Vol. 13, pp. 479-492, 1977.

[17] Hartranft, R. J. and Sih, G. C., "Effect of plate thickness on the bending stress distribution around through cracks", J. of Mathematics and Physics, Vol. 47, No. 3, pp. 276-291, September 1968.

[18] Sih, G. C., "Three-dimensional stress-state in a cracked plate", Proc. of the Air Force Conf. in Fatigue and Fracture of Aircraft Structures and Materials, AFFDL, TR 70-144, pp. 175-191, 1970.

[19] Gross, B. and Srawley, J. E., "Stress intensity factors for single-edge-notch specimens in bending or combined bending and tension by boundary collocation of a stress function", NASA TN E-2801, 1965.

SECTION VI
FATIGUE CRACK PROPAGATION

ANALYSES OF MICROSTRUCTURAL AND CHEMICAL EFFECTS ON FATIGUE CRACK GROWTH

H. W. Liu

Syracuse University
Syracuse, New York 13210

ABSTRACT

Fatigue crack propagation is insensitive to micro-structure of a material in the intermediate ΔK region. But it is highly sensitive in the low ΔK near threshold region. da/dN decreases with ΔK. As ΔK decreases below a transition point, ΔK_t, da/dN decreases rapidly, and a crack ceases to propagate when ΔK is below ΔK_{th}.

A fatigue crack may propagate by the deformation or fracture mode or by a mixture of the two. Fatigue crack propagation in deformation mode is structurally insensitive if the cyclic plastic zone is several times the grain size or the size of the mean free path for dislocation movement when the crack tip deformation can be described in terms of the "average" yield strength of the material.

FCP in deformation mode is structurally sensitive, if the cyclic plastic zone is nearly the size of the grain or the mean free path of dislocation movement; and FCP in fracture mode is also structurally sensitive.

Intergranular fatigue crack growth by the fracture mode in detrimental chemical environment is structurally sensitive. The temperature and frequency effects of the crack growth in detrimental chemical environment as well as the relation between ΔK_t and ΔK_{th} are analyzed.

INTRODUCTION

In the low ΔK near the threshold region, fatigue crack growth is sensitive to the microstructure and the chemical environment. The effects of grain size [1-4] and yield strength [2,3,5-9] have been studied. More recently, Yoder et al [10] have shown that the effects of both cyclic yield strength and the mean free path for dislocation movement (i.e., grain size) can be combined in terms of cyclic plastic zone size. Fatigue crack growth in detrimental chemical environment has been investigated extensively. The temperature and frequency effects have been analyzed in a manner of a kinetic process [11-14]. Convincing evidence of micro-structural effects on da/dN has been demonstrated in the near threshold region and in detrimental chemical environments.

A crack may grow by the deformation or fracture processes. In this paper, quantitative and qualitative analyses are made of the microstructural effects on the mechanical aspects of these two crack growth processes in the near threshold region and in detrimental chemical environments.

THE UNZIPPING MODEL OF FATIGUE CRACK GROWTH

Neumann [15] observed that under a cyclic load, a crack grows by shear slip process, Figure 1. The shear deformation concentrates in narrow slip bands sepa-

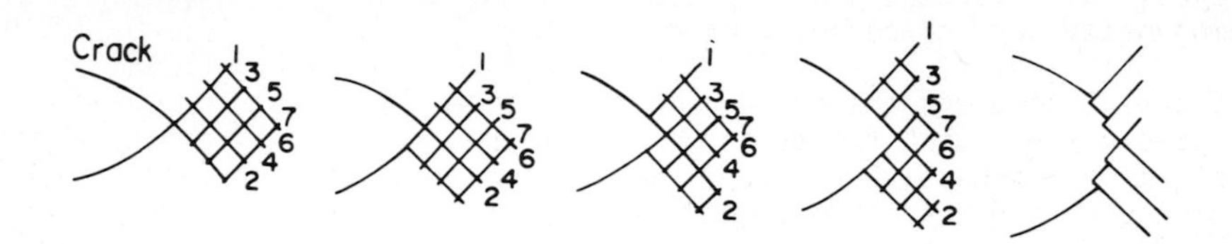

Fig. (1) - Unzipping model for crack opening and crack growth

rated by lightly deformed or non-deformed slabs. As the load increases, the slab at the advancing crack tip move away from the tip, one at a time, like the teeth in a zipper during the unzipping process.

Kuo and Liu [16] modelled this unzipping crack growth process, using the finite element method. First, they applied a small load, K, to a cracked solid in the range of small scale yielding. The plastic zone at the applied K was delineated. A slip band, such as AB in Figure 2 was chosen. Each of the nodal points

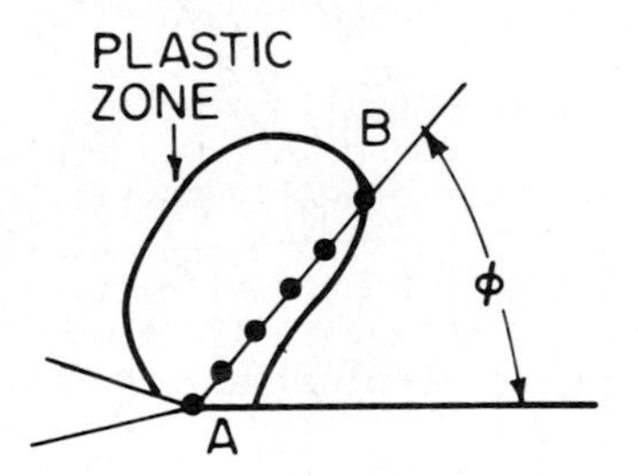

Fig. (2) - Crack tip plastic zone and slip band AB

along the slip band were branched into two, one on each side of the slip band. They then applied a small increment δK. During the increment, they allowed the branched nodal points along AB to slip, but not to move away from the slip band. The unzipping crack increment, δa, during δK was calculated. During the increme δK, all the elements are allowed to deform plastically according to the constitu tive relation, while the shear stress along AB remained unchanged. The calculat δa is related to da/dN. The details of the modelling are given in [16].

When a load is applied, the shear deformation in the crack tip region blunts the tip, causing the crack tip to open. But only a small portion of the crack t opening displacement contributes toward crack growth. Kuo and Liu's calculation takes into consideration only the opening displacement that causes crack growth. The calculated crack growth rate is

$$\frac{da}{dN} = 0.018 \frac{\Delta K^2}{E\sigma_{Y(c)}} \tag{1}$$

where $\sigma_{Y(c)}$ is cyclic yield stress; and E, the Young's modulus. $\sigma_{Y(c)}$ in equation (1) is the cyclic yield strength of polycrystals. Therefore, equation (1) applies only if the cyclic plastic zone is several times the grain size. When the cyclic plastic zone size is close to the grain size, the effects of material inhomogeneity have to be considered.

Figure 3 compares the calculated crack growth rate with the measured striation spacings and with the measured crack growth rates in the intermediate ΔK structurally insensitive region. The agreement between the calculated and the measured growth rates is very good.

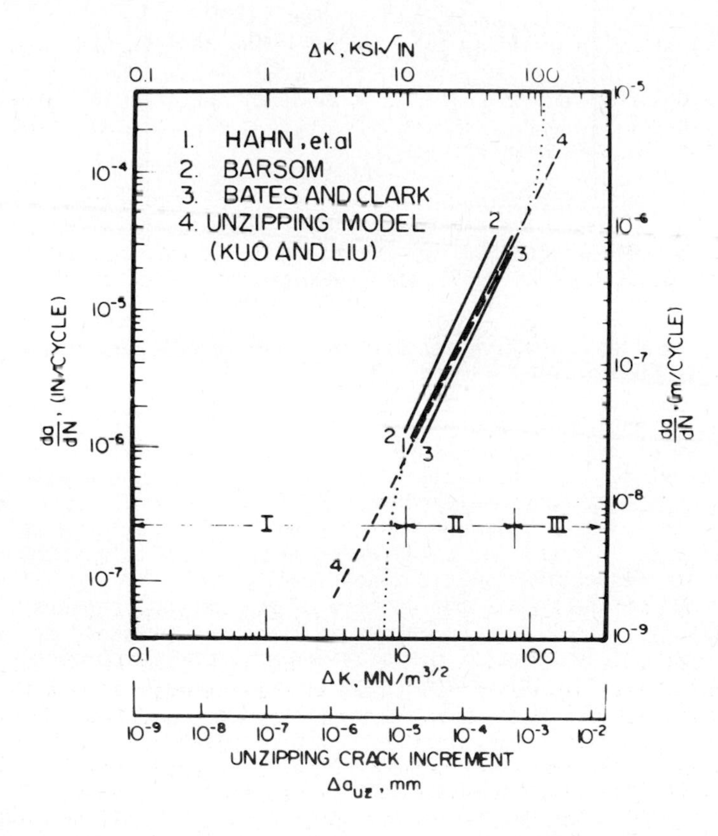

Fig. (3a) - Calculated and measured crack growth rate ([16b])

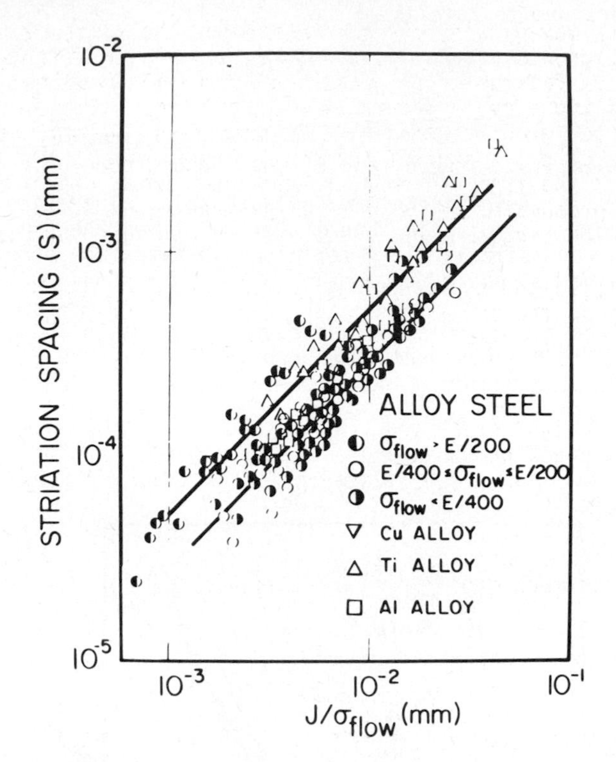

Fig. (3b) - Measured striation spacing and calculated crack growth rate ([16b])

In modelling the unzippings process, we use the macromechanical finite element method because of the large number of dislocations involved in the process. According to Yoder et al [9], the cyclic plastic zone size in the near threshold regions of the steels they have analyzed, ranges from 0.4 to 70 μm. The corresponding sizes of the plastic zone are 10^{-8} cm^2 to 3 x 10^{-4} cm^2. With an assumed dislocation density of 10^{12}/cm^2 for cold worked metals, the total numbers of dislocations within the respective plastic zones are 10^4 and 3 x 10^8. If the mobile dislocations are 10% of the total, the numbers of mobile dislocations are 10^3 and 3 x 10^7, respectively. These are the values in the near threshold region. At a higher ΔK-value, the total number of dislocations increases accordingly. Because of the large number of dislocations involved, the application of the continuum mo is justified.

The calculated growth rate is applicable only in the intermediate ΔK region, where da/dN is insensitive to microstructures. In the lower ΔK region, where da/dN is highly sensitive to microstructure, the calculated rate can serve as a reference to evaluate the microstructural effects.

FATIGUE CRACK GROWTH THRESHOLD, ΔK_{th}

A fatigue crack may grow by the deformation or the fracture mode. The unzipping fatigue crack growth process is a deformation mode of fatigue crack growth. The fracture mode of fatigue crack growth may occur in embrittled solids, such as intergranular fatigue crack growth at elevated temperature or in hydrogen embrittlement and crack growth across brittle particles. Often a fatigue crack grows in dual mode, i.e., a part of the crack front grows in the deformation mode and a part grows in the fracture mode. When these crack growth processes stop, a crack ceases to propagate. No single unique crack growth process exists. Therefore, we have no single unique model for the fatigue crack growth threshold.

Lal and Weiss [17] and subsequently, Weertmann [18] proposed the maximum fracture stress theory as the model for ΔK_{th}. When the crack tip stress is below the theoretical fracture strength of the material, i.e., E/10, a crack stops propagating.

The unzipping fatigue crack growth process is the deformation mode of crack growth. When this crack growth process stops, a crack ceases to propagate. Liu and Liu [19] proposed a ΔK_{th} model based on the unzipping fatigue crack growth process.

The fatigue crack growth rate decreases with ΔK. Beyond a transition point, ΔK_t, da/dN decreases much more rapidly, shown in Figure 4. As ΔK further decreases,

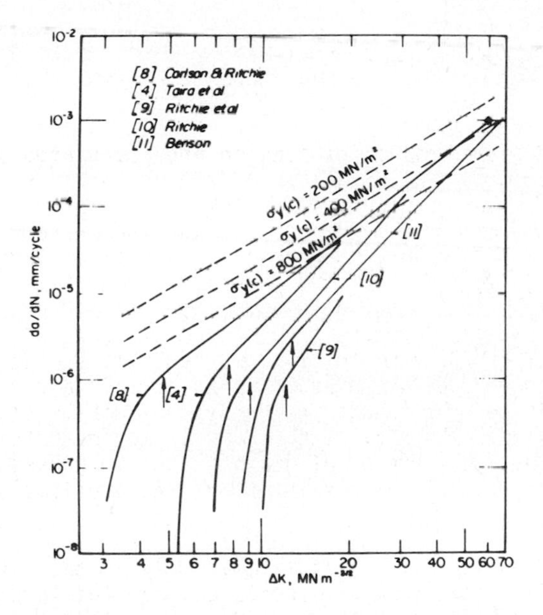

Fig. (4) - Crack growth data of several steels. Arrows indicate the transition point, ΔK_t ([19])

da/dN slows down so much that the crack practically stops propagating. The lowest value of ΔK, when a crack is on the threshold of propagation, is defined as ΔK_{th}.

Liu and Liu [19] have examined fifty sets of crack growth data and have determined the values of ΔK_t and ΔK_{th}, Figure 5. The empirical relation is

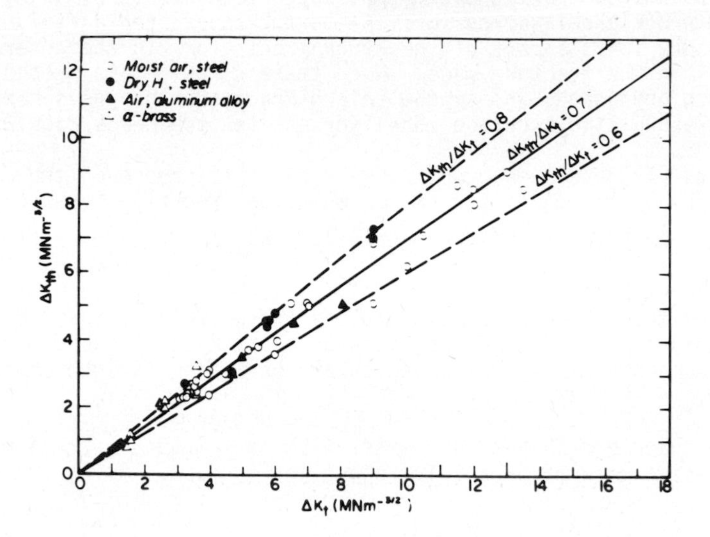

Fig. (5) - The relation between ΔK_t and ΔK_{th} ([19])

$$\Delta K_{th} = 0.7\ \Delta K_t \quad (2$$

Plastic deformation is a discrete process. The thickness of the "elastic" slab must have a minimum finite size ℓ_m. When crack growth rate is larger than ℓ_m, a crack may grow according to the unzipping process. When crack growth rate is less than ℓ_m, the crack front "retreats" during the unloading half cycle. The retreat of the crack front during the unloading half cycle may account for the drastic decrease in da/dN, when ΔK is below ΔK_t. At ΔK_t, $da/dN = \ell_m$.

Figure 6 shows the back and forth movement of the crack front under a cyclic load, when da/dN is less than ℓ_m. The numbered locations of the crack front cor

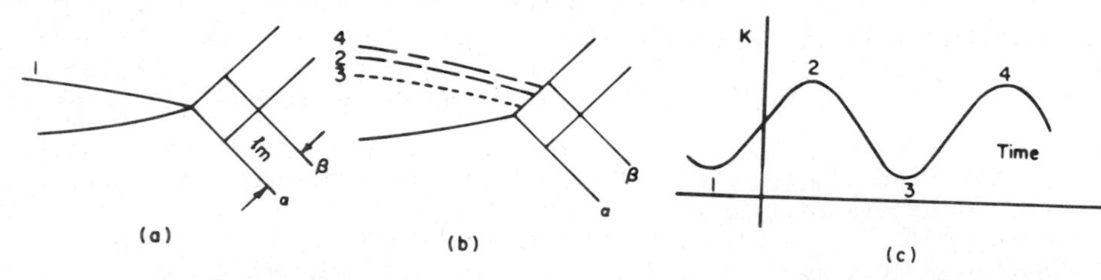

Fig. (6) - Cyclic creep movement of crack front ([19])

respond to the numbered positions during the loading cycles. As the cyclic load continues, the crack front creeps forward toward the intersecting conjugate slip plane β. As it does so, the mean stress relaxes. The unzipping process continues when the crack front reaches β. The rate of the forward movement of the crack relates to the rate of cyclic creep. If the crack front still cannot reach the conjugate slip band when the maximum cyclic creep shear slip is realized and when the mean stress completely relaxes to zero, the crack stops propagating. The corresponding ΔK is ΔK_{th}.

Cyclic creep and stress relaxation take place when a positive mean stress exists. Figure 7 shows the stress relaxation when the imposed strain range $\Delta\varepsilon$ is

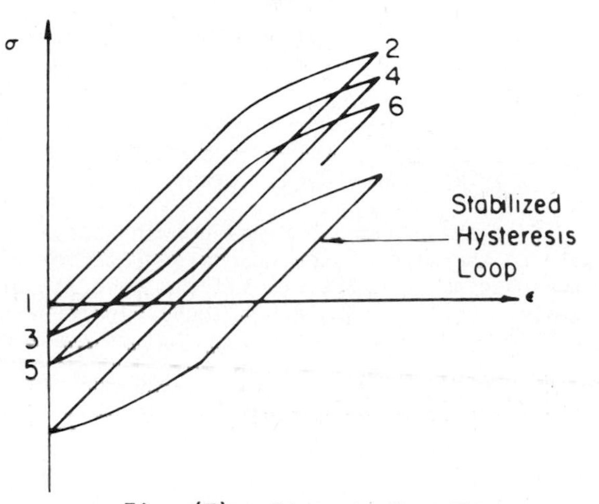

Fig. (7) - Stress relaxation

fixed. The σ_{max} decreases, cycle by cycle, until the mean stress relaxes to zero. A stabilized hysteresis loop is shown with $\sigma_{mean} = 0$. The σ_{max} of the stabilized hysteresis loop is reduced to $\Delta\sigma/2$. Without stress relaxation, $\sigma_{max} = \Delta\sigma$.

Figure 8 shows the cyclic creep, when $\Delta\sigma$ is fixed and σ_{mean} is positive. The cyclic creep strain is positive and increases every cycle. At a crack tip, the cyclic plastic deformation combines both of these two processes.

If we consider $\sigma_{Y(c)}$ in equation (1) as the macro-cyclic-yield-strength, the equation applies only if the cyclic plastic zone is several times the grain size or several times the size of the mean free path for dislocation movement, so that the cyclic plastic deformation at the crack tip can be described in terms of $\sigma_{Y(c)}$. Figure 4 shows the crack growth of several steels. When the da/dN is above 10^{-4} mm/cycle, the measured rate (solid curves) seems to merge with the calculated rate given by equation (1) (dashed lines). This happens because the cyclic plastic zone is large enough for the macromechanical calculation to agree with the crack tip deformation behavior.

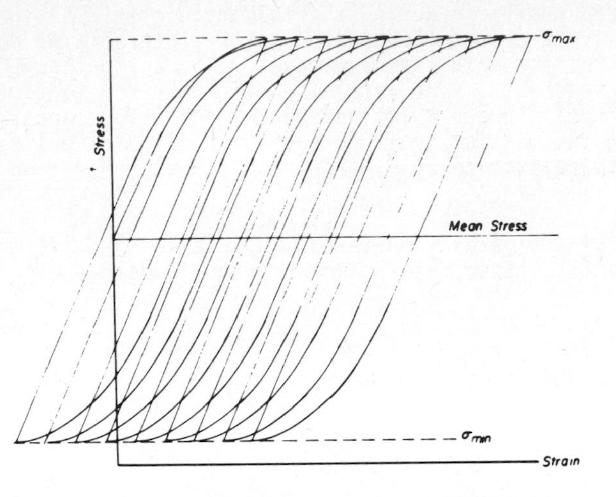

Fig. (8) - Cyclic creep

As da/dN decreases from this range of crack growth rate, the cyclic plastic zone size is comparable to the size of the mean free path for dislocation movement; the microstructure begins to exert its effects on crack growth rate; and the data start to deviate from the calculated lines. In the region immediately above ΔK_t, crack growth rate is much slower than that given by equation (1), and we can write the growth rate as

$$\frac{da}{dN} = A\left(\frac{\Delta K^2}{2E\sigma_{Y(c)}}\right)^m \quad (3a)$$

At the transition point,

$$\frac{da}{dN} = \ell_m = A\left(\frac{\Delta K_t^2}{2E\sigma_{Y(c)}}\right)^m \quad (3b)$$

where $2\sigma_{Y(c)}$ is the cyclic yield stress range. A material must experience the stress range before plastic deformation takes place. m and A are constants. m is close to one, but A is much smaller than the coefficient in equation (1) because of the microstructural effects. If the crack tip deformation under a monotonic load follows the same relation as given by equation (3a), the monotonic crack increment by the shear mode is

$$\Delta a_s = A\left(\frac{K_{max}^2}{E\sigma_Y}\right)^m \quad (4a$$

When the mean stress is fully relaxed, the total crack increment by the shear mod is

$$\Delta a_t = A\left(\frac{K_{max}^2}{E(\sigma_Y/2)}\right)^m \tag{4b}$$

The cyclic creep shear decohesion is the difference between these two quantities. With the approximation of $\sigma_Y \simeq \sigma_{Y(c)}$, $\Delta K \simeq K_{max}$ and $m \simeq 1$, the cyclic creep crack increment by the shear mode is

$$\Delta a_c = \ell_m = A\left(\frac{\Delta K_{th}^2}{E\sigma_{Y(c)}}\right) \tag{5}$$

At ΔK_{th}, Δa_c is equal to ℓ_m.

Combining equations (3b) and (5) together with the assumption m = 1, we have

$$\Delta K_{th} = 0.7\ \Delta K_t \tag{2}$$

This is the solid line in Figure 5.

In equation (3), the value of A is much smaller than the coefficient of equation (1). Microstructural effects reduce the value of the coefficient. When applying the linear elastic fracture mechanics, we assume that the crack front is straight and the crack surface is a plane. A real crack is not flat and its front is not straight. When one studies the microstructural effects, only the local stress intensity factor at the microstructural feature is relevant. Yoder et al [10] observed bifurcation at the crack tip in the low ΔK region. Bifurcation reduces the local stress intensity factor. If a material is not homogeneous, the crack front in the "weaker" material grows faster, and the crack front in the "stronger" material grows more slowly. As a result, a crack front is not a straight line, and the local K fluctuates up and down as one moves along the crack front. Both of these processes may effectively reduce the local K for the crack growth processes. Therefore, we can consider "A" as a geometric correction factor for the local K-value, which is caused by the irregularity of the crack front that is due to the microstructural crack growth features. Since the values of A at ΔK_t and ΔK_{th} may differ, equation (2) is valid only if the difference is small, i.e., if the pertinent geometric features at ΔK_t and ΔK_{th} are essentially the same.

Yoder et al [10] have shown that a crack greatly reduces its growth rate when the cyclic plastic zone equals the mean free path of dislocation movement between neighboring barriers. They certainly base their model for the transition point on the deformation mode of fatigue crack growth.

As pointed out by Yoder et al [10], grain boundary, the boundaries of pearlite colony and the packet of martensite laths, and the inter-lath boundary of martensite can serve as dislocation barriers. In the case of ferritic-martensitic dual phase steel, the interface is the barrier. Taira, Tanaka and Nakai [20a] have analyzed the crack tip slip band blocked by grain boundary. The grain boundary does reduce the size of the shear slip step at a crack tip.

Yang and Liu [20b] studied the crack tip shear slip in a ferritic-martensitic dual phase steel, Figure 9. When a crack in a soft ferrite phase approaches the ferritic-martensitic interface and when the plastic zone impinges upon the interface, the hard martensite constrains crack tip deformation in the ferrite, and the

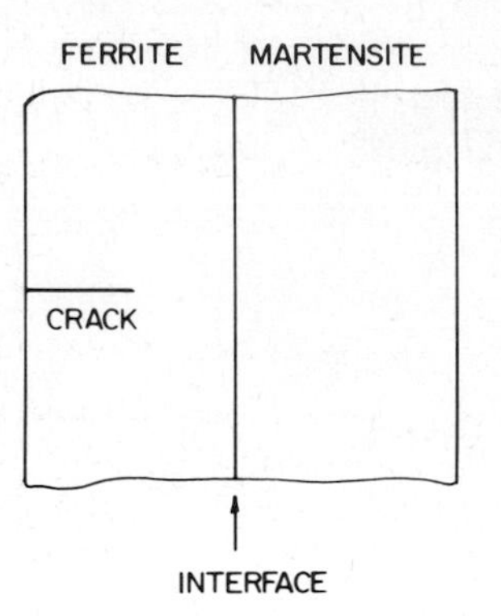

Fig. (9) - A crack in a ferritic-martensitic two phase steel ([20])

shear slip step decreases, as shown in Figure 10. The martensite acts like the dislocation barrier. When the unzipping crack increment decreases to ℓ_m, the crac stops propagating.

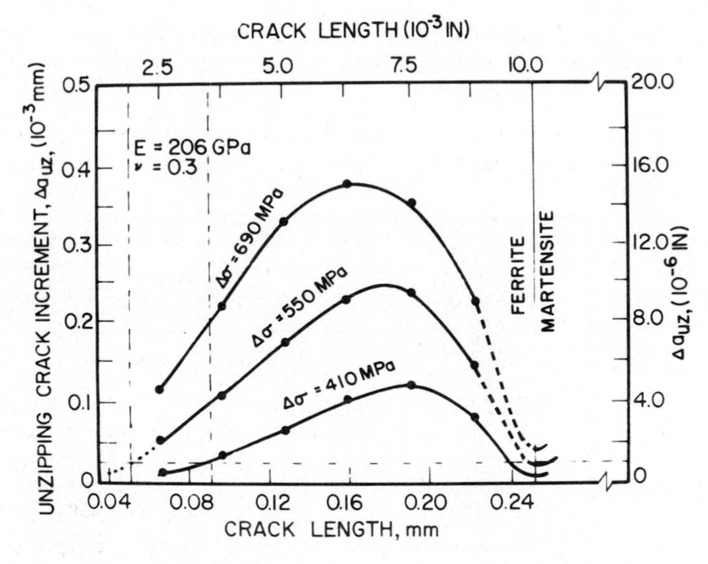

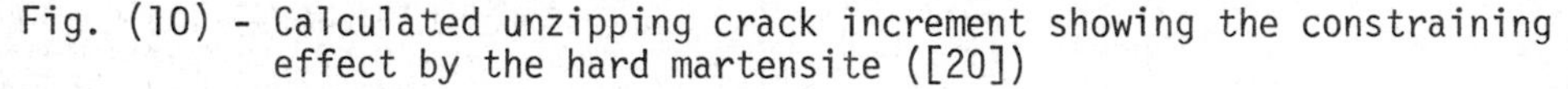

Fig. (10) - Calculated unzipping crack increment showing the constraining effect by the hard martensite ([20])

ENVIRONMENTAL EFFECTS

Fatigue crack growth at elevated temperatures and/or in detrimental chemical environments has been analyzed as a kinetic process [11-14]. In a like manner, Liu and McGowan [21] analyzed fatigue crack growth of IN-100, Waspaloy, and 304 stainless steel at elevated temperatures. They attributed the increased crack growth rate at the elevated temperature to chemical effects and found that the temperature and frequency effects follow an Arrhenius relation. In addition, they recognized two limit lines in a crack growth diagram, shown in Figure 11. At these two limit lines, da/dN is independent of temperature and frequency. Between these two limit lines, da/dN is temperature and frequency dependent.

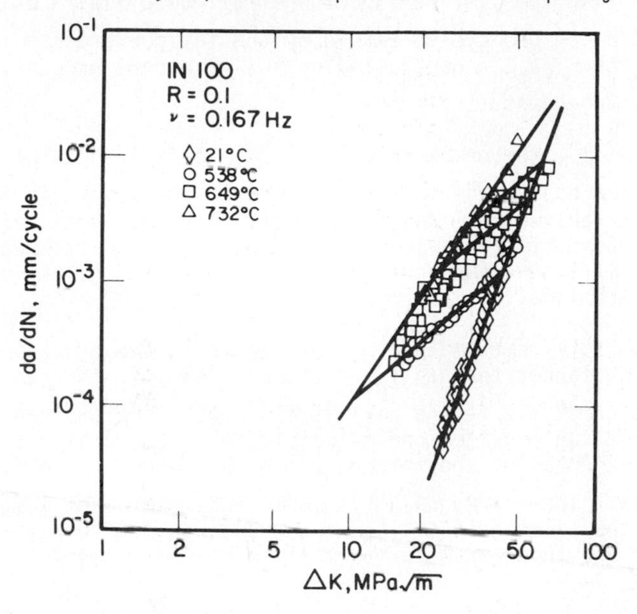

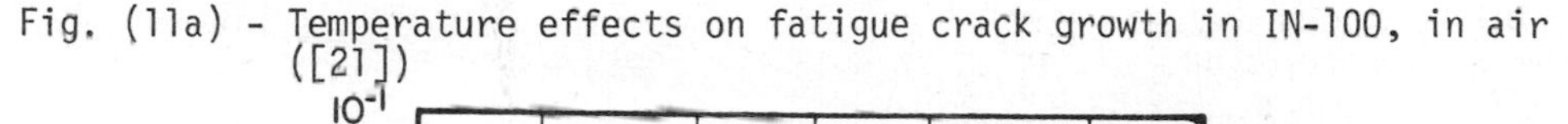
Fig. (11a) - Temperature effects on fatigue crack growth in IN-100, in air ([21])

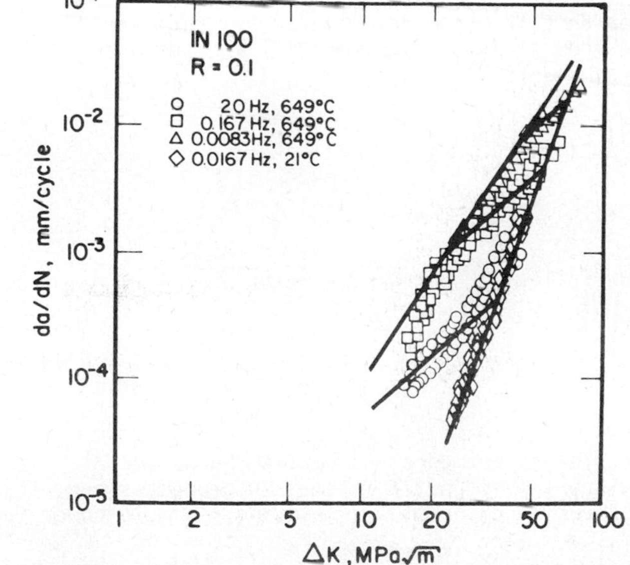

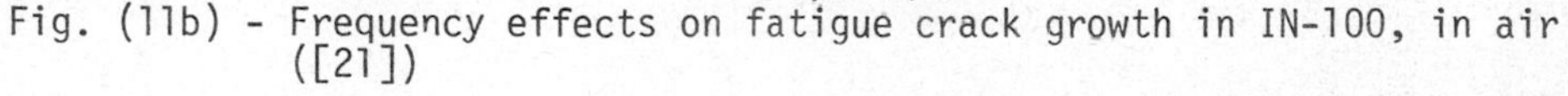
Fig. (11b) - Frequency effects on fatigue crack growth in IN-100, in air ([21])

When a cracked sample is stressed in a detrimental chemical environment, the crack grows much faster than in an inert atmosphere. For the detrimental effects to occur, the chemical specie has to reach the crack tip. Let's define the "penetration depth" as the crack area, Δa_p, that can be reached by the detrimental chemical specie during the time period of one fatigue cycle. Let $\Delta a_{f(n)}$ be the fatigue crack increment per cycle, caused by the fatigue process alone without the chemical effects. Δa_p is controlled by the transport process, which is temperature and time dependent.

When $\Delta a_{f(n)} >> \Delta a_p$, the crack growth rate is so fast that the detrimental chemical specie does not have enough time to reach the crack tip in large enough quantities to measurably affect da/dN. In this case, da/dN is neither frequency nor temperature dependent. This is the limit case of high frequency and low temperature. The limit line in the crack growth diagram is the nil-environmental-effect line, i.e., NEE line.

In a material "fully embrittled" by the chemical, the crack increment per cycle $\Delta a_{f(f)}$, is much larger than $\Delta a_{f(n)}$. If $\Delta a_p \geq \Delta a_{f(f)}$, the penetration depth runs ahead of the crack tip at the next stress cycle; the environment has affected da/dN fully; and we can consider the material "fully embrittled". Any further increase in Δa_p, by increasing the temperature or by reducing the frequency, will not increase da/dN. Therefore, da/dN is again independent of temperature and frequency. This is the limit case at low frequency and high temperature. The limit line in the crack growth diagram is the full-environmental-effect line, i.e., FEE line.

In the region between these two limit lines, the rate of the kinetic process becomes rate limiting, and da/dN is both temperature and frequency dependent.

The crack growth data in a number of materials indicate that the effects of ΔK can be separated from those of temperature T and frequency ν; therefore, we have

$$\frac{\Delta a}{\Delta N} = F(\nu,T)f(\Delta K) \tag{6a}$$

and

$$\frac{\Delta a}{\Delta N} = \frac{\Delta a}{\Delta t}\frac{1}{\nu} \tag{6b}$$

We assume that da/dt follows the Arrhenius relation. During the time period, $\Delta t = 1/\nu$, the crack growth is

$$\Delta a = \int_0^{\Delta t} \frac{da}{dt}\, dt \tag{7}$$

The rate da/dt is certainly not constant during the period Δt. It is an unknown function of the pertinent transport and reaction processes and, therefore, da/dN probably is not a linear function of ν. Assuming a simple power function for ν, we have

$$\frac{da}{dN} = \beta\left(\frac{\nu_o}{\nu}\right)^m \exp\left[-\frac{Q}{R}\left(\frac{1}{T} - \frac{1}{T_o}\right)\right]\Delta K^n \tag{8}$$

where ν_o and T_o are the reference state. Liu and McGowan examined the data of IN-100, Waspaloy and 304 stainless steel, (Figures 11, 12 and 13). The data fit

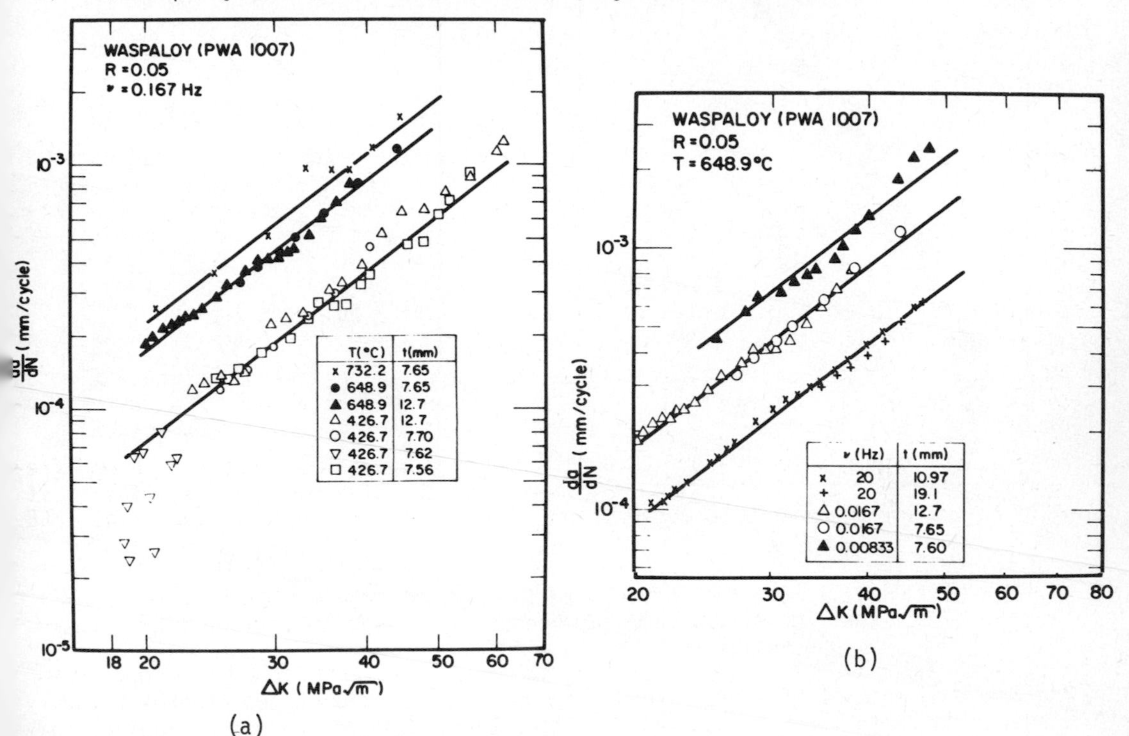

Fig. (12) - (a) Temperature effects on fatigue crack growth in Waspaloy in air ([21]); (b) Frequency effects on fatigue crack growth in Waspaloy in air ([21])

the solid lines of the equation very well. The data of IN-100 clearly shows the two limit lines. The data of 304 stainless steel indicate strongly the existence of the limit lines, but the temperature and frequency ranges for the Waspaloy are not wide enough to show any of the limit lines.

When a solid is cyclically loaded in a detrimental chemical environment, fatigue crack growth is often intergranular, and the crack grows in a fracture mode, primarily controlled by maximum tensile stress. Near the threshold, the plastic zone size could be much smaller than the grain size, and the crack plane remains relatively flat. When the grain boundary at the tip is not oriented in the plane of maximum tensile stress, the crack has to propagate in a transgranular manner by the shear unzipping process. When such a mixed mode of crack growth takes place, the fracture surface will be partly intergranular and partly transgranular as often observed. In general, the fracture mode of crack growth is much faster than the

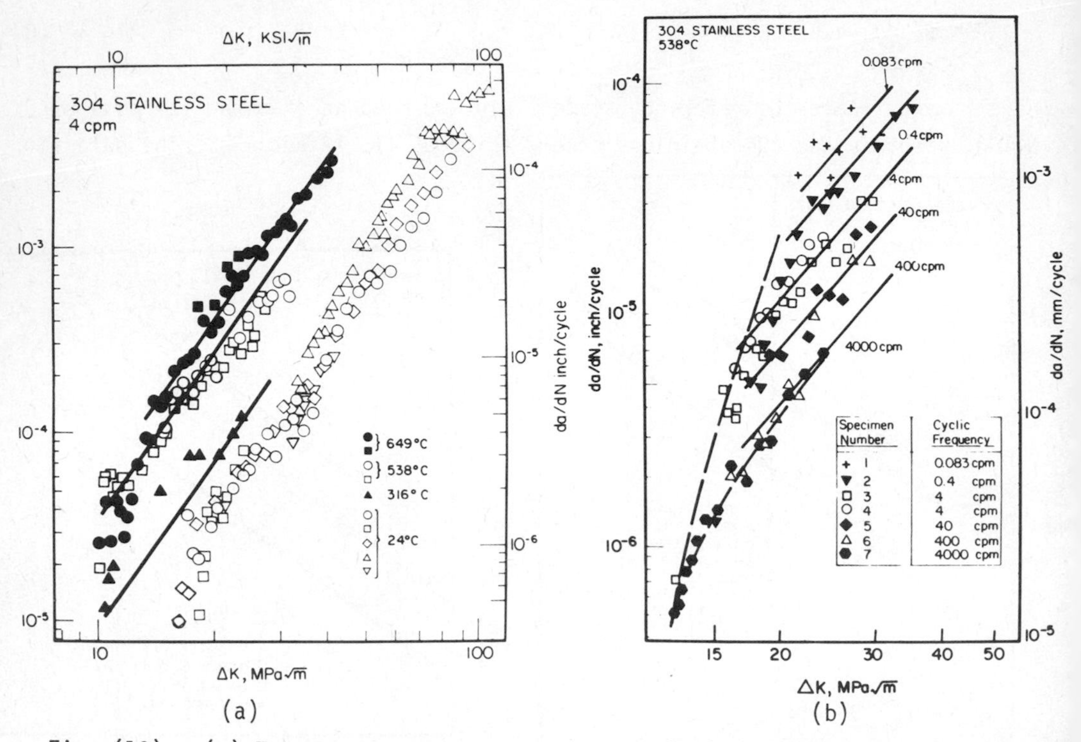

(a) (b)

Fig. (13) - (a) Temperature effects on fatigue crack growth in 304 stainless steel on air ([21]); (b) Frequency effects on fatigue crack growth in 304 stainless steel in air ([21])

deformation mode. Therefore, a crack front is not a straight but rather a wavy line. The crack front of the fracture mode runs ahead of the crack front of the deformation mode as shown in Figure 14. The local stress intensity factor at the crack front that runs ahead is lower than that of the crack front that lags behind. In Figure 14, the fatigue crack growth rate in the fracture and deformation modes are shown schematically. The middle curve is the measured crack growth rate. The crack configuration adjusts itself so the local da/dN will approach the average rate.

The crack front in the region of the fracture mode runs ahead. Consequently, the local stress intensity factor will be reduced and da/dN will decrease. While the crack front in the deformation mode of crack growth lags behind, the local stress intensity factor increases. And, as a result, the da/dN increases. To illustrate simply: we assume that the local crack growth rates in fracture and deformation modes have adjusted themselves to the average value. The corresponding local stress intensity factors for fracture and deformation modes, ΔK_f and ΔK_d, are shown.

With some simplifying assumptions, we will be able to analyze the local stress intensity factor. The fraction of crack front in the deformation mode is ϕ, and the local stress intensity factor K_d is assumed to be uniform along the segment in

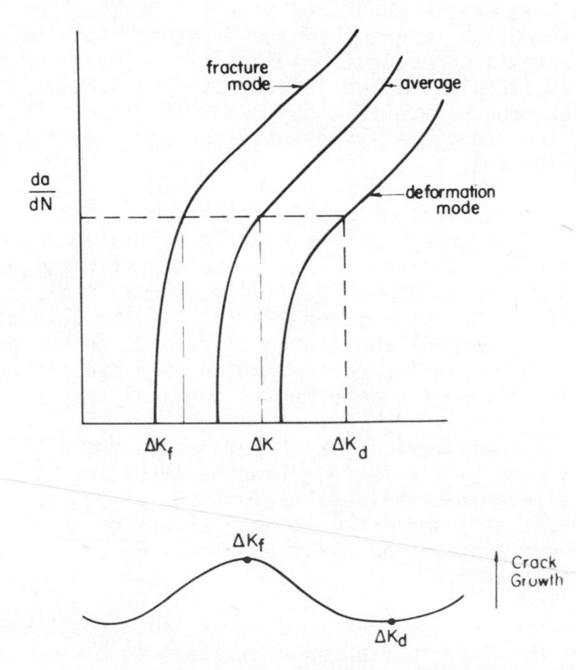

Fig. (14) - Schematical fatigue crack growth diagram showing the effects of wavy crack front

the deformation mode. This assumption is reasonable in view of the fact that the growth rate along the deformation mode crack front is reasonably constant. The fraction in the fracture mode is $(1-\phi)$, and the local stress intensity factor is K_f. With a virtual crack displacement δa, the principle of virtual work gives

$$G = \phi G_d + (1-\phi)G_f \tag{9}$$

The virtual work of a straight crack front is equal to the virtual work of the wavy crack front. G is the crack extension force of a straight crack. G_d and G_f are crack extension forces of the crack front in deformation and fracture modes respectively. Equation (9) corresponds to

$$\Delta K^2 = \phi \Delta K_d^2 + (1-\phi)\Delta K_f^2 \tag{10}$$

If ϕ is 2/3 and ΔK_f is 0.5 ΔK_d. The second term is only 12%. Thus, we have

$$\Delta K \simeq \sqrt{\phi}\Delta K_d \tag{11}$$

The constant factor of $\sqrt{\phi}$ shifts the crack growth curve in the near threshold region to the left, and the relation between ΔK_t and ΔK_{th} of crack growth in detrimental environments remains unchanged, as shown by the data in Figure 5.

DISCUSSIONS

A fatigue crack grows by the deformation or the fracture mode. If a crack grows by deformation mode and if the crack tip plastic zone size is several times larger than the grain size or the mean free path for dislocation movement, we can describe crack tip deformation in terms of the "average" tensile yield stress, and the crack growth rate is structurally insensitive. Assuming a typical grain size of 0.05 mm, the cyclic plastic zone size should be 0.20 mm to be structurally insensitive. The crack growth rate is about 0.5 x 10^{-3} times the cyclic plastic zone size; therefore, the structurally insensitive region is above da/dN = 10^{-4} mm/cycle as shown in Figure 4.

The fatigue crack growth is structurally sensitive if a crack grows by the fracture mode, such as crack growth across brittle particles and crack growth alonc embrittled grain boundaries. In the case of brittle particles, da/dN relates to th size, spacing, coherent strength of the particles, and to the strength of the inter face between the brittle particles and the matrix. In the case of grain boundaries embrittled by detrimental chemical specie, da/dN depends on the grain size and the strength of the embrittled grain boundaries, which is a function of the chemical composition of the material and the detrimental chemical specie.

When a crack grows by deformation, da/dN relates to the cyclic yield strength and is not greatly affected by the maximum tensile stress at the crack tip region. Therefore, it is speculated that the deformation mode of fatigue crack growth in the state of plane strain is insensitive to mean stress or R-ratio. Liu and McGow [22] have studied the macromechanical R-ratio effects on crack closure which are beyond the theme of this paper.

When one studies microstructural effects, only the "local" stress intensity factor is relevant to the "local" crack growth rate. According to Liu [23], the validity of the linear elastic fracture mechanics lies in the capability of the stress intensity factor to characterize the crack tip stress, strain, and displace ment fields even within the plastic zone. In other words, a given K-value uniquel corresponds to a specific state of crack tip stresses and strains. The apparent stress intensity factor often loses its capability to characterize the local crack tip field that is relevant in analyzing microstructural effects.

To illustrate the point, let us examine a two phase material, shown in Figure 9. A crack is in the weaker ferrite phase. For the sake of simplicity, we assume that the stronger martensitic phase is elastic and incapable of plastic deformatio and that both phases have the same elastic constants. Before the crack tip plastic zone reaches the interface boundary, the martensitic phase does not affect the crack tip field, and the value of the apparent K or J can characterize the crack tip field. As soon as the plastic zone reaches the interface boundary, the stronger martensitic phase constrains the crack tip deformation field in the weaker fer rite. As the crack grows further, the value of the apparent K increases, but the crack tip deformation is further constrained. Yang and Liu [20b] clearly show thi constraining effect in their analysis of the ferritic-martensitic two phase steel. As the crack grows longer, the apparent K-value and the unzipping crack increment, increase. As the crack extends further and the crack tip plastic zone reaches the interface boundary, the unzipping crack increment decreases because of the constraining effects, Figure 10. No unique relation exists between the value of the apparent K and the state of crack tip deformation. Any value of the apparent K corresponds to a great number of different crack tip fields. Therefore, the apparent K or J, as given by the conventional linear elastic fracture mechanics, fails to predict the crack growth rate.

When we use ΔK to correlate with da/dN, we use ΔK to characterize a state of crack tip field which causes the crack to grow at a specific rate. We use ΔK as an indirect correlation parameter.

When a crack grows in the structurally insensitive region, the calculated unzipping crack increment, Δa_{uz}, is the direct physical parameter of fatigue crack growth. In the structurally sensitive region, the calculated Δa_{uz}, based on the macromechanical yield stress and without the consideration of the microstructural effects on crack tip deformation process, no longer equals the crack growth rate directly and it becomes an indirect correlation parameter. However, the calculated Δa_{uz} does reflect the microstructural constraining effects. Therefore, if deformation is the mode of crack growth, we prefer to use Δa_{uz} for crack growth correlation than the apparent ΔK.

Obviously, mechanical analyses must consider the microstructural features of deformation and fracture in order to analyze the microstructural effects. The mechanical analyses that take the metallurgical features of deformation and fracture processes can be called metallurgical mechanics. This paper illustrates some of the applications.

SUMMARY

1. A fatigue crack may propagate by the deformation or fracture mode or by a mixture of the two.

2. When the deformation mode causes fatigue crack growth, the growth is structurally insensitive only if the cyclic plastic zone size is several times the grain size or several times the mean free path for dislocation movement, so that the crack tip deformation can be described in terms of macromechanical-tensile yield strength.

3. The measured rates of the deformation mode of fatigue crack growth in the structurally insensitive region agree well with theoretical calculations based on the unzipping mode.

4. When a fatigue crack grows by deformation, but the cyclic plastic zone size is in the order of grain size or less, the crack tip deformation and da/dN must be closely related to the size of the plastic zone relative to the mean free path for dislocation movement as described by Yoder et al.

5. When a fatigue crack grows by fracture mode, da/dN is structurally sensitive. In the case of crack growth across brittle particles, da/dN must be a function of the size and spacing of the brittle particles, the fracture strength of the particles, and the fracture strength of the interface between the particle and the matrix.

6. When a fracture mode propagates a fatigue crack along the grain boundaries embrittled by detrimental chemicals, da/dN must be a function of the grain size and the fracture strength of the embrittled grain boundaries.

7. We speculate that when a crack grows by the deformation mode, da/dN is insensitive to the mean stress or the R-ratio. If a crack propagates by the fracture mode, da/dN is probably sensitive to mean stress or R-ratio.

ACKNOWLEDGEMENT

The extensive discussion with Dr. George Yoder is greatly appreciated. The author acknowledges the assistance of Mr. H. Turner, Mrs. K. Szymanski and Mr. R. Ziemer in preparing this manuscript.

REFERENCES

[1] Kitagawa, H., Nishitani, H. and Matsumoto, J., Proc. 3rd Int. Congr. on Fracture, Vol. 5, Paper V-444/A, Dusseldorf, Verein Deutscher Eisenhuttenleute, 1973.

[2] Masounave, J. and Bailon, J.-P., Scr. Metall., 10, p. 165, 1976.

[3] Ritchie, R. O., J. Eng. Mater. Technol. (Trans. ASME, H), 99, p. 165, 1977.

[4] Ritchie, R. O., Met. Sci., 11, p. 368, 1977.

[5] Ritchie, R. O., "Near-Threshold Fatigue-Crack Propagation in Steels", Int. Metals Reviews, Nos. 5 and 6, p. 205, 1979.

[6] Taira, S., Tanaka, K. and Hoshina, M., in Fatigue Mechanisms, ASTM STP 675, J. T. Fong, ed., p. 135, 1979.

[7] Suzuki, H. and McEvily, A. J., Metallurgical Transactions A, 10A, No. 4, p. 475, 1979.

[8] Cooke, R. J. and Beevers, C. J., Materials Science and Engineering, 13, No. 3, p. 201, March 1974.

[9] Benson, J. P., Metal Science, 13, p. 535, September 1979.

[10] Yoder, G. R., Cooley, L. A. and Crooker, T. W., "A Critical Analysis of Grain-Size and Yield-Strength Dependence of Near-Threshold Fatigue-Crack Growth in Steels", Naval Research Laboratory, NRL Memorandum Report 4576, July 1981.

[11] Wei, R. P., Int. J. Fract. Mech., 4, p. 156, 1968.

[12] Rider, J. T. and Gallagher, J. P., J. Basic Eng., 92, p. 121, 1970.

[13] Yokobori, T., Physics of Strength and Plasticity, A. S. Argon, ed., MIT Press, Cambridge, Mass., p. 327, 1969.

[14] James, L. A. and Schwenk, E. B., Jr., Met. Trans., 2, p. 491, 1971.

[15] Neumann, V. P., "New Experiments Concerning the Slip Processes at Propagating Fatigue Cracks - I, II", Acta Met., 22, p. 1155, 1974.

[16a] Kuo, A. S. and Liu, H. W., "An Analysis of Unzipping Model for Fatigue Crack Growth", Scripta Met., 10, p. 723, 1976.

[16b] Liu, H. W. and Kobayashi, H., "Stretch Zone Width and Striation Spacing - The Comparison of Theories and Experiments", Scripta Met., 16, May 1980.

[17] Lal, D. N. and Weiss, V., "An Analysis of Non-Propagating Fatigue Cracks", Metallurgical Transactions A, 6A, p. 1623, 1975.

[18] Weertman, J., "Fatigue Crack Growth in Ductile Metals", Mechanics of Fatigue, T. Mura, ed., ASME, AMD-47, 1981.

[19] Liu, H. W. and Liu, D., "Near Threshold Fatigue Crack Growth Behavior", Script. Met., May 1982.

[20a] Taira, S., Tanaka, K. and Nakai, Y., "A Model of Crack Tip Slip Band Blocked by Grain Boundary", Mech. Res. Comm. 5(6), p. 375, 1978.

[20b] Yang, C. Y. and Liu, H. W., "The Application of Unzipping Model of Fatigue Crack Growth to a Two-Phase Steel", Int. J. Fracture, 17, No. 2, April 1981.

[21] Liu, H. W. and McGowan, J. J., "A Kinetic Analysis of High Temperature Fatigue Crack Growth, Scripta Met., 1981.

[22] McGowan, J. J. and Liu, H. W., "The Role of Three-Dimensional Effects in Constant Amplitude Fatigue Crack Growth Testing", J. of Eng. Materials and Tech., 102, p. 341, October 1980.

[23] Liu, H. W., "On the Fundamental Basis of Fracture Mechanics", to be published in Int. J. of Engineering Fracture Mechanics.

A FRACTOGRAPHIC STUDY OF CORROSION-FATIGUE CRACK PROPAGATION IN A DUPLEX STAINLESS STEEL

M. Ait Bassidi, J. Masounave

Institut de génie des matériaux, C.N.R.C.
Montréal, Québec, Canada H4C 2K3

J.-P. Bailon and J. I. Dickson

Ecole Polytechnique
Montréal, Québec, Canada H3C 3A7

ABSTRACT

A fractographic study of fatigue propagation in air and in white water indicates that the ferritic phase of KCR 171, a duplex stainless steel is susceptible to corrosion-fatigue which occurs by hydrogen embrittlement. The crystallography of the cleavage-like facets is identified. The strong tendency for crack propagation in a crystallographic direction results in twinning in the vicinity of the crack tip when the local crack propagation direction must change.

INTRODUCTION

Few detailed studies [1-5] of the fractographic aspects of corrosion-fatigue cracking have been reported. The fractographic features are generally similar to those produced during stress-corrosion cracking, which have been studied in more detail [6-11], especially for f.c.c. metals. Transgranular fracture surfaces produced during stress corrosion cracking generally have a very crystallographic aspect, suggestive of cleavage fracture even in ductile f.c.c. metals, such as austenitic stainless steels and α-brass. Moreover, as far as can be determined fractographically, opposing surfaces match perfectly [7,8,11]. These fractographic features have led to a reexamination [11,12] of whether such stress-corrosion cracking stems from an anodic corrosion process (dissolution) or from a cathodic corrosion process (hydrogen absorption). If such fracture surfaces are produced by dissolution, the dissolution must be very fine and crystallographic, Scamans and Swann [13] and Silcock and Swann [14] have concluded that their high voltage electron microscopy observations on austenitic stainless steels indicate such very fine crystallographic dissolution along {110} planes.

The objective of this paper is to present and interpret some detailed aspects of fracture surfaces produced during the corrosion-fatigue of KCR 171, an austenitic-ferritic stainless steel employed for suction press rolls in the pulp and paper industry.

EXPERIMENTAL PROCEDURE

The chemical composition and the mechanical properties of the steel employed are presented in Table 1. The microstructure consists of approximately 55% austenite, as islands within a ferritic matrix, Figure 1. The size of most of the ferritic grains varied between 1 and 4 mm.

TABLE 1 - CHEMICAL COMPOSITION AND MECHANICAL PROPERTIES OF KCR 171

Element	C	Si	Mn	P	S	Cr	Ni	Mo
wt%	0.07	1.55	0.85	0.028	0.015	24.4	8.1	1.1

Yield Stress	Ultimate Tensile Stress	Elongation
470 MPa	700 MPa	31%

Fig. (1) - Austenitic-ferritic microstructure (X 50)

Compact tension specimens (W = 50.8 mm, B = 12.7 mm) were machined from a section of cast roll, with the sample taken tangentially and oriented for crack growth in the circumferential direction. Fatigue crack propagation measurements, based on optical crack length measurements, were performed in air and in artificial white water at different temperatures and frequencies. A sinusoidal wave form and an R-ratio of 0.1 were employed. The composition of the artificial white water is given in Table 2. For the tests in white water, a plexiglass cell was attached to the sample around the fatigue precrack with a silicone glue free of acetic acid. A beaker of white water was controlled at a previously calibrated temperature on a hot plate, and white water was continuously circulated into the plexiglass cell. After stabilization, the white water in the cell was maintained within ±2.5°C of the desired temperature, which was a few degrees lower than the beaker temperature.

The fracture surfaces were studied by scanning electron microscopy, after ultrasonic cleaning in a solution of 5% HCl, 45% methanol, 50% distilled water and 0.5 gm of hexamethylene tertramine, a corrosion inhibitor. The orientation of the crystallographic fracture surfaces in the ferritic phase were determined from etch-pits [8], the sides of which correspond to {110} or {112} planes. Fig-

TABLE 2 - COMPOSITION OF ARTIFICIAL WHITE WATER

NaCl	(3.29 gm/ℓ)	10 mℓ
HCl	(.01 N)	3.16 mℓ
$Na_2S_2O_3$	(1.412 gm/ℓ)	100 mℓ
Na_2SO_4	(1.479 gm/ℓ)	100 mℓ
H_2O	(demineralized) 1ℓ	

ure 2 indicates the different etch-pit shapes possible on a {100} fracture surface.

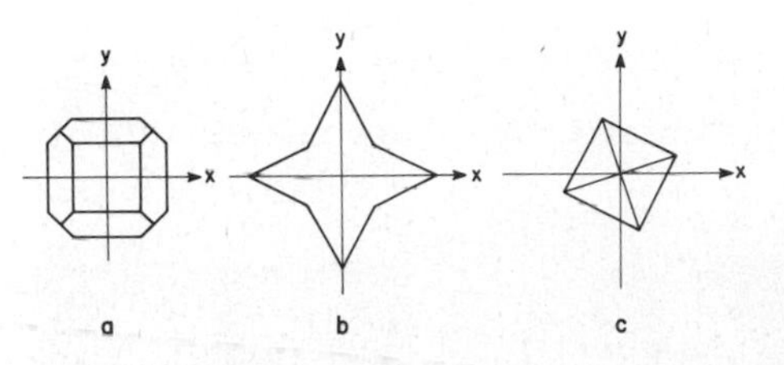

Fig. (2) - Etch-pit shapes possible on a {100} plane for sides parallel to a) {110} planes; b) 8 {112} planes and c) 4 {112} planes

RESULTS

A. Crack Rate Measurements

As seen in Figure 3, fatigue crack propagation for a cycling frequency of 1 Hertz is more rapid at 90°C in white water than at room temperature in air. The crack propagation results will be presented in detail elsewhere [15]. Faster crack propagation rates were obtained in white water at all temperatures studied (23°C to 90°C). Figure 3 also shows that the crack propagation in air is similar at frequencies of 1 and 20 Hertz.

B. Macrofractographic Observations

Visual examination of the fracture surfaces is sufficient to observe the striking fractographic difference between samples tested in white water and in air, Figure 4. Fracture surfaces formed in air have a dull aspect, with individual grains generally not distinguishable. In contrast, fracture surfaces formed in white water have a very crystallographic aspect, with the ferritic grains clearly distinguishable. The regions of the fracture surface to have been exposed the longest to the white water environment including the region precracked in air are covered with a thick porous film of corrosion product, whose thickness decreases as the exposure time decreases.

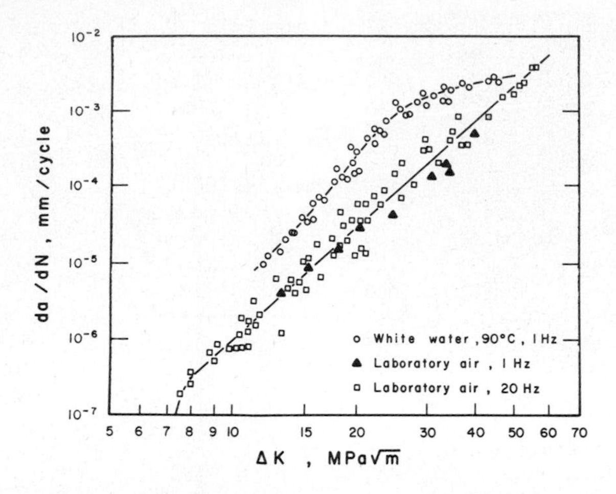

Fig. (3) - Fatigue crack growth rate curves in air and white water

Fig. (4) - Macrofractographic aspect a) in air; b) in white water

C. Microfractographic Observations - Fracture in Air

Microfractographic observations indicated that crack propagation in air produced ductile striations, with similar interstriation spacing in both phases, as shown in Figure 5. Near the edges, there is an appreciable angle which can approach 90° between the microscopic and macroscopic propagation directions, and the interstriation spacing is irregular, Figure 6, but generally larger than at mid-thickness. At all values of ΔK where ductile striations were produced in air, the occasional ferrite grain could be found presenting crystallographic facets with brittle striations and river lines, Figure 7, very similar to those observed for fatigue crack propagation in white water, which will be described next. In the austenitic phase however, the striations were still of the ductile type with an interstriation spacing approximately four times smaller than for the brittle striations.

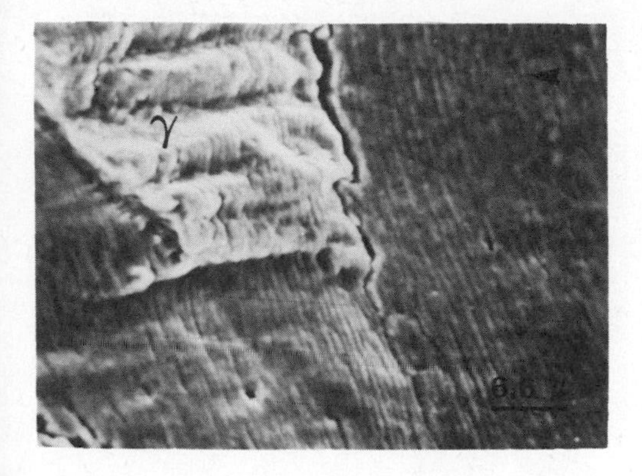

Fig. (5) - Ductile striations in air near mid-thickness (G = 1500X, ΔK = 40 MPa$\sqrt{m}$)

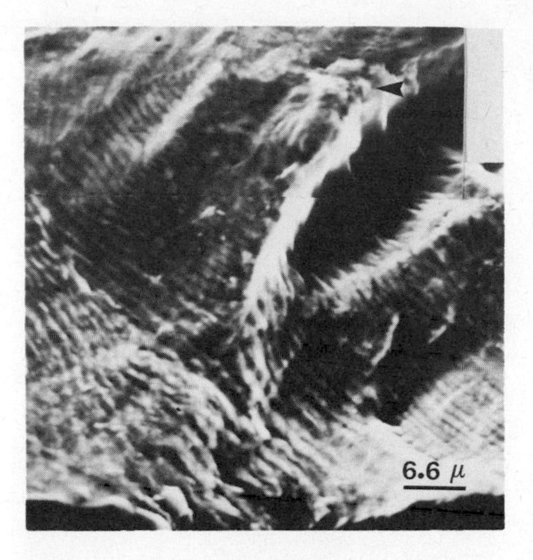

Fig. (6) - Irregularly spaced ductile striations produced in air observed at the edge (G = 1500X, ΔK = 40 MPa$\sqrt{m}$)

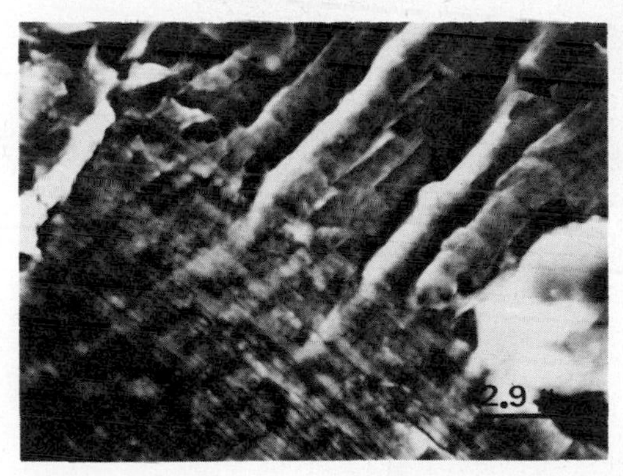

Fig. (7) - Transition from a region of brittle striations to a region of ductile striations in air (G = 3500X, ΔK = 40 MPa$\sqrt{m}$)

D. Microfractographic Observations - Fracture in White Water

Figure 8 shows typical fractographic features produced at 90°C in white water at ΔK levels of 22 and 45 MPa$\sqrt{m}$. Near the sides of the samples, Figure 8a, ductile striations are found with a variable interstriation spacing and with an important angle, often close to 90°, between the microscopic and macroscopic crack propagation directions. In this region, the fracture surface is very similar to that obtained in air. The width of the region of ductile striations obtained in white water increases as ΔK increases.

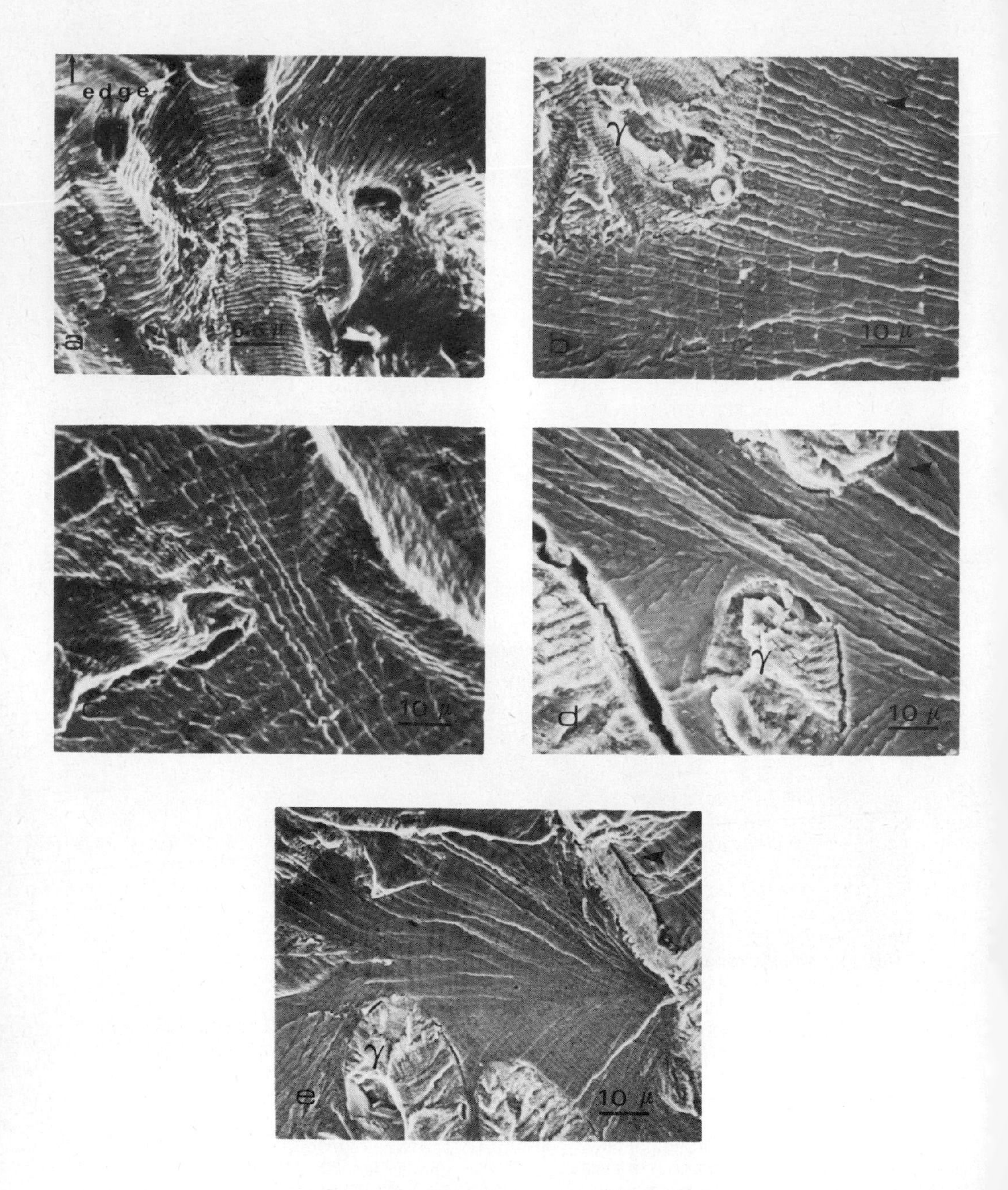

Fig. (8) - Typical fractographic features produced in white water a) near the edge, b-d) near mid-thickness. a-c) $\Delta K = 45\ MPa\sqrt{m}$; d-e) $\Delta K = 22\ MPa\sqrt{m}$

Away from the lateral surfaces, brittle striations are present in the ferritic phase, Figure 8b-8e. Depending on the relative orientation between the ferritic grain and the macroscopic crack plane, the fracture surface can have either a flat or a stepped appearance. In the islands of austenite, ductile striations are observed. The interstriation spacing in the austenitic grains is generally 4-10 times smaller than in the neighboring ferritic phase. Because of this difference in resistance to fatigue crack propagation, the crack front often bypasses austenitic grains. In such grains, crack propagation then proceeds from several initiation sites along the circumference, with the ductile striations converging towards the centre and with the interstriation spacing increasing. Before this convergence is achieved, necking and ductile tearing can occur.

The distance between the brittle striations in the ferrite is not uniform, Figures 8b-8e, and appears to be influenced principally by the distance between and the size of the neighboring austenite grains. In particular, a decrease in interstriation spacing can occur when the crack passes between two closely-spaced austenite grain, Figures 8c and 8d. As well, when brittle striations radiate from an initiation point at a grain boundary, the interstriation spacing generally increases for some distance as this initiation point is left behind, Figure 8e. A few observations suggested that at times brittle striations initiate at an austenite-ferritic interface ahead of the main crack front, e.g., Figure 8e.

At intermediate values of ΔK, a triangular pattern, Figures 9b and 10, often forms on the crystallographic fracture surface, with the base of each triangle corresponding to a brittle striation and the sides to two sets of river lines. The brittle striations become less marked as ΔK decreases, Figure 9a, and eventually disappear. At high values of ΔK, the brittle striations change direction to become parallel to the slip traces, Figure 9c, and have greater amounts of plasticity associated with them. Eventually, ductile striations appear in the ferritic phase.

Both large and fine twins can be observed on the crystallographic fracture surface. The very fine twins cause microscopic jogs in the brittle striations and occur often in regions in which the triangular pattern forms. Large twins form to permit the crack front to change direction over a larger distance within a ferritic grain. Figures 11 shows some small jogs in the brittle striations close to a larger change in direction. The presence of a fine twin, indicated by an arrow, can be seen by a change in the direction of the slip traces. Large twins form preferentially when corrosion-fatigue in a ferritic grain initiates at a point rather than along a front at the grain boundary, Figure 8e, and crack propagation must then radiate from this point, which requires several crack propagation directions. Large twins also form preferentially after crack propagation in the ferrite has passed an island of austenite, Figure 8d, 8e and Figure 12a, and a change in direction is required. Figure 9b shows a boundary of a large twin that acts as a mirror plane for slip lines (center of the fractograph) and brittle striations.

By comparing the etch-pit shapes of Figures 12 and 2, it can be seen that the crystallographic fracture facets produced in the ferrite correspond to {100} surfaces. From a study of a number of fracture facets containing triangular patterns, it was concluded that the two sets of river lines correspond to <210> directions and the brittle cleavage striations to a <110> directions. As well, slip lines which make an angle of approximately 19° with the striations are also parallel to a <210> direction and correspond to a {211} slip traces. The twins have a

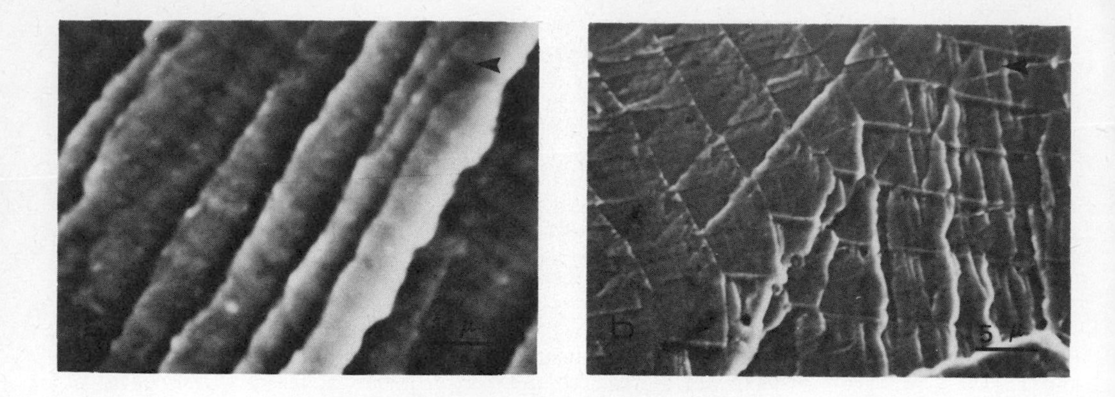

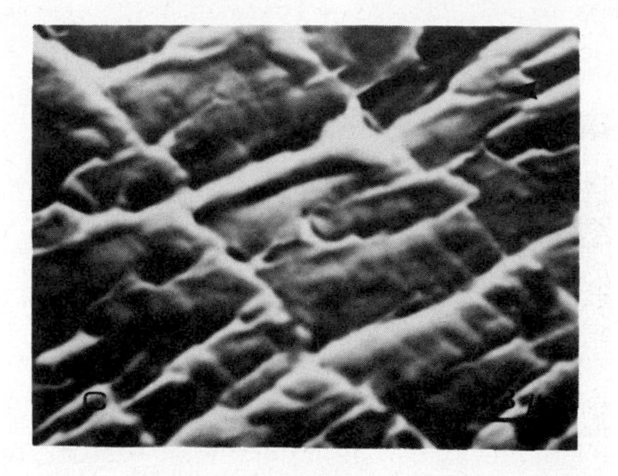

Fig. (9) - Transition in striation appearance with ΔK, a) ΔK = 18 MPa$\sqrt{m}$; b) ΔK = 45 MPa$\sqrt{m}$; c) ΔK = 50 MPa$\sqrt{m}$

{211} orientation and for the large twins containing triangular patterns one set of river lines in both the twin and the matrix are parallel to the twin boundary. The change in the orientation of the brittle cleavage striations and the slight change in orientation of the fracture plane across the twin is consistent with the twin orientation. In most facets, the angles measured between the different lines are not the ideal angles because of differences in the manner the facets studied were oriented in the scanning electron microscope.

DISCUSSION

A. Fractography - Crack Propagation in Air

Two fractographic observations obtained on the samples tested in air appear particularly interesting. The first is the occurrence of occasional corrosion-fatigue facets similar to those produced in white water. Clearly, a slight

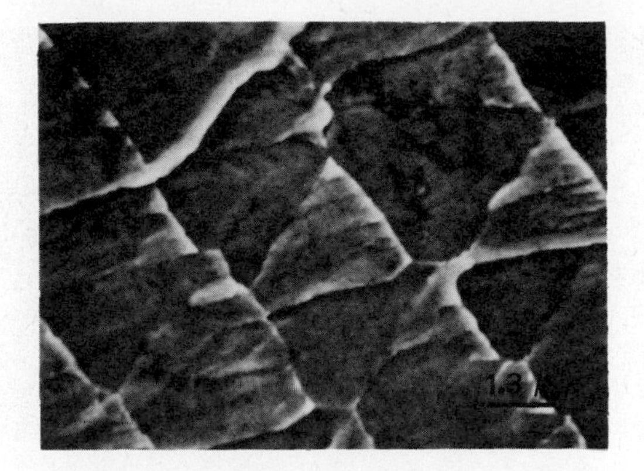

Fig. (10) - Detail of triangular pattern

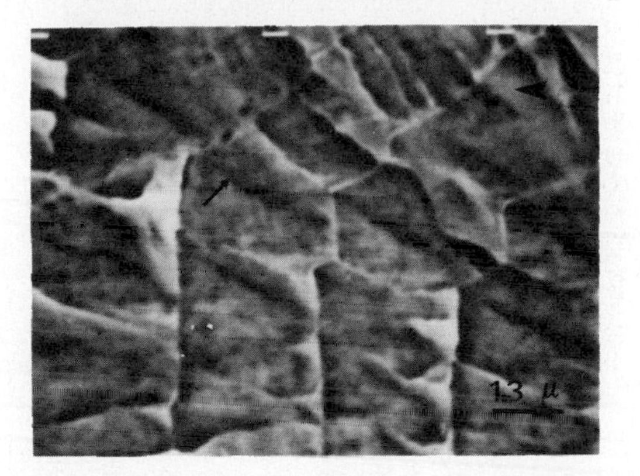

Fig. (11) - Changes in direction of the brittle striations associated with small and large twins. Note the mirror image (arrow) in the slip traces at the boundary of a small twin

amount of corrosion-fatigue in the ferritic phase can occur in laboratory air. The second is the strongly curved crack front and the aspect of the ductile striations observed near the edges of the fracture surfaces. The generally larger but more variable interstriation spacing in this region indicates that near mid-thickness the crack propagates in a continuous manner but near the edges, where the conditions tend to plane stress, it propagates discontinuously with periods of temporary crack arrest followed by periods of crack acceleration. Robin and Pelloux [16] previously observed the opposite effect on a 2214-T351 aluminum alloy. They attributed their smaller interstriation spacing near the edges to a more important crack closure effect in this region. In the present case, this more important closure effect can explain the occurrence of crack arrests, but when the crack breaks away from its pinning point, the important curvature of the crack front and the narrow width of the uncracked ligament at the edges results in rapid local propagation.

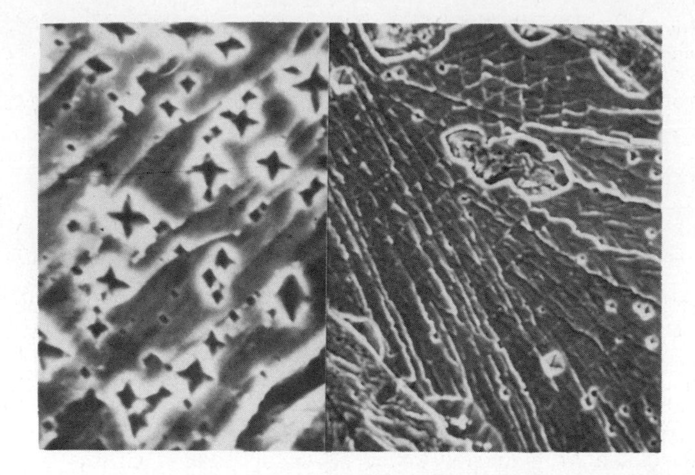

Fig. (12) - Typical etch-pit shapes observed on the ferritic fracture surface

B. Fractography - Crack Propagation in White Water

Moscovitz and Pelloux [5] previously observed the tendency for the formation of cleavage-like facets in the ferritic phase of VK-A271 during fatigue in white water. The present results show that these brittle striations are absent near the edges of the fracture surfaces. Away from the edges, where the ferritic phase is susceptible to corrosion-fatigue, unbroken ligaments of austenite can be left behind the main crack front. These then behave as individual low-cycle fatigue samples with cracks initiating at several sites and with striations converging towards the center at an accelerating rate. Final fracture of these ligaments generally occurs by a combination of necking and ductile tearing, as a result of the critical strain required for this fracture mode being surpassed.

At values of ΔK below 20 MPa $(m)^{\frac{1}{2}}$, the brittle striations generally are very faint even though the interstriation spacing in the ferrite can be important, Figure 9a. These observations are consistent with a cleavage-type fracture mechanism occurring with little plasticity. As the amount of plasticity at the crack tip increases, the brittle striations become better delineated. At a sufficiently high value of ΔK, ductile striations are obtained and the environment stops influencing the crack propagation rate. Two sets of river lines, neither of which is parallel to the crack propagation direction, are often obtained in the ferritic phase for intermediate values of ΔK, thus giving rise to the triangular patterns. At higher and lower values of ΔK, there is generally only one site of river lines and these are perpendicular to the striations.

Although the crystallography of the crack plane and propagation direction in the ferrite is in possible agreement with a decohesion by alternate slip model, originally proposed for ductile striations [17,18] and extended by Lynch [19] to some cleavage type fractures, the principal slip traces, for intermediate values of ΔK are not parallel to the striations, which clearly questions the applicabilit of this model.

The other striking fractographic observation is the deformation twinning that accompanies crack propagation. From the sites at which the large twins occur it is clear that these are produced in the vicinity of the crack tip to permit the crack propagation direction to change so as to fracture ferritic regions which could not be cracked easily otherwise. The occurrence of such twinning attests to the very strong tendency for crack propagation in a definite crystallographic direction.

C. The Corrosion-Fatigue Mechanism

There are several fractographic aspects which indicate that the corrosion-fatigue mechanism occurring in the ferritic phase is hydrogen embrittlement. The most evident is the cleavage-like fracture. The twins that form at the crack tip in order to permit the crystallographic crack propagation direction to change suggests strongly that the crack propagation in the ferritic phase occurs mechanically, and appears inconsistent with fatigue crack propagation aided by fine crystallographic dissolution. During fatigue crack propagation in laboratory air, a few ferritic grains have corrosion-fatigue facets essentially identical to those produced in white water. In air as suggested previously for iron [20,21], and other metals [1], the source of hydrogen would be associated with the oxidation of the fresh fracture surface in the presence of water vapor.

Another fractographic aspect suggesting a hydrogen embrittlement mechanism in white water are the brittle striations in the ferritic phase observed away from the edges of the specimen while ductile striations are found near the edges. This result agrees with the tendency of atomic hydrogen to concentrate in the region of high tensile hydrostatic stress ahead of the portion of the crack front that is in plane strain. The thickness of the region of ductile striation moreover increased with increasing ΔK, in qualitative agreement with the increase in thickness of the region subjected to quasi-plane stress conditions. This result, however, is also consistent with an adsorption mechanism, which requires [22] a high local value of stress at the crack tip and little ductility as favored by plane strain conditions. It also does not totally exclude a possible explanation associated with an electrolyte in contact with the crack tip which is significantly different near mid-thickness than near the sides. It is well-known in stress-corrosion cracking studies that the electrolyte in contact with the crack tip can often have a pH significantly different than that of the bulk electrolyte [23] and this transition must occur over a certain thickness. It can even be argued that the variation of the thickness of this zone with ΔK may be related to changes in electrolyte pumping conditions caused by opening and closing of the crack.

The observation of striated cleavage-like facets in the ferritic phase which initiate at an austenite-ferrite interface and apparently ahead of the main crack front also suggests hydrogen embrittlement. These observations appear to rule out the possibilities of an embrittlement by adsorption of an ion at the crack tip or of crack advance aided by fine crystallographic dissolution. Since decohesion often occurs at the austenite-ferrite interface, there remains some possibility that such initiation sites were in contact with the white water. While the combination of the different fractographic aspects strongly indicate hydrogen embrittlement of the ferritic phase, however, they do not permit to differentiate with complete certainty between an absorption or an adsorption mechanism. The decohesion observed at the ferritic-austenite interface may also be associated with a hydrogen embrittlement mechanism as this interface can be expected to act as a trap for hydrogen. This decohesion was observed less frequently at higher values of ΔK, which suggests that it is not simply produced by incompatibility between the plastic strains occurring on opposite sides of this interface.

Lastly, while there are some clear similarities between the corrosion-fatigue facets observed in the ferritic phase in KCR 171 and those produced in f.c.c. metals by stress-corrosion cracking, further work is required to determine whether the latter are also caused by hydrogen embrittlement.

CONCLUSIONS

From the present study, it can be concluded the ferritic phase of KCR 171 is embrittled by hydrogen during corrosion-fatigue in white water. The embrittlement appears to occur by absorption but an adsorption mechanism cannot be completely ruled out. It can also be concluded that the twinning observed on the fracture surfaces occurs to permit the crystallographic crack front to change directions.

ACKNOWLEDGEMENTS

Financial assistance from the Natural Sciences and Engineering Research Council of Canada and the Ministry of Education of Quebec (FCAC program) is gratefully acknowledged.

REFERENCES

[1] Meyn, D. A., Met. Trans., Vol. 2, p. 853, 1971.

[2] Stoltz, R. E. and Pelloux, R. M., Met. Trans., Vol. 3, p. 2433, 1972.

[3] Hioki, S. and Mukai, Y., ASTM STP 645, pp. 144-163, 1978.

[4] Amzallag, C., Mayonobe, B., Rabbe, P., Lieurade, H. P. and Truchon, M., Cahier d'Information Techniques, Vol. 5, p. 1015, 1978.

[5] Moscovitz, T. A. and Pelloux, R. M., ASTM STP 642, pp. 133-154, 1978.

[6] Scully, J. C., "The theory of stress corrosion cracking in alloys", NATO, Brussels, pp. 127-166, 1971.

[7] Dickson, J. I., Russell, A. J. and Tromans, D., Can. Met. Quart., Vol. 19, p. 161, 1980.

[8] Mukai, Y., Watanable, M. and Murata, M., ASTM STP 645, pp. 164-175, 1978.

[9] Liu, R., Narita, N., Altstetter, C., Birnbaum, H. and Pugh, E. N., Met. Trans., Vol. 11A, p. 1563, 1980.

[10] Marek, M. and Hochman, R. F., Corr., Vol. 29, p. 361, 1971.

[11] Bursle, A. J. and Pugh, E. N., Environment-Sensitive Fracture of Engineering Materials, edited by Z. A. Foroulis, pp. 18-47, 1979.

[12] Staehle, R. W., "Mechanisms of environment sensitive cracking of materials", Guildford Conference, The Metals Society, London, pp. 574-601, 1977.

[13] Scamans, G. M. and Swann, P. R., Cor. Sci., Vol. 18, p. 983, 1978.

[14] Silcock, J. M. and Swann, P. R., Environment-Sensitive Fracture of Engineering Materials, edited by Z. A. Foroulis, pp. 133-152, 1979.

[15] Ait Bassidi, M., Masounave, J., Bailon, J.-P. and Dickson, J. I., to be published in Can. Met. Quarterly.

[16] Robin, C. and Pelloux, R. M., Mater. Sci. Eng., 44, p. 115, 1980.

[17] Pelloux, R. M., Fracture 1969, Proceedings of the Second Int. Conf. on Fracture, Brighton, pp. 731-744, 1969.

[18] Neumann, P., Acta. Met., 22, p. 1115, 1974.

[19] Lynch, S. P., Advances in Fracture Research, ICF5 Proceedings, Pergamon Press, Vol. 1, pp. 245-254, 1981.

[20] Pelloux, R. M., La Fatigue des Matériaux et Structures, edited by S. A. Maloine, pp. 271-289, 1980.

[21] Wei, R. P., cited in [20].

[22] Liu, H. W., Private communication.

[23] Hartt, W. H., Tennant, J. S. and Hooper, W. C., ASTM STP 642, pp. 5-18, 1978.

EFFECTS OF ENVIRONMENT ON FATIGUE DEFORMATION OF IRON

D. Majumdar and Y.-W. Chung

Department of Materials Science and Engineering, Northwestern University
Evanston, Illinois 60201

ABSTRACT

Samples of commercially pure iron were fatigued in the reverse bending mode in different environments including ultrahigh vacuum, oxygen and water vapor. Fatigue in ultrahigh vacuum produced a rumpled surface (without any prominent slip line) with cracks initiated along grain boundaries. In the presence of oxygen, fatigue generated well developed slip lines. Fatigue cracks were observed more along slip lines than grain boundaries. When the same sample was fatigued in both ultrahigh vacuum and oxygen in succession, the surface structure produced was dependent on the order in which the two environments were employed. Fatigue in water vapor suppressed the surface deformation to a large extent, and gave rise to grain sliding with occasional void formation along the grain boundaries. Possible reasons for these observations are discussed.

INTRODUCTION

It has been known that the fatigue behavior of metals depends on the environment. One of the earliest studies in this context was made by Gough and Sopwith [1] who with a partial vacuum of only 10^{-3} Torr (0.133Pa) observed an increase in fatigue life of some metals. Since then, a number of studies [2-7] have been devoted to determine the environment-dependence of fatigue, and various mechanisms have been proposed. Oxygen and water vapor are some of the components of the atmosphere which have been studied extensively for their effects on fatigue [8,9]. Premature fatigue failure of some materials was attributed to oxide films or oxygen adsorption in some reports [2,5,6] whereas in others [3,4,7] water vapor was held responsible for deteriorated fatigue life. In some cases [3], however, the combined action of oxygen and water vapor was found to be most damaging to fatigue life. In general, the fatigue life of most engineering metals and alloys is improved significantly in vacuum [8]. Although of great practical importance, the determination of the nature of this phenomenon is yet to be complete. Most previous studies involving fatigue in vacuum have been performed at pressures not lower than 1×10^{-6} Torr (1.33×10^{-4} Pa). At this pressure only a few seconds are needed for the residual gas molecules to form a monolayer on the surface of the metal. This poses an unknown and uncontrollable factor in fatigue experiments. The objectives of the present research are to characterize the fatigue deformation behavior of iron under ultrahigh vacuum condition, and to study systematically the various changes brought about by different environments of oxygen and water vapor.

EXPERIMENTAL

Samples of dimension 72 x 7.5 x 0.7 mm with a gauge section, as shown in Figure 1, were machined from polycrystalline commercially pure iron (see Table 1) supplied by Inland Steel Company. Machined samples were annealed at 850°C for half an hour at a pressure of 1 x 10^{-6} Torr (1.33 x 10^{-4} Pa), polished to 0.3μ and lightly etched in 5% nital solution.

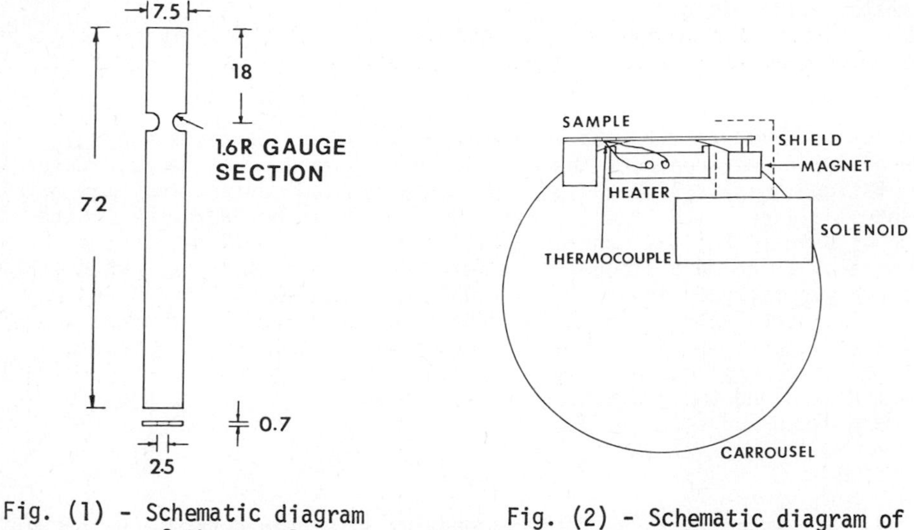

Fig. (1) - Schematic diagram of the sample

Fig. (2) - Schematic diagram of the fatigue drive

TABLE 1 - COMPOSITION OF THE COMMERCIALLY PURE IRON

Element	C	Mn	P	S	Si	O	Al	N
Wt. %	0.017	0.01	0.008	0.003	0.017	0.0156	<0.008	0.002

A schematic diagram of the electromagnetic drive used to fatigue the samples is shown in Figure 2. The sample was clamped at one end with the free end attached to a permanent magnet located in front of a solenoid coil. The coil was energized by a variable frequency AC power supply, thereby producing a reverse bending fatigue on the sample at the set frequency. The magnet was shielded by μ-metal. A tungsten filament was mounted behind the sample to facilitate heating, and the temperature was measured by a chromel-alumel thermocouple. The entire assembly was placed inside a Physical Electronics 590A scanning Auger microprobe (SAM) chamber with a base pressure less than 1 x 10^{-9} Torr (1.33 x 10^{-7} Pa). The chamber was equipped with an electron gun placed coaxially inside a cylindrical mirror analyzer, an argon ion gun, secondary electron detector and gas handling facilities.

Samples were fatigued at room temperature inside SAM at a frequency of 15 Hz and a total strain of ±0.09% measured at the gauge section. The natural oxide on the sample surface could be removed prior to fatigue by repetitive cycles of sputtering for 20 minutes using a 2keV argon ion beam and subsequent annealing at 450°C for 15 minutes. The different environments in which samples were fatigued included ultrahigh vacuum (UHV) and various pressures of oxygen and water vapor. The surface deformation could be observed in situ by operating the SAM in the secondary electron detection mode or ex situ under a standard scanning electron microscope.

RESULTS AND DISCUSSIONS

The impurities detected on the surface of a typical sample in the as-polished condition were carbon, oxygen and chlorine, as shown in Figure 3.

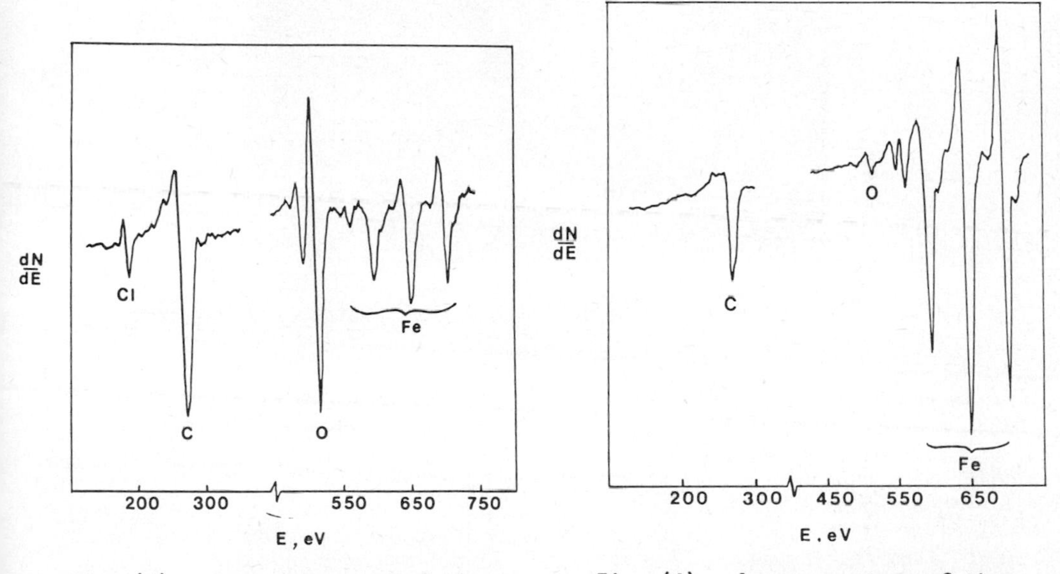

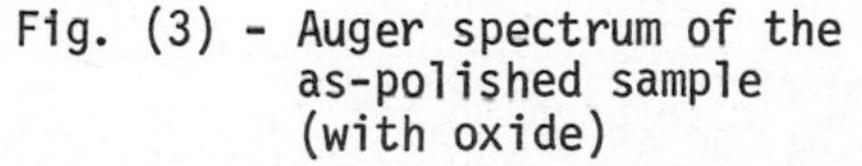

Fig. (3) - Auger spectrum of the as-polished sample (with oxide)

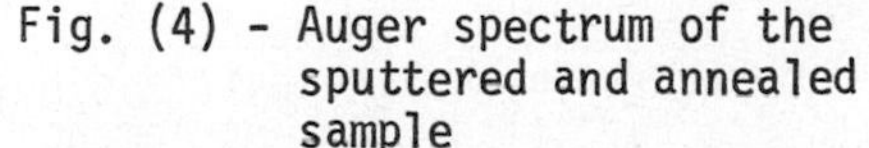

Fig. (4) - Auger spectrum of the sputtered and annealed sample

Presence of such impurities on iron samples has been reported before [10,11]. All these impurities except carbon were reduced to traces through repetitive cycles of ion-sputtering and annealing, as indicated by Figure 4. Although, not known for certain, the persistent carbon peak could accrue from some impurities in the argon gas used for sputtering.

A. Fatigue in UHV

Samples fatigued in UHV in the as-polished condition (with the natural oxide on) showed a generally rumpled surface without any prominent slip line, as shown in Figure 5. Surface rumpling observed during fatigue in vacuum has been reported in literature [12]. This rumpling may come from numerous fine slip lines distributed evenly throughout the grain.

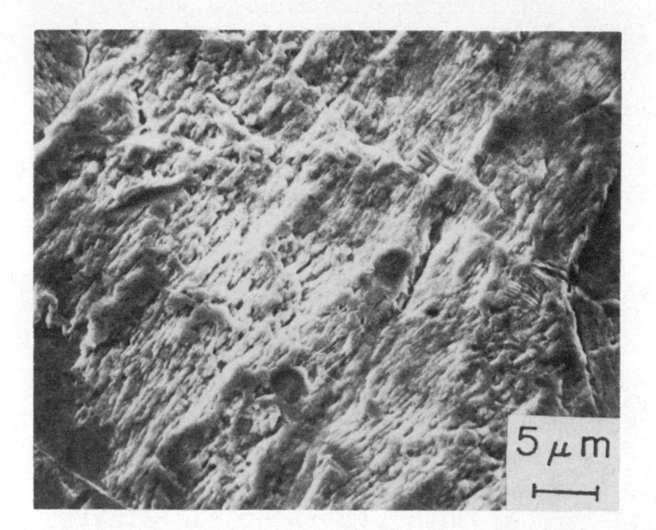

Fig. (5) - As-polished sample (with oxide) fatigued in UHV for 9×10^4 cycles. Notice the surface rumpling.

The removal of the natural oxide from the sample prior to fatigue did not change the post-fatigue appearance of the surface. Figure 6 shows the surface of one such oxide-free sample after UHV fatigue, revealing a rumpled structure similar to Figure 5. It was therefore concluded that there was no significant difference in the surface structures of samples with and without the natural oxide after UHV fatigue. Hence, in subsequent studies, the oxide was not removed from the sample surface for experimental convenience.

Contrary to the present findings, Grosskreutz [13] observed slip suppression due to dislocation entrapments underneath the surface oxide during fatigue of aluminum in UHV. The apparent discrepancy could, however, be resolved from the following consideration. The trapping of dislocations under the oxide layer is dependent on the elastic properties of the oxide and the substrate as well as the thickness of the oxide [13]. So, the reason for no slip suppression in the present case could be twofold: (1) the natural oxide on the iron samples was too thin (~30Å as estimated by Auger depth profile technique) to provide any effective barrier to the dislocations emerging through the surface, and/or (2) the Young's modulus of the oxide was not high enough (relative to that of the metal) to provide a strong repulsive force on the dislocations.

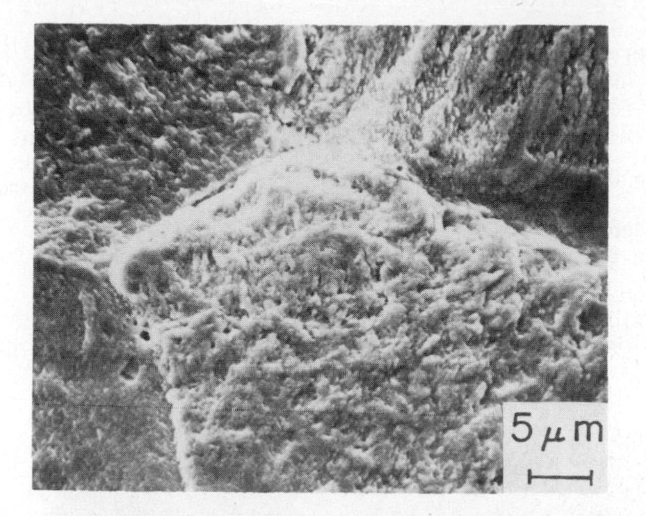

Fig. (6) - Oxide-free sample fatigued in UHV for 9 x 10^4 cycles. Surface shows similar rumpling as in Figure 5.

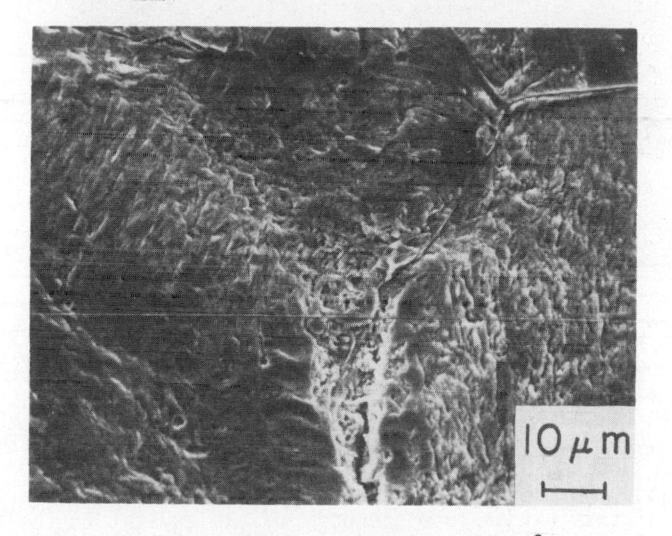

Fig. (7) - Sample fatigued in UHV for 1 x 10^6 cycles. Notice the surface rumpling and the grain boundary crack.

Another important aspect of fatigue on UHV is the formation of cracks along grain boundaries, as one shown in Figure 7. Transgranular cracks were rarely found under UHV conditions. It is believed that the fine slip lines forming the rumpled surface led to a uniform distribution of plastic deformation throughout the grain. This "slip homogenization", as mentioned by Verkin and Grinberg [14], would eliminate the possibility of high stress concentrations along the slip lines and thereby impede the formation of slip line cracks. Under such conditions, grain boundary cracks would form with accumulating plastic strain before slip line cracks. This may explain the present observation of cracks initiated along grain boundaries under UHV conditions.

B. Fatigue in Oxygen

In an oxygen environment, fatigue produced well developed slip lines, as shown in Figure 8, instead of a rumpled surface. Oxygen pressure as low as 2×10^{-6} Torr (2.66×10^{-4} Pa) was sufficient to cause noticeable slip steps to appear on the surface, as indicated in Figure 9. The crack morphology also changed in the oxygen environment. An increasing fraction of fatigue cracks was initiated along slip lines, like a few shown in Figure 10.

Fig. (8) - Sample fatigued in 5×10^{-2} Torr of oxygen for 1×10^{6} cycles. Prominent slip lines are formed during fatigue.

The initiation of cracks along slip lines in the presence of oxygen may occur in a way suggested by Thompson et al. [2]. According to this mechanism, a freshly exposed surface in the slip band adsorbs oxygen during one half cycle of the fatigue test. On the following half cycle, slip in the reverse sense draws some of this oxygen into the body of the crystal in the form of dissolved atoms. Repetition of this process builds up a high oxygen concentration near the slip band thus weakening the crystal and giving rise to a crack there. Alternatively, the formation of an oxygen layer on an exposed slip surface may prevent the rewelding of slip steps during the reverse half cycle thus initiating micro-cracks along slip lines. The observation of slip line cracks in oxygen in the present work can be rationalized by either one of these mechanisms.

The proposed rewelding mechanism has been further studied by many investigators. Laird and Smith [15] performed tension-compression fatigue on copper

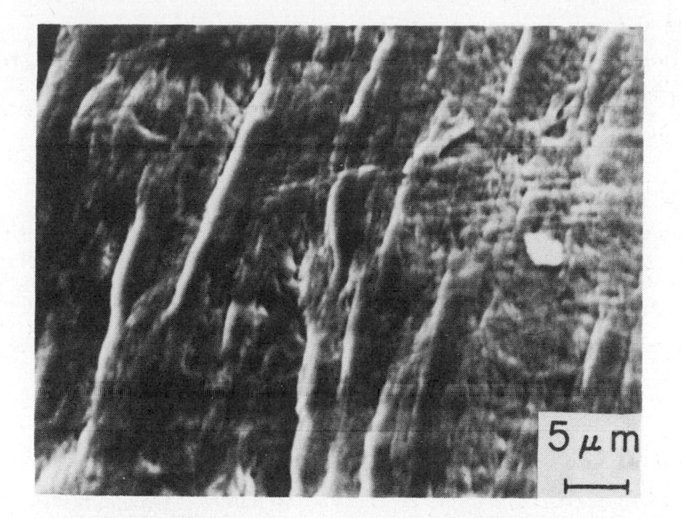

Fig. (9) - Sample fatigued in 2 x 10^{-6} Torr of oxygen for 1 x 10^{6} cycles. Well-developed slip steps can be observed over a rumpled background.

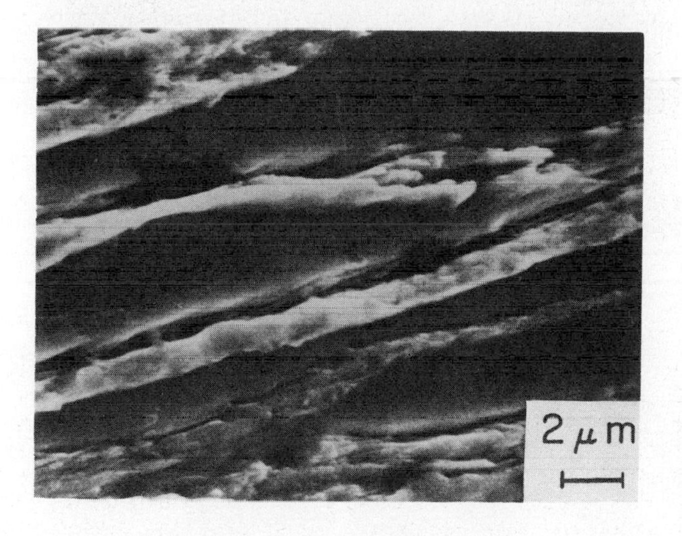

Fig. (10) - Sample fatigued in 1 x 10^{-2} Torr of oxygen for 1 x 10^{6} cycles. Notice the slip line cracks.

and aluminum under a pressure of 5 x 10^{-6} Torr (6.65 x 10^{-4} Pa). They did not observe any evidence of welding of the cracks after stopping the test during the compression cycle. However, working with a lower pressure (1 x 10^{-6} Torr (1.33 x 10^{-4} Pa)), Martin [16] found some indication of possible reweldment of fatigue cracks in stainless steel, copper and aluminum. More recently, Miyauchi et al. [17] conducted tension-compression and tension-tension modes of fatigue on commercially pure aluminum in different environments. They found that the environmental effect on fatigue life is larger in the tension-compression loading as compared to the tension-tension one. These results suggest some sort of welding crack surfaces during the compressive cycles of fatigue. The rewelding mechanism is also consistent with the present observation that a low oxygen pressure (2 x 10^{-6} Torr (2.66 x 10^{-4} Pa)) could cause well-developed slip steps on the surface since small amounts of gas adsorbates can affect the surface energy which in turn determines rewelding of two surfaces. One can show that at a cycling frequency of 15 Hz and under an oxygen pressure of 2 x 10^{-6} Torr (2.66 x 10^{-4} Pa), about 2% of a fresh surface in a crack will be covered with oxygen every cycle. Such an oxygen coverage may significantly change the surface energy and influence the deformation behavior, as it did in the present case. As mentioned before, many fatigue studies in vacuum have been performed in the past at pressures not lower than 1 x 10^{-6} Torr (1.33 x 10^{-4} Pa), commonly achieved through a diffusion pump. The present work indicates that such a pressure may not be low enough to reveal the true fatigue characteristics in the absence of any gaseous medium.

C. Fatigue in Mixed Environments

Interesting results were obtained when the same sample was fatigued in both oxygen and UHV in succession. As shown in Figure 11, a mixture of

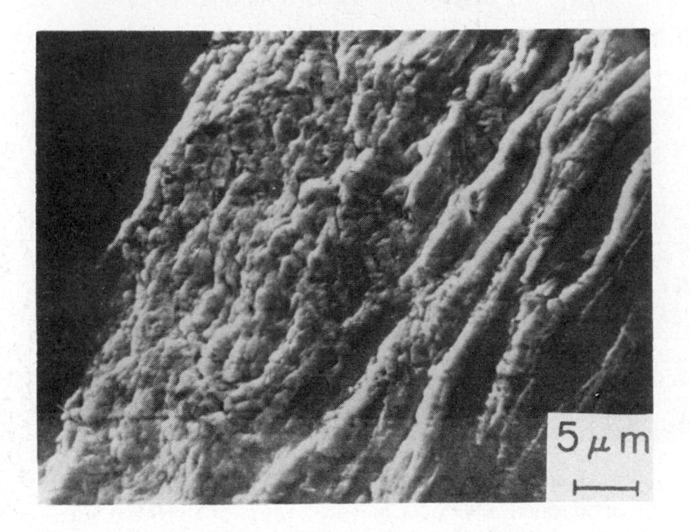

Fig. (11) - Sample fatigued first in 5 x 10^{-2} Torr of oxygen for 9 x 10^4 cycles and then in UHV for another 9 x 10^4 cycles. Observe the mixture of surface rumpling and slip lines.

surface rumpling and prominent slip lines appeared on the surface when the sample was fatigued first in 5×10^{-2} Torr (6.65 Pa) of oxygen and then in UHV for the same number of cycles. But if the order of the two environments was reversed, i.e. first half of fatigue in UHV and the following half in 5×10^{-2} Torr (6.65 Pa) of oxygen, the sample showed a predominantly rumpled appearance with much fewer and less prominent slip lines, as shown in Figure 12. In the

Fig. (12) - Sample fatigued first in UHV for 9×10^4 cycles and then in 5×10^{-2} Torr of oxygen for another 9×10^4 cycles. It shows a rumpled surface with faint slip lines.

latter case, the fatigue cracks were also found mostly along grain boundaries as in the case of UHV fatigue. These results indicate that the surface deformation depends on the order in which the environment is changed. Such a dependence may arise if two different types of dislocation configuration, one being energetically more favorable than the other, are responsible for the surface structure produced in the two environments. In that case, a change in the environment will cause a change in the dislocation structure only if it is from the less favorable to the more favorable one but not vice versa. Some TEM work is in progress for a better understanding. Although this hypothesis is yet to be confirmed, these results are of practical importance. They show that the fatigue damage produced in one environment is dependent on the prior fatigue history of the sample. In the present case, the surface structure produced in UHV did not change much during subsequent fatigue in a reactive medium like oxygen. This suggests a possible means of reducing fatigue damage of machine components in a corrosive medium, by subjecting them to prior fatigue treatments in inert environments like ultrahigh vacuum.

D. Fatigue in Water Vapor

Preliminary experiments were carried out at room temperature to study the effects of water vapor on fatigue deformation behavior of iron. Fatigue in an environment of 0.6 Torr (79.8 Pa) of water vapor produced a surface structure as shown in Figure 13. The following features are apparent: (1) the

Fig. (13) - Sample fatigued in 0.6 Torr of water vapor for 1×10^5 cycles. Note the reduction in surface deformation, grain boundary sliding and void formation.

overall surface deformation is much reduced when compared with UHV fatigue (Figure 5); (2) grain boundary sliding could be detected and (3) the nucleation of several voids could be observed along the grain boundaries.

Thermodynamically [18] water vapor can react with the clean iron surface to produce hydrogen. In fact, evolution of hydrogen during fatigue of carbon steel in moisture has already been reported [12].. It is this hydrogen which is believed to be responsible for the fatigue characteristics observed in water vapor.

Effects of hydrogen on the mechanical behavior of iron have been extensively studied. Several investigators [19-21] observed hydrogen-induced hardening effects in iron. In the present case, a similar hardening effect may take place, manifesting itself in the observed suppression of plastic deformation.

The observed grain boundary sliding and void formation could also be direct effects of the hydrogen released from the reaction between water vapor and iron. Hydrogen atoms, which could be readily transported to grain boundaries by dislocations [22], are believed [23,24] to reduce the cohesive energy of grain boundaries. This makes grain sliding a likely phenomenon during fatigue. Also, the coalescence of hydrogen atoms and/or hydrogen-bearing dislocations in the grain boundaries could give rise to the observed voids. Such void formation due to hydrogen has been reported [25].

In general, preliminary results suggest that the main effect of water vapor on fatigue of iron is by the way of incorporating hydrogen into the metal. More work is in progress for better understanding of the phenomenon.

CONCLUSIONS

(1) Fatigue of iron under UHV causes rumpling of the surface (with no prominent slip lines) with cracks formed along grain boundaries. The removal of the natural oxide does not change the post-fatigue appearance of the surface.

(2) Fatigue in oxygen produces well-developed slip lines. Cracks are initiated along slip lines instead of grain boundaries.

(3) In case of fatigue in both UHV and oxygen, the surface deformation depends on the order in which the environments are administered.

(4) Fatigue in water vapor reduces surface deformation, causes grain sliding and gives rise to voids along grain boundaries.

ACKNOWLEDGEMENTS

The authors gratefully acknowledge numerous discussions with Drs. S. Bhat and R. Cline of Inland Steel Research Laboratory and Professor M. E. Fine, Mr. C. V. Cooper and Mr. K. S. Shin of Northwestern University. This work was supported by NSF.

REFERENCES

[1] Gough, H. J. and Sopwith, D. G., J. Inst. Metals 52, p. 55, 1935.

[2] Thompson, N., Wadsworth, N. and Louat, N., Phil. Mag., 1, p. 113, 1956.

[3] Wadsworth, N. J. and Hutchings, J., Phil. Mag., 3, p. 1154, 1958.

[4] Broom, T. and Nicholson, A., J. Inst. Metals, 89, p. 183, 1960.

[5] Snowden, K. U., Acta Met., 12, p. 295, 1964.

[6] Shen, H., Podlaseck, S. E. and Kramer, I. R., Acta Met., 14, p. 341, 1966.

[7] Bradshaw, F. J. and Wheeler, C., Appl. Mater. Res., 5, p. 112, 1966.

[8] Duquette, D. J., Fatigue and Microstructure, p. 335, ASM, Ohio, 1978.

[9] Achter, M. R., ASTM STP 415, p. 181, Am. Soc. Testing Mats., 1967.

[10] Leygraf, C. and Ekelund, S., Surface Sci., 40, p. 609, 1973.

[11] Simmons, G. W. and Dwyer, D. J., Surface Sci., 48, p. 373, 1975.

[12] Hudson, C. M. and Seward, S. K., Eng. Fract. Mech., 8, p. 315, 1976.

[13] Grosskreutz, J. C., Surface Sci., 8, p. 173, 1967.

[14] Verkin, B. I. and Grinberg, N. M., Mat. Sci. Eng., 41, p. 149, 1979.

[15] Laird, C. and Smith, G. L., Phil. Mag., 8, p. 1945, 1963.

[16] Martin, D. E., Trans. ASME, 87 (D), p. 850, Am. Soc. Mech. Engnrs., 1965.

[17] Miyauchi, J., Nishioka, K., Fujiwara, H., Nakamura, S. and Komatsu, T., Scripta Met., 11, p. 617, 1977.

[18] Gaskell, D. R., Introduction to Metallurgical Thermodynamics (2nd. Ed.), p. 586, McGraw-Hill Book Co., New York.

[19] Asano, S. and Otsuka, R., Scripta Met., 10, p. 1015, 1976.

[20] Tobe, Y. and Tyson, W. R., Scripta Met., 11, p. 849, 1977.

[21] Oguri, K. and Kimura, H., Scripta Met., 14, p. 1017, 1980.

[22] Donovan, J. A., Met. Trans., 7A, p. 1677, 1976.

[23] Shin, K. S. and Meshii, M., to be presented in 3rd Int. Congress on Hydrogen and Metals, Paris, France.

[24] Oriani, R. A. and Josephic, P. H., Acta Met., 22, p. 1065, 1974.

[25] Oriani, R. A. and Josephic, P. H., Acta Met., 27, p. 997, 1979.

MICROSTRUCTURAL PROCESSES PRIOR TO AND DURING FATIGUE SOFTENING OF STRUCTURAL STEELS

H. Veith

Bundesanstalt für Materialprüfung (BAM)
Unter den Eichen 87, D-1000 Berlin 45 FRG

ABSTRACT

Induced by local stress concentration and/or favored grain orientation a few grains become already plastified in the first cycles. The mobility of the free dislocations seems to be controlled by the thermally activated overcome of obstacles. The process of fatigue softening in normalized structural steels by stress amplitudes below the yield strength is explained by the stress induced diffusion of pinning point atoms in the pipe of pinned dislocations. By the components of the line tension the pinning point atoms will move in direction of smaller free dislocation segments as well as in direction of the applied shear stress. So critical free dislocation segments will be produced in the course of cycling and FRANK-READ sources can be activated even by stress amplitudes far below the yield strength. Larger oscillations of pinning point atoms with the dislocation in direction of the applied shear stress mark the experimental observed transition from a time dependent start of softening to a cycle dependent beginning with increasing temperature and/or decreasing cyclic frequency. The rise of the deformation amplitude during softening under stress cycling is caused by the successive and accelerated plastification of originally elastic grains, starting on grains already plastified in the first cycles.

INTRODUCTION

In normalized unalloyed structural steels during cycling with stress amplitudes below the initial yield strength a softening process takes place in the first stage of fatigue. Under constant stress amplitude the process starts by an increase of the deformation amplitude after a characteristic number of cycles which rises with decreasing stress amplitude [1-3]. Figure 1 demonstrates the effect of fatigue softening on the German standard steel St 37-3 at room temperature and a load frequency of 0.1 cps.

The softening process is caused by the successive plastification of originally elastic grains and therefore seems to be fundamental for the failure of structural steels under cyclic loading at stress amplitudes below the original yield strength. A few grains seem to originate this process by becoming already plastic in the first cycles as a consequence of local stress concentrations and being assisted by favored grain orientation [2,4].

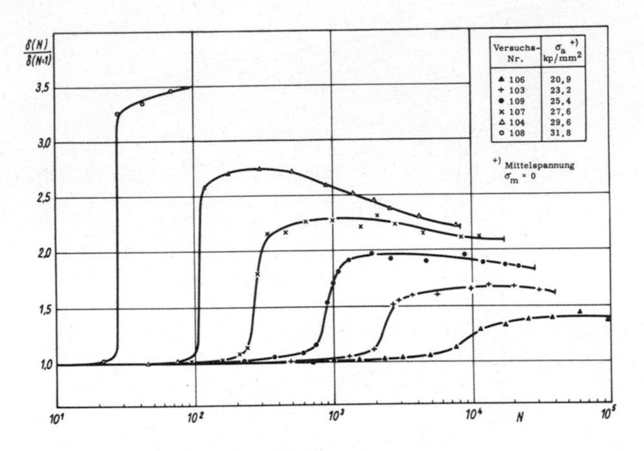

Fig. (1) - Fatigue softening under constant stress amplitude, steel St 37-3

MICROSTRUCTURAL PROCESSES PRIOR TO FATIGUE SOFTENING

In the cycle interval before softening starts no change of the quasi elastic macroscopic behavior is observed. However, microplasticity is already observed within the first stress cycles, when strain measurements were performed using strain gauges of a registration sensitivity for the plastic strain amplitude of 10^{-6}. Thus the micromechanical behavior in dependence on various parameters can be presented by the microplastic strain amplitude ε_{pa} as a function of the stress amplitude σ_{am}. Functions $\varepsilon_{pa} = f(\sigma_{am})$ can experimentally be verified by increasing σ_{am} step by step from cycle to cycle starting at zero stress amplitude. Typical results for the change of $\varepsilon_{pa} = f(\sigma_{am})$ in the course of stress cycling up to the start of fatigue softening of structural steels are illustrated in Figure 2 for the German standard steel Ck 10 at -43°C, a load frequency of 1 cps and a stress amplitude of $\sigma_a = 280$ Nmm^{-2}. The first microplastic strain in the order of 10^{-6} is observed above a stress amplitude σ_{am} of about 30 Nmm^{-2}. At the beginning of stress cycling, especially at lower temperature, an increase of σ_{am} above 30 Nmm^{-2} results only in a small amount for ε_{pa} in the range of 5 to 10 $\cdot 10^{-6}$. But when σ_{am} exceeds the limit of about 200 Nmm^{-2}, ε_{pa} increases again with rising stress amplitude. The determination of $\varepsilon_{pa} = f(\sigma_{am})$ in dependence on the number of stress cycles N demonstrates that the range of σ_{am} where only a very small increase of ε_{pa} is observed, decreases successively. Therefore, for instance, before stress cycling (N = 0) the plastic strain amplitude ε_{pa} is about $5 \cdot 10^{-6}$ at

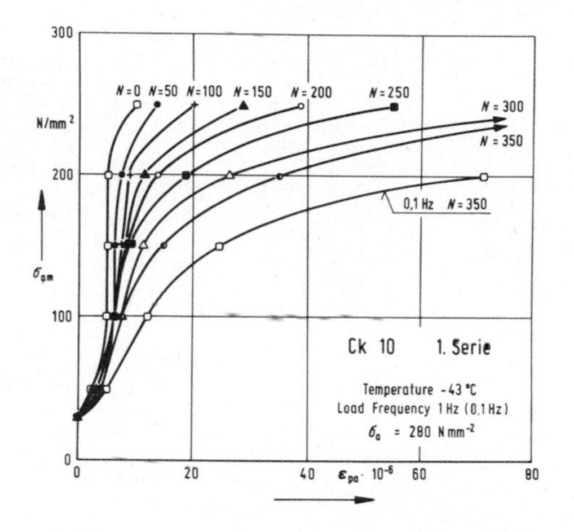

Fig. (2) - Microplastic behavior in dependence on the number of stress cycles N, steel Ck 10

σ_{am} = 200 Nmm^{-2}, and after 250 cycles under a stress amplitude of 280 Nmm^{-2} ε_{pa} has been increased to $20 \cdot 10^{-6}$ at the same level of σ_{am}.

The softening process observed in the microplastic range seems to be influenced also by the cycling frequency as a change of the frequency from 1 cps to 0.1 cps at N = 350 demonstrates in Figure 2. So a decrease of the frequency by a factor of 10 results in an increase of ε_{pa} by a factor of about 2. From Figure 3 it can be seen that a similar softening effect can be achieved by raising the temperature. The deformation barrier of about 200 Nmm^{-2} at low temperature will be reduced continuously with increasing temperature. So the change of the shape of the curve $\varepsilon_{pa} = f(\sigma_{am})$ caused by a temperature elevation from -43°C up to 390°C at 1 cps cycling frequency is comparable to that within the first 100 stress cycles under a stress amplitude of 280 Nmm^{-2} at -43°C and 1 cps frequency, as the relevant curve of Figure 2 and Figure 3 demonstrates.

In normalized unalloyed structural steels the dislocations are pinned by interstitial atoms such as C or N which might be concentrated in clusters [3-8]. Hence by stress amplitudes above the fatigue limit, but below the original yield strength, only the free parts of the dislocations between the pinning points can move over small distances by bowing out. By local stress concentrations or by favored grain orientation free glide dislocations will be generated in a few grains within the first stress cycles [3,9]. But a similar arrangement of C- or N-clusters in the matrix as assumed along the pinned dislocations allows the free dislocation in the plastified grains also to move only between the clusters in the matrix at low stress amplitudes. Thus ε_{pa} observed at low σ_{am} before macroscopic softening starts, should be produced mainly by oscillations of free dislocation parts between pinning points [10].

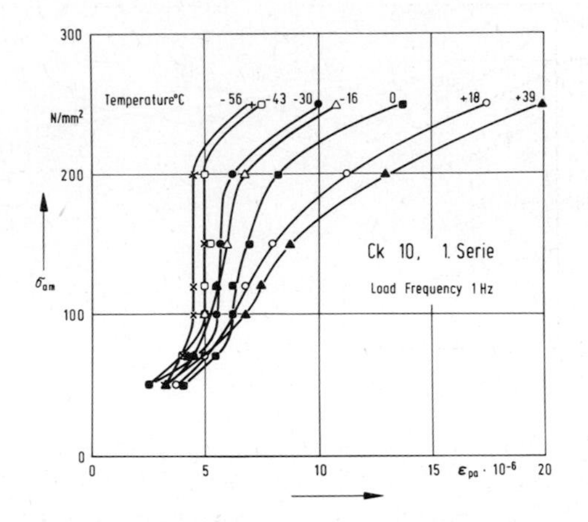

Fig. (3) - Microplastic behavior in dependence on temperature, steel Ck 10

Assuming an average distance of pinning points in the range of 150 b up to 200 b [3,6] and a grown-in dislocation density of about 10^8 cm^{-2} a plastic strain amplitude of about $5 \cdot 10^{-6}$ will be produced. This result is in a quite good agreement with ε_{pa} in Figure 2 at σ_{am} below the barrier of about 200 Nmm^{-2}. Thus this low temperature deformation barrier seems to be controlled by the restricted mobility of the pinned grown-in dislocations.

As it is demonstrated in Figure 2 the barrier can be overcome by an applied stress about 200 Nmm^{-2} and then a reduction of the barrier occurs in the further course of cycling. The rise of deformation amplitude up to 10^{-4} should primarily be caused by the movement of free dislocations in grains already plastified in the first stress cycles, because the height of the barrier is far below the yield strength and the threshold for macroscopic softening at -43°C and 1 cps cycling frequency as it can be seen from Figure 4. Already 10% plastified grains having a density of mobile dislocations of about 10^8 cm^{-2} will produce a plastic strain amplitude of about 10^{-4} if the glide dislocations can move over a distance in the order of the grain radius. In order to obtain this the glide dislocations have to overcome obstacles such as interstitial clusters and forest dislocations. The low plastic strain amplitude of about $5 \cdot 10^{-6}$ below the deformation barrier should indicate that the dislocation motion in the plastified grains will be mainly controlled by obstacles of clusters as they have an essentially higher density than that produced by the forest dislocation [6,8,10]. Hence from Figure 2 it would follow that cutting of particles by glide dislocations seems to be possible above a stress of 200 Nmm^{-2} at -43°C and 1 cps cycling frequency. As the dislocations moving

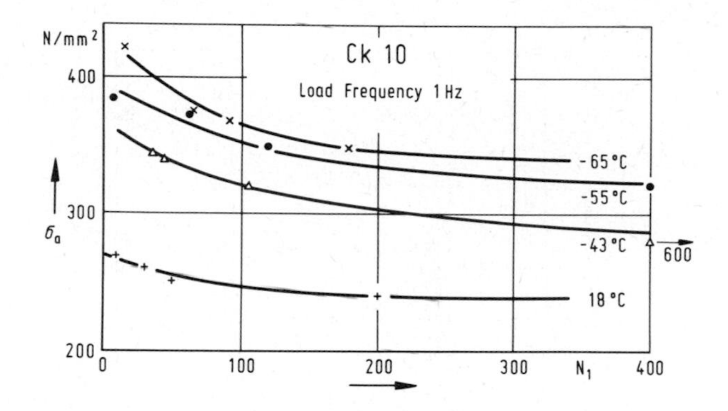

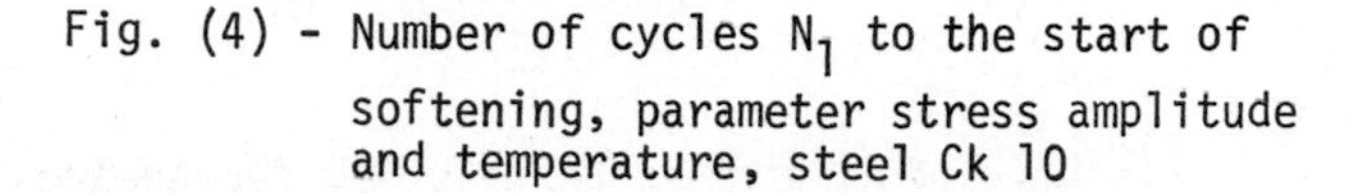
Fig. (4) - Number of cycles N_1 to the start of softening, parameter stress amplitude and temperature, steel Ck 10

through the particles reduce the resistance of the obstacles proportionally to the number of passing dislocations the deformation barrier will be successively removed in the course of cycling, as can be seen from Figure 2. The barrier to dislocation motion can also be lowered by increasing the temperature or by decreasing the strain rate because cutting of particles is a thermally activated process. This behavior is observed in the experiment by the weakening of the deformation barrier as the decrease of the deformation resistance with increasing temperature in Figure 3 and the change of the load frequency from 1 to 0.1 cps at N = 350 in Figure 2 demonstrate.

START OF FATIGUE SOFTENING

To elucidate the origin and the start of fatigue softening the question has to be raised how a major plastification process in originally elastic grains can occur by applied stress amplitudes far below the initial yield strength [5]. During fatigue cycling the corresponding alternating shear stress in the slip planes of pinned dislocations will bow out the free dislocation segments between the pinning points formed by C- or N-clusters. So the pinning points experiences individual forces by the line tension of the dislocation as is demonstrated schematically for the i-th pinning point between two free dislocation segments L_{ci} and L_{ci+1} in Figure 5. The superposition of the force components in x- and y-direction results in a y-component F_{yi} alternating within the limits $\pm[F_{yi}]_{max}$ and a x-component F_{xi} changing between zero and $[F_{xi}]_{max}$ and is always directed towards the smaller dislocation segment. Thus in principle the pinning point atoms will oscillate in the direction of the acting shear stress vector and are forced to move towards the smaller dislocation segment in direction perpendicular to the applied shear stress vector, if their mobility and the cycling period are large enough. The drift rate of the atoms can be verified by the EINSTEIN-relation and so the change of the length of the dislocation segments is evaluated by integration with incremental time steps over the time of cycling.

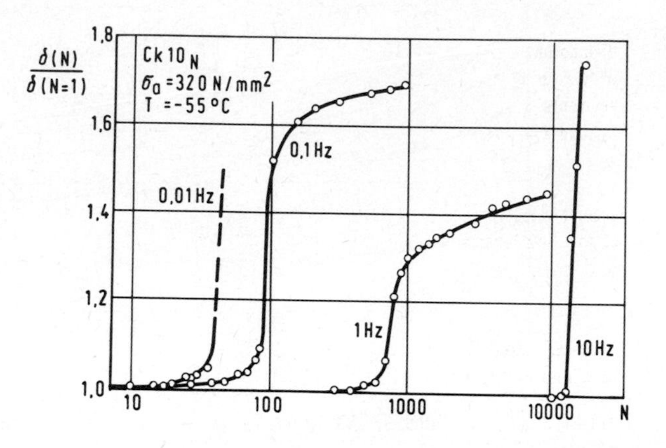

Fig. (10) - Influence of frequency on the start of fatigue softening, steel Ck 10

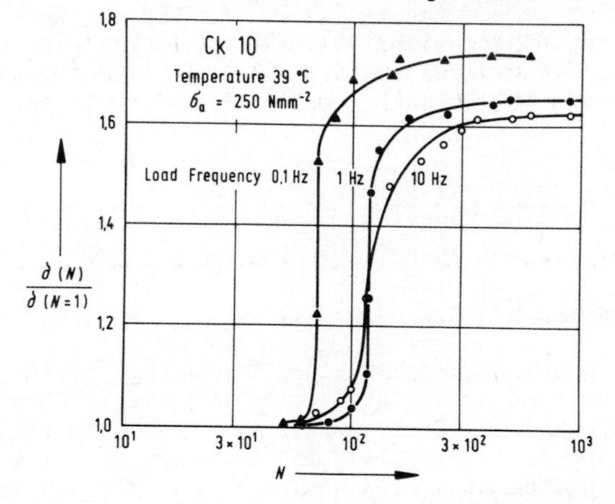

Fig. (11) - Start of softening independent of load frequency, steel Ck 10

a higher stress $\tau_V(x,t)$ in the vicinity of their boundaries than that resulting from the applied stress. This means that from the beginning of cycling there will exist two groups of elastic grains as presented in Figure 12: First a group 1 including all grains which are neighbors of piled up dislocations in plastified grains and therefore exposed to higher stresses than the applied stress, and a second group 2 including all other elastic grains which are cycled by the applied stress only. Hence the critical free dislocation segment will be reached in grains of group 1 much earlier than in those of group 2 and so the spread of plastic ranges will start from grains already plastified in the first stress cycles. The plastification rate is steered by the process that in grains of group 2 before they become grains of group 1 an increase of their maximum free dislocation

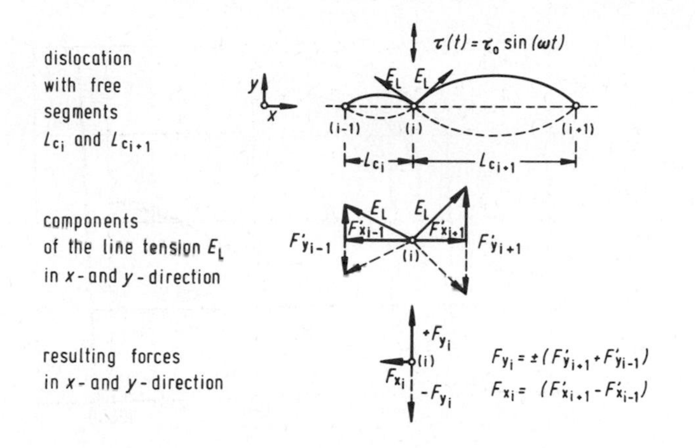

Fig. (5) - Forces on the i-th pinning point of a dislocation

The results of a computer simulation for a dislocation network length of 4000 b and initial free dislocation segments of 150 ± 15 b in Figure 6 demonstrate that in the course of cycling major free dislocation segments will be

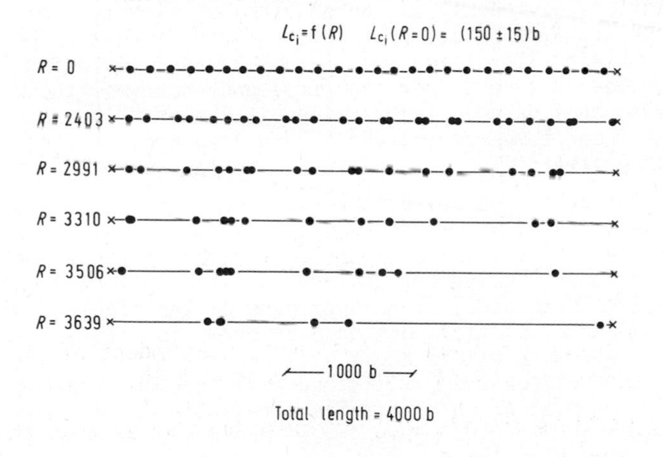

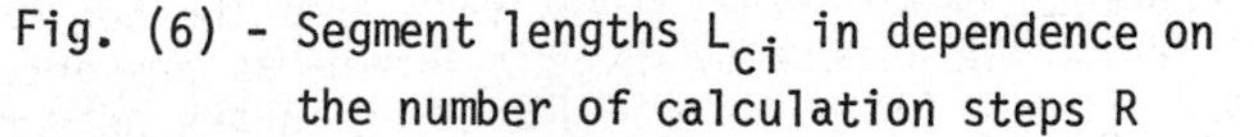

Fig. (6) - Segment lengths L_{ci} in dependence on the number of calculation steps R

produced. Thus the critical free dislocation length even for an applied stress amplitude below the yield strength will be reached and then the network lengths can be activated to a FRANK-READ-source. As the increase of the free dislocation segments by stress induced diffusion is a time dependent process, the start of fatigue softening should also be time dependent. This means that the number of cycles up to the start of softening should increase with rising load frequency. This is confirmed by experimental results as shown in

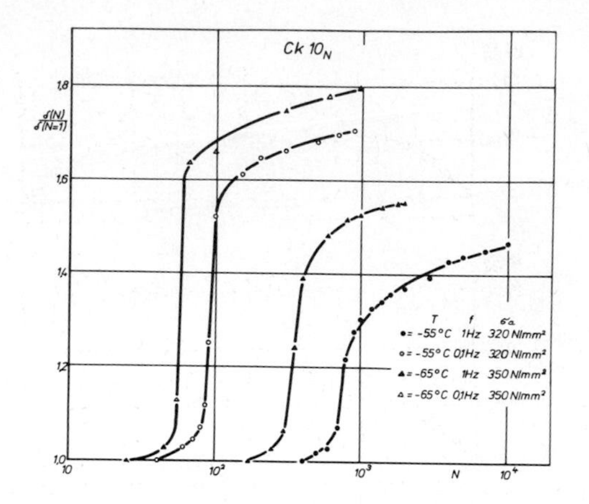

Fig. (7) - Influence of temperature and frequency on the deformation amplitude in the low temperature-high frequency region

Figure 7 for example on steel Ck 10 for test temperatures of -55°C and -65°C if the load frequency is changed from 0.1 cps to 1.0 cps.

As the oscillations of the pinning point atoms in direction of the applied shear stress vector are also caused by stress induced diffusion the oscillation amplitudes will increase with rising temperature and/or decreasing load frequency. So at low temperature and/or high frequency the amplitudes will be very small, however, considerably large values in the order of 1000 b and more can be reached at higher temperatures and/or low frequency as a computer simulation for a ratio of the diffusion coefficient to the load frequency of 10^{-14} cm^2, corresponding to the performance of tests for instance at -80°C and 0.0016 cps or at 40°C and 0.16 cps, demonstrates in Figure 8. If a similar arrangement of clusters or particles exists in the matrix as assumed for pinning the dislocations a bowing out dislocation will contact those particles in the matrix when the mean oscillation amplitude of the pinning point appears in the order of the medium particle distance as shown in Figure 9 schematically. By these contacts particle atoms will migrate into the dislocation pipe by stress induced diffusions [11,12] and will then be arranged to new pinning points, the removal of which entails a retardation of reaching the critical free dislocation length. This delay should increase proportionally with the mean oscillation amplitude of the pinning point atoms and hence will depend linearly on the ratio of the diffusion coefficient to the load frequency. As the number of cycles for reaching the critical free dislocation segment rises proportionally to the reciprocal of the above ratio a change from the time dependent start of softening at low temperature and/or high frequency to a cycle dependent start of softening should be expected when the oscillations of pinning point atoms in direction of the applied shear stress vector will exceed the medium particle distance in the matrix by raising the temperature and/or lowering the load frequency. This transition can be observed on steel Ck 10 for example at -55°C by the reduction of the load frequency from 0.1 cps down to 0.01 cps in Figure 10 or in addition in

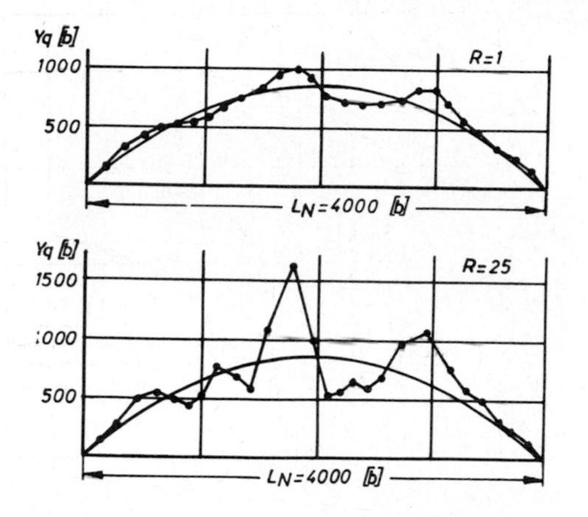

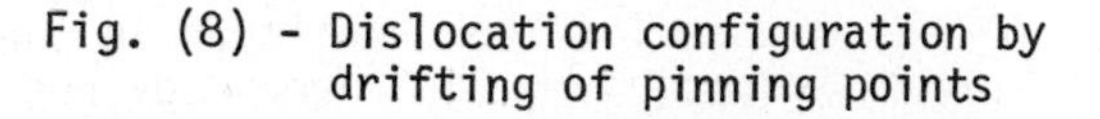

Fig. (8) - Dislocation configuration by drifting of pinning points

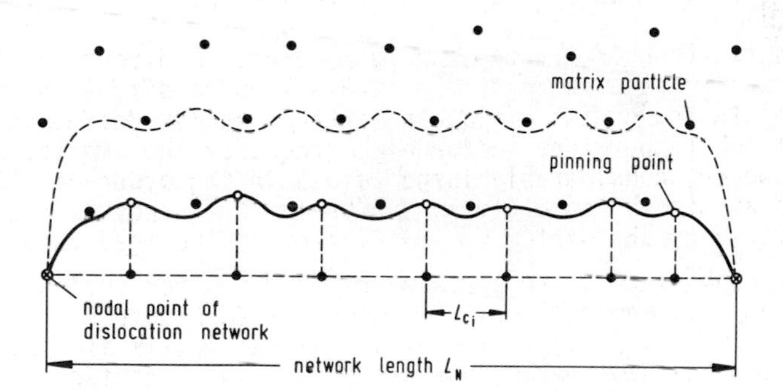

Fig. (9) - Dislocation contact with lattice particles by drifting of pinning points

Figure 11 where it can be seen that at 39°C already below 10 cps a cycle dependent start of softening occurs.

FATIGUE SOFTENING PROCESS

From the microplastic behavior prior to fatigue softening it is evident that already in the first stress cycles a few grains will become plastified by local stress concentrations or by favored grain orientation. Hence in these grains free dislocations can move over larger distances and piled up dislocations will be produced on major obstacles as for instance grain boundaries. So neighbor grains of the piled up dislocations will experience

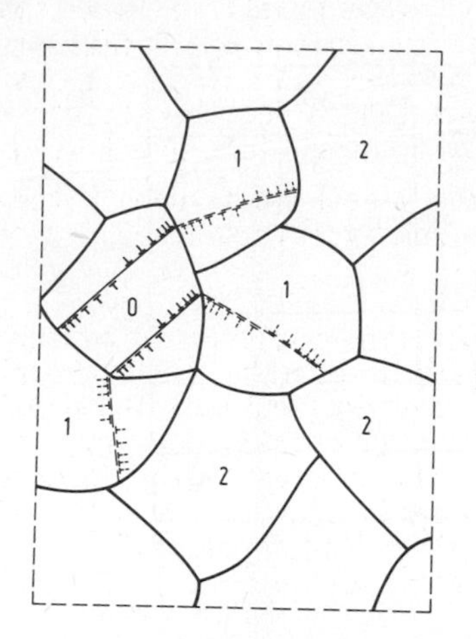

0 = Grain plastified in the First Cycle
1 = Grains of the First Grain Series
2 = Grains of the Second Grain Series

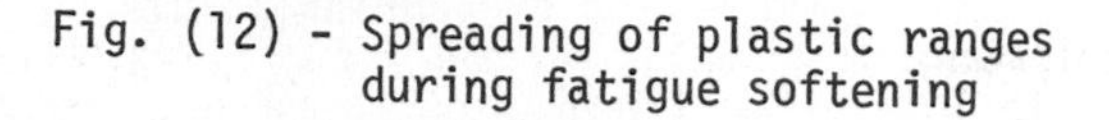
Fig. (12) - Spreading of plastic ranges during fatigue softening

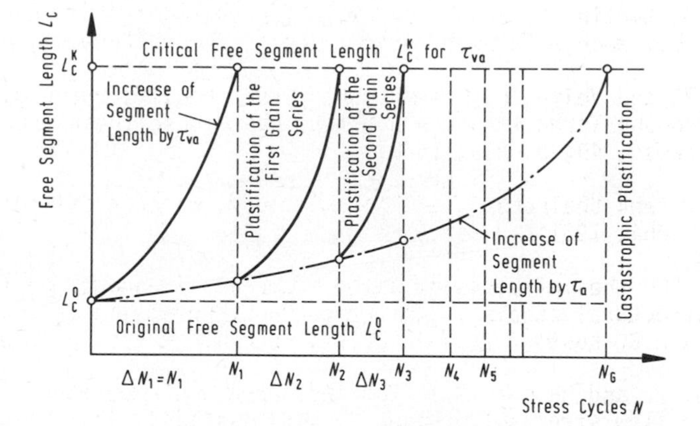

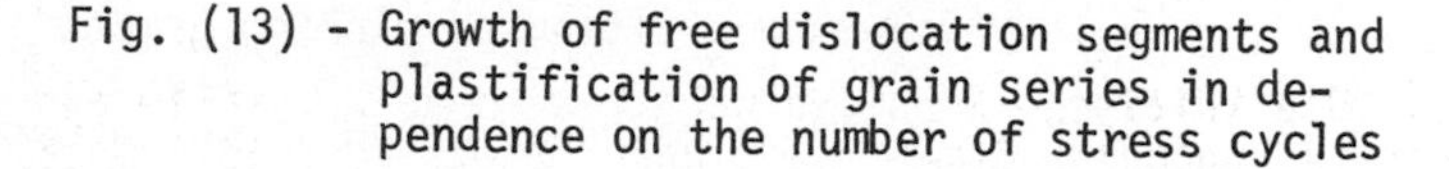
Fig. (13) - Growth of free dislocation segments and plastification of grain series in dependence on the number of stress cycles

segments by the applied stress amplitude τ_a takes place as demonstrated in Figure 13 by the dashed curve. So the further segment growth for reaching the critical free dislocation length by the higher stress amplitude τ_{va} after the grains have changed from group 2 to group 1 decreases with increasing number of stress cycles. The periods ΔN_1 between the change of grains from group 2 to group 1 will accordingly be successively reduced, indicating that the cycle intervals between the plastification of grain series will decrease in the course of the softening process. This accelerated plastification is continued until all elastic grains have become plastified. A catastrophic plastification might occur at the end of the softening process if there even will exist elastic grains belonging to group 2 when their dislocations have already reached the critical free segment length for the activation of the dislocation to a FRANK-READ source at N_G cycles. This event is observed in the experiment by a rapid increase of the strain amplitude in the main phase of fatigue softening as it can be seen from experimental results in Figures 10 and 11.

When a grain series become just plastified the plastic amount of their individual deformation amplitude will have a maximum. This amount will be reduced in the course of further cycling by cyclic hardening until a saturation amplitude has been reached. Thus the macroscopic observed deformation amplitude during the softening process will be produced by the superposition of the individual momentary deformation amplitudes of the plastified grain series, characterized by their individual points of plastification and individual stages of hardening after plastification.

REFERENCES

[1] Klesnil, M., Holzmann, M., Lukás, P. and Rýs, P., "Some aspects of the fatigue process in low-carbon steel", J. Iron Steel Ins., 203, p. 47, 1965.

[2] Löhberg, K., Martin, E. and Veith, H., "Zun Spannung-Dehnung-Verhalten von Baustählen bei Wechselbeanspruchung", Arch. Eisenhüttenwes., p. 798, 1972.

[3] Seiffert, R. and Veith, H., "Temperatur- und Frequenzabhängigkeit des Spannung-Dehnung-Verhaltens von Baustahl bei Wechsellbeanspruchung", Arch. Eisenhüttenwes., 49, p. 279, 1978.

[4] Suits, J. G. and Chalmers, B., "Plastic microstrain in silicon-iron, Act. Met., 9, p. 854, 1961.

[5] Veith, H., "On the influence of temperature and frequency on fatigue softening of structural steels". In: Strength of metals and alloys, P. Haasen, V. Gerold and G. Kostorz (eds.), Oxford Pergamon Press, p. 1151, 1979.

[6] Mordike, B. L. and Haasen, P., "The influence of temperature and strain rate on the flow stress of α-iron single crystals", Phil. Mag., 7, p. 459, 1962.

[7] Frank. W., "Die kritische Schubspannung kubischer Kristalle mit Fehlstellen tetragonaler Symmetrie", Phys. Stat. Sol. 19, p. 239, 1962.

[8] Lütjering, G. and Hornbogen, E., "Mechanische Eigenschaften der Mischkristalle des α-Eisens", Z. Metallkde., 59, p. 29, 1968.

[9] Suits, J. G. and Chalmers, B., "Plastic microstrain in silicon-iron", Acta Met., 9, p. 854, 1961.

[10] Pilo, D., "Zum Wechselverformungsverhalten normalisierter unlegierter Stähle mit Kohlenstoffgehalten von 0.01 bis 1.02 Gew %," Dissertation, TU Karlsruhe, 1979.

[11] Gleiter, H., "Die Formänderung von Ausscheidungen durch Diffusion im Spannungsfeld von Versetzungen", Act. Met., 16, p. 455, 1968.

[12] Gleiter, H., "Ausscheidungshärtung durch Diffusion im Spannungsfeld einer Versetzung", Act. Met., 16, p. 857, 1968.

[13] Eshelby, J. D., Frank, F. C., and Nabarro, F. R. N., "The Equilibrium of Linear Arrays of Dislocations", Phil. Mag. 42, p. 351, 1961.

SECTION VII

STATISTICAL FRACTURE AND FATIGUE

ON A PREDICTIVE FATIGUE CRACK INITIATION RELIABILITY FOR LARGE STEEL CASTINGS

U. S. Chawla, Y. Carmel, P. H. B. Hamilton

Dominion Engineering Works, Ltd.,
Montreal, Quebec, H3C 2S5, Canada

and

J. W. Provan

McGill University
Montreal, Quebec, H3A 2K6, Canada

ABSTRACT

For over a century, large steel and cast iron castings have been and are being successfully produced and operated in fatigue loading situations. More recently, however, customer requirements as to the power transmitted and reliability of heavy machinery have increased along with a corresponding increase in both component size and the effect of inherent casting defects. The long-term objectives of the research whose current status is reported in this paper are therefore twofold. The first is to establish a reliability and decision making algorithm based on probabilistic concepts which takes into consideration the nature, geometry and location of casting defects, their severity, the local fatigue material properties of metals commonly used in cast form and the variety of fatigue loading situations. The second is to accumulate an expanding data bank to aid in the formulation of future design decisions in an ever increasing number of situations which involve large cast steel components. Although this predictive technique applies in general to any large casting situation, the research effort reported here is directed toward establishing crack initiation reliability estimates for 6 to 12 m diameter ring gears used to drive ore grinding mills.

THE LOCAL STRAIN APPROACH

With reference to Figure 1, the major aspects of the current research into predicting the fatigue crack initiation reliability of ore grinding mill ring gears may be described in the following manner.

A. Metallurgical Characterization

The material used in these gears is a low-alloy cast steel similar to 4340 wrought steels. The carbon content has been raised, however, to between 0.35 and 0.45%, so as to obtain the desired hardness and durability of the gear teeth. Manganese runs from 0.5 to 1.0% and silicon from 0.4 to 0.6%, as required

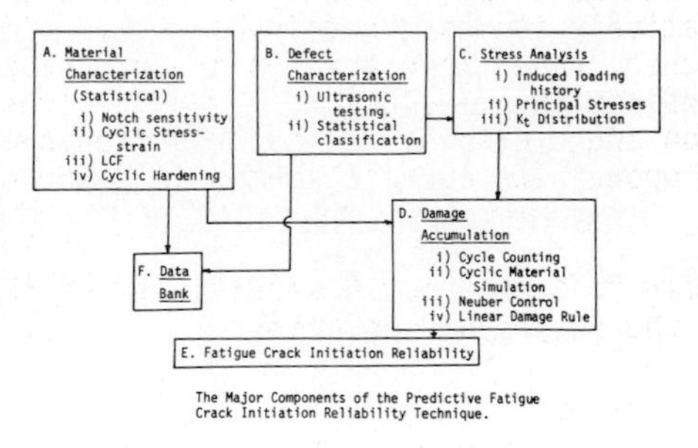

Fig. (1) - The local strain philosophy

by the casting process. Nickel has been increased to 1.75 to 2.25% to increase toughness, while chromium and molybdenum are held at the 4340 limits of 0.6 to 0.9% and from 0.2 to 0.3%, respectively. The levels of porosity and micro-shrinkage are above those of wrought steels while the manganese sulphide inclusions are globular spheroids as opposed to the linear stringers normally found in rolled and forged steels. Whereas 4340 is usually oil quenched to obtain the optimum engineering properties, this is impossible with large cast segments weighing in the vicinity of 30 tonnes. These castings are therefore normalized at 900°C for eight hours, air quenched, and tempered at about 290°C for another eight hours. The mechanical properties thus obtained are:

Brinnell hardness:	350-300 BHN	
Yield strength:	620 MPa	(1)
Ultimate strength:	900 MPa	

The effects of slack quenching is shown in Figure 2 where a mixture of martensite and bainite with some ferrite separation is shown.

Fig. (2) - The cast 4340 microstructure (etched in 2% Nital)

Since the specific material used in this investigation has not previously been considered from a local strain and fatigue damage accumulation point of view, the pertinent material characterizations have to be established. Consequently, a version of the now relatively standard but involved and time-consuming local strain, smooth and notched, round specimen testing procedure [1,2] designed to establish the notch sensitivity, monotonic and cyclic stress-strain, low-cycle-fatigue crack initiation and cyclic hardening or softening properties of the cast 4340 was embarked upon. Its current status is described in the following paragraphs.

Notch Sensitivity "a". The notch sensitivity of the material in question can be determined from the Peterson's equation:

$$K_f = 1 + \frac{K_t - 1}{1 + a/r} \tag{2}$$

where K_t is the elastic stress concentration factor, r is the notch radius and K_f is the fatigue concentration factor defined as:

$$\frac{\Delta S/2 \text{ unnotched at } 10^6 \text{ cycles}}{\Delta S/2 \text{ notched at } 10^6 \text{ cycles}}$$

where $\Delta S/2$ is the nominal stress amplitude involved in the rotating cantilevered beam experimentation. The geometry of both specimens and the unnotched Woehler "S-N" curve are shown in Figures 3 and 4, respectively. As can be observed, the

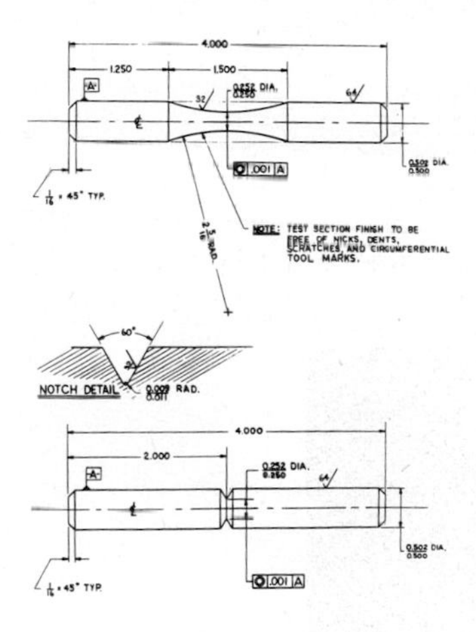

Fig. (3) - Unnotched and notched specimen geometry

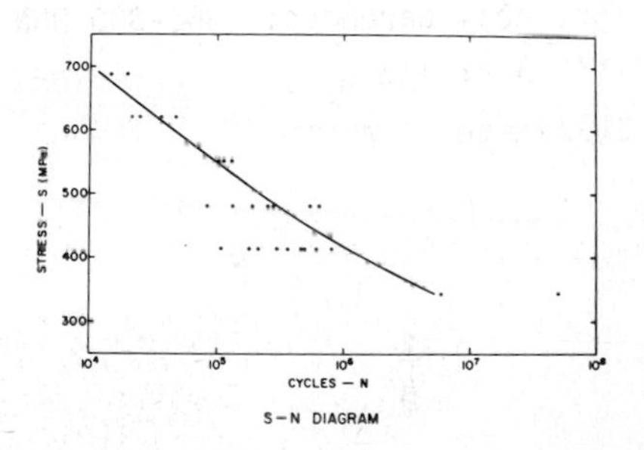

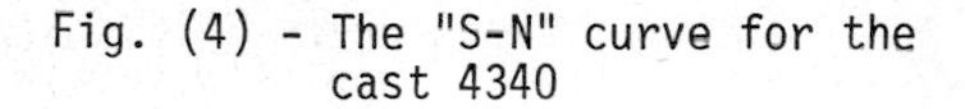

Fig. (4) - The "S-N" curve for the cast 4340

scatter is large and will eventually be incorporated into a statistical description of the notch sensitivity. The major cause of this scatter is the material itself as is evidenced by the fractographs 5a and 5b indicating the influence of the aforementioned manganese spheroids on the fatigue characteristics of the cast 4340 steel.

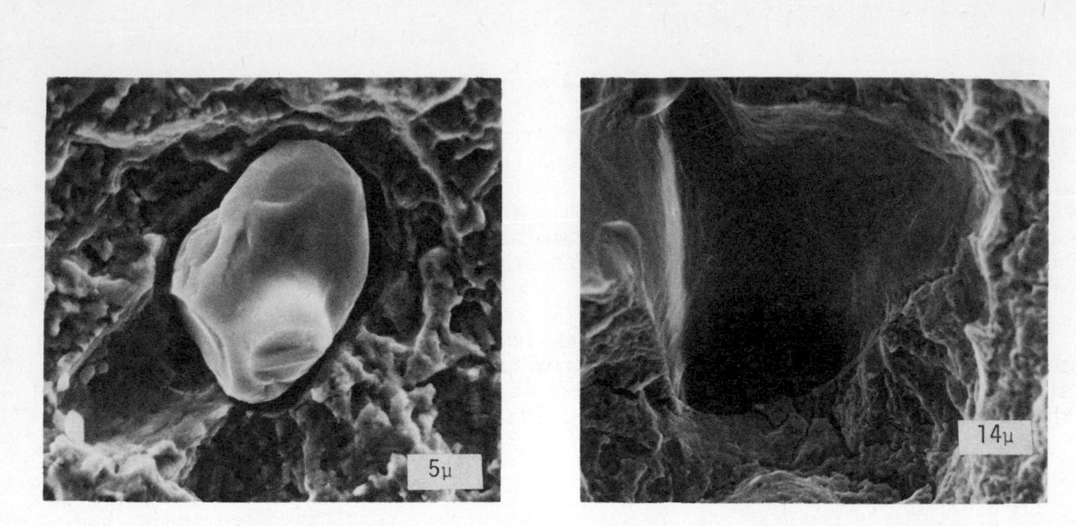

Fig. (5) - Fractographs of manganese spheroids

The Cyclic Stress-Strain Behavior. The cyclic stress-strain behavior of the cast steel is established by using a Landgraf incremental strain procedure [3], which involves joining the locus of the tips of the stable hysteresis loops. The specimen geometry for both the cyclic stress-strain and the following LCF investigation is shown in Figure 6. The cyclic stress-strain behavior of the 4340

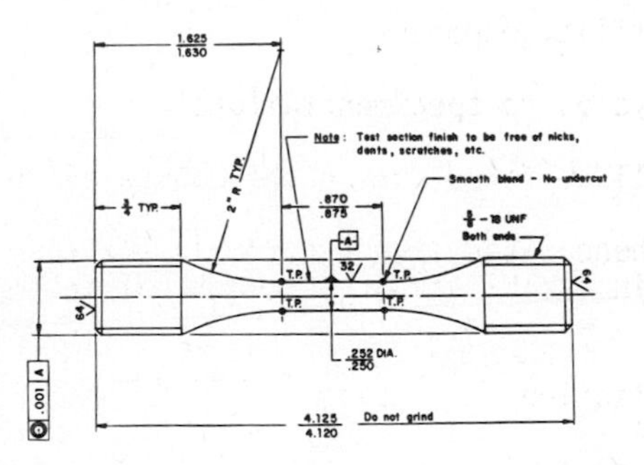

Fig. (6) - Cyclic stress-strain and LCF specimen geometry

steel is then characterized by the equation:

$$\frac{\Delta\varepsilon}{2} = \frac{\Delta\sigma/2}{E} + \left(\frac{\Delta\sigma/2}{K'}\right)^{1/n'} \tag{3}$$

where:

$\Delta\varepsilon/2$ = local strain amplitude

$\Delta\sigma/2$ = local stress amplitude

K' = cyclic strength coefficient (4)

n' = cyclic strain hardening exponent

E = Young's modulus

The Low-Cycle-Fatigue (LCF) Material Characterization. The remaining material characterization, which is required for the implementation of the local strain approach, is the cast 4340 steel's low-cycle-fatigue properties. By using the standard ASTM E466 [4] procedure and the Morrow equation:

$$\frac{\Delta\varepsilon}{2} = \frac{\sigma_f'}{E}(2N_f)^b + \varepsilon_f'(2N_f)^c \quad (5)$$

where:

σ_f' = the fatigue strength coefficient

b = the fatigue strength exponent

ε_f' = the fatigue ductility coefficient (6)

c = the fatigue ductility exponent

N_f = the number of cycles to specimen failure

These characteristics will be found during the course of this research program.

Again, in both these cases, the scatter in the results will be carefully documented and utilized in establishing the crack initiation reliability of the entire structure.

B. Defect Characterization

To establish for cast ring gears the required statistical description of flaw geometries, such as those shown in Figure 7 and characterized in Figure 8, ultrasonic testing procedures are employed. Both straight and angular beam methods are used and a strict procedure is reported which ensures a full investigation of the gear segment in question. This procedure involves a reference 3.2 mm diameter flat bottomed hole and a correlation between ultrasonic indication and flaw geometry. This latter correlation is established by a companion test which involves a thorough ultrasonic test of a cast specimen coupled with a sectioning procedure which results in the characterization of the defects existing in the actual specimen. Other procedures commonly used in the non-destructive evaluation of castings of this type are industrial radiography, for internal flaws, and magnetic particle and/or dye penetrant techniques, for surface flaws.

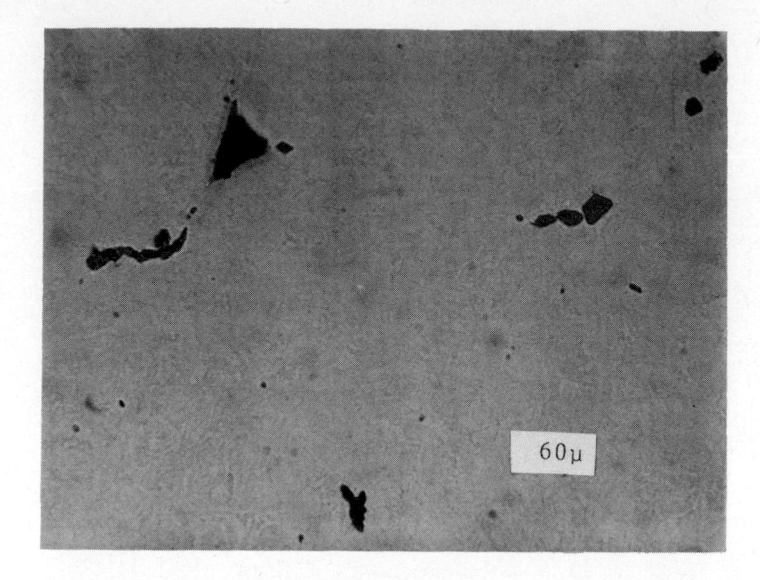

Fig. (7) - Small casting flaws

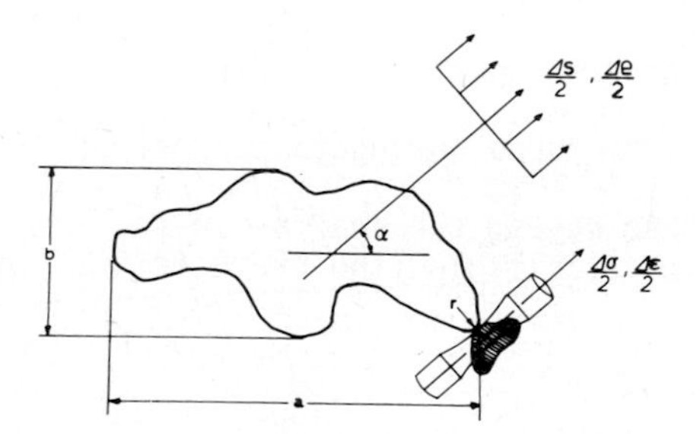

Fig. (8) - Flaw characterization

C. Stress Analysis

Preliminary field trial tests using "in-situ" strain gage and radio telemetry techniques have been performed on an 8.3 m diameter ring gear at the locations shown in Figure 9. Representative strain gage outputs are shown in Figure 10. By combining the findings of the strain gage investigation with a combined analytic and finite element analysis, both the location, magnitude and direction of the severest nominal stresses and strains are determined along with a representative fatigue loading cycle or block of fatigue cycles.

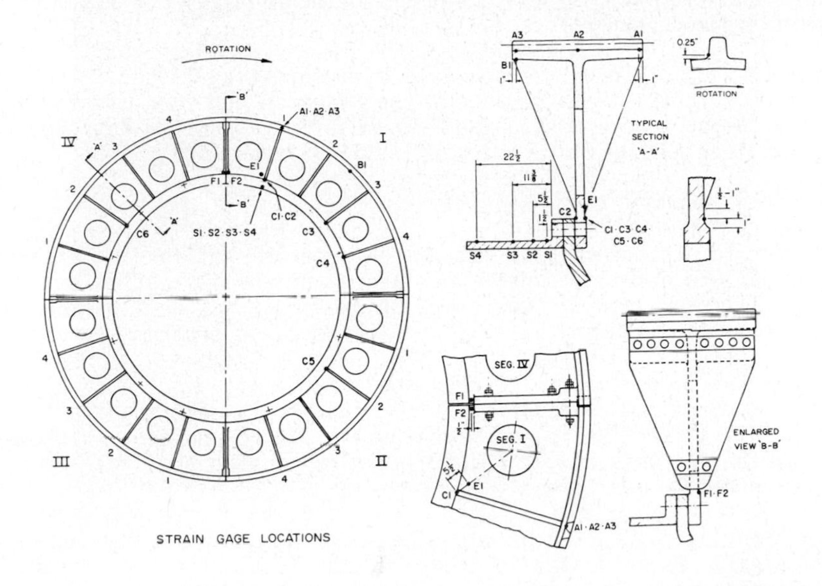

Fig. (9) - The strain gage locations on an 8.3 m diameter ring gear

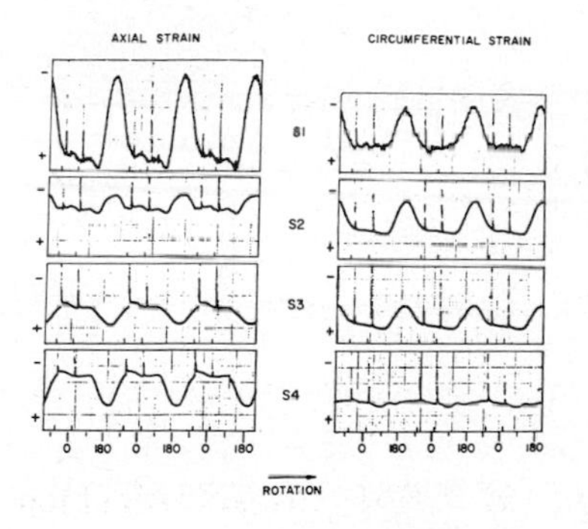

Fig. (10) - Representative strain gage outputs

A third facet of the stress analysis contribution to this research involves the determination of the statistical distribution for the stress concentration factor, K_t. This is determined by implementing B, the statistical characterization of defects and their geometries, in conjunction with above stress analysis C. This distribution can then be combined with the statistical expressions for the notch sensitivity established in the manner described in the section, Notch Sensitivity "a", so as to determine the required distribution function for the fatigue concentration factor, K_f.

With the tabulation of the statistical information listed in subsections A, B and C of this local strain approach, the research now turns to the implementation of a computer dependent damage accumulation algorithm designed to establish the required estimate of ring gear reliability.

D. The Damage Accumulation Algorithm

In order for the previously determined material and loading distributions to be integrated into a fatigue crack initiation reliability estimate for ore grinding ball mill ring gears, computer techniques must now be implemented. Briefly, the steps involved in this part of the research program are as follows.

Fatigue Cycle Counting. The representative fatigue cycle or block of fatigue cycles, determined as described in Section C, is described in terms of either a range-pair, rainflow [5] or Markov matrix [6] technique, whichever is the most suitable. In this way, the actual cycles causing the cycle-by-cycle damage can be correctly monitored.

Simulation of the Cyclic Stress-Strain and Hardening/Softening Properties of Cast 4340 Steel. As indicated in Figure 8, the local strain philosophy considers the material near the critical radius of a defect to have properties determined by the laboratory testing of smooth round specimens. This dictates that these responses be correctly modelled so that the amount of fatigue damage per cycle can be assessed. Again, this aspect is now well-documented in the literature [7] with modifications to the Jahnsale [8] sliding rule control rheological model being, at present, the most appropriate. Care will again be taken to incorporate the statistical aspects of these properties.

Neuber Control. With both the block of fatigue cycles and the cyclic material response being correctly described, the Neuber control of damage developed by Topper [9] will be implemented using the distribution function for K_f. This involves the basic equation:

$$K_f \Delta S = K_f (\Delta S \Delta e E)^{1/2} = (\Delta\sigma \Delta\varepsilon E)^{1/2} = 2E[\frac{\sigma_f'}{E} (2N_f)^b \frac{\Delta\varepsilon}{2}]^{1/2} \quad (7)$$

which both relates the nominal to the local material responses and enables the statistical defect behavior to be predicted from the elastic portion of the Morrow smooth specimen fatigue data, equation (5).

The Linear Damage Rule. The required fatigue crack initiation reliability curve will be generated in two stages. The first involves the implementation of the Palmgren-Miner linear damage accumulation rule for *each and every* representative combination of defect geometry and material response characteristic. This will generate a considerable amount of data in the form of number of cycles to initiate fatigue cracks. The second stage is to weigh this data according to the statistical descriptions of all the parameters, be they geometric or material, involved. This statistical weighting procedure, which amounts to practical implementation of the multidimensional Stieltjes integral concept, generates the desired estimate of the fatigue crack initiation reliability of the ball mill ring gears described in this paper.

E. Fatigue Crack Initiation Reliability

Thus, from:

1. the geometry of a ring gear,

2. the nature of its operation,

3. a strain gage and stress analysis investigation,

4. a statistical estimate, based on NDE, of flaws and their representative geometries,

5. a knowledge, again in statistical terms, of the cast 4340 steel's
   - notch sensitivity,
   - cyclic stress-strain properties, and
   - its LCF properties,

the entire component's reliability against crack growth initiation is found.

F. Data Bank

A long term and important aspect of the research project outlined in this paper is the accumulation of the statistical characterizations of flaws and material properties. In this way, the reliability of other cast steel systems can be ascertained during the design stage and suitable modifications made prior to the manufacture of the large scale casting.

ACKNOWLEDGEMENTS

The financial assistance of the National Research Council of Canada through an IRAP grant to The Dominion Engineering Works, Montreal, is gratefully acknowledged.

REFERENCES

[1] Mitchell, M. R., "A unified predictive technique for the fatigue resistance of cast ferrous-based metals and high hardness wrought steels", SAE, SP-79/448, pp. 31-66, 1979.

[2] Dowling, N. E., Brose, W. R. and Wilson, W. K., "Notched member fatigue life predictions by the local strain approach", in Fatigue Under Complex Loading, SAE, AE-6, 1977.

[3] Landgraf, R. W., Morrow, JoD. and Endo, T., "Determination of the cyclic stress-strain curve", J. of Matls., 4, pp. 176-188, 1969.

[4] 1980 Annual Book of ASTM Standards, 10, American National Standard, ANSI/ASTM E466-76, pp. 614-619, 1980.

[5] Dowling, N. E., "Fatigue failure predictions for complicated stress-strain histories", J. of Maths., 7, pp. 71-87, 1972.

[6] Argyris, J. H., Aicher, W. and Artelt, H. J., "Analysis and synthesis of operational loads", Library translation 2008, Royal Aircraft Establishment, Ministry of Defence, Farmborough, Hants, U.K., 1979.

[7] Mattos, R. J. and Lawrence, F. V., "Estimation of the fatigue crack initiation life in welds using low cycle fatigue concepts", SP-424, Society of Automotive Engineers, Warrendale, Pa., 1977.

[8] Jhansale, H. R., "A new parameter for the hysteretic stress-strain behavior of metals", ASME Transactions, p. 33, January 1975.

[9] Topper, T. H., Wetzel, R. M. and Morrow, JoD., "Neuber's rule applied to fatigue of notched specimens", J. of Matls., 4, pp. 200-209, 1969.

RELIABILITY OF STRUCTURES WITH TIME DEPENDENT PROPERTIES

G. F. Oswald

Technische Universität Munchen
D-8000 Munchen 2, Federal Republic of Germany

and

G. I. Schuëller

Universität Innsbruck
A-6020 Innsbruck, Austria

ABSTRACT

A reliability analysis approach for structures is presented taking into consideration the deterioration effects due to random loading. The deterioration of the resistance in critical cross sections of structures is formulated as a function of the underlying physical mechanism, i.e., the growth of fatigue cracks. In order to treat the physical process consistently over the entire range of the expected load spectrum, the limit load range is also included in the analysis. Finally, the concept is applied to an offshore structure and the reliabilities with respect to various failure modes are compared.

INTRODUCTION

For the case of weakly damped dynamically excitable structures with stochastic loading, the problem of a possible deterioration of the resistance is of major importance. Typical examples are structures of nuclear power plants under operational and earthquake loading [1,2], offshore structures under wave loading and tall, slender chimneys under wind loading. In the static case, a structure may fail by (brittle or ductile) fracture of one or more components or by other failure modes such as plastic collapse (plastic hinges) or by buckling, etc. For the dynamically loaded structure, an additional failure mode might be fatigue failure, i.e., the fracture of components deteriorated by the fatigue process. Therefore, in the reliability concept, not only the loading but also the resistance is to consider as a time dependent random process. The deterioration of the resistance due to the load history is modeled by fracture mechanical considerations which are then introduced in the reliability concept. In order to carry out a consistent treatment of the physical process over the whole range of the expected load spectrum, the plastic range is also included. The advantage of a fracture mechanical approach - as compared with the fatigue approach (S-N curves) - is, that the deterioration of the critical cross section is related to the crack length. Therefore, a relation between the initial flaw size distribution and the reliability can be obtained. As a further practical application, this procedure can be utilized to determine acceptable initial flaw sizes. This can be achieved by comparison of the reliability level for fracture with that for other failure modes.

METHOD OF ANALYSIS

A. Reliability Concept

The objective of a reliability analysis considering time variant system properties is the determination of the probability of survival* at any point of time within the design or service life of the structure. The reliability is time dependent, because on one hand, there are loads with random intensity, duration and occurrence, whereby the expected maximum load becomes an increasing function of time [3] and on the other hand, the resistance of the critical cross section decreases during the service life. The reliability concept is schematically shown in Figure 1 which indicates the distributions of the most important param-

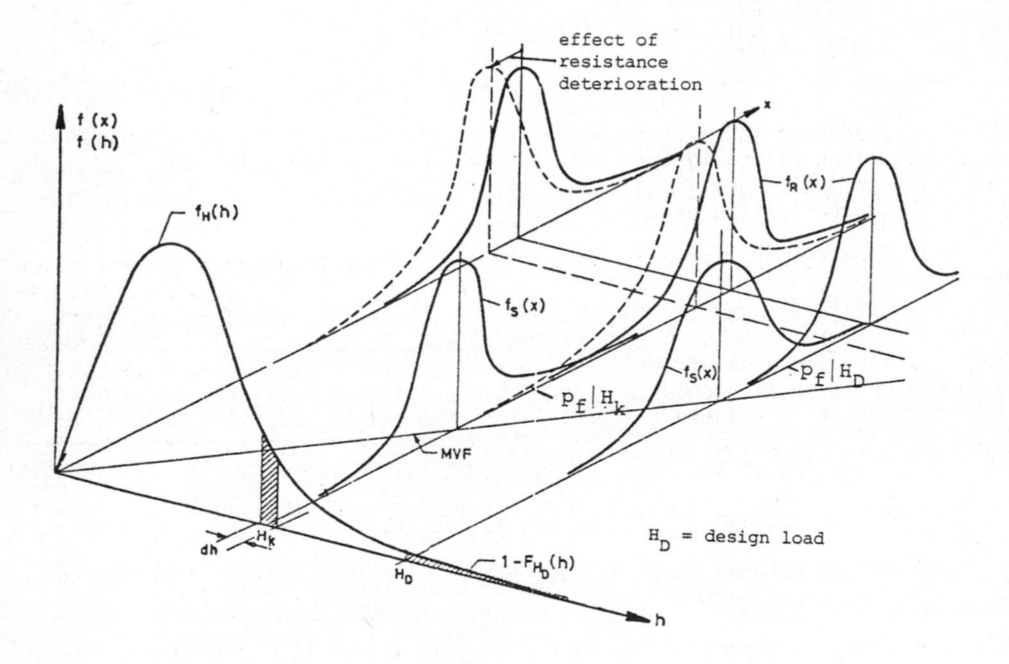

Fig. (1) - Schematic presentation of the reliability concept

eters. In this figure, the intensity of the load events (external loads acting on the structure), e.g., storms characterized by the expected maximum wave height, is described by $f_H(h)$. For each realization of $f_H(h)$, e.g., H_K, by

*The complement of the reliability is the probability of failure.

means of structural analysis, a distribution $f_S(x)$ for the load effects of the critical cross section can be determined. The convolution of $f_S(x)$ with $f_R(x)$ yields the probability of failure given this particular load intensity H_K. The time dependent reliability is calculated by the following equation (4), which is extended in the following development:

$$L_T(t) = \int_o^\infty \sum_{k=o}^\infty \frac{(\nu t)^k e^{-\nu t}}{k!} \{ \prod_{j=o}^{k} F_{S_j}(x_j)\} f_{R_o}(x_o) dx_o \tag{1}$$

$F_{S_j}(x_j)$ herein is the distribution function of the cross sectional load at the j-th load event, $f_{R_o}(x_o)$ is the density function of the initial resistance (at t=0) and ν is the average occurrence rate of the load events (Poisson's law). The deterioration of the resistance of the critical cross section is implied, as x_j denotes thr realization of the resistance at the j-th load event which is the equivalent of the residual resistance after j-1 load events. In [4], for the residual resistance x_j, only the following *qualitative* equation is given:

$$x_j = x_o \cdot \psi(t) \tag{2}$$

where $\psi(t)$ is assumed to be a monotonically decreasing function of time respectively, a function of the load history of the cross section $\psi(t) = \psi[\{S(t)\}]$. The objective of this work is to show a *quantitative* formulation of $\psi(t)$ for modeling the deterioration of the resistance.

B. Elastoplastic Fracture Mechanical Model for the Resistance Deterioration

The physical cause for the decrease of the resistance of a considered cross section is the extension of a fatigue crack, and as a result, the decrease of the load carrying area of the cross section. Furthermore, the stress distribution in the cross section changes producing a greater stress intensity. Recognizing reality, it is assumed that initial cracks are unavoidable, even in case of careful construction, as, e.g., for reactor- or offshore structures. Regarding the deterioration of the resistance, two major problems have to be solved:

- Relation between crack length and resistance
- Relation between crack propagation and the load history experienced by the cross section

The first problem is approached by introducing an appropriate fracture criterion; the second one by utilizing realistic models for calculating crack propagation.

When plastic flow, i.e., yield in larger regions of the cross section is expected - depending on the respective load, crack length and yielding capacity of the material - linear elastic fracture mechanics (LEFM) is no longer applicable and methods of post yield fracture mechanics have to be used. In Figure 2, case (a) corresponds to the range of the LEFM, as the size of the plastic zone r_{pl} is much smaller than the crack length ($r_{pl} \ll a$). In case (b) of $r_{pl} < 0.4a$, an approximate application of the LEFM is possible using the Irwin's plastic zone

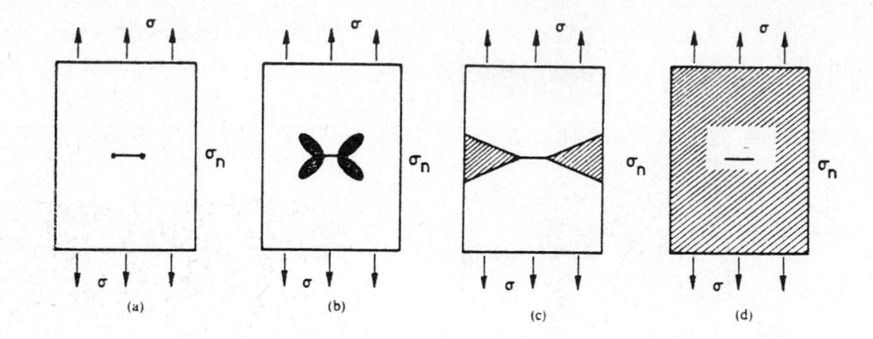

Fig. (2) - Schematic presentation of several grades of plastic flow [5]

correction. Case (c), however, requires the application of post yield fracture mechanics, whereas the fully plastic case (d) can be treated on the basis of the nominal stress in the remaining cross sectional area, as the stress distribution over the cross section is not likely to be affected by the crack.

1. *Elastoplastic failure diagram.* Beyond the limits of LEFM, generalized parameters as the J-integral or the crack opening displacement (COD) are frequently used. As there are still scarce data for the respective material parameters J_c and COD_c for fitting distributions, the Feddersen Scheme [6], which is an empirical elastoplastic concept, is chosen as fracture criterion. This concept - like the Two Criteria Approach - represents a link between the linear elastic and the fully plastic failure (case (a) and (d) in Figure 2). Actually, the different functions $K_r = f(S_r)$ proposed by several authors for the Two Criteria Approach can be transferred to the form $a = f(\sigma)$ and are comparable to the inverse function $\sigma = f(a)$ represented by the Feddersen Scheme. For the present purpose, the Feddersen Scheme is given preference because a functional relationship of the type $\sigma = f(a)$ is required. As shown in Figure 3, the Feddersen-diagram indicates the resistance σ_c of the cross section in relation to the crack length, the yield stress σ_y and the critical stress intensity factor K_{c_1} (for strain hardening materials the yield stress σ_y is replaced by the ultimate strength σ_u). The failure curve is divided into 3 regions. The central region (region II) corresponds to LEFM, whereas in the other regions (I and III), the failure curve deviates from the hyperbola (which represents the inverse basic equations of LEFM) due to plastic effects. From the equations for the straight line and the hyperbolic section, the failure stresses for the respective regions can be determined as follows:

$$\text{region I} \quad \sigma_c = \sigma_y \left[1 - \frac{4\pi}{27} \left(\frac{\sigma_y}{K_{c_1}}\right)^2 \cdot a(t)\right] \tag{3}$$

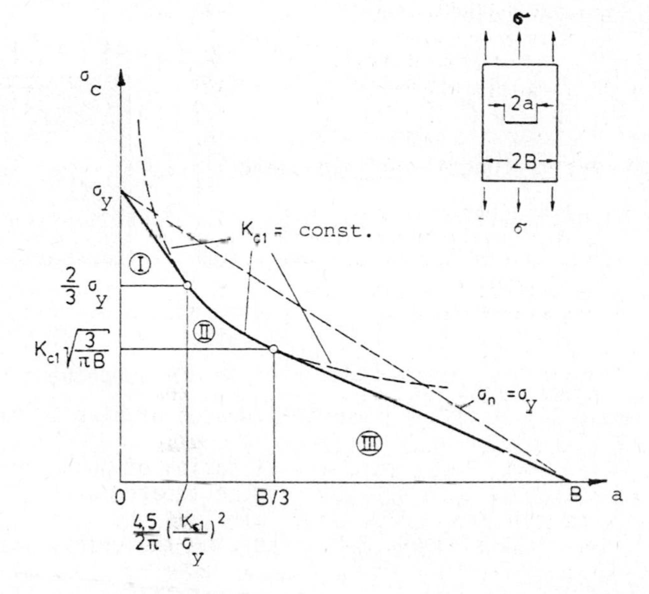

Fig. (3) - Feddersen Scheme [7]

$$\text{region II} \quad \sigma_c = K_{c_1} \sqrt{\pi a(t)} \tag{4}$$

$$\text{region III} \quad \sigma_c = \frac{3}{2} K_{c_1} \sqrt{\frac{3}{\pi B}} \left[1 - \frac{a(t)}{B}\right] \tag{5}$$

For the fully plastic case (high toughness material), the diagram shows ($\sigma_u = \sigma_y$) for the three regions identical relations, i.e.,

$$\text{region I, II, III} \quad \sigma_c = \sigma_y \left(1 - \frac{a(t)}{B}\right) \tag{6}$$

Using equations (3) to (6), the function $\psi(t)$, which represents the time dependent deterioration of the resistance of the cross section, can be formulated quantitatively as

$$\psi(t) = \frac{\text{resistance at } t}{\text{resistance at } t=0} = \frac{\sigma_c}{\sigma_{c_o}} = \frac{\sigma_c(a(t))}{\sigma_c(a_o)} \tag{7}$$

Therefore, the initial crack length a_o and the crack length $a(t)$ - as a function of the load history have to be determined.

2. *Distribution of the initial crack length.* As mentioned before, the initial crack length is considered to be randomly distributed. Consequently, equation (1) has to be extended by a further convolution, i.e.,

$$L_T = \int_{a_m=0}^{\infty} L_T(t|a_m)\, f_{A_m}(a_m) da_m \tag{8}$$

where $f_{A_m}(a_m)$ is the density function of the *largest* initial crack length* and $L_T(t|a_m)$ is the conditional reliability function. The first step for determining $f_{A_m}(a_m)$ is to establish $f_a(a)$, which is the crack length distribution confined to one crack only. This may be regarded as the product of the distribution in size of defects remaining in the component after fabrication, $g_A(a)$ with the probability that defects of a given size are not detected in the subsequent ultrasonic or any other inspection, $h_A(a)$. Thus,

$$f_A(a) = g_A(a) \cdot h_A(a) \tag{9}$$

where $f_A(a)$ has to be renormalized. Generally, for $g_A(a)$ and/or for f_A, an exponential type distribution is assumed [8,9]. For the following, let $f_A(a)$ be

$$f_A(a) = c \cdot e^{-c(a-a^*)} \tag{10}$$

where a^* is the minimum crack length detectable by the respective inspection procedure. As the distribution of the flaw size $f_A(a)$ and the mean rate of the occurrence of detectable flaws are measurable parameters, the following way to determining the distribution $f_{A_m}(a_m)$ was chosen. The number k of the flaws occurrin in the welding length l considered, can be modeled by the Poisson law

$$p(k) = \frac{(\lambda \cdot l)^k \cdot e^{-\lambda l}}{k!} \tag{11}$$

where λ is the average number of flaws per unit length. Utilizing the relation $F_{X_m}(x) = F_X(x)^n$, which is well-known from extreme value theory, the probability distribution of the maximum flaw size results in

$$F_{A_m}(a_m) = \sum_{k=0}^{\infty} F_A^k(a) \cdot p(k) \tag{12}$$

---

*The claim that only the distribution of the *largest* crack is relevant is based on the assumption of a uniform stress condition along the weld length. In the case of greater stress variation, the weld length is to subdivide in intervals of approximately uniform stress.

The density function $f_{A_m}(a_m)$ can then be obtained by derivation

$$f_{A_m}(a_m) = \frac{dF_{A_m}(a_m)}{da_m} = \sum_{k=0}^{\infty} k \cdot F_A(a)^{k-1} \cdot f_A(a) \cdot p(k) \qquad (13)$$

In Figure 4, the distributions $f_A(a)$ and $f_{A_m}(a_m)$ respectively are presented (assumption: λ = 10 flaws/m, c=1, a* = 2 mm). One can see in Figure 4, that, when

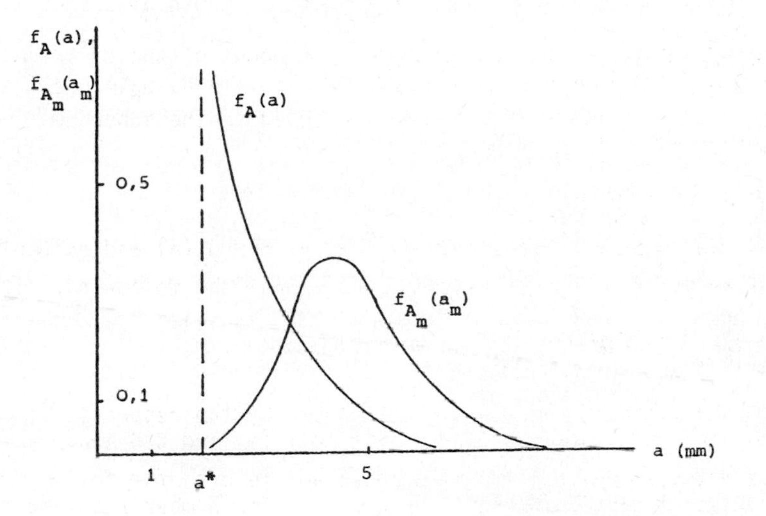

Fig. (4) - Distribution $f_A(a)$ of the size of one crack and $f_{A_m}(a_m)$ of the size of the largest crack

more than one crack of detectable size is expected in a weldment, the shape and position of the distribution of crack length changes shape and central tendency, whereby larger crack lengths become more likely.

3. *Crack propagation.* The crack propagation models of the LEFM are mostly based on the range of the stress intensity factor ΔK. Due to larger plastic flow at the crack tip, displacement or energy parameters are relevant beyond the limits of the LEFM. Here, an approach using the parameter ΔJ - as proposed by Dowling et al [10,11] - was chosen:

$$\frac{da}{dN} = C_1(\Delta J)^{C_2} \qquad (14)$$

where C_1, C_2 are material parameters. The parameter ΔJ can be determined from the crack length and the parameters of the cyclic stress strain hysteresis [11] by approximate methods. Good correlation with experimental data are reported for this approach, although it has been shown [12] that the quantity ΔJ might be misleading, because it cannot be interpreted as the range of the J-integral.

4. *Sequence effects on crack propagation.* For the load history considered here, basically two types of sequence effects can be distinguished. When the load history, over the entire service life of the structure (macro-aspect) consists of several load events of different load intensities and durations (e.g., storms, earthquakes, etc.), effects due to the given sequences of these events can be determined. These effects might be of interest when one or several events of high intensity in the beginning of the service life cause a relatively large fatigue crack, so that afterwards, also events of smaller intensities can contribute significantly to the fatigue process. This sequence effect can be taken into account by performing the calculations for crack propagation for the particular load events separately, considering the sequence. The treatment of the second type of sequence effects, which results from the grouping of the amplitudes within a load event (micro-aspect) is not straightforward. These effects are known as acceleration or retardation of the crack propagation, which is caused by the interactions between these amplitudes (residual stresses, crack closure, etc.). It can be shown [13,14] that a significant sequence effect is to be expected only if the sequence patterns typical for increase or decrease of the crack propagation occur at larger intervals and, moreover, the ratios of amplitudes are sufficiently large. It should be noted, that randomly applied tensile overloads will lead to a significant decrease only if these overloads are followed by 10^2 - 10^4 amplitudes required to be at least 30% smaller. Whereas sequences with these characteristics frequently appear on aircraft components, this is not expected on structures dynamically excited by environmental forces. As a consequence, in this investigation, the calculation of the crack propagation can be carried out neglecting the interaction effect of the various amplitudes.

5. *Application of Markoff theory to crack propagation.* In the light of the previous discussion on sequence effects, the crack growth at any load event will depend only on the previous crack length and the acting load. This situation can be modeled by Markoff chains, which have already been applied for modeling the micromechanism of propagation [15]. By applying the Markoff model, the range of the crack length is discretized into b intervals of the length $a = B/b$, (where B denotes the component width) which one considers as "stages" of crack length. Following a load event, the new stage will be equal or higher than the previous one. Thus, the transition-probability matrix can be expressed as follows.

$$P = \begin{bmatrix} p_{0,0} & p_{0,1} & \cdots\cdots & p_{0,b} \\ 0 & p_{1,1} & \cdots\cdots & p_{1,b} \\ \vdots & & & \vdots \\ 0 & 0 & \cdots\cdots & p_{b-1,b} \\ 0 & 0 & \cdots\cdots & 1 \end{bmatrix} \tag{15}$$

The transition probabilities $p_{i,k}$ for the matrix P can be defined as

$$p_{i,k} = \text{Prob.}\{(k - i - \frac{1}{2})\Delta a < Z < (k - i + \frac{1}{2})\Delta a\} \tag{16}$$

where Z is the crack increment during the load event. As the crack increment Z is a random variable, equation (16) can be expressed as

$$p_{i,k} = \int_{(k - i - \frac{1}{2})\Delta a}^{(k - i + \frac{1}{2})\Delta a} f_Z(z)dz \tag{17}$$

The crack increase Z includes the crack propagation law considered. The discrete distribution of the crack length after j load events, which is the probability of being in state k = 1,2, ... b at the j-th event can be expressed in vector form:

$$p(j) = \{p_1(j), p_2(j), \ldots, p_b(j)\} \tag{18}$$

and from Markoff theory follows for homogeneous chains [16]

$$p(j) = p(0) \cdot P^j \tag{19}$$

in which p(0) denotes the discrete distribution of the initial stage, i.e., crack length and P is the transition matrix. An alternative method, which is, in fact, used more frequently to obtain the distribution of the actual crack length after j load events is the application of the following equation

$$a(j) = a_o + \sum_{i=1}^{j} Z_i \tag{20}$$

where the initial crack length a_o, the crack increment Z and the resulting crack length a(j) are considered to be randomly distributed. Except for normal distributed parameters, equation (20) cannot easily be solved in a closed form. Therefore, Monte Carlo Simulation procedures are applied. For comparison of the two different approaches, a numerical example is carried out. To simplify the problem, all input parameters were assumed to be log-normally distributed. Figure 5 shows the random crack growth due to three load events, starting from a log-normal distribution for the initial crack length. The good agreement of the results indicates the compatibility of the two methods. Moreover, it can be seen how the mean value and the shape of the distributions changes with random crack propagation. As cracks confined to higher stages of crack length will grow faster than the smaller ones, the distributions become considerably flatter at each step.

NUMERICAL EXAMPLE

In order to obtain a first estimate of the expected influence of the deterioration due to dynamic loads, the concept described in the preceding sections is

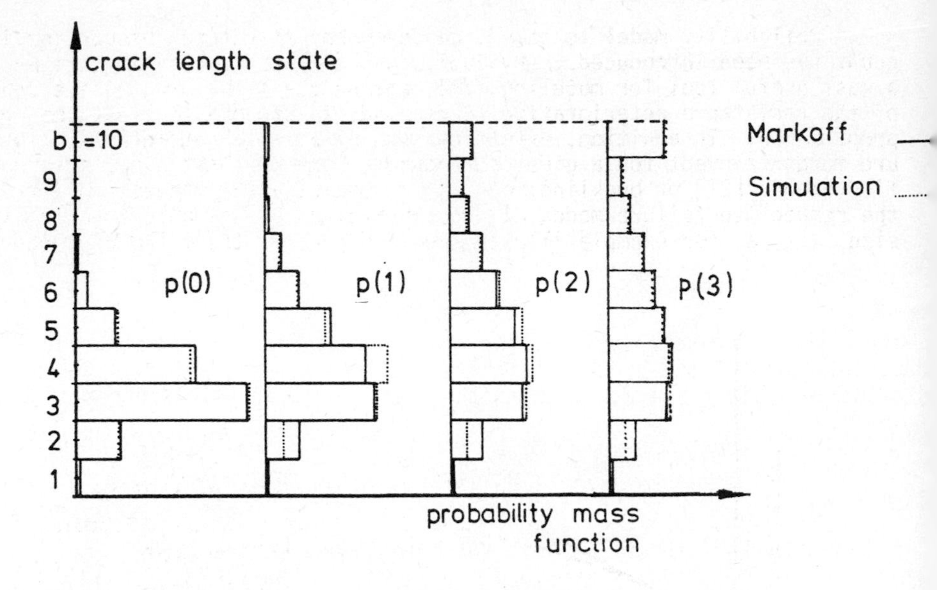

Fig. (5) - Comparative presentation of random crack growth

now applied to a practical example. The calculations are carried out for a welded joint, which can be considered as a tubular joint of an offshore steel construction. The cyclic load action on the cross section is modeled by a Gumbel distribution based on previous structural calculations [17,18]. The initial crack length is modeled by a distribution as shown in Figure 4. The information of [19] concerning the expected size of flaws in platform constructions is also taken into account. Steel of the quality St 520-3 is considered and the statistical data as presented in [20] are utilized. Assuming a natural period of T = 3 sec and a mean occurrence rate of 1 storm/year (duration 12 hours), the results for the time dependent failure probabilities are presented in Table 1.

TABLE 1 - FAILURE PROBABILITY AS A FUNCTION OF SERVICE LIFE

service life failure probability	1 year	15 years	30 years
without deterioration of resistance	$1.84 \cdot 10^{-4}$	$2.03 \cdot 10^{-3}$	$3.91 \cdot 10^{-3}$
with deterioration of resistance	$1.90 \cdot 10^{-4}$	$3.44 \cdot 10^{-3}$	$6.36 \cdot 10^{-2}$

As the present calculation is based on some simplifying assumptions, the result should be considered as a first estimate. Nevertheless, the trend is obvious that the influence of the deterioration of the resistance of the critical cross section increases progressively during the service life. For an average service life of 30 years, the influence may increase up to one order of magnitude or more, Table 1. Due to the influence of corrosion, which is yet to be investigated, this trend is likely to be of even greater importance.

DISCUSSION OF RESULTS AND CONCLUSIONS

A reliability model to take time dependent structural properties into account has been introduced. Obviously, the application of fracture mechanics is a most useful tool for modeling time dependency as the progressive characteristics of the resistance deterioration corresponds to the physical phenomenon of crack propagation. In addition, extending the reliability consideration to other failure modes relevant for a given structure - such as plastic collapse due to plastified hinges [21] or buckling, etc., - a comparison of the reliability levels of the respective failure modes allows a statement on the allowable initial flaw size. If, as for example illustrated in Figure 6, the failure probability for

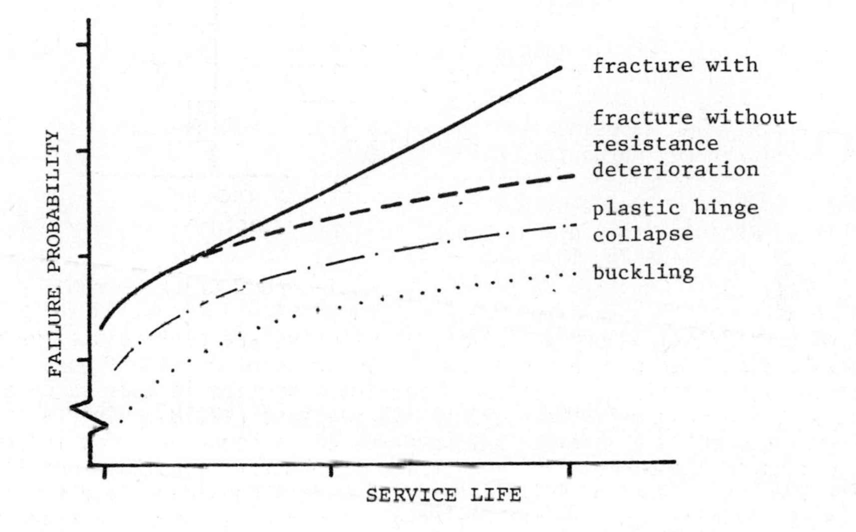

Fig. (6) - Time dependent failure probabilities

the fracture mode is considerably higher than for other modes, the requirement of an equilibrated reliability level would require an improvement of the welds reducing the initial crack length by a more stringent inspection and repairing procedure respectively.

REFERENCES

[1] Schüeller, G. I., Kafka, P. and Schmitt, W., "Some aspects of the interaction between systems and structural reliability", Invited Paper (M8/1*), Fourth Int. Conf. on Struct. Mech. in Reactor Technology, Berlin, Aug. 12-19, 1979.

[2] Benjamin, J. R., Schüeller, G. I. and Witt, J., (ed.), Proc. 2nd Int. Sem. Struct. Rel. Mech. Comp. Subassembl. Nucl. Power Plants, Special Volume, Nuclear Engr. and Design, Vol. 59, 1980.

[3] Freudenthal, A. M. and Schueller, G. I., "Risikoanalyse von Ingenieurtragwerken", Konstruktiver Ingenieurbau, Berichte, Report No. 25/26, pp. 7-95, Univ. Bochum, 1976.

[4] Freudenthal, A. M., Garrelts, J. M. and Shinozuka, M., "The analysis of structural safety", J. Struct. Div., Proc. ASCE, Vol. 92, No. ST1, pp. 267-325, 1966.

[5] Turner, C. E., "Methods for post-yield fracture safety assessment", in: Post Yield Fracture Mechanics, D. G. H. Latzko, ed., Applied Science Publishers, London, 1979.

[6] Feddersen, C. E., "Evaluation and prediction of the residual strength of center cracked tension panels", ASTM STP 486, pp. 50-78, 1970.

[7] Schwalbe, K.-H., "Bruchmechanik metallischer Werkstoffe", Carl Hanser Verlag, München, Wien, 1980.

[8] Marshall, W., "An assessment of the integrity of PWR pressure vessels", U.K. Atomic Energy Authority, Harwell, 1976.

[9] Becher, P. E. and Pedersen, A., "Application of statistical linear elastic fracture mechanics to pressure vessel reliability analysis", Nucl. Eng. Design 27, pp. 413-425, 1974.

[10] Dowling, N. E. and Begley, J. A., "Fatigue crack growth during gross plasticity and the J-integral", in Mechanics of Crack Growth, ASTM STP 590, pp. 82-103, 1976.

[11] Dowling, N. E., "Crack growth during low-cycle fatigue of smooth axial specimens", ASTM STP 637, 1976.

[12] Wüthrich, C., "The extension of the J-integral concept to fatigue cracks", to be published in Int. Journ. of Fracture.

[13] Broek, D. and Leis, B. N., "Similitude and anomalies in crack growth rates", Proc. of Fatigue 81 SEE, Warwick University, England, pp. 129-146, March 1981

[14] Oswald, G. F., "Über den Rißfortschritt bei variabler Belastung", Internal Working Report No. 2, Project No. SFB96/B8-8, Techn. Univ. München, March 1981.

[15] Provan, J. W., "The micromechanics approach to the fatigue failure of polycrystalline metals", in Voids, Cavities and Cracks in Metallic Alloys, J. Gittus, ed., ARES MECHANICA Monograph Elsevier's Applied Science Publishers Ltd., England, 1980.

[16] Schueller, G. I., "Einführung in die Sicherheit und Zuverlässigkeit von Tragwerken", Verlag Wilhelm Ernst & Sohn, Berlin - München, 1981.

[17] Oswald, G. F., "Zuverlässigkeitsbeurteilung von Strukturen unter Berücksichtigung zeitabhängiger Festigkeitseigenschaften", Berichte zur Zuverlässigkeitstheorie der Bauwerke, Heft 47, pp. 43-61, München, 1980.

[18] Oswald, G. F. and Schüeller, G. I., "Fracture mechanical aspects in the reliability assessment of structures", Seminar on Stochastic Structural Mechanics and Reliability, SFB 96, No. 49, München, pp. 30-33, September 1980.

[19] Pan, R. B. and Plummer, F. B., "A fracture mechanics approach to non-overlapping, tubular K-joint fatigue life prediction", Journal of Petroleum Technology, pp. 461-468, April 1977.

[20] Hawranek, R., "Verteilung der Festigkeitseigenschaften der Baustähle St 37-2 und St 52-3", Arbeitsbericht zur Sicherheit der Bauwerke, LKI, Heft 5, TUM, January 1974.

[21] Schüeller, G. I. and Kappler, H., "Zuverlässigkeitsorientierte Bemessung von Meeresplattformen", Berichte zur Zuverlässigkeitstheorie der Bauwerke, Heft 59, Munchen, 1981.

AN EXPERIMENTAL INVESTIGATION OF FATIGUE RELIABILITY LAWS

J. W. Provan and Y. Theriault

McGill University
Montreal, Quebec, H3A 2K6, Canada

ABSTRACT

This paper begins with a review of the more familiar fatigue reliability laws, such as the normal, log normal, Weibull, etc., and the not so familiar laws of Birnbaum and Saunders [1] and Provan [2], which have been proposed as suitable descriptions of the scatter in fatigue data and the reliability of components being subjected to fatigue loading situations. Except for the latter two, all of the above laws are, to some extent, based on empiricism with little or no effort being made to relate the form of these laws to the microstructural behavior of the material being investigated. The Birnbaum-Saunders model, on the other hand, is based on a probabilistic description of the microstructural fatigue crack growth process but is, unfortunately, based on the known to be false premise that the fatigue crack grows per fatigue cycle an amount which is independent of the size of the crack. The reliability law of Provan, however, does depend on the crack size in the sense that it is based on the stochastic fatigue crack growth being described by a Markov linear birth process.

The main purpose of the present paper is, then, to present the results of an experimental program developed to ascertain the validity of these laws, especially in relation to the "Provan" law. The program involved fatigue failing a statistical number of specimens whose manufacture and preparation were carefully controlled thereby being able to infer that the resulting scatter in the fatigue data is due, in part, to the microstructure itself rather than to variations in the surface and testing conditions. The program further involved the viewing of a selected number of specimen fracture surfaces using a scanning electron microscope to determine the basic material crack growth transition intensity required for the implementation of the "Provan" law. In this way, the applicability of both the empirical reliability laws and the microstructural laws is assessed.

INTRODUCTION

In the coming years, many design considerations will be related to an assessment of the reliability of a particular component or system. As is well-known, the determination of the fatigue reliability of large engineering structures by full scale testing is, in many cases, just not feasible. Hence, reliability estimates, as they pertain to such systems, are primarily based upon:

i) the fracture and fatigue material characteristics obtained by small scale laboratory testing,

ii) the characterization, usually by non-destructive evaluation techniques, of material and manufacturing defects,

iii) stress analysis, fracture mechanics and fatigue damage characterizations, and

iv) the assessment of size effects.

It is the desire of the on-going research being carried out at McGill to estimate the mechanical reliability of large components and/or structures based upon probabilistic and stochastic descriptions of both microstructural defects and a material's fracture and fatigue degradation mechanisms.

In the past, this research endeavor has led to mathematical descriptions of internal stress statistical distributions [3], fatigue crack initiation [4], propagation [5] and reliability [2], as they apply to polycrystalline metals. The aim of the current research program is to experimentally assess the validity of existing reliability laws along with the "Provan" law, whose derivation is detailed in the Proceedings of the first Defects and Fracture Symposium held in Tuczno, Poland [2].

THE RELIABILITY LAWS

The more common empirical and semi-empirical laws, along with some of their characteristics and applications, are tabulated in Table 1, where: erfc stands for the complementary error function, I the incomplete gamma function and Γ the gamma function. All of these reliability laws have previously been utilized to describe safety levels for all types of systems. Since these laws are, for the most part, described in the literature pertaining to mechanical reliability, (see [6], for example), they will not be dealt with in detail in the present situation.

A further two laws whose derivations are based on the modelling of fatigue crack growth in polycrystalline metals are presented in Table 2, where: erf stands for the error function. The Birnbaum and Saunders law [1] is based on the false premise that the incremental fatigue crack extension during any fatigue cycle has a probability distribution depending *only* on the amplitude of the applied load. This is not the case and although the law, as we will see, when used in an empirical manner does give a recommendable description of the scatter in fatigue data, it is fundamentally based on a crack growth model which does not describe the observed physics of the situation.

The "Provan" law, first presented in [2] and where all the symbols are defined, is based on the assumption that fatigue crack growth does depend on both the nominal load amplitude and the crack size and does indeed predict an exponential mean crack growth rate which is in keeping with the dimensionability and similarity descriptions of fatigue crack growth [7]. Since the development of this new reliability law has been fully described in [2], it will not be repeated here.

TABLE 1 - THE COMMON RELIABILITY LAWS

Distribution	Parameters	Cumulative Distribution Function	Mean	Variance
Exponential	$\delta > 0$	$F(x) = 1 - e^{-x/\delta}$	$\frac{1}{\delta}$	$\frac{1}{\delta^2}$
Normal	$-\infty < \mu < \infty$ $\sigma > 0$	$F(x) = \frac{1}{2}\,\mathrm{erfc}[\frac{-1}{\sqrt{2}}(\frac{x-\mu}{\sigma})]$	μ	σ
Log Normal	$\infty < \mu' < \infty$ $\sigma' > 0$	$F(x) = \frac{1}{2}\,\mathrm{erfc}[\frac{-1}{\sqrt{2}}(\frac{\ell n\, x-\mu'}{\sigma'})]$	$e^{\mu'+\sigma'^2/2}$	$e^{2\mu'+\sigma'^2} \times (e^{\sigma'^2}-1)$
Gamma	$\alpha > 0$ $\eta > 0$	$F(x) = I[\frac{\alpha x}{\sqrt{\eta}}, (\eta-1)]$	$\frac{\eta}{\alpha}$	$\frac{\eta}{\alpha^2}$
Weibull (2 parameters)	$\theta > 0$ $\beta > 0$	$F(x) = 1-\exp[-(\frac{x}{\theta})^{\beta}]$	$\theta\Gamma(\frac{1}{\beta}+1)$	$\theta^2\{\Gamma(\frac{2}{\beta}+1) - [\Gamma(\frac{1}{\beta}+1)]^2\}$
Weibull (3 parameters)	$\theta>0,\ \theta_0>0$ $\beta > 0$	$F(x) = 1-\exp[-(\frac{x-\theta_0}{\theta-\theta_0})^{\beta}]$	$\theta_0 + (\theta+\theta_0)\Gamma(\frac{1}{\beta}+1)$	"
Gumbel (largest value)	$-\infty<\alpha<\infty$ $\delta > 0$	$F(x) = \exp\{-\exp[-(\frac{x-\alpha}{\delta})]\}$	$\alpha + 0.5776\delta$	$1.645\ \delta^2$
Gumbel (smallest value)	$-\infty<\alpha<\infty$ $\delta > 0$	$F(x) = 1-\exp[-\exp(\frac{x-\alpha}{\delta})]$	$\alpha - 0.5776\delta$	"

a) - Empirical and semi-empirical reliability laws

Distribution	Comments
Exponential	- probability distrib ution of life when a constant conditional failure (or hazard) rate, λ, is assumed - more appropriate for complex systems or assemblies - special case of Weibull and Gamma distributions
Normal	- applicable as a time to failure model if $\mu \geq 3\sigma$ otherwise, density function must be truncated - symmetric distribution without shape parameter
Log Normal	- derived from the consideration of analytical process wherein failure is due to the growth of a fatigue crack
Gamma	- time-to-failure distribution of a system if system failure occurs as soon as K independent subfailures have taken place at a constant rate α.
Weibull (2 parameters)	- widest applicability of all failure distributions - more appropriate to represent the life distribution of parts or components
Weibull (3 parameters)	- same comments as for 2-parameter Weibull - also known as the Type III asymptotic distribution for minimum values - for cases where the lower bound life is non-zero.
Gumbel (largest value) Gumbel (smallest value)	- also known as the Type I asymptotic distribution for maximum (largest) or for minimum (smallest) values - applicable whenever failure depends on the largest or smallest value of a variable (e.g., strength or flaw size) whose distribution is of the exponential type, such as normal, gamma, or exponential. - no shape parameter

b) - A brief review of their characteristics

TABLE 2 - FATIGUE CRACK GROWTH RELIABILITY LAWS

Distribution	Parameters	Cumulative Distribution Function	Mean	Variance
Birnbaum-Saunders	$\alpha > 0$ $\beta > 0$	$F(x) = \frac{1}{2}\,\mathrm{erfc}\{\frac{-1}{\alpha\sqrt{2}}\,[(\frac{x}{\beta})^{1/2} - (\frac{x}{\beta})^{-1/2}]\}$	$\beta(1 + \frac{\alpha^2}{2})$	$(\alpha\beta)^2(1 + \frac{5\alpha^2}{4})$
'Provan'	$\lambda > 0$ $Vac > 0$	$F(x) = \frac{1}{2}\,\{\mathrm{erf}(\frac{\mu_{a_0} e^{\lambda x} - \mu_{a_f}}{\sqrt{2\,Vac}}) + \mathrm{erf}\,(\frac{\mu_{a_f}}{\sqrt{2\,Vac}})\}$	known complex expression	known complex expression

a) - Fatigue crack growth laws.

Distribution	Comments
Birnbaum-Saunders	- probabilistic model based on the restrictive assumption that fatigue crack growth is independent of crack length.
'Provan'	- derived from a probabilistic micromechanics model of fatigue crack propagation - more applicable for components - no shape parameter

b) - A brief review of their characteristics.

THE EXPERIMENTAL INVESTIGATION

A high purity oxygen-free-high-conductivity (O.F.H.C.) brand copper was used in the experimental validation procedure. Each specimen was meticulously prepared in accordance with the low-cycle-fatigue standard test procedures designated in ASTM E466 [8]. The specimen dimensions are as shown in Figure 1 with the final surface finish being predominantly in the longitudinal direction and each specimen being heat treated so as to eliminate any residual machining effects.

All but two specimens were tested under strain control at a total strain amplitude of:

$$\frac{\Delta\varepsilon}{2} = 0.003 \qquad (1)$$

at a frequency of 0.5 Hz. After eliminating those specimens whose mode of failure clearly showed the influence of spurious testing conditions, such as the fatigue crack initiating in the vicinity of the clip-on-gage's knife edge, the results of 18 tests were collected and are presented in Figure 2.

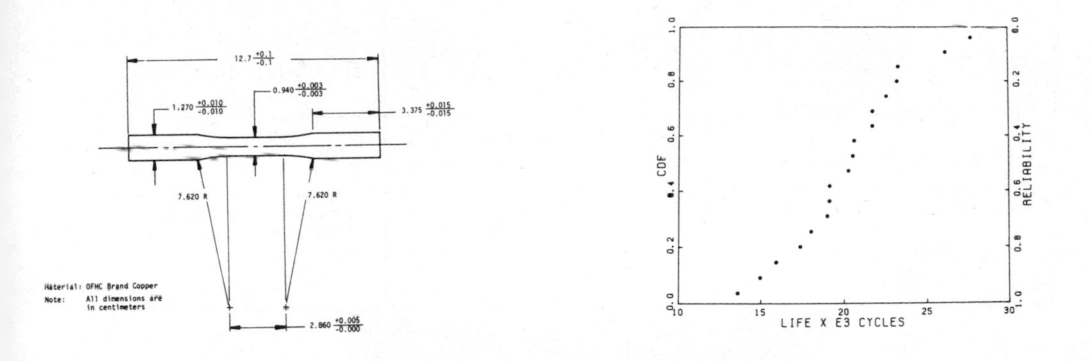

Fig. (1) - Specimen configuration

Fig. (2) - The raw experimental results

However, before discussing the implications of this data on the previously described reliability laws, it is advantageous to return for a moment to a consideration of both the main objective of this research endeavor and the basic Markovian stochastic description of the fatigue crack growth process.

THE FATIGUE TRANSITION INTENSITY

Detailed in [5] and not repeated here are two methods, one theoretical and one experimental, of assessing the basic material characteristic parameter which must be determined if the stochastic fatigue crack growth mechanism is to be successfully modelled in terms of Markovian processes. This basic property is termed the *transition intensity* and is given the symbol λ. It depends on the load amplitude and material and both completely characterizes the Markov linear birth interpretation of fatigue crack growth and is a major parameter in the "Provan" reliability law given in Table 2. The entire development of this reliability law would be vindicated in the case where both, or either, of these methods of assessing λ were to give reliability estimates bourne out first by laboratory experimentation, the current paper being the first such example, and then subsequently confirmed by an accumulation of in-service large component data. Alternatively stated: gaining confidence in the Markov description of the fatigue process and the laboratory interpretation of the transition intensity is tantamount to gaining confidence in the fatigue reliability of large engineering structures manufactured from the material in question.

A. λ_{th}

As noted in [5], the theoretical interpretation of the transition intensity, λ_{th}, is determined by an iteration procedure which utilizes two points on the Woehler "S-N" data curve pertaining to the polycrystalline metal being investigated. For the O.F.H.C. being examined here, one of these points corresponded to the strain amplitude of 0.003 while the other was obtained by testing the remaining two specimens at a strain amplitude of 0.0015. By implementing the procedure detailed in [5], the value of λ_{th} in the present case was determined as:

$$\lambda_{th} = 0.12 \times 10^{-3}/\text{cycle}; \ \frac{\Delta\varepsilon}{2} = 0.003 \tag{2}$$

Its validity is assessed in the following section.

B. λ_{exp}

Again, as detailed in [5], an experimental procedure using a SEM (scanning electron microscopic) technique is also available for determining the transition intensity; in this case designated by λ_{exp}. Briefly, this fractographic procedure counts striations and their spacings, an example of which is shown in Figure 3, and from the statistics generated in this way is able to find λ_{exp}. Following this procedure, for the O.F.H.C. copper being viewed here, the experimental value of the transition intensity was found to be:

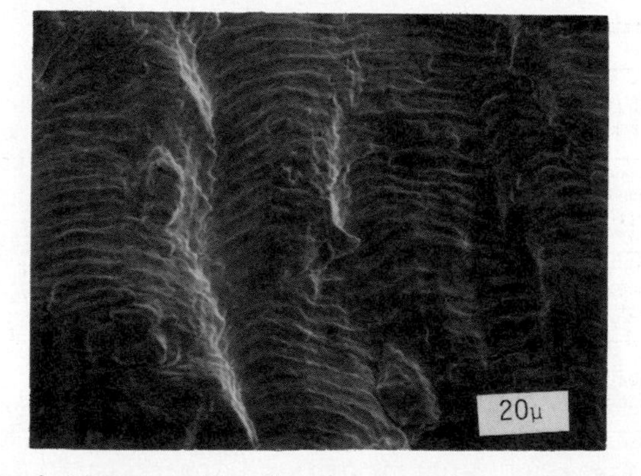

Fig. (3) - Striations on the fracture surface of an O.F.H.C. specimen

$$\lambda_{exp} = 1.89 \times 10^{-3}/\text{cycle}; \ \frac{\Delta\varepsilon}{2} = 0.003 \tag{3}$$

The discrepancy between the theoretical and experimental values of λ was previously noted [5] and will be further discussed in the concluding remarks.

VERIFICATION OF THE RELIABILITY LAWS

The ten laws briefly listed in Tables 1 and 2 were computer curve fitted to the experimental data presented in Figure 2. The results are shown in Figure 4. It should be noted that the Weibull 3-parameter curve does not appear since it gave results completely in keeping with those found using the 2-parameter form.

As can be seen from these comparisons all, with the possible exception of the exponential law, describe the experimental data to varying degrees of goodness-of-fit. An idea of just how good this comparison is, is given in Table 3, where the simple error estimate shown for each comparison is listed.

TABLE 3 - A LISTING OF RELIABILITY LAW PARAMETERS AND ERRORS

Distribution	Parameters	Error = $\left(\frac{1}{18}\sum_{1}^{18} E_i^2\right)^{1/2}$
Exponential	α = 16,060 cycles δ = 4419 "	0.222
Normal	μ = 20,280 cycles σ = 3,653 "	0.029
Log Normal	μ' = 9.902 σ' = 0.184	0.036
Gamma	α = 0.00152/cycle η = 30.82	0.032
Weibull (2 parameters)	θ = 21,770 cycles β = 6.269	0.037
Gumbel (largest value)	α = 18,550 cycles δ = 3,187 "	0.049
Gumbel (smallest value)	α = 22,030 cycles δ = 3,228 "	0.053
Birnbaum-Saunders	β = 19,960 cycles α = 0.179	0.036
'Provan'	λ = 2.0E-4/cycle Vac = 1.8E-5 m^2	0.048

Of primary interest to the current investigation is, however, the empirically determined value of the Markovian transition intensity, λ_{emp}. From Figure 4i and Table 3, the value obtained in the present circumstances is:

$$\lambda_{emp} = 0.20 \times 10^{-3}/\text{cycle}; \ \frac{\Delta\varepsilon}{2} = 0.003 \tag{4}$$

CONCLUDING REMARKS

As can be seen from Figure 4 and Table 3, the utilization of any of the reliability laws tested in this investigation, with perhaps the exception of the exponential law, will provide an adequate description of the bulk of the scatter in fatigue data. The differences mainly arise near the "tails" of each distribution with the most appropriate being chosen on the basis of the particular application being contemplated. The applicability of both the Birnbaum and Saunders and "Provan" distributions *in an empirical fashion* has also been confirmed. The previously noted [2] skewness to the left of the "Provan" distribution may be of interest in situations where this non-normal characteristic has been observed in fatigue data.

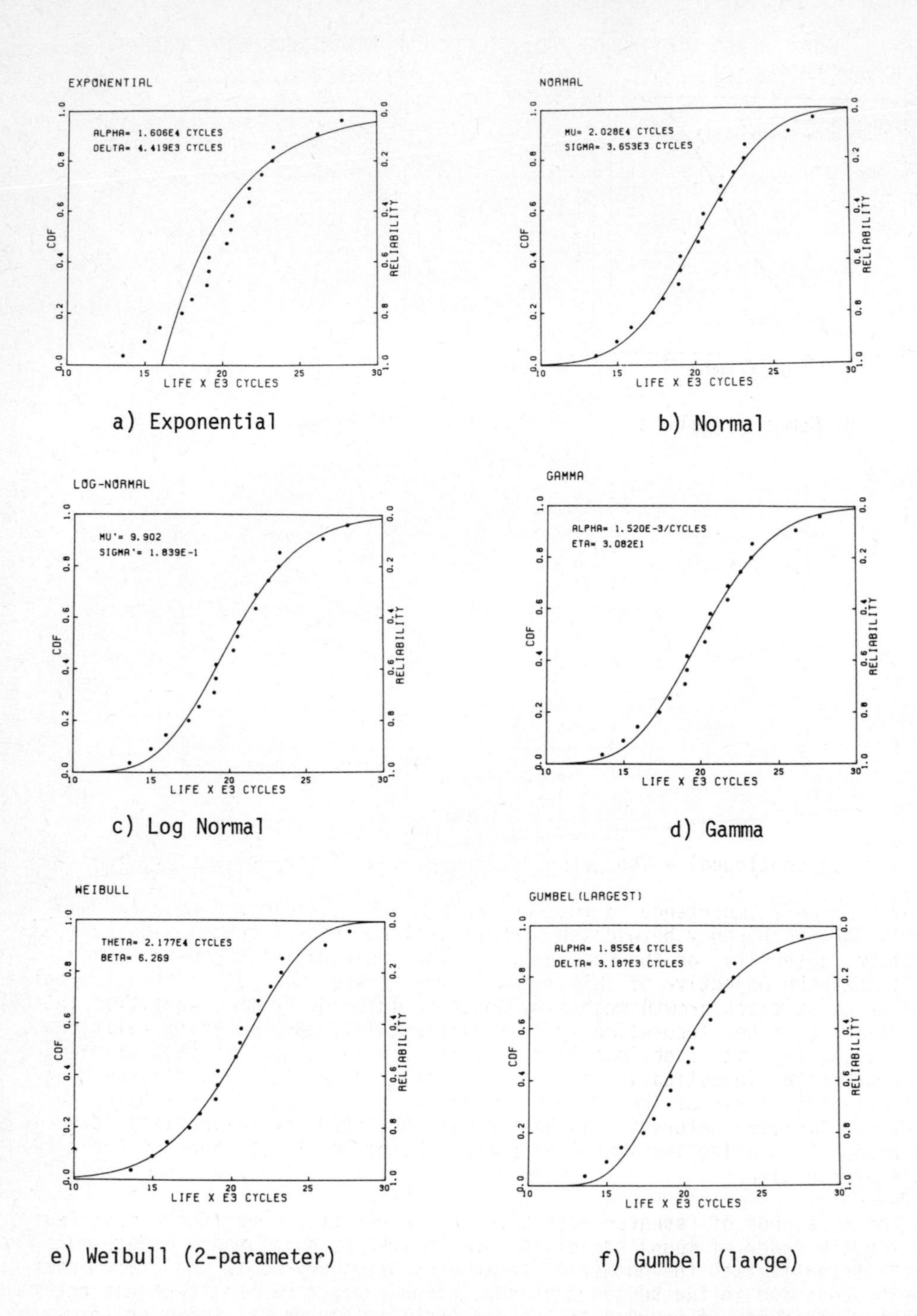

Fig. (4) - The graphical comparison of the reliability laws

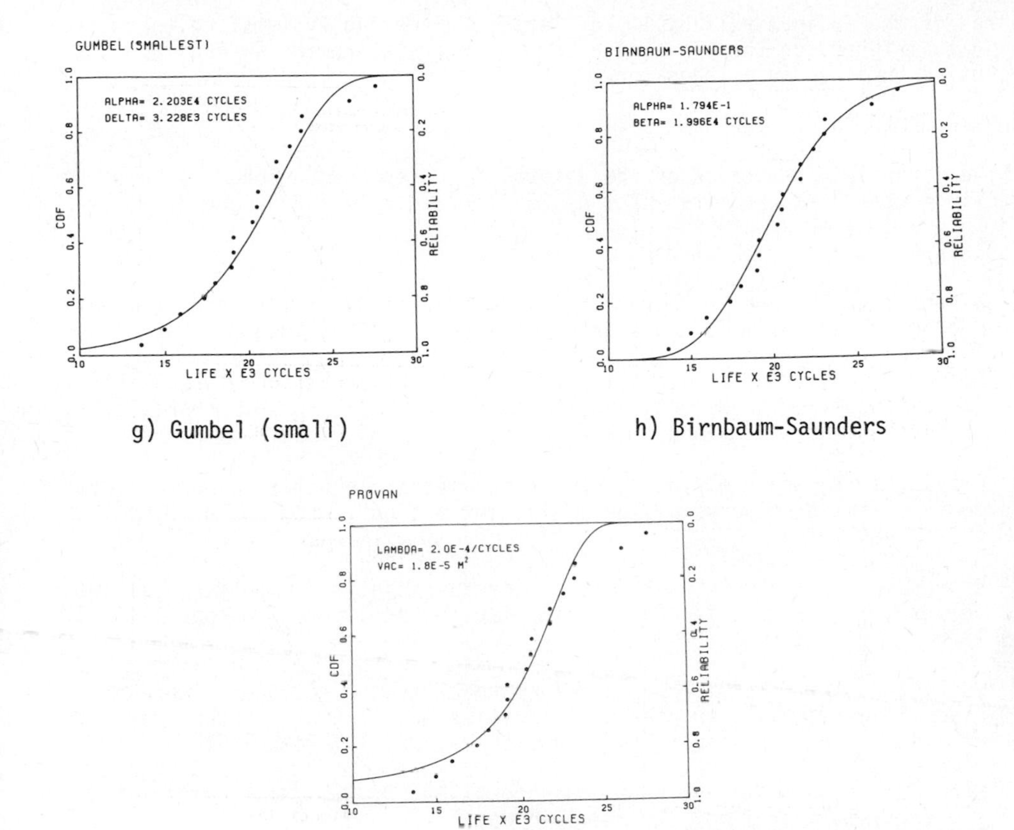

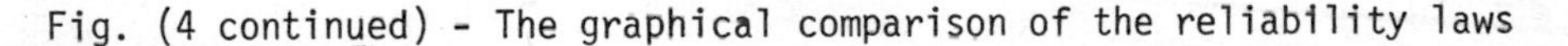
Fig. (4 continued) - The graphical comparison of the reliability laws

Of much more importance to the long-term objectives of this research program is the discrepancy between the theoretical and experimental values of the transition intensity, on the one hand, and the empirical value on the other. Recall that the objective of this research was to see if a reliability law based on a Markovian crack growth mechanism could be inferred from a transition intensity measured either theoretically or experimentally. By comparing relations (2), (3) and (4), it is obvious that this conclusion cannot, as yet, be drawn. The experimental investigation described in this paper clearly indicates that the theoretical basis of the "Provan" reliability law has to be seriously reviewed and improved before it can be used in the hoped for reliability form. This research is being instigated at present based on the findings of the current investigation.

Another avenue of research must also now be pursued. One of the main reasons for the order of magnitude difference in the three values of λ determined in this paper is that the number of fatigue cycles involved in the crack initiation process and in the subsequent propagation process has not been correctly assessed. This is, of course, an age old problem but one which can no longer be ignored if the fatigue reliability of large engineering systems are to be successfully inferred on the basis of a Markovian crack growth law and material properties obtained in a laboratory.

ACKNOWLEDGEMENTS

The financial assistance of the Natural Sciences and Engineering Research Council of Canada through its Grant A7525 is gratefully acknowledged.

REFERENCES

[1] Birnbaum, Z. W. and Saunders, S. C., "A probabilistic interpretation of Miner's rule", SIAM J. Appl. Math., 16, pp. 637-652, 1968.

[2] Provan, J. W., "A fatigue reliability distribution based on probabilistic micromechanics", in Defects and Fracture, G. C. Sih and H. Zorski, eds., Martinus Nijhoff Publishers, The Hague, pp. 63-69, 1982.

[3] Provan, J. W. and Axelrad, D. R., "The effective elastic response of randomly oriented polycrystalline solids in tension", Arch. Mech. Stos., 28, pp. 531-547, 1976.

[4] Provan, J. W., "A model for fatigue crack initiation in polycrystalline solids", in Fracture 77, 2, D. M. R. Taplin, ed., Waterloo, pp. 1169-1176, 1977.

[5] Provan, J. W., "The micromechanics approach to the fatigue failure of polycrystalline metals", Chapter 6 of Cavities and Cracks in Creep and Fatigue, J. Gittus, ed., Applied Science Publishers, pp. 197-242, 1981.

[6] Mann, N. R., Schafer, R. E. and Singpurwalla, N. D., "Methods for statistical analysis of reliability and life data", John Wiley, 1974.

[7] Liu, H. W., "Fatigue crack propagation and applied stress range - an energy approach", J. Basic Engng., Trans. ASME, 85, pp. 116-122, 1963.

[8] 1980 Annual Book of ASTM Standards, 10, American National Standard, ANSI/ASTM E466-76, pp. 614-619, 1980.

SECTION VIII
FRACTURE TESTING AND CONTROL

ON CRACKING INSTABILITY IN PLATES CONTAINING CIRCULAR HOLES

C. L. Chow

The BFGoodrich Research & Development Center
Brecksville, Ohio 44141

ABSTRACT

For functional reasons, most engineering structures contain geometrical discontinuities such as notches, fastener/rivet holes, etc. This paper describes an investigation into the effect of multiple circular holes on the stability condition of cracked plates.

The cracked plates chosen for the investigation contained two circular holes placed immediately above and below a crack. The plate height (H) and width (W) ratio used was 0.730. At this ratio an isolated crack without the presence of holes had been found to be unstable. For the cracked geometry with the holes, the stability conditions were found to be governed not by the geometrical plate ratio (i.e., H/W) as in the case of an isolated crack, but by the ratio of the crack length and hole radius (a/R). The prediction that stable crack growth can be attained for the a/R ratio, about 1.3, agreed fully with the experimental observations.

INTRODUCTION

In the past decades, fracture mechanics has evolved as a quantitative aid to safe design by focusing on crack initiation characteristics (or sometimes known as local instability) as a means of fracture control. Design engineers have gained increasing awareness of the need for utilizing existing materials to their fullest potential and of learning to "live" with real-life cracked/flawed elements.

In addition to crack initiation aspects of design analysis, engineers have also been aware of the development of crack instability phenomenon during the crack propagation process (which is sometimes known as global instability) [1-10]. The development of global instability can cause a crack to propagate in an unstable manner resulting in catastrophic failure of an entire structure. This has led to the increasing interest by structural engineers in the characterization and prediction capability of fracture instability. Therefore, the prevention of possible catastrophic failure of a structure may also be included in the overall design analysis.

The cracking instability described in this paper refers to conditions of global rather than local instability. Cracking instability analysis is sometimes related by some investigators to the crack path and multiple crack interaction problems [11-14]. No attempt is made in this paper to relate the stability of cracking criteria for characterizing those problems.

In fracture toughness testing, the cracking instability behavior of test specimens have long been observed. Irwin, in 1958, was the first investigator who successfully expounded the observed instability phenomenon through the crack-extension force concept. This concept was extended by Krafft, et al [2], Srawley [3], Clausing [5] and Glucklich [6] to explain the cracking behavior of a wide range of crack geometries under different modes of loading. Gurney [4] and Sih [15,16], on the other hand, independently derived stability criteria which have been found equally valid in characterizing a wide range of fracture problems under both linear and nonlinear elastic structural behavior.

Examination of the stability criteria proposed by Irwin, Gurney and Sih reveals that their criteria are, in principle, equivalent. They are all based on the balance between the absorbing fracture surface energy of the material under consideration and the elastic strain energy stored in the test system. The energy concept postulates that a stable crack propagation is attained when the ability of the test material to absorb strain energy to create a new surface is balanced by the release of stored elastic strain energy in the system. Otherwise, the released excess energy converts into kinetic energy, resulting in uncontrol or unstable crack growth.

Due primarily to mathematical complexities in the analysis of the fracture surface and stored elastic strain energies, information characterizing the stability conditions for crack propagation is limited and confined mainly to cracking stability of an isolated embedded crack in simplified finite or rectangular plates [17]. For functional reasons, most engineering structures contain geometrical discontinuities such as notches, fastener/rivet holes, etc. This paper describes an investigation into the effect of multiple circular holes on the stability conditions of cracked plates.

A. Stability Conditions of Crack Propagation

While the stability criteria proposed by Irwin [1], Gurney [4] and Sih [15] are equally valid, the quasistatic energy analysis due to Gurney was chosen for the present investigation. Gurney's approach has been particularly valuable in characterizing crack stability for both linear and nonlinear elastic structures [4,7]. The stability criteria based on the quasistatic energy analysis are expressed as [7]

$$\frac{dG_c}{dA} \geq \left(\frac{\partial G}{\partial A}\right)_Q = GSF \tag{1}$$

for $dQ/Q > 0$ under load-control testing condition, and

$$\frac{dG_c}{dA} \geq \left(\frac{\partial G}{\partial A}\right)_u = GSF \tag{2}$$

for du/u > 0 under displacement-control testing conditions. In equations (1) and (2), G_c denotes fracture toughness; G, strain energy release rate; A, crack surface area; GSF, geometrical stability factor; Q and u are respectively load and displacement at the loading point. It is clear from the equations that the governing parameters controlling the crack stability are the material property G_c and the geometrical stability factor (GSF). The term $\partial G/\partial A$ was denoted as GSF because it is related to the relative difference between the stored elastic strain energies at different crack lengths, and these stored energies are in turn dependent upon the specimen geometry and mode of loading.

The physical implications of stability conditions described in equations (1) and (2) may be examined with a graphical method. Figures 1a and 1b depict a "family" of G curves in a load-displacement diagram characterizing fracture be-

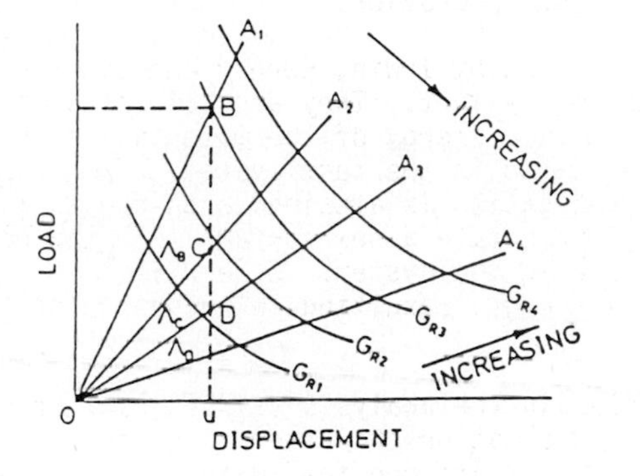

Fig. (1a) - Load-displacement diagram of stable crack propagation

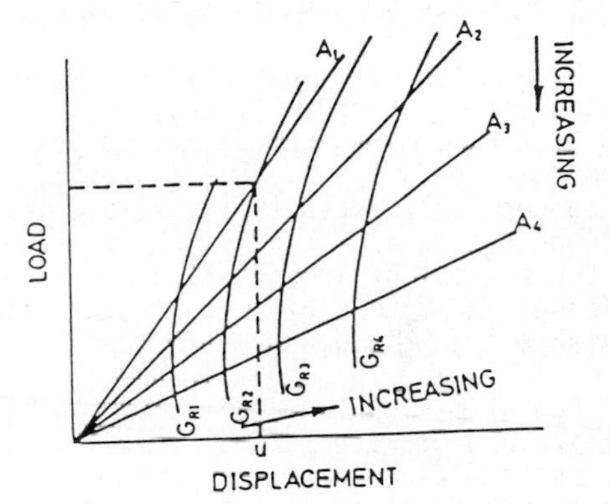

Fig. (1b) - Load-displacement diagram of unstable crack propagation

havior for two different types of specimen geometry and/or loading. It is evident from examining the G-loci depicted in Figure 1a that the term $\partial G/\partial A$ has negative values, thus satisfying the stability criteria of equations (1) and (2). On the other hand, positive values of $\partial G/\partial A$ can be deduced from Figure 1b, indicating a situation in which a crack propagates unstably. The G-loci depicted in Figure 1a are typical of test specimens having low height/width (H/W) ratios while the G-loci shown in Figure 1b represent the unstable behavior of test specimens with high H/W ratios. The interpretation of stability conditions from the family of G-curves will be discussed further and verified experimentally in the next section.

Another graphical interpretation, based on the load-displacement-toughness (Q-u-G) diagram shown in Figure 1, may be made by considering the changes of strain energy in a test specimen with respect to the increase of crack length.

Since

$$\left(\frac{\partial G}{\partial A}\right)_u = \left(\frac{\partial^2 \Lambda}{\partial A^2}\right)_u$$

then

$$(\frac{\partial G}{\partial A})_u = [\frac{(\Lambda_B-\Lambda_C)-(\Lambda_C-\Lambda_D)}{\Delta A}]_u$$

where Λ_B, Λ_C and Λ_D are the strain energies of crack lengths A_1, A_2 and A_3, respectively at a constant displacement (Figure 1a). Since the denominator in equation (3) is negative, stable crack propagation can only be achieved if the numerator is positive, or the area of triangle OBC must be greater than that of OCD. It can be observed from Figure 1a that the triangle OBC is greater than OCD, thereby implying that the test specimen exhibits stable crack propagation. Similarly, we can easily deduce that the test specimen displacing Q-u-G in Figure 1b exhibits unstable crack growth. These conclusions agree with those with an earlier graphical interpretation based on equation (2).

EXPERIMENTS AND DISCUSSION OF RESULTS

For the sake of illustration, the analysis and subsequent discussions on the stability conditions of crack propagation in plates containing an isolated embedded crack with or without the presence of multiple circular holes, Figure 2, are con-

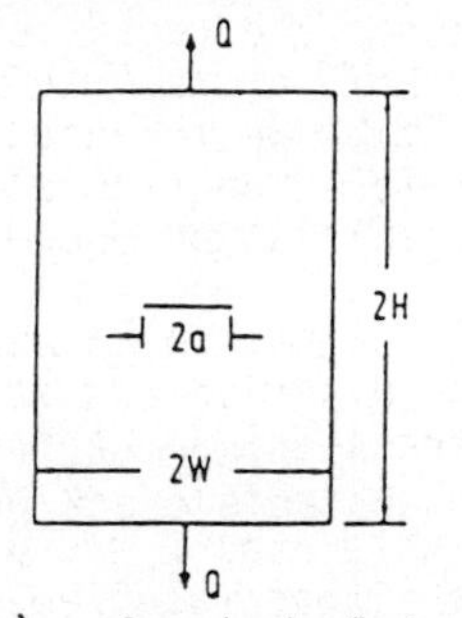

Fig. (2a) - Cracked plate under concentrated tensile load

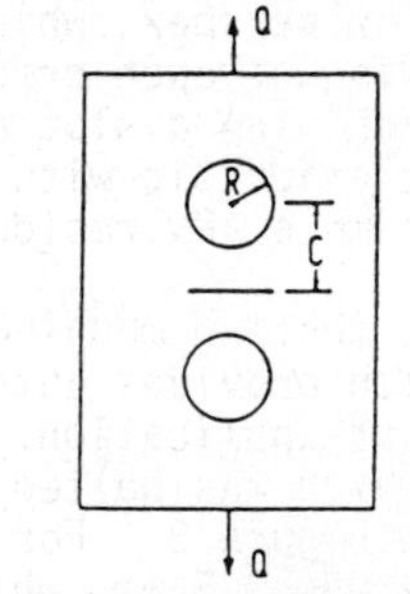

Fig. (2b) - Cracked plate with circular holes under concentrated tensile load

fined to the displacement control condition ($du/u > 0$) based on equation (2). Similar analysis can be easily extended for the load control testing condition ($dQ/Q > 0$). Based on the stability criterion of equation (2), it is evident that positive and negative GSF values characterize respectively unstable and stable crack growth. Alternatively, the stability criterion may be expressed in terms of normalized geometrical stability factor (NGSF).

Since the fracture toughness is expressed as [18]

$$G = -(\frac{\partial \Lambda}{\partial A})_u = -(\frac{u^2}{2})(\sum_{i=1}^{n} iC_i A^{i-1}) \qquad (4)$$

We can express the NGSF in an alternative form by combining equations (2) and (4) as

$$\frac{1}{G_c} \frac{dG_c}{d(a/W)} \geq - \frac{BWu^2}{2G} \sum_{i=2}^{n} i(i-1)(C_i A^{i-2}) = NGSF \quad (5)$$

where a denotes crack length; B, specimen thickness; W, specimen half width; C_i, polynomial coefficients; G, strain energy release rate; G_c, fracture toughness.

A. Rectangular Plate with an Isolated Embedded Crack

In order to examine the predictability of the stability of crack propagation based on equation (5), a series of experiments was conducted with rectangular cracked specimens of varying plate length (H) and width (W) with or without the presence of holes, Figure 2. The experiments to verify the validity of cracking stability for an isolated crack as depicted in Figure 2a were performed with a constant width of 90 mm and thickness of 6 mm. However, the plate height varied from 27 mm to 256 mm in steps of 25.4 mm, making a total of 9 different specimen geometries. For each specimen geometry, the crack length was made to vary from 6.35 mm to 44.5 mm in steps of 6.35 mm, providing a total of six compliance measurements.

The test material was polymethylmethacrylate (PMMA), commonly known as plexiglass or perspex, which was chosen for its brittleness and positive visual identification of each crack increment. An initial crack in a specimen was produced by machining a slot at which a "natural" crack or flaw was created by tapping at the crack tip with a sharp knife. The specimen was then carefully annealed to remove any residual stresses during the crack initiation process.

The testing machine used for the experiment was an Instron Universal Tester, which provides autographic recording of load and displacement at the point of load application. Two holes were drilled in the specimen at the loading point which was bolted to the Instron tester by pins placed in the holes, as shown in Figure 3. For the experiments reported in this paper, the diameter of the pins was 9.5 mm, which was designed to sustain the load required for the tests. The loading arrangement was intended to create the theoretical point-load system, although a closer approximation to the system may be achieved by using, if possible, smaller diameter loading pins. It may be worth noting that, provided that the pin holes are not located in the immediate vicinity of the crack tips and there is no irreversible deformation of the pin holes, the size of pin hole used should not adversely affect the fracture toughness measurement. This is because the fracture toughness is evaluated based on the differential strain energy stored in the entire test specimen for crack increment, rather than a single strain energy value.

Figure 3 depicts typical experimental results of a test specimen (H = 27 mm, W = 89 mm) exhibiting stable crack propagation. The compliance curves for seven different crack lengths were measured from the specimen, whereas the lines of constant fracture toughness were computed based on equation (4). Stable crack propagation was observed for all the crack lengths from the specimen. This agrees with the theoretical prediction based on the stability conditions derived from equations (2) and (3) as described in an earlier paragraph.

Figure 4 depicts typical experimental results of a test specimen (H = 65 mm, W = 89 mm) exhibiting unstable crack propagation. The Q-u-G diagram

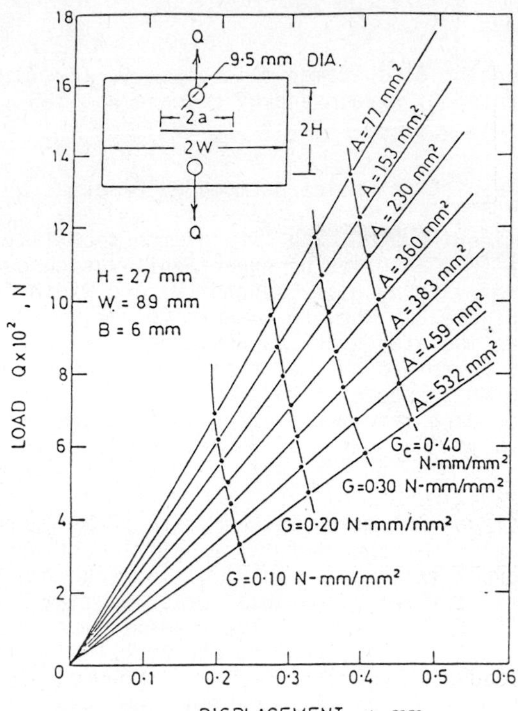

Fig. (3) - Fracture toughness G with quasistatic energy method

appears similar in shape to that depicted in Figure 1b, which we predicted theoretically would display unstable crack growth.

Figure 5 summarizes the stability conditions of crack propagation for all nine specimen geometries examined. As plotted, the figure reveals the effect of the H/W ratio on the stability of cracking of the point-load center-cracked rectangular plate. It can be observed from the figure that there is clearly a geometrical transition at H/W ratio of about 0.525, above which stable crack propagation cannot be attained. This information is important for the fracture researchers who are often interested in the evaluation of fracture toughness within the quasistatic range [4,7-9]. The prediction described in Figure 5 agrees well with the observed cracking instability conditions.

B. Cracked Rectangular Plates with Multiple Circular Holes

To examine the effect of multiple holes on the stability conditions of cracked rectangular plates, two circular holes were placed immediately above and below an embedded crack for a chosen geometrical H/W ratio of 0.73 (see Figure 2b). This ratio was chosen because a cracked plate at this H/W ratio without the presence of holes had found, from Figure 5, to be unstable.

Another series of experiments were performed for rectangular plates having plate width W of 89 mm, plate height H of 65 mm, thickness B of 5.84 mm, cir-

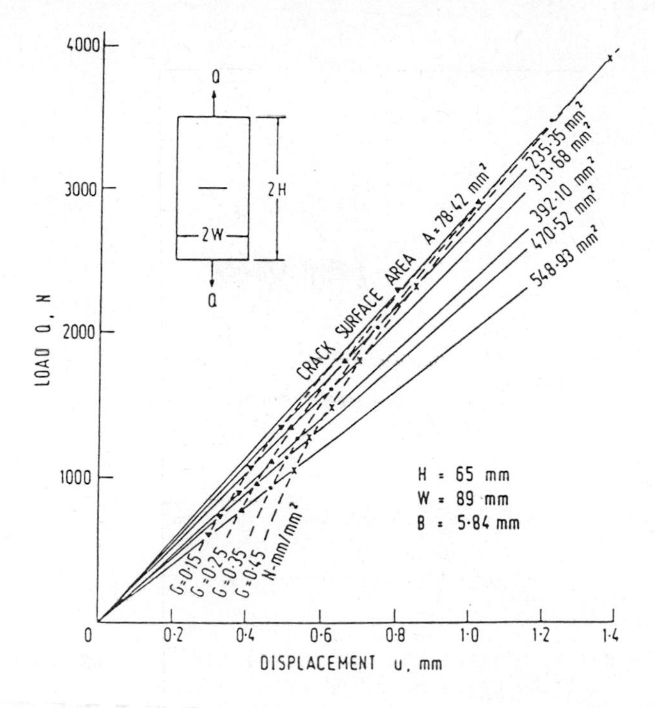

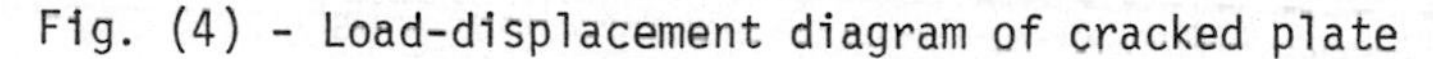
Fig. (4) - Load-displacement diagram of cracked plate

cular hole radius of 12.7 mm and the distance between the hole center and an embedded crack C of 25.4 mm (see Figure 2b). The crack length, a, varied from 12.7 mm to 89 mm. Figure 6 depicts the compliance curves measured from the autographic records of the Instron testing machine, and the evaluations of lines of constant G were based on equation (4). Compliance curves were also used to predict the normalized geometrical stability factor (NGSF) based on equation (5), and the results are summarized in Figure 7. Continuous (solid) and discontinuous lines in Figure 7 are the predicted NGSF values for the cracked plates without and with circular holes, respectively. Crack propagation for the cracked plate without the holes is, as expected, unstable throughout the entire crack length range. For the cracked plate with two circular holes, stable crack propagation can be observed from Figure 7 for the crack length to the hole radius ratio of about 1.3. Above this point, the crack propagation remains unstable. Circular holes demonstrate the ability to convert unstable crack propagation to a stable propagation. This may be attributed to the fact that compressive stresses are developed in the immediate region between the two circular holes [19]. Maximum compressive stresses are located along the line between the hole centers and their magnitudes decrease with the increase in distance from the center line. When the crack tip is located within the compressive region, the presence of holes can convert an unstable crack growth to a stable crack growth up to the a/R ratio of about 1.3. If the crack tip is remotely located from the compressive zone, the crack propagates unstably as if the holes are not present. This prediction agrees well with the results we observed experimentally when the cracked plates containing two circular holes were tested using a Universal Instron testing machine.

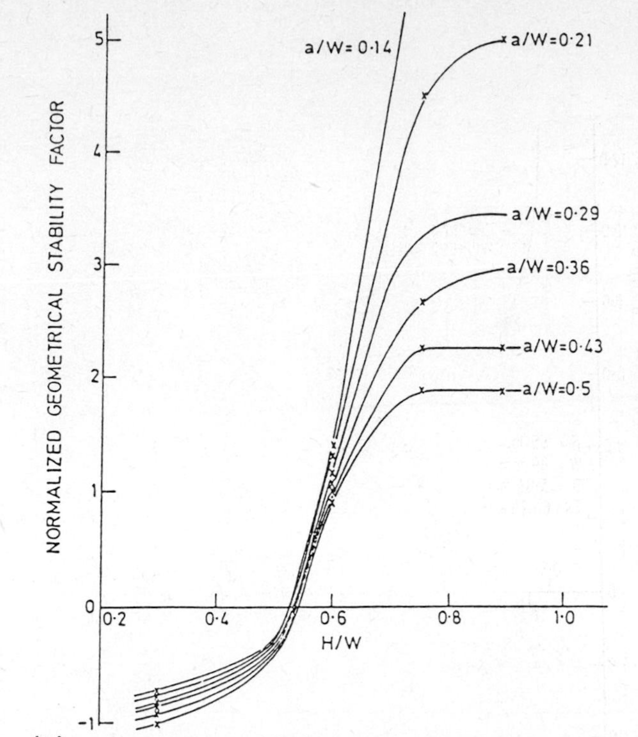

Fig. (5) - Fracture stability of rectangular plate with an isolated crack at different H/W ratios

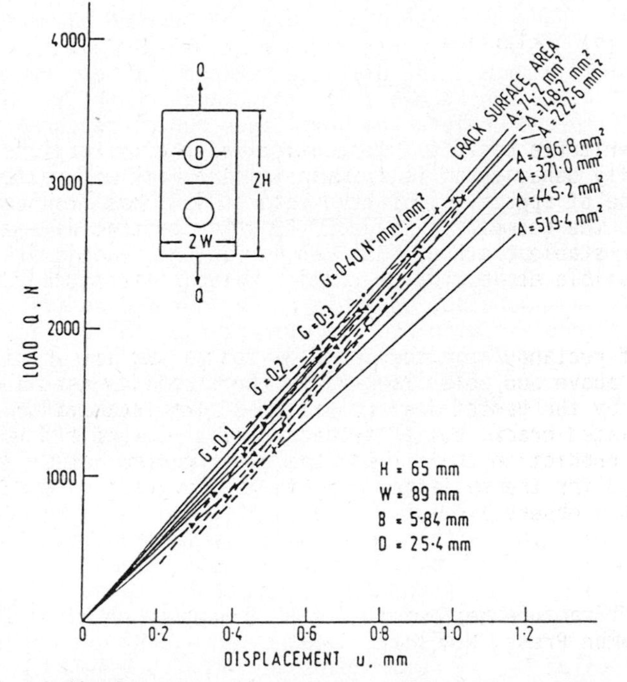

Fig. (6) - Load-displacement diagram of cracked plate with circular holes

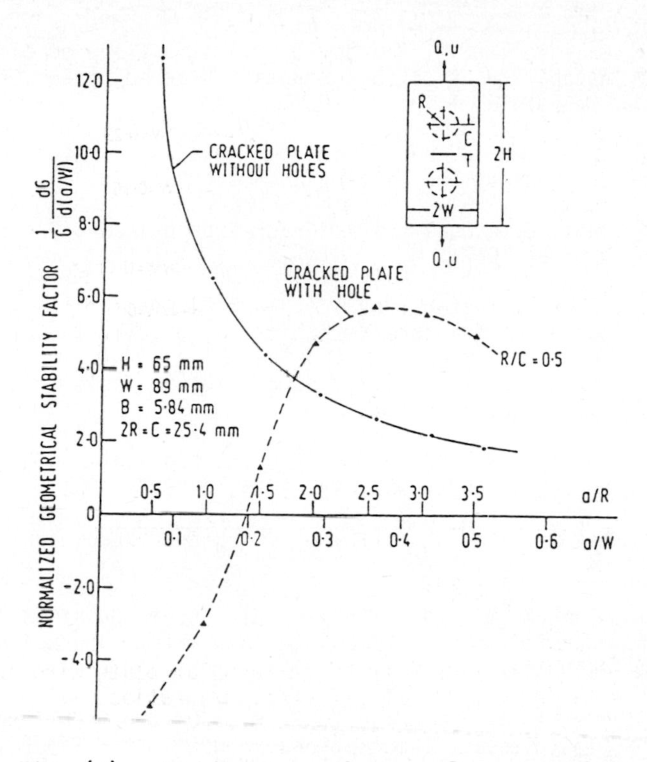

Fig. (7) - Stability conditions of cracked plates

CONCLUSIONS

The stability criteria based on the quasistatic energy analysis were used to examine the stability behavior of rectangular plates with and without circular holes. For the case of specimens embedded with an isolated crack only without the circular holes, the geometrical transition ratio of specimen height and width (H/W) for attaining stable crack propagation was 0.525. No stable crack growth was found to be possible above this transition ratio, irrespective of the crack length.

For the case of rectangular cracked plates containing two circular holes placed immediately above and below the crack, the stability conditions were found to be governed not by the geometrical H/W ratio of the rectangular plate as in the case of an isolated crack, but by the ratio of the crack and hole at a radius of about 1.3. The prediction that the stable crack growth can be attained for the a/R ratio of 1.3 for the rectangular plate of H/W ratio of 0.73 agrees fully with the experimental observations.

REFERENCES

[1] Irwin, G. R., "Fracture Mechanics", First Symposium on Naval Structural Mechanics, Pergamon Press, New York, 1958.

[2] Krafft, J. M., Sullivan, K. M. and Boyle, R. W., "Effect of Dimensions on Fast Fracture Instability of Notched Sheets", Proceedings of Crack Propagation Symposium, Cranfield, 1, p. 8, 1961.

[3] Srawley, J. E. and Brown, W. F., "Fracture Toughness Testing and Its Applications", American Society for Testing and Materials, 381, p. 133, 1965.

[4] Gurney, C. and Hunt, J., "Quasistatic Crack Propagation", Proceedings Royal Society, A299, p. 508, 1967.

[5] Clausing, D. P., "Crack Instability in Linear Elastic Fracture Mechanics", International Journal of Fracture Mechanics, 5, p. 211, 1969.

[6] Glucklich, J., "On Crack Stability in Some Fracture Tests", Engineering Fracture Mechanics, 3, p. 333, 1971.

[7] Gurney, C. and Ngan, K. M., "Quasistatic Crack Propagation in Nonlinear Structures", Proceedings Royal Society, A325, p. 207, 1971.

[8] Gurney, C. and Mai, Y. W., "Stability of Crackings", Engineering Fracture Mechanics, Vol. 4, p. 853, 1973.

[9] Chow, C. L. and Lam, P. M., "Stability Conditions in Quasistatic Crack Propagation for Constant Strain Energy Release Rate", Journal of Engineering Materials and Technology, 96, p. 41, 1974.

[10] Mai, Y. W., Atkins, A. G. and Caddell, R. M., "On the Stability of Cracking in Tapered DCB Testpieces", International Journal of Fracture, 11, p. 939, 1975.

[11] Cotterell, B., "On Fracture Path Stability in the Compact Tension Test", International Journal of Fracture Mechanics, 6, p. 189, 1970.

[12] Andersson, H., "A Finite Element Representation of Stable Crack Growth", J. Mech. Phy. Solids, 21, p. 337, 1973.

[13] Broberg, K. B., "On Stable Crack Growth", J. Mech. Phy. Solids, 23, p. 215, 1975.

[14] Nemat-Nasser, S., Sumi, Y. and Keer, L. M., "Unstable Growth of Tension Cracks in Brittle Solids: Stable and Unstable Bifurcations, Snap-Through, and Imperfection Sensitivity", Int. J. Solids Structures, 16, p. 1017, 1980.

[15] Sih, G. C., "Mechanics of Fracture", Vol. 1, Noordhoff, Leyden, Holland, 1973.

[16] Sih, G. C., "Mechanics of Fracture", Vol. 4, Noordhoff, Leyden, Holland, 1977.

[17] Mai, Y. W. and Atkins, A. G., "Crack Stability in Fracture Toughness Testing", Journal of Strain Analysis, 15, p. 63, 1980.

[18] Chow, C. L. and Lau, K. J., "Fracture Studies of Point-Loaded Center-Cracked Plates", Journal of Strain Analysis, 12, p. 286, 1977.

[19] Ling, C. B., "The Collected Papers in Elasticity and Mathematics of Chih-Bing Ling", Institute of Mathematics, Academia Sinica, Taiwan, 1963.

EFFECT OF INCLUSIONS ON DYNAMIC TOUGHNESS OF LINE-PIPE STEEL

W. R. Tyson, J. D. Boyd and J. T. McGrath

Physical Metallurgy Research Laboratories, Canada Centre for Mineral and Energy Technology, EMR, Ottawa, Canada K1A 0G1

ABSTRACT

As part of a comprehensive evaluation of line-pipe steels, toughness measurements have been made by drop-weight-tear testing (DWTT) of commercial steels. Desulphurization and shape control of inclusions have been found to exert dramatic effects on toughness anisotropy and energy absorption. Without such controls, HAZ toughness can be vanishingly small.

The results of DWTT tests are combined with information from quantitative metallography and Charpy tests to illuminate the significance of inclusion control in improving dynamic toughness.

INTRODUCTION

The demand for large-diameter pipelines for exploitation of frontier resources has stimulated a rapid evolution in pipemaking technology. Steelmakers have produced material of higher strength and better cleanliness, and spiral-welding processes of high productivity have been developed.

PMRL has monitored these developments in an ongoing project of evaluation of commercial line pipe, in parallel with laboratory-scale work to point the way to future technologies. In the course of this work, a great deal of data has been gathered on a variety of pipes which has enabled us to identify some of the effects of changes in commercial practice on pipe properties. The purpose of this paper is to demonstrate the marked effect of one such change, namely in inclusion control, on the toughness of pipe and of mill weldments.

STEELMAKING AND PIPE FABRICATION

Two sample pipes were investigated in this work, designated "AB" and "AG". Both were 1000 mm diameter, 13.7 mm wall thickness, CSA Z245 Grade 483 line pipe. Their chemical compositions are given in Table 1. Both pipes were produced from electric-furnace, ingot cast, controlled-rolled, coiled skelp. However, pipe AB was made from semikilled steel, whereas pipe AG is representative of the current practice of aluminum deoxidation and ladle desulphurization by injection of calcium silicide. Both pipes were spiral welded by the double-submerged-arc process, with a helix angle of 70° measured from the pipe axis.

TABLE 1 - COMPOSITIONS OF STEELS (WT. PCT.)

	C	Mn	Si	Al	S	P	Nb	Mo	Ni	Cr	Cu
AB	0.06	1.80	0.02	<0.002	0.017	0.009	0.058	0.27	0.11	0.04	0.22
AG	0.08	1.70	0.15	0.02	0.004	0.014	0.075	0.21	0.20	0.09	0.23

EXPERIMENTAL INVESTIGATIONS

Metallographic observations of the microstructure and inclusion distribution of the base metal were made, and quantitative measurements were obtained using a Cambridge Model 720 image analysis system.

Tensile and toughness measurements were made on the base (parent) metal and the weldments. A pipe section is shown in Figure 1, indicating one of the

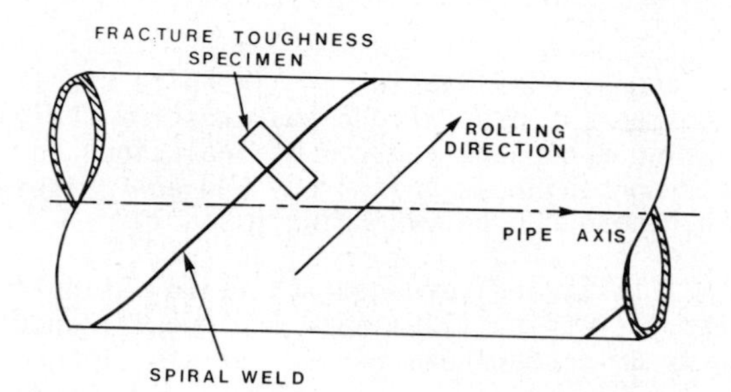

Fig. (1) - Pipe section, showing orientation relationships. Specimen is transverse to rolling direction, T (RD).

specimen orientations. Conventional full-size Charpy specimens were cut with their long axes either parallel (longitudinal, L) or perpendicular (transverse, T) to the rolling direction (RD) or pipe axis (PA), and notched parallel to the thickness direction. DWTT specimens (ASTM E436) were cut with long axes in T(PA) or T(RD) using the above notation; a T(RD) specimen including a mill weld is shown in Figure 1. Notches parallel to the thickness direction were pressed in the base plate (parent metal, PM), weld metal (WM), or heat-affected zone (HAZ).

Charpy samples of the HAZ were machined in the T(RD) orientation, and precracked into the HAZ. In order to have the crack tip entirely within the HAZ, it was necessary to orient the notch parallel to the surface rather than the thickness.

The energy absorbed to fracture the DWTT samples was measured in a pendulum impact machine of 11 kJ capacity.

PROPERTIES

A. Microstructure, Strength

Through-thickness sections parallel (L) to the rolling direction are shown in Figure 2. Pipe AB had a duplex microstructure consisting of fine grained

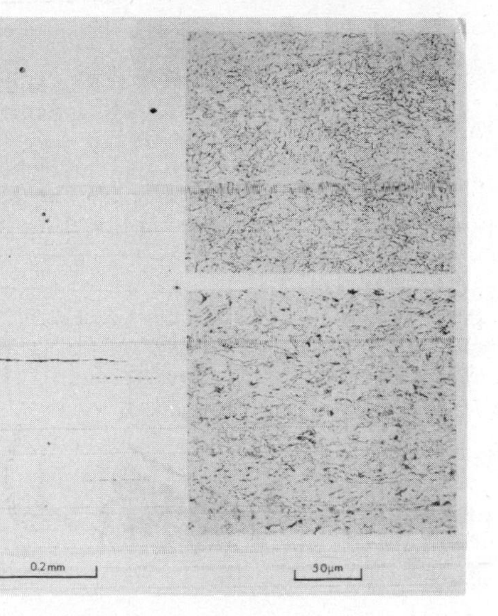

Fig. (2) - Micrographs of unetched (left) and etched (right) sections of AG (top) and AB (bottom)

acicular ferrite with a small fraction (<20%) of polygonal ferrite, whereas the structure of pipe AG was composed exclusively of acicular ferrite. The number and the distribution of inclusions was significantly different for the two pipes. Pipe AB had many long inclusion stringers, whereas pipe AG had a much lower area fraction of inclusions which were predominantly equi-axed. Quantitative measurements supported these observations (Table 2).

TABLE 2 - QUANTITATIVE METALLOGRAPHIC DATA

	Inclusion area pct., f	Mean aspect ratio $\bar{R}$
AB, L(RD)	0.134 ± 0.002	7.96
AB, T(RD)	0.176 ± 0.002	3.02
AG, L(RD)	0.08 ± 0.01	1.00
AG, T(RD)	0.07 ± 0.01	0.99

Tensile properties are given in Table 3. Anisotropy in strength and difference in strength between the two steels is small, varying by a maximum

TABLE 3 - TENSILE PROPERTIES

	0.2% offset yield stress, MPa	UTS MPa	% Elong. (50 mm)
AB, L(RD)	534	623	34.7
AB, T(RD)	559	616	26.7
AG, L(RD)	563	655	35.3
AG, T(RD)	606	681	31.3

of about 8% in both cases. Elongation in the L(RD) direction is almost identical for AB and AG, but the elongation of AB in the T(RD) orientation is significantly less than in the L(RD) orientation while AG shows much less anisotropy. These differences in anisotropy of ductility are reflected and greatly magnified in toughness anisotropy.

B. Charpy V-notch Impact Energy

Charpy impact results are shown in Figures 3 and 4 for pipes AB and AG respectively. Anisotropy in toughness is markedly greater for AB than AG.

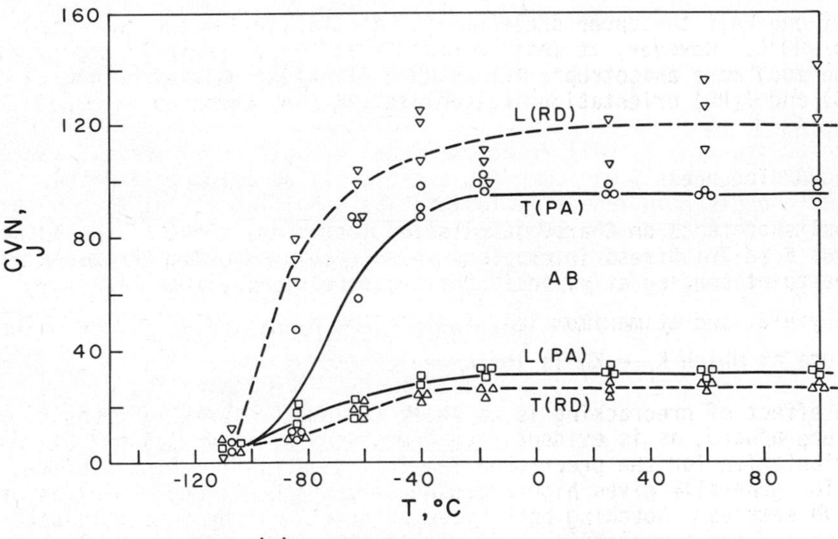

Fig. (3) - CVN impact energies for AB

As expected, the L(RD) orientation gives highest energy absorption and T(RD) the least (since the crack propagates parallel to the "fibre direction" of the microstructure in this orientation), with L(PA) and T(PA) falling between the extremes. The ductile/brittle transition temperature, defined at a CVN

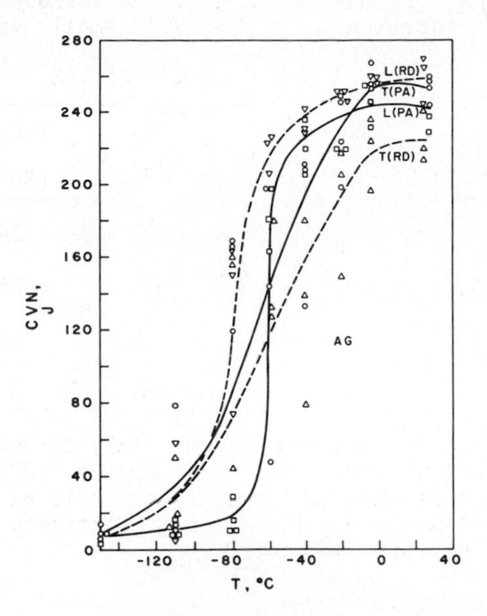

Fig. (4) - CVN impact energies for AG.
(Symbols same as for Figure 3)

energy of one-half the upper shelf energy, is similar for the two steels at -60°C to -80°C. However, it must be noted that the upper shelf energy is much lower and much more anisotropic for AB than AG, being only 120 J and 25 J for the L(RD) and T(RD) orientations respectively for AB compared with 260 J and 225 J for AG.

C. HAZ Toughness

Results of tests on Charpy samples precracked into the HAZ are shown in Figures 5 to 7. Stress intensities were calculated using the usual formula for three-point bending at yield (K_Y) or at fracture (K_F) when fracture preceded yield, and at maximum load (K_M) following general yielding. The temperature at which $K_F = K_Y$ is indicated as T_F.

The effect of precracking is to shift the ductile/brittle transition temperature upward, as is evident upon comparison of Figures 4 and 5. Although notch orientation for the precracked samples is parallel to the surface, this orientation generally gives higher toughness than the through-thickness notch of the CVN samples. Notching both types of sample in the same orientation would increase the temperature shift resulting from sharpening the notch still more.

The toughness of the HAZ is lower than that of the base metal and T_F is higher for the HAZ (Figures 5,6). However, in pipe AG above T_F the value of K_M is comparable for base metal and HAZ, indicating that propagation of

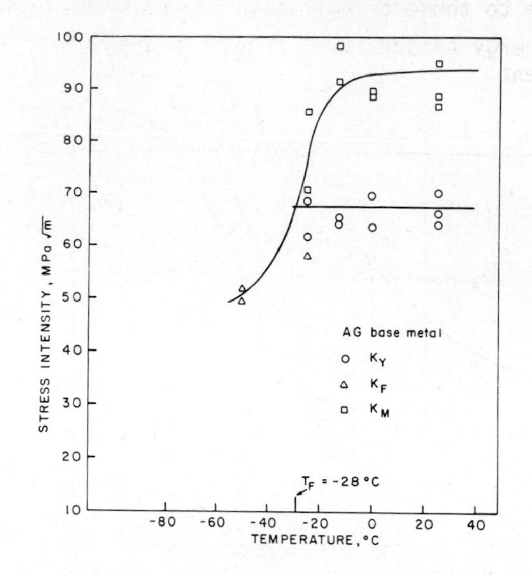

Fig. (5) - Precracked Charpy results for AG base metal. T(RD) orientation, notched parallel to plate surface.

ductile fracture is accompanied by extensive plasticity and high energy absorption for both the base metal and HAZ. This was confirmed by comparing measured energy absorption for the two cases.

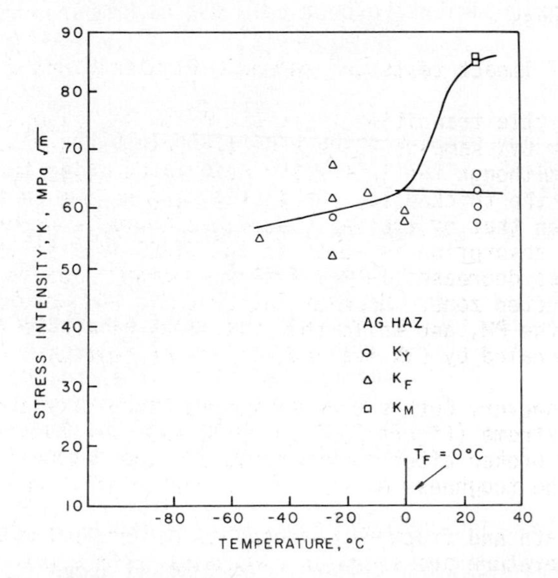

Fig. (6) - Precracked Charpy results for HAZ of AG T(RD) orientation.

Turning to results for AB, Figure 7 shows that while values of K_F for the HAZ are comparable to those of AG, above T_F the value of K_M remains small. Very little energy is absorbed in crack propagation, again confirmed by measurement.

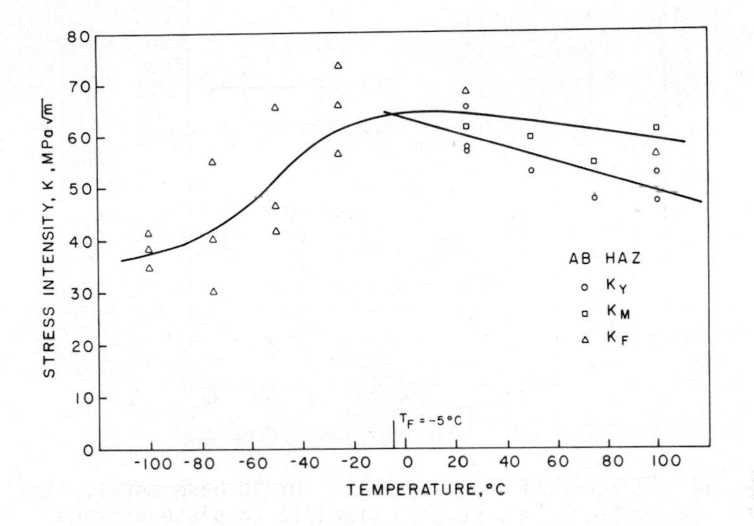

Fig. (7) - Precracked Charpy results for HAZ of AB. T(RD) orientation.

D. DWTT Toughness

Results of DWTT impact tests are shown in Figures 8 and 9.

The ductile/brittle transition temperature for AG, Figure 8, is much higher than that for CVN samples, Figure 4, and for the precracked Charpy samples, Figure 5 (although the latter are notched parallel to the surface rather than through the thickness). Similarly, the HAZ has a higher transition temperature than that of the HAZ precracked Charpy specimens (Figure 6). As expected, energy absorption is lower in the T(RD) than in the T(PA) orientation, and toughness decreases in the following order: parent metal, weld metal, and heat-affected zone. However, at 20°C the HAZ absorbs 25% of the energy absorbed by the PM, and while the anisotropy of the PM toughness is larger than that revealed by CVN testing, it is not severe.

For pipe AB, however, both the anisotropy of toughness and the level of HAZ toughness are extreme (Figure 9). In the T(PA) orientation, specimens were not completely broken after absorbing 11 kJ; the reason for this will be discussed later. The toughness of the HAZ is vanishingly small.

The fracture path and fracture surfaces of the HAZ-notched samples broken at room temperature are shown in Figure 10. The crack ran parallel to the rolling direction, and remained in the HAZ. AG showed brittle cleavage

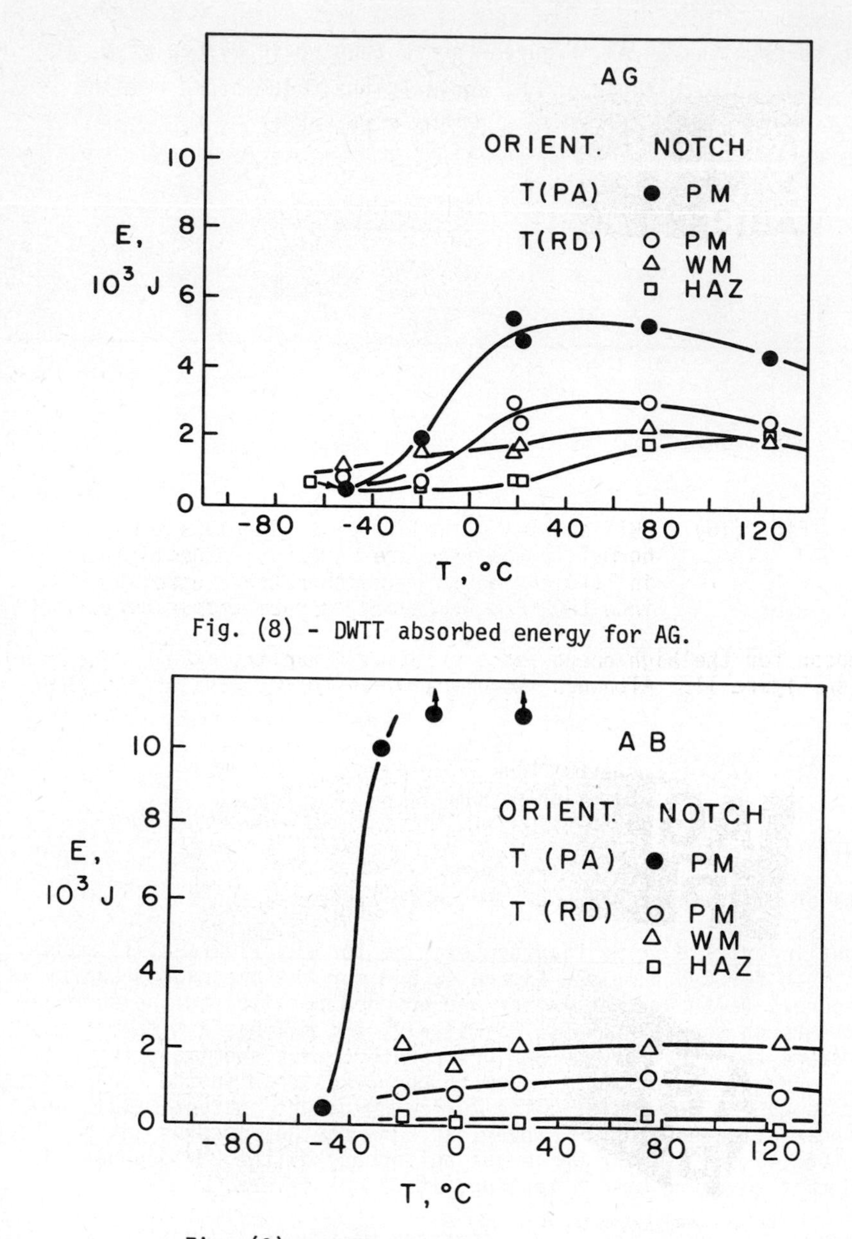

Fig. (8) - DWTT absorbed energy for AG.

Fig. (9) - DWTT absorbed energy for AB.

with some shear lip, while AB exhibited a woody, fibrous fracture as well as cleavage. The fibrous fracture of the HAZ of AB occurred with very little energy absorption (Figure 9), even at temperatures as high as 120°C. At 120°C, the HAZ of AG fractured by 100% shear with almost the same energy absorption as the parent metal (Figure 8).

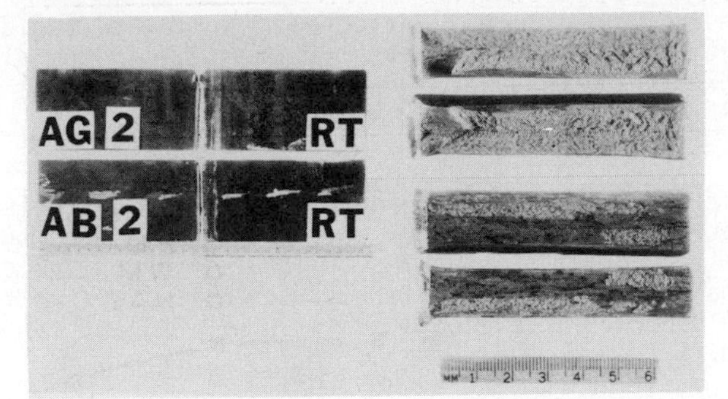

Fig. (10) - DWTT samples viewed in plan (left), and normal to the fracture surfaces. Specimens in T(TD) orientation, with notch pressed into the HAZ, fractured at room temperature.

The reason for the high energy absorption of AB in the T(PA) orientation is evident in Figure 11. Although the crack began to run parallel to the

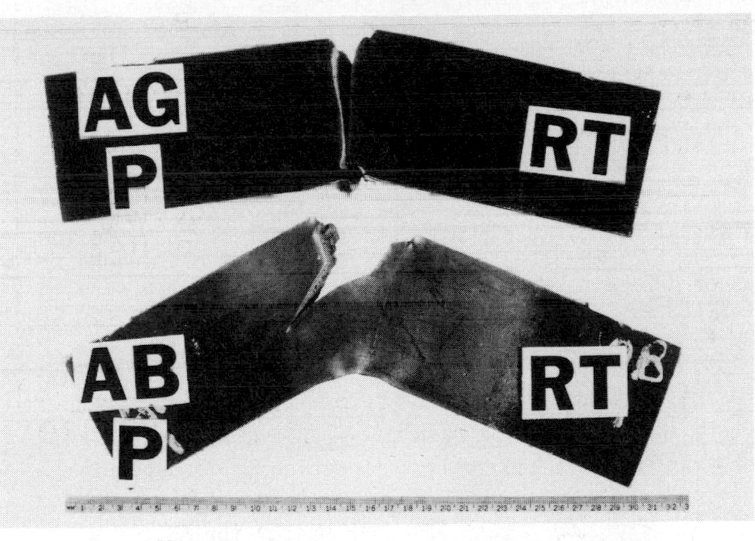

Fig. (11) - DWTT samples viewed in plan. Specimens in T(PA) orientation, with notch pressed into parent metal, fractured at room temperature.

direction of the pipe axis, it soon was deflected into the rolling direction. Pipe AG, because of its much smaller anisotropy, fractured in the direction of highest driving force (Figure 11).

DISCUSSION

The principal microstructural difference between the two steels in this study is the area fraction and the shape of the inclusions. Consistent with the relative S levels (Table 1), AB contains ~0.16 area percentage of inclusions while AG has less than half this amount. More importantly, the inclusions in AB were deformed into elongated stringers while those of AG were nearly equi-axed.

Inclusion stringers present an easy path for crack propagation. Hence, impact energy absorption is highly anistropic in pipe AB for both CVN and DWTT specimens. The lowering of energy absorption is more pronounced for DWTT specimens of pipe AB notched in the HAZ, where the inherent low toughness of the coarse-grained matrix in combination with the inclusion stringers makes possible the propagation of fracture with almost negligible energy absorption, Figure 9.

There is a significant influence of both notch sharpness and specimen size on transition temperature. Transition temperatures of pipe AG are approximately -60°C, -30°C, and 0°C respectively for CVN (Figure 4), pre-cracked Charpy (Figure 5), and pressed-notch DWTT (Figure 8) specimens. This underlines the necessity for material specification and design to be based on the toughness testing of full-thickness specimens. However, there is correlation between tests using full-size and Charpy-size samples as demonstrated in this report, notably for anisotropy of properties, showing that quality control can be based on smaller specimens. Nevertheless, there are important quantitative differences when the section size is increased, for example in the transition temperature as discussed above.

The very low energy absorption for HAZ fracture in pipe AB is very disconcerting at first sight. However, the driving force for pipe fracture following a spiral path is certainly less than for fracture following the pipe axis, and may in fact lead to fracture arrest by cut-off around the weld. The implications of HAZ brittleness are of more concern in fracture initiation, since defects are more likely to occur in this region, than for fracture propagation.

CONCLUSIONS

1. This work has emphasized the importance of inclusion control in avoiding low-energy fracture in the rolling direction of line-pipe.

2. There is good qualitative correlation between toughness and anisotropy using Charpy-size and full-size specimens. However, there are quantitative differences, notably in transition temperature, underlining the importance of basing design data on full-section tests.

ACKNOWLEDGEMENTS

This work is extracted from an extensive and on-going study on line pipe steel that has involved a number of scientists and technicians at PMRL. Special acknowledgement must be made to P. Trudeau, J. Ellis, and E. Smith for DWTT testing, B. Casault and M. Shehata for quantitative metallography, and R. Narraway and G. Weatherall for Charpy testing.

DISCONTINUITY SOURCES IN MANUFACTURING PROCESSES

H. J. McQueen

Concordia University
Montreal, Quebec H3G 1M8

ABSTRACT

The goals of processing are to produce objects meeting specifications in dimensions and surface finish, in microstructure (100-0.01 μm) and properties and in sound condition without macrostructural defects (1000-10 μm) to serve as crack initiators. The objective of this paper is to examine the important manufacturing processes to point out the sources of incipient cracking and the inadequate control of processing which give rise to them. A great many types of defects are well-known causes of failure so that in proportion to the required confidence in the product, process and quality controls can be applied to eliminate them. There are also certain conditions in mechanical forming which give rise to small discontinuities which could serve as initiators for service failures and which are not so readily recognized and avoided.

INTRODUCTION: PROPERTIES - MICROSTRUCTURE - PROCESSING

The deployment and safe utilization of a mechanical system depends upon the cooperation of the research metallurgist, the designer, the fabricator and the user. The prevention of component failure depends on the avoidance of discontinuity introduction and the provision of a matrix microstructure resistant to failure propagation. This is not always accomplished either because of improper specification or inadequate process control.

Proper materials specification depends on the establishment of the relationship between properties and microstructure. Even extensive research programs cannot always determine the optimum for each microstructural constituent and the best of models is usually more simple than the reality it represents. The organization of the macroscopic processing to achieve the microscopic structure, as well as the shape, suffers from the limits of available control technique and, in many cases, from inherent inhomogeneity that makes it difficult to keep within the specified range throughout the entire component. In addition, the microstructural transformations during the processing may follow an entirely different set of principles from those immediately governing the final properties.

CLASSIFICATION OF DISCONTINUITIES

Processing can lead to a variety of discontinuities which are quite discrete from the specified matrix structure. The discontinuities to be considered here are cracks, voids and second phase inclusions which usually have a larger size, less uniform distribution and poorer mechanical properties than the intended precipitates. Their location relative to the external dimensions are dependent on the nature of the processing and have an important influence on the degree to which they augment stress concentrations. These defects will be described without discussing the toughness of the matrix despite its acknowledged role in determining the ease with which fissures propagate from the discontinuities. Variations in crystal size, orientation, composition or precipitate distribution are considered to be part of the matrix characteristics.

Defects play very different roles depending on their size so three groups are distinguished. The large ones range from those which lead to failure during the process, such as hot short in rolling, to those which are just visible in the product thus requiring repair or rejection. They reduce the profitability of the manufacturing operation but do not become responsible for high costs in service failures. The intermediate size are not visible in the product but can be detected by non-destructive techniques. The quality control system required by the probability of their presence increases the cost of the processing; moreover, early inspection and rejection of defective parts save processing costs compared to those scrapped only at the end. The high expense of quality assurance leads to its omission in the manufacture of low security products in which this class of defects causes the vast majority of service failures (where they increase the cost of maintenance and the chance of accident). It is of interest that the best source of information on these are not research journals but handbooks (Table 1). Most research is devoted to fundamental mechanisms or to matrix characterization so that specimens containing such defects by chance are deliberately rejected.

The third class of defects are too small to be easily detected and generally only pose a dangerous problem in high security components. It is not certain that their contribution to fracture initiation has been clearly distinguished from matrix characteristics. Their scale is of the same magnitude as that of the grains and their fractographic appearance makes it difficult to determine whether they existed prior to fracture or were created by it. While these small defects may arise as minor manifestations of the earlier classes, they can also be produced as a result of certain distinct process conditions to be described. Before doing this, a summary will be given of the larger sizes of defects, particularly to point out those shared by different processes and how they can be carried from one class of processing to another, often becoming exacerbated and hence rendered visible. The ductility limit in metal forming is examined as an indicator of damage generation.

COMMON MACROSCOPIC DEFECTS

The large processing defects, which are easily detectable by standard non-destructive techniques, yet still the causes of many service failures, are listed in Table 2 according to their geometric characteristics for the three major classes of metal shaping processes. The list is not exhaustive and particularly omits defects peculiar to a specific process, such as electron beam welding or die casting. The column on forming mainly lists defects that flow from those in the ingot or concast slab; in general, the latter has a much lower probability of defects.

TABLE 1 - INFORMATION ON DEFECT FORMATION AND CONTROL

SOURCES	WELDING BRAZING		CASTING		FORMING	
ASM Metals Handbook						
Forging and Casting Vol. 5, 1970			Quality control for specific processes	149-444	Quality control for specific processes	1-148
Welding and Brazing Vol. 6, 1971	Quality control for specific processes					
Failure Analysis and Prevention Vol. 10, 1973	Weldments Brazed Joints	333-368 369-372	Iron and Steel Castings	315-332	Cold Formed Pts. Forging Mech. Fasteners Springs	285-290 291-314 470-487 487-499
Non Destructive Inspection and Quality Control Vol. 11, 1976	Weldments Brazed Assemblies Soldered Joints	340-355 355-358 359-366	Castings	308-320	Forgings Bars, Billets Wrought Tubes Threaded Fast. Powder Parts	287-307 320-327 327-339 373-379 396-398
Fractography and Atlas of Fractographs Vol. 9, 1974	Discontinuities Leading to Fracture that are Revealed by Fractography			93-103		

TABLE 2 - DISCONTINUITIES COMMONLY ARISING IN MANUFACTURING PROCESSES

DISCONTINUITY	WELDING BRAZING		CASTING	FORMING
Internal				
Grain Structure	1 columnar grains 2 grain growth 3 martensite	(W) (HAZ) (HAZ)	1 dendritic, oriented, columnar grains	1 fibrous, oriented grains 2 large annealed grains, secondary grain growth
Second Phase	4 slag, W-electrode 5 oxide on base 6 grain boundary precipitates, eutectic	(W) (W-B) (W)	4 entrapped slag, sand 5 non-metal inclusions 6 grain boundary precipitates, eutectic colonies	4 fibrous stringers 5 inclusions 6 precipitates
Voids	7 gas porosity 8 H_2O cracking 9 shrinkage 11 incomplete fusion	(W) (HAZ) (W-B)	7 blow and worm holes 8 hydrogen flakes 9 shrinkage cavities, centerline porosity	7-9 fibrous, unhealed cavities due to oxidation or insufficient working 10 center bursts, splits
External				
Voids	12 crater cracks 13 lack of penetration undercutting	(W)	12 pipe, shrinkage 14 surface blow holes	12 unhealed centerline defects 14 seams, laminations
Cracks	16 hot tearing 17 hot cracking 18 liquation cracking 19 ductility dip cracks	(W) (W) (HAZ) (HAZ)	15 cold shuts, splatter, scabs 16 hot tearing (2 phases) 17 hot cracking (solid) 19 hot cracking	15-19 seams (⊥ surface) laminations (∥ surface) 18 burning, hot shortness 19 pores, fissures in hot work 20 laps, flow through; hoop, edge cracking; delamination cracks
NDT	die penetrant, magnetic particle X-ray or γ ray radiography ultrasonics helium leak or pressure test			die penetrant, magnetic particle ultrasonics eddy current, magnetic permeability

The defects produced by deformation are presented in more detail in Table 3. The common means of NDT are indicated for each category, but again there is not space to give the shape or process limitations.

In castings and weldments, certain defects arise from the solidification mechanism: columnar grains (Table 2), intergranular segregates of substances dissolved in the liquid (5,6), blow holes from dissolved gases (7,8,14) and cavities from liquid-to-solid shrinkage (9,12). These defects can be reduced by control of melt composition, rate and direction of solidification through mold design, or welding power, speed and preheat, and proper gating and risering or welding edge preparation, run-out tabs, etc. In casting, certain defects arise from the process techniques and can be eliminated by improved practice: slag, refractory or sand (4) from furnace, ladle or mold, cold shuts and splatter (15) from turbulent mold filling and scabs (15) from patches of sand detached from the mold. In welding, poor control of the shielding atmosphere, of the protective slag, of the edge cleanliness, of the electrode feed rate, or of the power-speed balance, can give rise to inclusions (4), cavities and cracks (7,8,11,13) [1]. Constraints from the mold, or from the fabricated assembly generate stresses leading to tearing (16) of the partially solidified metal, and cracking after solidification (17) or in the HAZ (18,19). The HAZ is subject to special problems: grain growth (2), overaging, transformation to martensite (3) and even melting at the grain boundaries (18), which can be alleviated by the controlling the thermal cycle or by a heat treating the entire assembly. In welding, severe residual stresses can give rise to cracks of diverse geometries; transverse longitudinal, toe, root, fusion line, underbead in the weld or the HAZ.

In mechanical forming, unnoticed faults in the cast feed are greatly extended in the elongation direction, which helps to expose and eliminate them giving a superior product. Inclusions and precipitate are distributed as stringers (4,5,6) throughout the metal reducing the transverse properties and generating cavities and cracks. Shrinkage porosity (7,9,12,14) and severe segregation (4,5,6) give rise to defects along the centerline which can lead to processing failures when tension stresses are generated there. Surface defects (14-19) on the ingot give rise to seams and laminations (15). Improper conditioning to remove surface defects can produce grooves or grinding cracks which give rise to laps and seams respectively. Powder metallurgy avoids segregation, shrinkage and inclusions in cast material but inadequate control can leave residual porosity, nonuniform density and interparticle oxides.

Additional defects can be generated in the mechanical working process which includes preheating or annealing as required (Table 3). Thermal stresses alone can cause cracking in large ingots (Table 3:C1); overheating or an oxidizing atmosphere (C2) can induce embrittling damage at the grain boundaries. Hot working processes are noted for nonuniformity of temperature because the rolls or dies are usually cooled for their protection (D1,2) and of strain and strain rate arising from geometry and temperature variation which leads to highly sheared zones with substantial heat generation which further concentrates the deformation (D3,4). Thus, on one hand, the temperature may become so low at some stage that cracking ensues (D1,2) whereas on the other, local temperature may rise sufficiently due to adiabatic or frictional heating that grain boundary melting leads to a hot short (D3,4). Most primary working operations which involve heavy reductions of cast structures are compressive in nature; however, nonuniformity of deformation; particularly in finishing passes where reductions are light can give rise to regions of tensile stress, which may cause center bursts (E3,4) or surface

TABLE 3 - DEFECTS ARISING IN MECHANICAL FORMING

	CAUSAL CONDITION	RESULTING DEFECT
A.	DEFECTS CARRIED OVER FROM CAST INGOT OR SLAB	
1.	internal porosity, inclusions	center cracking
2.	external porosity, cracks, cold shuts	seams, laminations
B.	DEFECTS FROM CONDITIONING	
1.	grinding cracks	seams
2.	deep abrupt grooves	laps
C.	THERMAL EFFECTS ALONE	
1.	too rapid preheating	center cracks
2.	too high or oxidizing heating	grain boundary liquation
3.	too rapid cooling	cracking especially of martensite
4.	rapid heating and cooling in grinding	surface cracking
D.	THERMO MECHANICAL	
1.	low T, inadequate heating	center cracks
2.	cold tooling	surface, flash cracks
3.	high T, adiabatic heating	center hot short (alligatoring in rolling)
4.	friction heating	surface hot short (fir-tree in extrusion)
E.	MECHANICAL - INHOMOGENEOUS TENSION	
1.	rolling without edge support	edge cracking at convex surface*
2.	open or closed die forging	hoop cracking in barrel, splitting of head*
3.	side forging of rounds	center splitting
4.	drawing, extrusion, low ε, high angle	center bursts
F.	MECHANICAL - DIRECT TENSION	
1.	deep drawing, stretching	localized necking (FLD), pores
2.	(low n, r<1, inclusions, friction)	at inclusions*
3.	punching, blanking, trimming	exfoliation cracking
4.	machining with built-up edge	cracks at surface
G.	MECHANICAL - FRICTIONAL FLOW CONSTRAINTS	
1.	forging or section rolling, mismatch	laps, folds
2.	of blank shape to die sequence	flow through defects
3.	leading to nonuniform friction	web holes in universal mill
H.	DEFORMATION MECHANISM	
1.	grain boundary sliding in hot working	boundary cracking and pores*
2.	hard inclusions	pores*
3.	superplastic deformation	pore formation*

*micro discontinuities

cracking (E1,2); of course, solidification segregation and low temperatures exacerbate this. The role of non-homogeneity in defect generation is considered later. Improper progress of sequential deformations can displace metal in the wrong direction giving rise to laps or flow-through cracks (G1-3). Some of the secondary operations are directly tensile and can give rise to necking or tearing in wire and tube drawing (F1,2). In sheet metal, working the situation is more complex since the proportion between the primary and secondary surface stresses affect the major strain at which localized necking begins as is discussed later. The distribution of strain ratio is dependent on die and blank geometry, blank surface finish and edge quality, die lubrication, and magnitude and distribution of hold down pressure. The processing is also influenced by material properties such as strain hardening rate, anisotropy, inclusion content, grain size and presence of yield point effect which gives rise to nonuniform Luder's strains.

The final stages of manufacturing are usually metal cutting and heat treating and even these can give rise to discontinuities. In machining, cutting ductile material with a built-up-edge on the tool can give rise to cracks when the chip tears-off (F4), and heat generated in grinding with inadequate lubricating coolant may result in thermal cracking (C4). In thermal treating, overly rapid heating can cause cracks, and overheating or inadequate control of the protective atmosphere can lead to oxide penetration of the grain boundaries. Unnecessarily severe quenching rates, or incorrect directions of quenchant immersion, can lead to cracking, or residual stresses in parts of complex geometry. Improper control of case depth and subsequent thermal cycling can lead to subcase cracking. Inadequate quenching or tempering can leave residual austenite which transforms in service to create residual stresses.

MICRO DEFECT GENERATION

After having looked at macro discontinuities which are common sources of failures, attention is now focused on internal mechanisms and deformation conditions which give rise to finer defects which are likely to be overlooked by both producer and purchaser. Hot working is usually noted for its high workability; however, intergranular pores are produced in some temperature ranges which will be described. Hydrostatic tension conditions can arise near the center line, or near the surface, in some circumstances of inhomogeneous flow, producing small discontinuities and in severe cases, obvious cracking. Difficulties are most likely to arise when a new stiffer alloy, or one with slightly more inclusions, is subjected to a tried shaping process, or when a well tried alloy is deformed to a more complex or severe shape.

INTERGRANULAR FISSURING IN HOT WORKING

In deformation at temperatures above half that of melting, the strain hardening is reduced by dynamic recovery which, in annihilating some dislocations, rearranges others in low density sub-boundaries [2-4]. In metals such as Al and bcc Fe, this can reach such a degree that steady state deformation ($\dot{\varepsilon}$,T,σ constant) is attained. In lower stacking fault, fcc metals such as Fe, Ni, Cu and their alloys, the accumulation of dislocations continues without attaining equilibrium and thus results in dynamic recrystallization at strains of the order of 0.5-2 dependent on the strain rate [2,3]. An additional deformation mechanism is grain boundary sliding which constitutes only a few percent of the total and declines as the strain rate is raised. The sliding gives rise to triple junction cracking and boundary pore formation to an extent dependent on the degree to which the lattice resists accommodating deformation, Figure 1, [4,5]. Al and bcc Fe dynamically recover to such a degree that intergranular cracking is very limited and strains of 50 to 100 are possible before rupture, Figure 2.

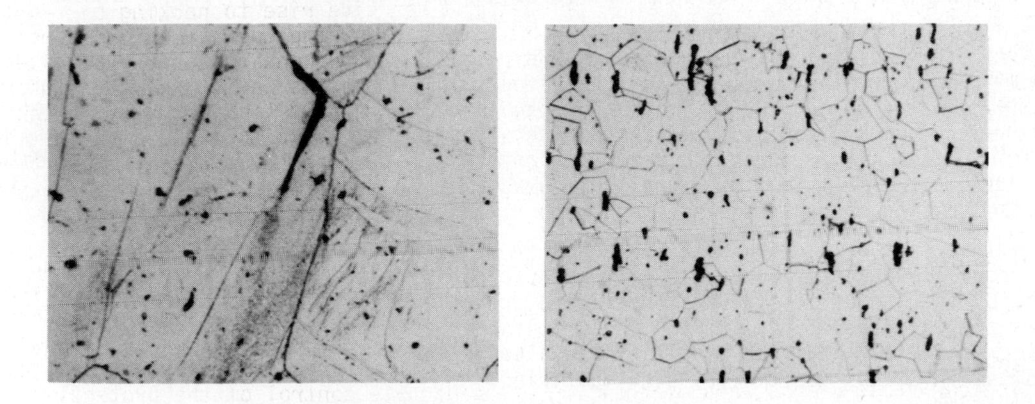

Fig. (1) - Waspaloy torsion specimens - a) grain boundary cracking at 815°C, $1s^{-1}$, $\bar{\varepsilon}_f = 0.6$; b) pores, which initially formed at grain boundaries, have been separated from them by dynamic recrystallization at 1040°C, $1.0s^{-1}$ and $\bar{\varepsilon} = 11.7$. X 225 [5]

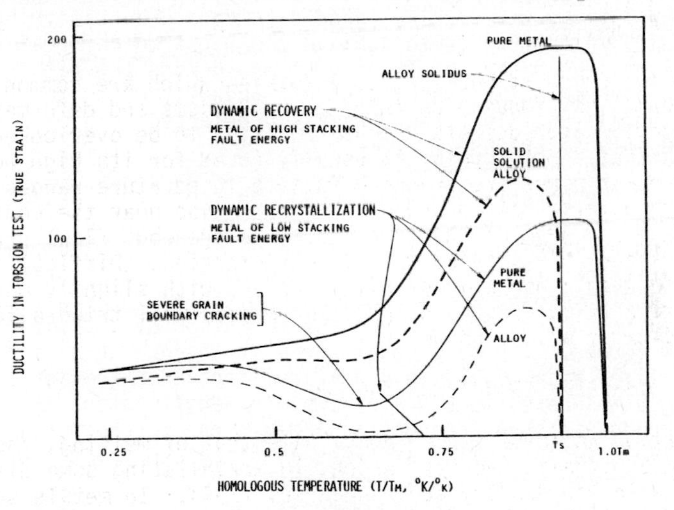

Fig. (2) - Schematic ductility - temperature curves comparing metals which undergo either dynamic recovery to a high degree or dynamic recrystallization. In the former, grain accommodation to intergranular sliding leads to ductility rising from room temperature; whereas, in the latter, there is a minimum just below the temperature at which recrystallization takes place [5]

The low stacking fault fcc alloys of Fe, Ni and Cu exhibit a minimum in ductility of the order of 0.5 at about 0.5-0.6 T_M in which failure occurs by intergranular cracking along the original grain boundaries, Figure 1a, 2 [4,5]. Below that temperature, ductility increases towards that at room temperature where a ductile cup-cone fracture is usual. As the temperature rises above 0.6 T_M, the ductility rises rapidly as a result of dynamic recrystallization taking place more rapidly. The pores already generated mainly become isolated within grains and, even if on boundaries, do not propagate readily because of differences in boundary orientation and character which alter the sliding, Figure 1b. As deformation proceeds, new pores are generated on the existing boundaries but are repeatedly separated from them by migration. Gradually at strains of the order of 10, the pores having increased in size and number, link together leading to fissures and failure. As the strain rate increases from the order of $10^{-4}s^{-1}$, the ductility-versus-temperature curve is higher because dynamic recrystallization is speeded up by the increased driving force; however, at still higher rates, it becomes lower as the lattice becomes stiffer and cracking is apparently speeded up relative to recrystallization. The ductility can also be improved by deforming in a series of small stages (up to 0.3) with sufficient time between them for static recrystallization.

With alloys of γFe, Ni, or Cu, the minimum broadens towards higher temperatures because solutes or precipitates hinder dynamic recrystallization. In C steels, the minimum is not clearly observed because of the transformation from γ to α which has higher ductility; moreover, the two phase region has poorer ductility than either phase. As the hot short, or nil ductility (solidus) temperature diminishes with increased alloying, the working range of adequate ductility becomes narrower and the difficulty in forming is to achieve the required deformation, often requiring several rolling passes or forging blows, before the temperature becomes so low that severe intergranular cracking occurs (usually close to the surface because of cooling by tools) [5]. This could remain undetected in a mild case or actually lead to failure in a severe one.

The growth of the pores is very dependent on the stress state. The behavior above is found in tension, in compression as barrel cracking, and in torsion where it is accelerated by longitudinal tension and diminished by compression. In most hot forming operations, there usually is a hydrostatic compression component, e.g., in extrusion, in die forging and in rolling due to the inward friction on both sides of the neutral point, which slows the formation of cavities and may even weld shut those formed. The healing of pores in the presence of hydrostatic compression is utilized effectively in powder rolling and extrusion (in the absence of an oxidizing atmosphere) to achieve complete densification. However, non-uniform deformation due to shape change, tool geometry, heterogeneous cooling and friction can give rise to tension stresses as discussed below. One may summarize by stating that pore formation is a fundamental mechanism occurring during hot working which is suppressed in most commercial shaping processes but not in all. In critical products where pores are suspected, they can be absolutely eliminated by subjecting the parts to an additional hot isostatic pressing.

INCLUSIONS AND DECOHESION

Recent studies of inclusion behavior during hot working show that the plasticity of different non-metallic inclusions vary with temperature in different ways from the metal matrix and each other [6,7]. When the inclusion is softer, it deforms more than the matrix giving rise to fibrous stringers whose effect on the properties have already been described. When the inclusion is much harder than the metal, it does not deform and may either decohere from the matrix to form conical holes pointing in the rolling directions or fracture forming rows of particles; the decrease in density due to such pores is measurable. In hot working, as the quantity of hard inclusions increases, they give rise to pores more rapidly than the grain boundary sliding, and thus dominate the cracking process and greatly reduce the ductility. Dynamic recrystallization gives little relief because the particles hinder the migration of the boundaries. Massive precipitates can have similar embrittling effects especially if they form on the boundaries and if the grain size is large. One may summarize that in critical temperature ranges for each ceramic compound, decohesion voids and cracked inclusions are produced by the hot shaping to serve as defects in service. The concentration of inclusions can be reduced by additional processing, e.g., steels can be deoxidized by vacuum degassing.

CENTERLINE DEFECTS

In industrial processes such as wire drawing, extrusion, strip drawing and strip rolling, center bursts have been observed under certain processing conditions which have been quite clearly explained by Rogers and others [8-10]. In slip line analysis of a plane strain forming process, a deformation zone is defined by a grid of lines which are tangent to the planes of maximum shear stress. By means of the Hencky equations, the mean stress can be calculated at every point in the field starting from a boundary region of known stress level, Figure 3a. Under

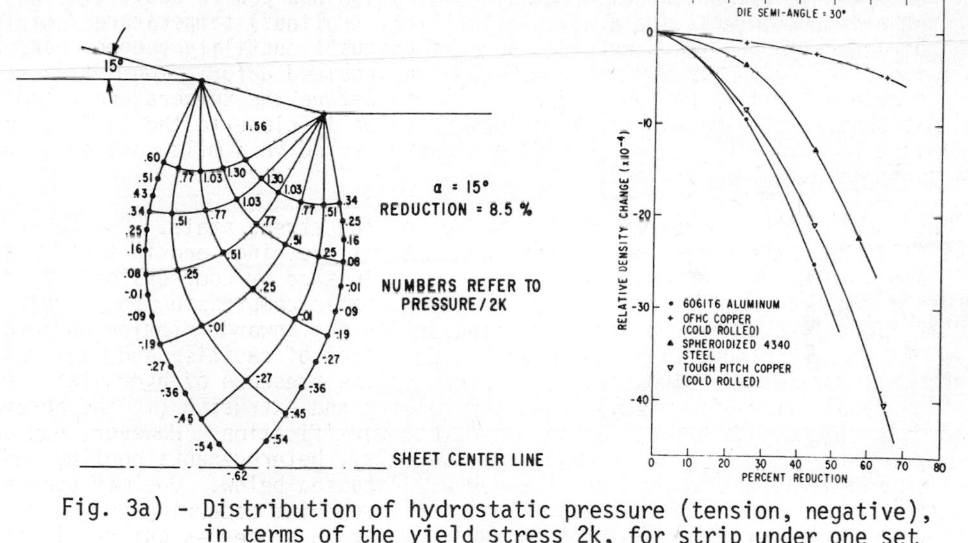

Fig. 3a) - Distribution of hydrostatic pressure (tension, negative), in terms of the yield stress 2k, for strip under one set of plane strain drawing conditions; b) Relative change in density during drawing through dies with a semi-angle of 30 degrees, of metals with different concentrations of hard particles [8]

conditions of high die angle and low reduction, the mean stress at the centerline becomes hydrostatic tension. As the number of passes through a sequence of dies of similar angle and reduction increases, Figure 3b, the damage (as measured by density and by metallography) progresses from holes at inclusions, to tubes elongated in the drawing direction, to planar fissures and to two eared fissures propagating along the planes of high shear strain, which finally become arrowhead defects, or center bursts, that can lead to fracture during drawing [10]. In terms of distribution of pores, there were a larger volume fraction in regions of larger hydrostatic tension stress [9]. The volume fraction of discontinuities increases faster per unit strain, Figure 3b, and the critical die angle is smaller for materials with greater concentration of inclusions, e.g., tough pitch copper compared to OFHC. It should be emphasized that the initial stages, when defects are small and undetectable, are the more dangerous for yielding poor quality product.

EDGE OR HOOP CRACKING

In upsetting, heading or rolling the unconstrained edge undergoes bulging as the surfaces in contact with the tools are restricted from enlarging as height is reduced. Tensile stresses are induced in the bulge because the circumference increases in length without having a pressure component which prevents void formation. Kuhn and colleagues have determined the conditions for crack opening from a study of surface strains as indicated by inscribed grids [11].

With no friction at the anvils, a cylinder remains such during compression so that the minor strain is one half of the major strain. As long as the cylinder expands uniformly, there is no surface cracking. With friction, the straining curve diverges to higher circumferential strains and the specimen fails by hoop cracking. Experimentally for a given alloy, the locus of the failure points is a straight line with the same slope as the uniform deformation line, Figure 4a.

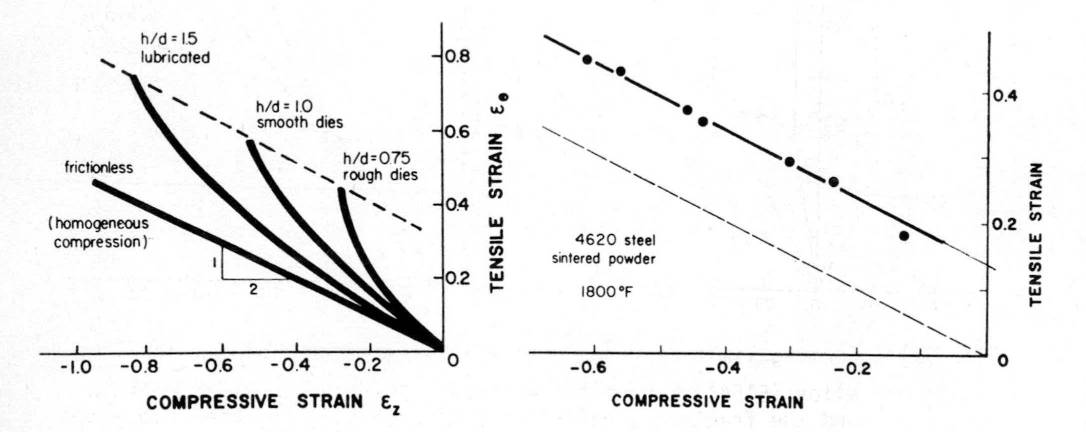

Fig. 4a) - The straining path to fracture diverges more rapidly from the uniform ratio as friction increases and height to diameter ratio decreases. The secondary strain at fracture exceeds half the pressure by a constant amount; b) The locus of hoop fracture strains in upset tests has a greater intercept when the metal is more ductile at the higher temperature [11]

The intercept on the secondary strain axis is larger for more ductile materials, for working at a higher temperature, Figure 4b, and for metals with less inclusions. Prior to actual cracking at the surface, pores form inside the bulge initiating from particles or grain boundary sliding [10,12]. Moreover, the failure criterion put forward is consistent with that of McClintock in which fracture occurs when the voids generated in a shear zone link up [10,11]. Once again, degrees of barrelling short of that causing fracture are the ones that generate discontinuities in the product.

CAVITATION IN BIAXIAL SHEET FORMING

A useful empirical methodology for sheet forming is the Goodwin-Keeler Forming Limit Diagram which shows the locus for formation of a localized constriction on a biaxial strain graph, Figure 5a, [13]. In a certain shaping operation, a

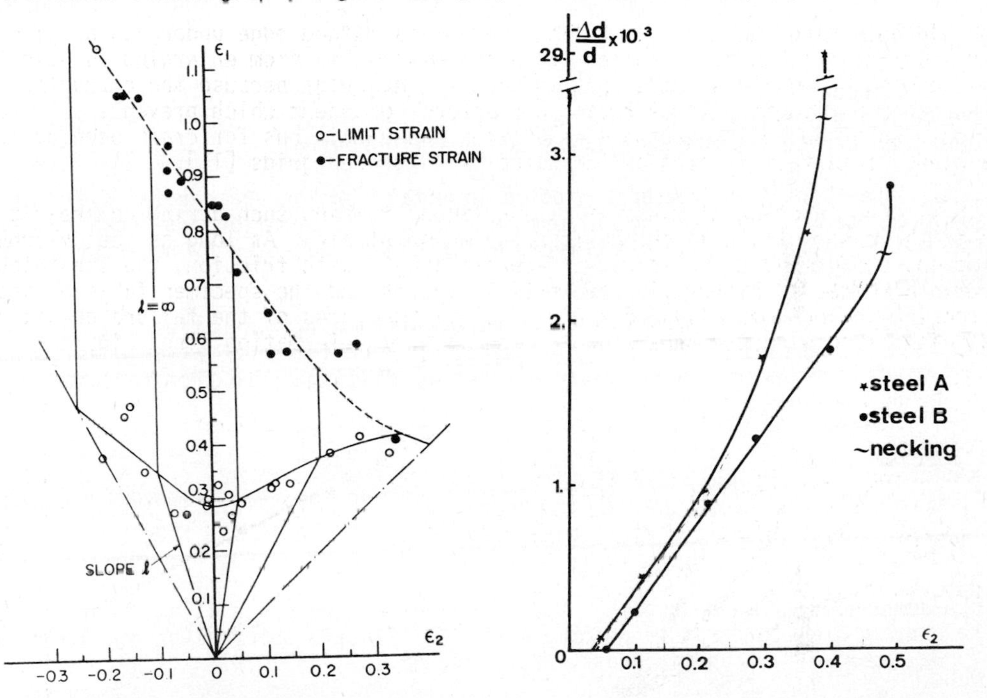

Fig. 5a) - Forming limit diagram (with ε_1 the major strain) for an Al-Mg alloy (5154) at room temperature. Both the localized necking and the fracture strains are indicated [13]; b) When subjected to equal biaxial straining steel A with a higher inclusion content than steel B suffers the greater void formation. The limit strain for A is 0.4 compared to 0.5 for B [14]

region of the sheet will follow a proportional straining path, that is with constant slope, up to the forming limit line determined experimentally. High levels of the forming limit are found for the following characteristics of the material: in radial drawing a high value of r (resistance to thinning), in stretching a high strain hardening rate and area reduction in tensile necking [13]. In determining the cause for the level being higher for metals with less inclusions, the density has been found to decrease continuously during equal proportioned straining and at a higher rate as the forming limit is approached [14]. Of two similar steels, the one with the higher inclusion level, had the greater rate of void formation, Figure 5b. Examination of the surface by SEM showed formation of pores at inclusions. It thus appears that biaxial stretching, even when the operation is successful, creates discontinuities in the product.

A sheet forming technology of growing importance is that based on superplastic deformation and is accomplished by vacuum or pressure against a single die. Extremely high strains are possible without unstable neck formation because of the high strain rate sensitivity. However, in the course of deformation, pores form at the interphase boundaries. The density of discontinuities in the product depends on the alloy, the initial grain size and the strain [15,16].

CONCLUSIONS

1. Discontinuities are generated in all common manufacturing operations and have been a major source of service failures. As required by the service situation, they can be avoided by process and quality control.

2. The majority of defects exposed in metal forming are carried over from cast feed stock. However, nonuniform heating and deformation are known to give rise to additional macroscopic defects which can be detected by non-destructive testing.

3. Under certain conditions of working, micro discontinuities are created which are not weeded out by current inspection techniques and, even when they have initiated or enhanced fracture, are not easily distinguished from features formed during the failure propagation.

REFERENCES

[1] Eaton, N. F., Glober, A. G. and McGrath, J. T., "Fracture in the production and service of welded structures", Adv. Res. Strength and Fracture of Materials, Vol. 1, D. M. R. Taplin, Pergamon Press, Toronto, pp. 751-773, 1978.

[2] McQueen, H. J. and Jonas, J. J., "Recovery and recrystallization of metals during high temperature deformation", The Plastic Deformation of Materials, R. Arsenault, Academic Press, New York, pp. 393-493, 1975.

[3] McQueen, H. J., "Dynamic recovery and its relation to other hot working mechanisms", Metalurgia I Odlewnictwo, Vol. 5, pp. 421-470, 1979.

[4] Tegart, W. J. McG., "The role of ductility in hot working", Ductility, ASM, Metals Park, Ohio, pp. 133-177, 1968.

[5] McQueen, H. J., Sankar, J. and Fulop, S., "Fracture under hot forming conditions", Mechanical Behavior of Materials, Vol. 2, Pergamon Press, London, pp. 675-684, 1979.

[6] Baker, T. J. and Gove, K. B., "Inclusion deformation and toughness anisotropy", Met. Tech., Vol. 3, pp. 183-193, 1976.

[7] Segal, A. and Charles, J. A., "Deformation characteristics of MnS inclusions in steel", Met. Tech., Vol. 4, pp. 177-182, 1977.

[8] Rogers, H. C., "Materials damage during deformation processing", Metal Forming Inter-relation Between Theory and Practice, A. L. Hoffmanner, Plenum Press, New York, pp. 453-474, 1971.

[9] Rogers, H. C., "Fracture initiation and propagation during strip drawing", Adv. Res. Strength and Fracture of Materials, Vol. 2, D. M. R. Taplin, Pergamon Press, Toronto, pp. 435-441, 1978.

[10] Spretnak, J. W., "Fracture initiation under metal working conditions", ibid., Vol. 2, pp. 431-441, 1978.

[11] Kuhn, H. A. and Dieter, G. E., "Workability in bulk forming processes", ibid., Vol. 1, pp. 307-324, 1977.

[12] Lee, P. W. and Kuhn, H., "Fracture in cold upset forging - a criterion and a model", Met. Trans., Vol. 4, pp. 969-974, 1973.

[13] Embury, J. D. and LeRoy, G. H., "Failure maps applied to metal deformation processes", Adv. Res. Strength and Fracture of Materials, D. M. R. Taplin, Pergamon Press, Toronto, Vol. 1, pp. 15-42, 1978.

[14] Jalinier, J. M., Baudelet, B. and Argemi, R., "Forming limit diagrams and damage in mild steel", J. Mat. Sci., Vol. 13, pp. 1142-1145, 1978.

[15] Ridley, N. and Livesey, D. W., "Factors affecting cavitation during superplastic flow", Adv. Res. Strength and Fracture of Materials, D. M. R. Taplin, Pergamon Press, Toronto, Vol. 2, pp. 533-540, 1978.

[16] Taplin, D. M. R. and Smith, R. F., "Fracture during superplastic flow of industrial Al-Mg alloys", ibid., Vol. 2, pp. 541-551, 1978.

A STOCHASTIC INTERPRETATION OF MATERIAL DEGRADATION PROCESSES

E. S. Rodriguez III, and D. R. Hay

Tektrend International, Inc. 750 Bel-Air St., Montreal H4C 2K3

J. W. Provan

Mechanical Engineering Department, McGill University Montreal H3A 2K6

ABSTRACT

The design of load-bearing structures and pressurized systems has advanced to the point where most structures and systems perform at high levels of reliability. However, the scale of many structures and systems and their functions are often such that it is essential to avoid the occurrence of even a very low probability of failure. Bridges, tower cranes, gantries, nuclear power plants, liquified gas containers, toxic fluid containers are all examples where our contemporary society will not tolerate threats to its physical safety. Likewise, industry imposes an economic reliability requirement in large-scale systems where failure costs can be very high.

An inescapable element in the reliability of such systems is the presence of flaws in the material. Unfortunately, there exists a high degree of uncertainty associated with the description of the flaw content and the way in which flaws grow as a function of time. This paper regards the growth of flaws as a random process and proposes Markov stochastic processes, in particular, the discrete state processes, to describe it. A generalized, qualitative model is developed which incorporates the stochastic interpretation of material degradation processes as well as non-destructive inspection. To illustrate its utility, the model is applied to a specific corrosion system involving pipelines.

INTRODUCTION

This paper has two main objectives.

i) To describe time-dependent material degradation processes such as corrosion, fatigue, stress corrosion, etc., using stochastic models, specifically a Markov stochastic process. While this paper deals with pitting corrosion, the same philosophy can be extended to other modes of degradation [1].

ii) To illustrate how one incorporates such a model into a failure control system designed to enhance the reliability of structures whose materials of construction are subject to time-dependent material degradation processes.

An important aspect in the reliable performance of any load-bearing structure is a failure control system. One such system has been devised by Hay, et al [2] which emphasizes the materials-related aspects of failure in load-bearing structures. A general characteristic of any failure control system is that it is multidisciplinary, integrating such disciplines as solid mechanics, strength of materials, fracture mechanics, materials engineering, and non-destructive inspection, among others. However, a system can be structured to show only the key input and the key output, i.e., estimates of reliability. Such a system is shown in Figure 1.

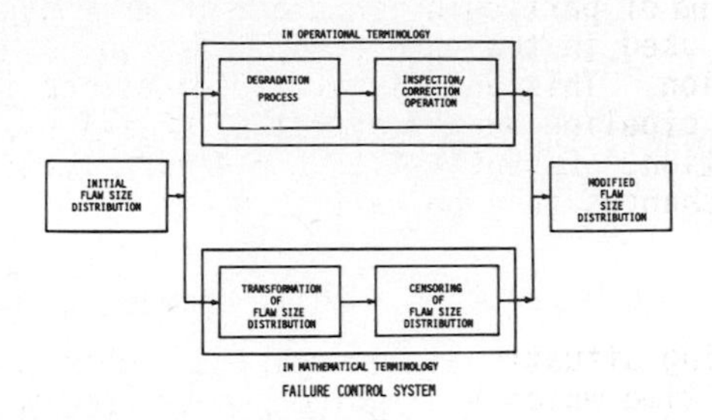

Fig. (1) - The basic elements of a failure control system

The extent of a structure's degradation is ultimately manifested in crack-like defects which propagate to a size at which the integrity of the structure, or its load-bearing capacity, is compromised. Hence, defects, characterized by probability functions, are the key input to any failure control system. Within the system are basically two models, namely: degradation and "upgradation". The degradation module describes the deterioration of the structure as a function of time. It is here that the flaw size distribution is generated and transformed with time. The upgradation module involves inspection and correction or repair operations and their effects on the flaw size distribution. These modules can be expressed both in operational and mathematical terms. The output of the system is a modified flaw size distribution from which reliability or risk estimates can be obtained. A theoretically infinite cycle of degradation and upgradation can be envisaged but, of course, in practice cost intervenes and the number of cycles is finite. For example, it may become more feasible to replace the structure than to inspect and repair it; or the target level of risk may no longer be obtainable.

In the sections which follow, each item of the system (Figure 1) will be discussed with special emphasis on the degradation module. Finally, the system will be applied to a structure consisting of 598 miles of 34 inch diameter crude oil pipeline whose extent of degradation was established after approximately ten years of service.

INITIAL FLAW SIZE DISTRIBUTION

In the context of this analysis, a flaw size distribution is a product of two random variables, i.e., the number and size of flaws. The "number" may be either represented by a distribution, such as the Poisson distribution, or by some deterministic average, while the "size" usually takes the form of a probability density function such as the log-normal, extreme value, exponential, etc.

A flaw size distribution is generally obtained in two ways: destructively or non-destructively. The first method involves sectioning the material and through metallographic techniques, tabulating the number of flaws falling into particular size intervals. Often, this method is not feasible as it involves destruction of the material, hence, the alternative method is through non-destructive inspection techniques. This involves an indirect measurement of the flaw by a probing medium and requires the interpretation of results. Some of the more common non-destructive testing techniques used in industry are radiography, ultrasonics, eddy current, dye penetrant, and magnetic flux leakage. As an example, and of particular interest in this paper, is a Linalog* instrument, which was used in the case study to survey corrosion of the pipeline under consideration. This unit is a self-contained instrument which is transported through a pipeline by the operating pressures, and does not disturb normal pipeline operation. Essentially, it records changes in the magnetic field as a result of changes in pipe wall thickness.

DEGRADATION PROCESS

Often in engineering situations, one considers observations of a phenomenon over a period of time which are influenced by random effects not just at a single instant but throughout the entire interval of time or the sequence of times under observation. In essence, a stochastic process is a phenomenon that varies to some degree unpredictably as time goes on. The unpredictability here implies that if an entire time-sequence of the process is observed on several different occasions under presumably identical conditions, the resulting sequence of observation would, in general, be different. Probability enters but not in the sense that each result of a random experiment determines only a single number, but a sequence or series of values, a function.

An accepted definition of a Markov process is that it is a stochastic process in which the system has no memory that would allow it to use past information to modify probabilities for the next stage. A comprehensive treatment of the theory of Markov processes is given in reference [3]. It is appropriate at this point to consider a specific degradation process, namely, pitting corrosion and see how it can be modelled by a Markov process.

Consider an idealized pitting model shown in Figure 2. The growth of the pits, $d(t)$, is the time-dependent process to be described. Hence, denote Ω, the thickness of the material, as the pitting space or state space, associated with the process $\{d(t), t \geq 0\}$, that is, Ω is the space of all the possible pit depths which the random variable $d(t)$ can assume. Thus,

$$d(t) \in \Omega \quad . \tag{1}$$

Because the pit depth can only be observed to within a certain degree of accuracy, $d(t)$ is contained in one of the observable zones, i, such that:

$${}^{i}X < d(t) \leq {}^{i}X + \Delta^{i}X \quad , \tag{2}$$

where i identifies the observable zones, 1, 2, 3, ... in the x-direction with each zone being equal to ΔX, the range of error in the measurement, Further-

*A trademark of AMF TUBOSCOPE, INC., Houston, Texas.

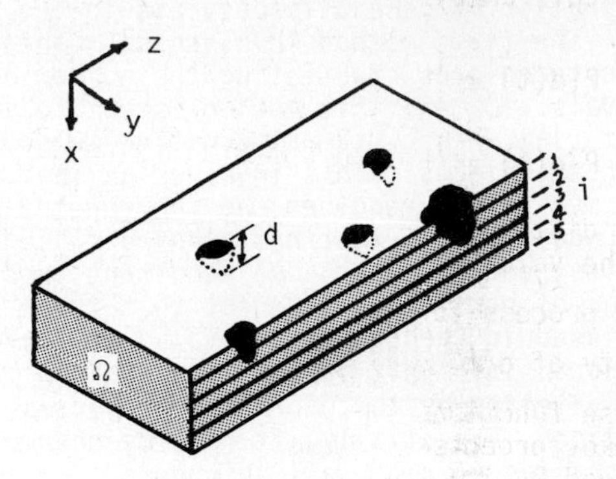

Fig. (2) - Idealized pitting corrosion model

more, these zones represent the statistical events ${}^{i}E$ which form the σ-algebra, F, necessary for the measure definition of the probability of d(t) being in ${}^{i}E$, i.e.:

$$\begin{aligned} P^{d}(t) &= P\{d(t) \varepsilon\ {}^{i}E\} \\ &= P\{{}^{i}X < d(t) \leq {}^{i}X + \Delta{}^{i}X\} \quad . \end{aligned} \tag{3}$$

Thus, the triad (Ω, F, P^d) defines the "probability pitting space" or the "abstract pitting system".

From the definition of the σ-algebra, the events ${}^{i}E$ form subsets of Ω, that is, ${}^{i}E \subset \Omega$. The probability of the event ${}^{i}E$ is a set function $P\{{}^{i}E\}$, called the probability function which satisfies the following conditions:

$$0 \leq P\{{}^{i}E\} \leq 1 \quad , \quad P\{\Omega\} = 1 \quad , \tag{4}$$

and

$$P\{\sum_{i=1}^{\infty} {}^{i}E\} = \sum_{i=1}^{\infty} P\{{}^{i}E\} \tag{5}$$

if the ${}^{i}E$ are disjoint, i.e., they have no common points. If for some time $t > 0$ the random variable d(t) sssumes a value $d\ \varepsilon\ {}^{i}E$, then it is said that event ${}^{i}E$ has occurred. The stochastic process $\{d(t), t \geq 0\}$ is said to be a Markov process if the conditional probability function $P({}^{i}E, t\ ; d, \tau)$ is such that:

$$P({}^{i}E, t\ ; d, \tau) = P\{d(t)\ \varepsilon\ {}^{i}E | d(\tau) = d\} \quad , \quad t > \tau \quad , \tag{6}$$

where $P\{d(\tau) = d\} > 0$, is uniquely determined. For the continuous time parameter case, as in this analysis, $\{d(t), t \geq 0\}$ is a Markov process if:

$$P({}^{i}E, t ; d, \tau) = P\{d(t) \in {}^{i}E | d(\eta) \quad , \quad 0 < \eta \leq \tau\}$$
$$= P\{d(t) \in {}^{i}E | d(\tau) = d\} \quad , \tag{7}$$

i.e., the probability distribution of $d(t)$ is completely determined for all $t > \tau$ by the knowledge of the value assumed by $d(\tau)$ and in particular is independent of the history of the process for all $t < \tau$. Since $P({}^{i}E, t ; d, \tau)$ gives the conditional probability of $d(t) \in {}^{i}E$ under the hypothesis that at a fixed time $\tau < t$, $d(\tau) = d$, these functions $P({}^{i}E, t ; d, \tau)$ are called "transition probabilities" of the Markov process $\{d(t), t \geq 0\}$. Analytically, a Markov process is completely determined by its transition probabilities. Since $d(t)$ assumes denumerable values, $P({}^{i}E, t ; d, \tau)$ which defines the Markov process, is reduced to a point function $P(j, t ; i, \tau)$ which can be written as $P_{ij}(\tau, t)$ where the subscripts i and j are referred to as the initial and final states, respectively. For the time homogeneous case:

$$P_{ij}(\tau, t) = P_{ij}(t - \tau) \quad , \tag{8}$$

i.e., the transition probabilities depend only on the duration of the time interval, and not on the initial time. In this case $\tau = 0$ and $P_{ij}(\tau, t)$ reduces to $P_{ij}(t)$.

For example, analytical expressions for a homogeneous linear birth process and a homogeneous Poisson process are given as follows: [3]:

$$P_{ij}(t) = \begin{cases} \binom{j-1}{j-i} e^{-i\lambda t}(1-e^{-\lambda t})^{j-i} & ; j \geq i \\ 0 & ; \text{otherwise} \end{cases} \quad , \tag{9}$$

and

$$P_{ij}(t) = \begin{cases} \frac{(\lambda t)^{j-i}}{(j-i)!} e^{-\lambda t} & ; j \geq i \\ 0 & ; \text{otherwise} \end{cases} \quad . \tag{10}$$

respectively. $P_{ij}(t)$ is the probability of going from state i to state j in the time interval t; i and j define the states in the state space, i being the earlier state and j, the later state; λ is a non-negative constant called the transition intensity which links the empirical process and the mathematical model.

A very important relation can now be established, that is:

$$P_j(t) = \sum_{i=0}^{\infty} P_{ij}(t) P_i(0) \quad , \tag{11}$$

which is the absolute probability relation. This states that if $P_i(0)$, the initial probability distribution, is known, such as one might obtain from the initial flaw size distribution, then the probability distribution can be calculated for any given subsequent time interval. This is very important as far as reliability estimates are concerned because it is from this distribution that one can obtain interference with a critical crack size or failure size distribution, the measure of reliability.

The object now is to find the appropriate transition probability expression that will describe the pitting corrosion process, and then can be done by experimentation. The ideal thing to do, of course, is to observe the degradation of actual structures undergoing pitting corrosion such as pipelines. Unfortunately, it takes time and resources to obtain such data, therefore, one can only resort to accelerated tests in the laboratory involving small samples. An example of a laboratory grown pit as shown in Figure 3 which resulted from immersion of

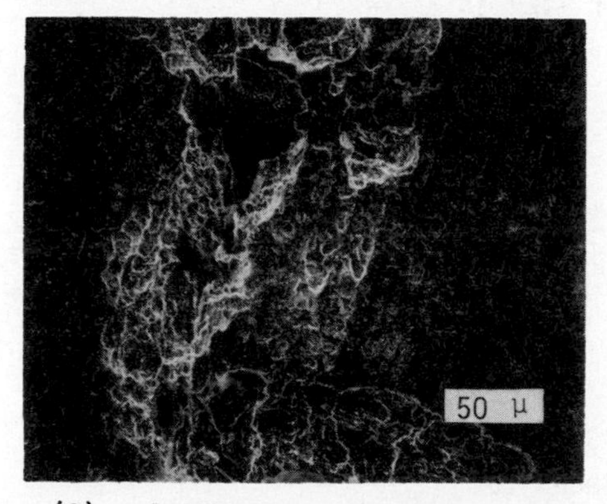

Fig. (3) - A SEM view of a corrosion pit

CA-15 stainless steel coupons in simulated "white water" [4] for 95 hrs at 50°C with no agitation. While the results of laboratory tests cannot be extrapolated with confidence to the field they do give a first estimate of the stochastic process to be used in the analysis. The eventual failure of the system is determined by the pits that penetrate the material.

UPGRADATION PROCESS

Regardless of the capabilities of designers combined with the recent developments in materials technology, it is neither economically possible nor technically possible to build and operate engineering systems with complete assurance that they will never fail. Therefore, one can only hope to achieve the lowest possible probability of failure. In many systems this is done by non-destructive inspection during fabrication, before the structure is put into service and periodically during its service life. Hence, inspection is an essential element in a failure control system.

The upgradation process consists of non-destructive inspection followed by the repair and/or replacement of components containing certain flaws. Each non-destructive testing procedure has limitations in its ability to detect small flaws and, to a certain extent, even large flaws. This uncertainty is conveniently expressed as probability of non-detection (PND). For example, an expression is given by Davidson [5]:

$$F_d(x) = \begin{cases} C_1(1-e^{-B_1(x-a_{th})}) & ; \ x \geq a_{th} \\ 0 & ; \ x < a_{th} \end{cases} , \quad (12)$$

where $F_d(x)$ is the detention probability, the complement of the probability of non-detection, as a function of the flaw size; x; a_{th} is the minimum detectable flaw size or the detection threshold; and C_1 and B_1 are constants with $C_1 < 1$. This formula applies for non-destructive testing in the laboratory. However, Yang [6] suggests that it can also be used for in-service conditions by choosing the appropriate constants. To obtain the probability of non-detection, the complement of equation (12) is taken, i.e.,:

$$PND(x) = 1-F_d(x) = \begin{cases} 1-C_1(1-e^{-B_1(x-a_{th})}) & ; \ x \geq a_{th} \\ 1 & ; \ x < a_{th} \end{cases} , \quad (13)$$

or

$$PND(x) = \begin{cases} (1-C_1) + C_1 e^{-B_1(x-a_{th})} & ; \ x \geq a_{th} \\ 1 & ; \ x < a_{th} \end{cases} , \quad (14)$$

The term $(1-C_1)$ is negligible compared to $C_1 e^{-B_1(x-a_{th})}$, hence, it is disregarded and equation (14) reduces to:

$$PND(x) = \begin{cases} C_1 e^{-B_1(x-a_{th})} & ; \ x \geq a_{th} \\ 1 & ; \ x < a_{th} \end{cases} , \quad (15)$$

The present paper acknowledges another aspect of the uncertainties of inspection, namely, sizing. Upon detecting a flaw, one can either underestimate, correctly estimate or overestimate this flaw; this must also be regarded as a random variable. This aspect will not be included in the present analysis, however, one must include it in a comprehensive risk analysis.

MODIFIED FLAW SIZE DISTRIBUTION

The failure control system in Figure 1 can be illustrated graphically, as shown in Figure 4. Suppose a certain structure has an initial flaw size density as in Figure 4(a). Before proceeding, two flaw sizes must be defined, namely, repair size, a_r, and failure size or critical size, a_f. a_r is a function of the repair policy while a_f depends on the failure criterion. At a time $t > 0$, the structure will have experienced deterioration, the extent of which is reflected in the shift of the flaw density to the higher range of flaw size,

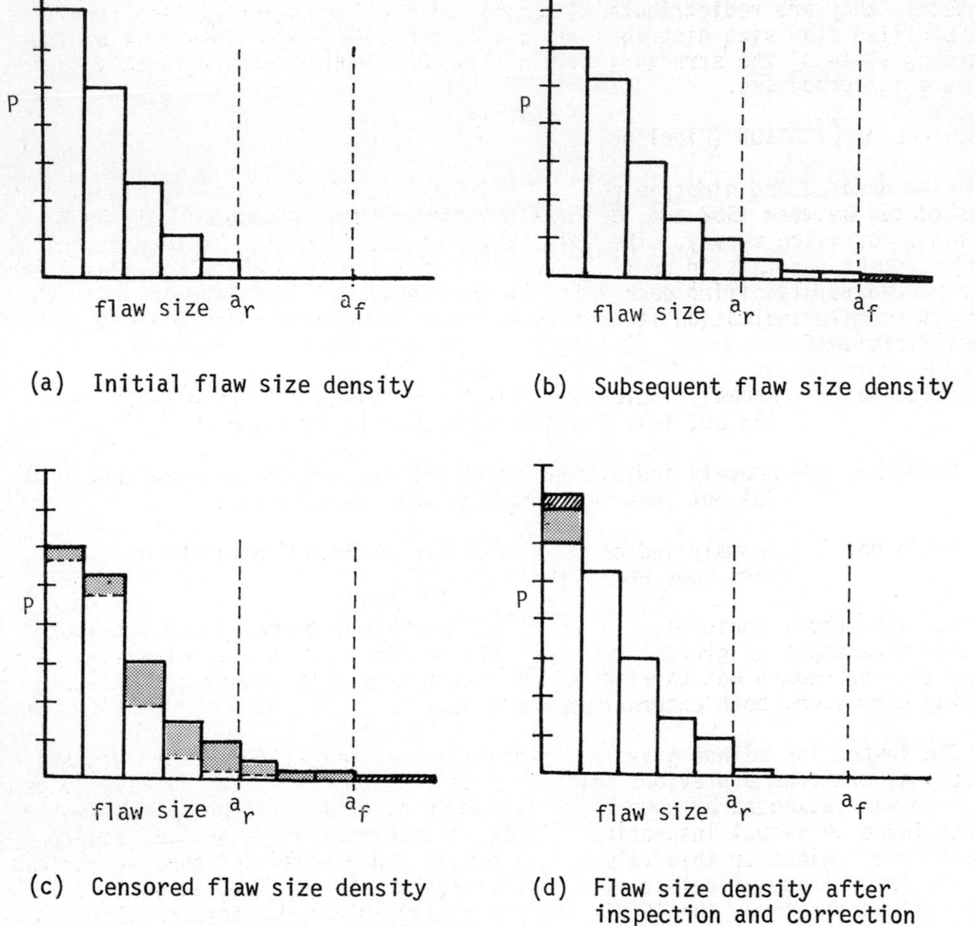

(a) Initial flaw size density

(b) Subsequent flaw size density

(c) Censored flaw size density

(d) Flaw size density after inspection and correction

Fig. (4) - Graphic representation of the failure control system

Figure 4(b). Note that a fraction of the population has exceeded a_f (shaded area). This is the measure of risk or the probability of failure. In this diagram, reliability is measured as the fraction of the population below a_f.

By inspecting at this time, some flaws will be detected and some will be missed, depending on the probability of non-detection. Mathematically, this operation involves multiplying the flaw density by the probability of non-detection. This is shown graphically in Figure 4(c) where the dotted areas represents the fraction of the population detected. As far as repair is concerned, one may repair all the flaws detected or only some of them, in which case only those greater than the specified a_r. Thus, even if a flaw is detected it remains in the structure if it is less than a_r.

It is assumed in this analysis that the components repaired and/or replaced are put back into service. Hence, depending on their quality (in terms of flaw content), they are redistributed according to the flaw density. The result is the modified flaw size distribution, shown in Figure 4(d). This will be the starting state of the structure that will undergo subsequent degradation and upgradation processes.

PRACTICAL APPLICATION (Pipelines)

Five-hundred and ninety-eight (598) miles of 34-inch crude oil pipeline constructed between 1962 and 1969 were non-destructively inspected in 1979 during a corrosion survey. The NDT method used was the flux leakage technique; the equipment for this inspection being the aforementioned Linalog. The inspection procedure classified each joint (approximately 40 feet) according to the maximum anomaly indication it contained. The following are the grade classifications:

Grade No. 1 - Anomaly indications which are believed to be more than 15% but less than 30% body wall penetration,

Grade No. 2 - Anomaly indications which are believed to be more than 30% but less than 50% body wall penetration,

Grade No. 3 - Anomaly indications which are believed to be 50% or more body wall penetration.

Hence, one anomaly indication in Grade No. 3 is in fact one pipe joint whose maximum flaw depth is greater than or equal to 50% body wall penetration. The inspection procedure was carried out in such a way as to detect generalized pitting corrosion, both external and internal.

The inspection method gives the probability of detection of the largest defects as 95% from a previous inspection of 24 miles of welded seamless pipe [7]. It was estimated by comparing the number of defects found by NDT to the number found by visual inspection. It is assumed that only the most severe defects are subject to this value of detection probability and that it is less for the lower range of defect size. Assuming that 95% applies to defects of full wall penetration, then 5% is the probability of non-detection. It is also reasonable to assume that below a lower limit, no defects can be detected. A value, 5% of body wall, is arbitrarily taken as the deepest penetration for which the probability of non-detection is unity. Using these as boundary conditions, equation (15) is used and the constants C_1 and B_1 are evaluated. In this case, $C_1 = 1$ and $B_1 = 0.46$, and a graph of this function is shown in Figure 5.

For this analysis, the population consists of pipe joints (assumed to be 40 feet long each) which were graded according to their maximum anomaly indication. Figure 6(a) shows the summary of the corrosion survey for the 598 miles of pipeline (approximately 78,500 joints). But because of the inspection method's finite probability of non-detection, these numbers must be adjusted in order to account for what have been missed. This is accomplished mathematically by dividing the number found in each cell by the average probability of detection for that cell. Figure 6(b) shows the adjusted number with the remaining joints being classified in the 0-15% wall thickness range.

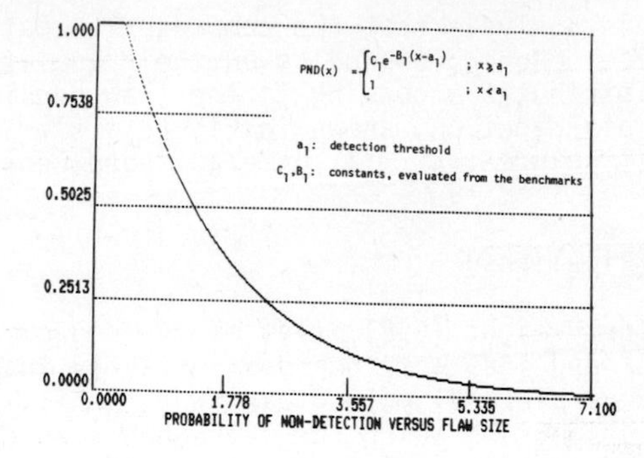

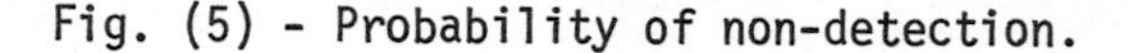
Fig. (5) - Probability of non-detection.

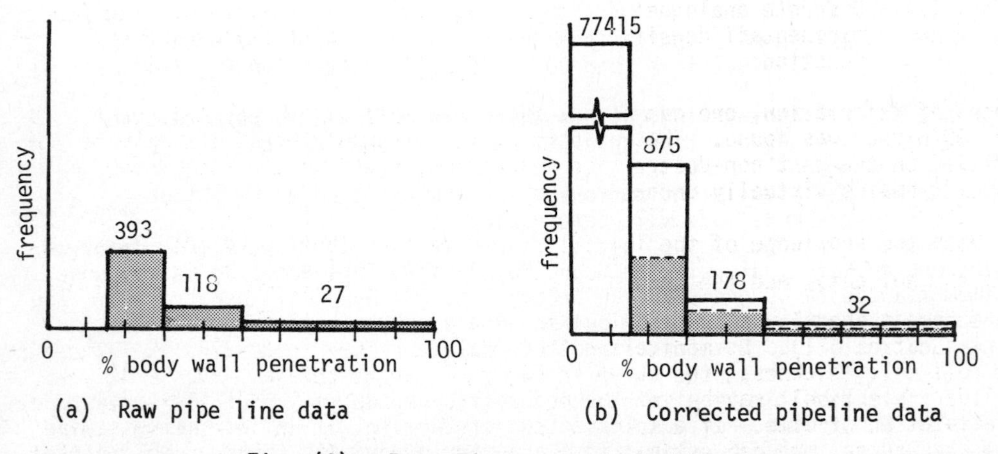

Fig. (6) - Raw and corrected pipeline data

An analytical probability density function can now be fitted to the data. A one-parameter exponential probability density function was assumed and a shape parameter equal to 0.455 was found to be a good fit. In order to be able to use the Markov discrete state processes, this function is normalized, and then quantized by dividing the flaw size range (in this case, the pipe wall thickness) with an appropriate number of cells, the result of which is the size for each cell. The choice of cell size is usually governed by the tolerance of the measuring equipment. For this analysis, a cell size 20% or 1.42 mm is taken. The discrete analogue of the exponential distribution function is shown in Figure 7. The probability for each cell or state can now be calculated.

The appropriate transition probability mechanism which describes the pitting corrosion process has not as yet been established but, for purposes of illustration, suppose it can be described as a homogeneous linear birth process as presented in equation (9). The state probabilities in Figure 7 were obtained from the NDT data after a time interval of approximately 10 years. Furthermore, it is reasonable to assume that at the beginning of service life the entire population is contained in the first state. Based on these two

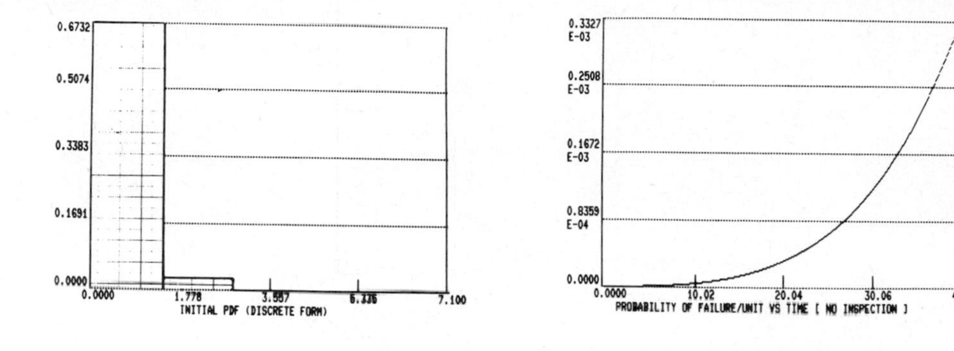

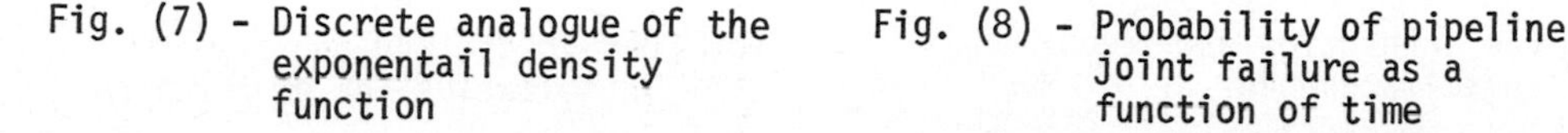
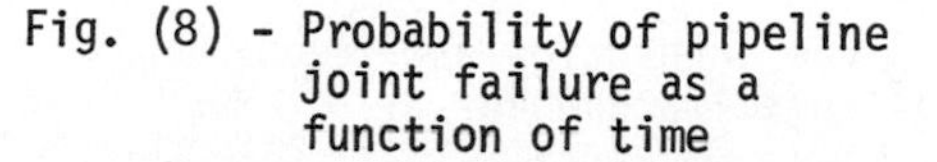

Fig. (7) - Discrete analogue of the exponentail density function

Fig. (8) - Probability of pipeline joint failure as a function of time

pieces of information, one can find a first estimate of λ. For this case, λ = .0045/year was found. The validity of the assumed process can only be verified on the next non-destructive inspection, provided the "corrosion system" remains virtually unchanged.

With the knowledge of the initial flaw size distribution, $P_i(0)$, inferred from the NDT data, and the knowledge of the process, described by $P_{ij}(t)$, one can forecast the flaw size distribution at any subsequent time interval by using equation (11). By monitoring the fraction of the population that exceeds the full wall thickness, the curve in Figure 8 can be obtained over a 40 year period. This result may be interpreted as the probability of failure per joint as a function of time. If a target risk level of operation is desired, then from this curve, one can estimate the time to upgrade the system in order that it remains in operation under the targetted risk level. On the other hand, if it is desired to minimize the expected number of failures by the end of, say, 40 years, one or more upgradations may be performed within this period. Suppose one inspection is planned within this period, there exists an optimum time to do it depending on the probability of non-detection, repair size level and the quality of the repaired or replaced units. Figure 9(a) shows the optimum time to do the inspection and the corresponding fraction of the population that have already failed at the time of inspection plus the estimated fraction of the population that would fail towards the end of the period. This curve was obtained by assuming a repair size level of 70% of wall thickness, that is, to repair all units whose anomaly indications exceed 70% of wall thickness, while the quality of the repaired or replaced units is assumed to be the same as the original. It can be seen that the expected number of failures can be reduced by at least 58% by inspecting after 25.6 years. Assuming a lower repair size level of 30% of wall thickness, Figure 9(b) shows that the expected number of failures by the end of the period can be further reduced by 37% with the corresponding optimum inspection time being at 20 years.

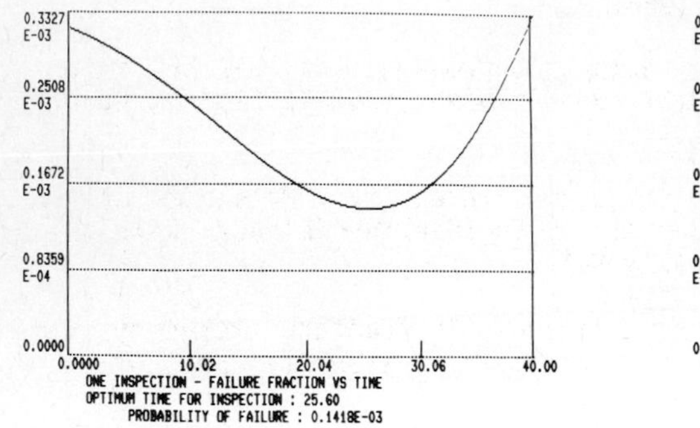

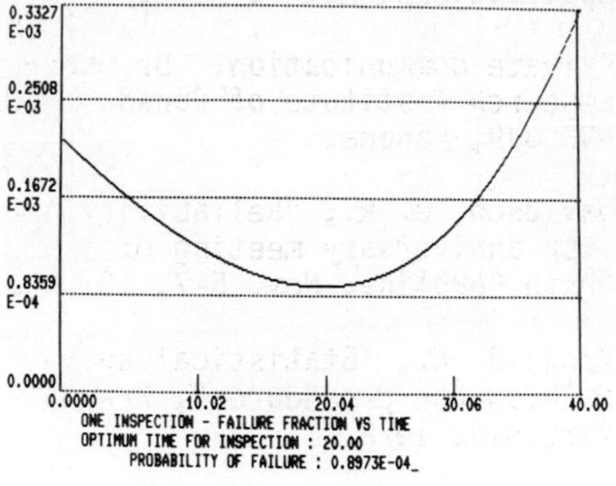

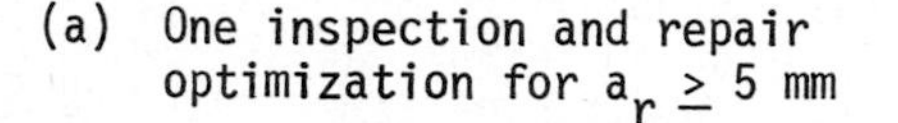

(a) One inspection and repair optimization for $a_r \geq 5$ mm

(b) One inspection and repair optimization for $a_r \geq 2.1$ mm

Fig. (9) - Inspection and repair optimization

CONCLUDING REMARKS

As has been shown in this paper the concept that material degradation mechanisms can be modelled in terms of stochastic processes, in particular Markov processes, has very powerful and far reaching implications in terms of component and/or system reliability. In particular, pitting corrosion modelled in this way has led to definite predictions of a pipeline system's reliability but, of course, only time will tell if these predictions are indeed correct. Any marginal outcome will, however, not detract from the viewpoint that very impressive reliability estimates, inspection and repair recommended schedules and general levels of confidence can be attained by considering time dependent degradation processes from this philosophical viewpoint. Indeed, all that remains is to determine the correct Markov law which describes a particular corrosion systems' time evolution. The model outlined in this paper is an example of its applicability and utility when incorporated into a failure control system.

ACKNOWLEDGEMENTS

The financial support of the Industrial Materials Research Institute - National Research Council Canada, through its contract with Tektrend International, and the NSERC Grant #A7525 is gratefully acknowledged.

REFERENCES

[1] Provan. J. W., "The micromechanics approach to the fatigue failure of polycrystalline metals", Cavities and Cracks in Creep and Fatigue, J. Gittus, ed., Applied Science Publishers, London and New Jersey, pp. 197-242, 1981.

[2] Hay, D. Robert, et al., "A systems approach to failure control in load-bearing structures", Canadian Metallurgical Quarterly, Vol. 19, Pergamon Press Ltd., pp. 79-85, 1980.

[3] Bharucha-Reid, A. T., Elements of the Theory of Markov Processes and Their Applications, McGraw Hill, Inc., 1960.

[4] Private communication. Dr. Andrew Garner, Corrosion, The Pulp and Paper Research Institute of Canada, 570 St. John's Blvd., Pointe Claire, Quebec H9R 3J9, Canada.

[5] Davidson, J. R., "Reliability and structural integrity", Presented at the 10th anniversary meeting of the Society of Engineering Science, Raleigh, North Carolina, Nov. 5-7, 1973, NASA-TM-X-71934.

[6] Yang, J. N., "Statistical approach to fatigue and fracture including maintenance procedures", Fracture Mechanics, The University Press of Virginia, 1978.

[7] AMF Tuboscope Pipeline Services Manual, AMF Tuboscope Inc., P.O. Box 808, Houston, Texas, 77001, USA.

LIST OF PARTICIPANTS

Achard, L.
McGill University
Montreal, Quebec H3A 2K6
Canada

Aifantis, E.
University of Minnesota
Minneapolis, Minnesota 55455

Ait Bassidi, M.
Institut de Genie des Materiaux
750, rue Bel-Air
Montreal, Quebec H4C 2K3
Canada

Aksogan, O.
Gannon University
Erie, Pennsylvania 16541

Alsem, W.
Northwestern University
Department of Materials Science
and Engineering
Evanston, Illinois 60201

Bailon, J. P.
Ecole Polytechnique
Montreal, Quebec H3C 3A7
Canada

Bandyopadhyay, K. K.
Gibbs & Hall, 11th Floor
393 7th Avenue
New York, New York 10001

Boutin, J.
Ecole Polytechnique
Montreal, Quebec H3C 3A7
Canada

Chawla, U.
Dominion Engineering Works Ltd.
Montreal, Quebec H3C 3S5
Canada

Chow, C. L.
BF Goodrich Company
9921 Brecksville Road
Brecksville, Ohio 44141

Cooper, C. V.
Northwestern University
Evanston, Illinois 60201

Cox, B. N.
Rockwell International Science
Center
P.O. Box 1085
Thousand Oaks, California 91360

Delph, T. J.
Lehigh University
Bethlehem, Pennsylvania 18015

Dickson, I.
Ecole Polytechnique
Montreal, Quebec H3C 3A7
Canada

Desroches, G.
Instron Canada Limited
Montreal, Quebec H1V 3M5
Canada

Eifler, D.
Universitat Karlsruhe (TH)
Postfach 6380
D-7500 Karlsruhe 1,
West Germany

Eringen, A. Cemal
Princeton University
Princeton, New Jersey 08544

Febres, A.
Ecole Polytechnique
Montreal, Quebec H3C 3A7
Canada

Gdoutos, E. E.
Institute of Fracture and
Solid Mechanics
Lehigh University
Bethlehem, Pennsylvania 18015

Geckinli, E.
Ecole Polytechnique
Montreal, Quebec H3C 3A7
Canada

Gerold, V.
Max-Planck-Institut fuer Metallforschung
Institut fuer Werkstoffwissenschaften
Seestrasse 92, D-7000 Stuttgart 1
West Germany

Handfield, L.
Ecole Polytechnique
Montreal, Quebec H3C 3A7
Canada

Harig, H.
Universitat Essen-GH
Institut fuer Werkstofftechnik
Universitatsstr. 15
4300 Essen 1
West Germany

Hsu, T.-R.
University of Manitoba
Winnipeg, Manitoba R3T 2N2
Canada

Imura, T.
Nagoya University
Chikusa-ku, Nagoya 464
Japan

Krzeminski, J.
Polish Academy of Sciences
00-049, Swietokrzyska St. 21
Warsaw, Poland

Liu, H.-W.
Syracuse University
Syracuse, New York 13210

Marchand, N.
Ecole Polytechnique
Montreal, Quebec H3C 3A7
Canada

Mayr, P.
Stiftung Institut fuer Harterei-
Technik
Lesumer Heerstr. 32
2820 Bremen 77
West Germany

Masounave, J.
Institut de Genie des Materiaux
750, rue Bel-Air
Montreal, Quebec H4C 2K3
Canada

McQueen, H.
Concordia University
Montreal, Quebec H3G 1M8
Canada

Mughrabi, H.
Max-Planck-Institut fuer Metal-
lforschung
Institut fuer Physik
Heisenbergstr. 1
7000 Stuttgart 80
West Germany

Mura, T.
Northwestern University
Evanston, Illinois 60201

Neimitz, A.
Technical University of Kielce
25-314 Kielce
Al. Tysiaclecia 7
Poland

Ohr, S. M.
Oak Ridge National Laboratory
Oak Ridge, Tennessee 37830

Oswald, G. F.
Techn. Universitat Munchen
Acrisstrasse 21
8000 Munchen 2
West Germany

Pardee, W. J.
Rockwell International Science Center
P.O. Box 1085
Thousand Oaks, California 91360

Provan, J. W.
McGill University
817 Sherbrooke Street West
Montreal, Quebec H3A 2K6
Canada

Rodriguez, E. S., III
Failure Control International
750, rue Bel-Air
Montreal, Quebec H4C 2K3
Canada

Scarth, D.
W. L. Wardrop & Assoc., Ltd.
77 Main Street
Winnipeg, Manitoba R3C 3H1
Canada

Shirai, H.
Northwestern University
Evanston, Illinois 60201

Sih, G. C.
Institute of Fracture and
Solid Mechanics
Lehigh University
Bethlehem, Pennsylvania 18015

Smith, C. W.
Virginia Polytechnic Institute
and State University
Blacksburg, Virginia 24061

Theriault, Y.
McGill University
817 Sherbrooke Street West
Montreal, Quebec H3A 2K6
Canada

Tyson, W. R.
PMRL, CANMET
568 Booth Street
Ottawa, Ontario K1A 0G1
Canada

Dr. H. Veith
Bundesanstalt fuer Materialprufung
(BAM)
Unter den Eichen 87
D-1000 Berlin 45
West Germany